Physics of Magnetism

WILEY SERIES ON THE SCIENCE AND TECHNOLOGY OF MATERIALS

Advisory Editors: J. H. Hollomon, J. E. Burke, B. Chalmers, R. L. Sproull, A. V. Tobolsky

---

PHYSICS OF MAGNETISM
Sōshin Chikazumi

PHYSICS OF III-V COMPOUNDS
Otfried Madelung

PRINCIPLES OF SOLIDIFICATION
Bruce Chalmers

APPLIED SUPERCONDUCTIVITY
Vernon L. Newhouse

THE MECHANICAL PROPERTIES OF MATTER
A. H. Cottrell

THE ART AND SCIENCE OF GROWING CRYSTALS
J. J. Gilman, editor

COLUMBIUM AND TANTALUM
Frank T. Sisco and Edward Epremian, editors

MECHANICAL PROPERTIES OF METALS
D. McLean

THE METALLURGY OF WELDING
D. Séférian—Translated by E. Bishop

THERMODYNAMICS OF SOLIDS
Richard A. Swalin

TRANSMISSION ELECTRON MICROSCOPY OF METALS
Gareth Thomas

PLASTICITY AND CREEP OF METALS
J. D. Lubahn and R. P. Felgar

INTRODUCTION TO CERAMICS
W. D. Kingery

PROPERTIES AND STRUCTURE OF POLYMERS
Arthur V. Tobolsky

MAGNESIUM AND ITS ALLOYS
C. Sheldon Roberts

PHYSICAL METALLURGY
Bruce Chalmers

FERRITES
J. Smit and H. P. J. Wijn

ZONE MELTING
William G. Pfann

THE METALLURGY OF VANADIUM
William Rostoker

INTRODUCTION TO SOLID STATE PHYSICS, SECOND EDITION
Charles Kittel

Physics of Magnetism

by Sōshin Chikazumi

PROFESSOR OF PHYSICS
INSTITUTE FOR SOLID STATE PHYSICS
UNIVERSITY OF TOKYO

ENGLISH EDITION PREPARED WITH THE ASSISTANCE OF

Stanley H. Charap

RESEARCH STAFF
IBM THOMAS J. WATSON RESEARCH CENTER
YORKTOWN HEIGHTS, NEW YORK

John Wiley & Sons, Inc., New York · London · Sydney

In 1959 substantial portions of this book were published by Syokabo Publishing Company, Tokyo, Japan, under the title *Physics of Ferromagnetism*.

Library of Congress Catalog Card Number: 64-14985
Printed in the United States of America

Preface

This book is intended as a textbook on magnetism for students and investigators who are interested in the physical aspects of magnetism. The level of presentation assumes only a fundamental knowledge of electromagnetic theory and atomic physics and a general familiarity with rather elementary mathematical exposition. Throughout the book the emphasis is primarily on explanation of physical concepts rather than on rigorous theoretical treatments which require a background of quantum mechanics or of high level mathematics.

Many books on magnetism have been published, as the reader may see in the General Reference list at the end of this book. The approach is different in different books, ranging from purely theoretical treatment of the magnetic structure of substances to plain description of magnetic materials. This is a simple reflection of the character of recent magnetics research which deals with fundamental physical problems including mechanisms of exchange interaction, origin of intrinsic magnetization, etc., and more secondary magnetic phenomena such as magnetic anisotropy, magnetostriction, domain structures, and technical magnetization. Engineering applications of magnetic materials to electromagnetic machines, permanent magnets, and electronic computers are also among the major interests within the scope of magnetics.

The purpose of this book is to give a general view of these magnetic phenomena, focusing its main interest at the center of such a broad field. As an outstanding textbook on magnetism which was written from this standpoint, Becker and Döring's *Ferromagnetismus*, published by Julius

Springer in 1939, can be cited. Unfortunately the content of that book has become outmoded owing to the rapid development of investigations, especially those on domain structures and magnetic oxides; but I believe that it is still a first rank textbook because it presents an interesting physical approach to various magnetic phenomena. It is always pleasant for me to recall the enjoyment experienced by a number of colleagues and me when, as students in the seminars of Professor S. Kaya, we read the Becker and Döring book. It has been my desire to reproduce this sort of approach in the description of more up-to-date magnetics.

The contents of this book are divided into six parts. After an introductory description of magnetic phenomena in Part 1, the origin and mechanism of para-, ferro-, antiferro-, and ferrimagnetism are treated in Part 2, and then magnetic domain structures and related subjects such as magnetic anisotropy, magnetostriction, and magnetostatic energy are dealt with in Part 3. Part 4 is devoted to discussions of magnetization processes on the basis of our knowledge of domain structures—from static to dynamic or resonance phenomena. In Part 5, special topics on magnetism, including induced magnetic anisotropy, thin films, helical spin configurations, rare earth metals, neutron diffraction, and the Mössbauer effect are discussed. The galvanomagnetic effect, magnetothermal effect, and magnetomechanical effect are also treated but in less detail. Engineering applications of magnetic materials are described in Part 6.

Throughout the book, the mksa system of units of *E-H* analogy is used in describing mathematical formulas. As is well known, this system is very convenient for describing all electromagnetic phenomena without introducing troublesome coefficients. This system also uses practical units of electricity such as volts, amperes, coulombs, and farads. This system is particularly convenient when we treat phenomena such as eddy currents or electromagnetic induction which relate magnetism to electricity. The use of the mksa system, however, requires a great deal of conversion of units of magnetic quantities—from gauss to Wb/m^2, oersteds to A/m, and so on. In most countries, I suppose, students of physics have been placed in this peculiar situation: they learn electromagnetic theory in the mksa system, and then they must switch their thinking from the mksa to the old cgs system when they enter a graduate course. The reason is that professors and senior investigators, having been familiar with the cgs system, may understand the mksa system but may not have a feeling for quantities expressed in that system. For such senior investigators, including myself, all quantities in the figures, tables, and text in this book are presented in both mksa and cgs units. It may be advantageous for those investigators to use mksa formulas for calculation and then convert the results obtained to cgs units. For this purpose the conversion table in Appendix 3 may be useful.

One of the difficulties which I faced during the preparation of the manuscript was the need to describe the most recent research on each item. Because so many papers have been published continuously, and some of them forced me to rewrite most of the paragraphs in the relevant section, I am afraid that I may have omitted the results of some important papers because of carelessness or ignorance. Even for referenced papers, the mode of explanation had to be modified sometimes to conform to the general form of presentation of this book. I apologize for these points in advance, and I shall be very glad if the reader will point out any unjustified statement or omission of important papers.

I borrowed a number of figures and tables from these books: R. Becker and W. Döring's *Ferromagnetismus* (Julius Springer, 1939), J. L. Snoek's *New Developments in Ferromagnetic Materials* (Elsevier, 1949), R. M. Bozorth's *Ferromagnetism* (D. Van Nostrand, 1951), C. Kittel's *Introduction to Solid State Physics* (John Wiley and Sons, 1956). I also borrowed a number of figures from original papers by Prof. E. Kneller, Dr. B. W. Roberts, Dr. I. S. Jacobs, Dr. W. H. Meiklejohn, Dr. E. W. Lee, Prof. C. G. Shull, Prof. G. W. Rathenau, Dr. A. V. Pohm, Dr. C. D. Olson, Prof. R. Street, Prof. L. Néel, Dr. F. G. Brockman, Dr. W. J. Carr, Dr. E. M. Pugh, Dr. G. H. Jonker, Prof. R. Pauthenet, Dr. E. W. Gorter, Prof. R. W. Hoffman, Dr. L. R. Walker, Dr. L. R. Bickford, Jr., Dr. G. K. Wertheim, Dr. P. E. Tannenwald, Dr. G. E. Bacon, Dr. R. F. Penoyer, Dr. M. K. Wilkinson, Dr. G. Squires, Dr. J. C. Slonczewski, Dr. H. Shenker, and Dr. J. Smit, whose papers appeared in *Reviews of Modern Physics*, *The Physical Review*, *Journal of Applied Physics*, *Proceedings of the IRE*, *Journal of the Physics and Chemistry of Solids*, *Proceedings of the Physical Society*, *Reports on Progress in Physics*, *Annalen der Physik*, *Zeitschrift für Physik*, *Zeitschrift für angewandte Physik*, *Physica*, *Philips Technical Review*, *Philips Research Reports*, and *Solid State Physics*, Vols. 2 and 5 (Academic Press). I thank these authors, editors, and publishers who kindly granted me permission to use their figures and tables with words of encouragement. Thanks are also due to Mr. Y. Takada, Prof. L. F. Bates, Dr. C. D. Mee, Dr. C. P. Bean, Dr. B. W. Roberts, Dr. R. W. De Blois, Dr. C. D. Graham, Jr., Dr. H. J. Williams, Dr. J. D. Remeika, Dr. R. C. Sherwood, Dr. J. B. Goodenough, Dr. E. E. Huber, Dr. D. O. Smith, Dr. M. Lamback, and Dr. T. Ichinokawa for their kind offer of beautiful photographs. Dr. K. Ohta kindly collected data of magnetic anisotropy and magnetostriction. Dr. J. C. Slonczewski, Dr. R. M. Bozorth, Dr. C. D. Graham, Jr., Prof. K. Yosida, Dr. H. Miwa, Dr. J. Kondo, Dr. M. Tachiki, Prof. S. Hoshino, Prof. K. Ōno, and Prof. Y. Ishikawa kindly read parts of the manuscript and gave me advice and many valuable suggestions.

This book was originally published in Japanese by Syokabo Publishing Company in Tokyo, and I have been assisted in its translation into English by Dr. Stanley H. Charap. Much new material, two completely new chapters, and three sections were added during the process of translation. For the realization of the English edition I am indebted to the kind offices of Mr. K. Endo of Syokabo Publishing Company and Mr. H. Tonami of Agne Publishing Company.

Finally, I express my sincere thanks to Professor S. Kaya who first introduced me to research work in magnetism.

Sōshin Chikazumi

Tokyo, Japan
June, 1964

Contents

PART 1 MAGNETIC PHENOMENA 1

1 Magnetic Substances, 3
1.1 Magnetostatics, 3
1.2 Classification of Various Kinds of Magnetism, 6
1.3 History of Magnetic Research, 12

2 Magnetization of a Ferromagnetic Body, 15
2.1 Magnetization Curve, 15
2.2 Demagnetizing Factor, 19
2.3 Magnetic Circuits, 24
2.4 Measurement of Magnetization and Magnetic Field, 29

PART 2 INTRINSIC MAGNETIZATION 37

3 Origin of Magnetism, 39
3.1 Orbital Motion and Spin, 39
3.2 Gyromagnetic Effect, 41
3.3 Ferromagnetic Resonance, 48
3.4 Vector Model of Magnetic Atoms, 52

4 Ferromagnetism, 60

4.1 Langevin Theory of Paramagnetism, 60
4.2 Weiss Theory of Ferromagnetism, 64
4.3 Intrinsic Magnetization of Alloys, 72

5 Ferrimagnetism, 79

5.1 Theory of Antiferromagnetism, 79
5.2 Theory of Ferrimagnetism, 88
5.3 Ferrimagnetic Oxides and Compounds, 99

PART 3 MAGNETIC DOMAIN STRUCTURE 111

6 The Concept of Magnetic Domains, 113

6.1 Discovery of Magnetic Domains, 113
6.2 General Features of Magnetic Domains, 117
6.3 Technical Magnetization, 123

7 Magnetic Anisotropy, 128

7.1 Phenomenology of Magnetic Anisotropy, 128
7.2 Pair Model of Magnetic Anisotropy, 147
7.3 One-Ion Model of Magnetic Anisotropy, 153

8 Magnetostriction, 161

8.1 Phenomenology of Magnetostriction, 161
8.2 Mechanism of Magnetostriction, 167
8.3 Volume Magnetostriction and Form Effect, 176
8.4 Inverse Magnetostrictive Effect, 182

9 Properties of Magnetic Domain Walls, 186

9.1 Structure of Domain Wall, 186
9.2 180° Domain Wall, 192
9.3 90° Domain Walls, 198

10 Magnetostatic Energy, 205

10.1 Fundamental Formulas, 205
10.2 Surface Distribution of Magnetic Poles, 209
10.3 μ^* Correction, 214

11 Domain Structure, 219

11.1 Domain Pattern and Its Interpretation, 219
11.2 Calculation of a Uniform Distribution of Magnetic Domains, 229
11.3 Domain Structure of Non-uniform Substances, 233

PART 4 THE MAGNETIZATION PROCESS 243

12 The Magnetization Curve and Domain Structure, 245

12.1 Mechanism of Technical Magnetization, 245
12.2 Distribution of Domain Magnetizations at Remanence, 248
12.3 Distribution of Domain Magnetization at the Coercive Force Point, 254

13 The Reversible Magnetization Process, 260

13.1 Reversible Rotation Magnetization, 260
13.2 Reversible Wall Displacement, 264
13.3 Law of Approach to Saturation, 274

14 The Irreversible Magnetization Process, 281

14.1 Irreversible Rotation Magnetization, 281
14.2 Irreversible Displacement of the Domain Wall, 285
14.3 Shape of Magnetization Curves, 291

15 Magnetic Aftereffect, 303

15.1 Phenomenology of Magnetic Aftereffect, 303
15.2 Mechanism of Magnetic Aftereffect, 309
15.3 Disaccommodation, 316

16 Dynamic Properties of Magnetization, 321

16.1 Eddy Current Loss, 321
16.2 Various Types of Loss and Resonance, 327
16.3 Spin Dynamics, 333
16.4 Various Modes of Ferro-, Ferri-, and Antiferromagnetic Resonance, 338
16.5 Equation of Motion for Domain Walls, 348

PART 5 SPECIAL TOPICS 357

17 Induced Magnetic Anisotropy, 359

17.1 Magnetic Annealing Effect, 359
17.2 Roll Magnetic Anisotropy, 373
17.3 Anisotropy Induced at the Crystal Transition Point, 387

18 Ferromagnetism of Thin Films, Thin Wires, and Fine Particles, 397

18.1 Ferromagnetism of Thin Films, 397
18.2 Ferromagnetism of Thin Wires and Fine Particles, 411

19 Various Phenomena Accompanying Magnetization, 419

19.1 Galvanomagnetic Effect, 419
19.2 Magnetothermal Effect, 427
19.3 Magnetomechanical Effect, 431

20 Special Spin Configurations and Rare Earth Metals, 440

20.1 Triangular and Helical Spin Configurations, 440
20.2 Electron Structure and Magnetic Properties of Rare Earth Metals, 446
20.3 Magnetic Alloys and Compounds Containing Rare Earth Elements, 454

21 Other Techniques for Investigating Internal Magnetic Structure, 459

21.1 Neutron Diffraction, 459
21.2 Measurement of Internal Field at Nucleus by Nuclear Magnetic Resonance and Mössbauer Effect, 472

PART 6 ENGINEERING APPLICATIONS OF MAGNETIC MATERIALS 489

22 Kinds of Magnetic Materials, 491

22.1 Metallic Core Materials, 491
22.2 Pressed Powder and Ferrite Cores, 498
22.3 Permanent Magnets, 500

23 Special Applications of Magnetic Materials, 507

23.1 Applications of Non-linear Characteristics, 507
23.2 Applications of Gyromagnetic Properties, 514

General References, 517

Solutions of Problems, 519

Appendix 1 Symbols Used in the Text, 521
Appendix 2 Fundamental Equations for Magnetism, 524
Appendix 3 Conversion Table of Magnetic Quantities—mksa and cgs Systems, 525
Appendix 4 Conversion Table for Various Units of Energy, 526
Appendix 5 Values of Some Constants, 527
Appendix 6 Periodic Table of Elements, 528

Indexes, 533

part I

MAGNETIC PHENOMENA

1

Magnetic Substances

1.1 Magnetostatics

In this section we describe various elementary magnetic phenomena in order to provide a background for those lacking familiarity with magnetism and to familiarize the reader with mksa system.

The fundamental magnetic phenomenon is the Coulomb interaction between magnetic poles. Consider two magnetic poles whose strengths are m_1 (weber) and m_2 (weber) respectively, separated by the distance r (meter). The force F (newton) exerted on one pole by the other is

$$F = \frac{m_1 m_2}{4\pi\mu_0 r^2}, \tag{1.1}$$

where μ_0 is the permeability of vacuum the value of which is, in units of henry per meter,

$$\mu_0 = 4\pi \times 10^{-7} \text{ H/m}. \tag{1.2}$$

Another method of exerting a force on a magnetic pole is to use electric current. Generally we call the space wherein a magnetic pole experiences an applied force a magnetic field, whether it is produced by other magnetic poles or by an electric current. A uniform magnetic field can be produced by a long, thin solenoid. When a current of i A (ampere) flows in the windings, having n turns per meter, the intensity of the field H is defined by

$$H = ni. \tag{1.3}$$

The unit of the magnetic field thus defined is the ampere per meter (1 A/m = $4\pi \times 10^{-3}$ Oe = 0.0126 Oe; 1 Oe = 79.6 A/m).

When a magnetic pole of strength m (Wb) is brought into a magnetic field of intensity H (A/m), the force F (N) acting on the magnetic pole is

$$F = mH. \tag{1.4}$$

[μ_0 was so defined as to avoid a coefficient in (1.4).] If a bar magnet of length l (m), which has magnetic poles m and $-m$ at its ends, is placed in a uniform magnetic field H, each pole is acted upon by forces as indicated by the arrows in Fig. 1.1, giving rise to a couple of force whose moment is

$$L = -mlH \sin \theta, \tag{1.5}$$

where θ is the angle between the direction of the magnetic field H and the direction of the magnetization ($-m \rightarrow +m$) of the magnet. Thus a

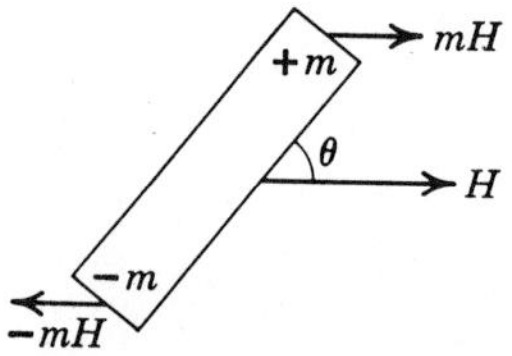

Fig. 1.1 A magnet under the action of a couple of forces in a uniform magnetic field.

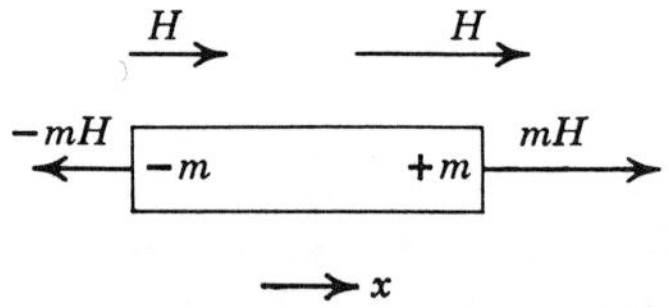

Fig. 1.2. A magnet under the action of a translational force in a gradient magnetic field.

uniform magnetic field can rotate the magnet but cannot translate it from its original position (cf. Fig. 1.2). For this purpose it is necessary to have a gradient of the field $\partial H_x/\partial x$, which gives rise to the translational force

$$F_x = ml \frac{\partial H_x}{\partial x} \tag{1.6}$$

in the x direction.

As seen in (1.5) and (1.6), any kind of force which acts on the magnet involves m and l in the form of the product ml. We call this product

$$M = ml \tag{1.7}$$

a magnetic moment the unit of which is the weber-meter (1 Wb-m = $(1/4\pi) \times 10^{10}$ gauss-cm^3).

In terms of M, the torque exerted on a magnet which is in a uniform field H is expressed as

$$L = -MH \sin \theta, \tag{1.8}$$

irrespective of the shape of the magnet. If no frictional forces are acting on the magnet, the work done by the torque (1.8) should be conserved, giving rise to the potential energy,

$$U = -MH \cos \theta. \tag{1.9}$$

This relation will be used frequently in the discussion of rotation of the atomic magnetic moment by an external field.

Next we discuss the interaction between two magnets. Suppose that two magnets having magnetic moments M_1 and M_2, respectively, are separated by the distance r_{12}. We assume the length of each magnet to be very small compared to r_{12}. We call such a small magnet a magnetic dipole. The

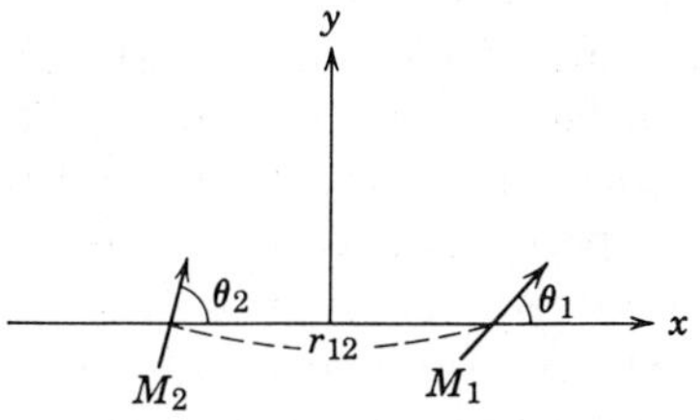

Fig. 1.3 Magnetic interaction between two dipoles.

Fig. 1.4. Two arrangements of parallel dipoles: (*a*) stable, (*b*) unstable.

field which magnet 2 produces at the site of magnet 1 is given by (cf. Fig. 1.3)

$$H_{1x} = \frac{M_2}{4\pi\mu_0}\frac{2\cos\theta_2}{r_{12}^3},\qquad H_{1y} = -\frac{M_2}{4\pi\mu_0}\frac{\sin\theta_2}{r_{12}^3}, \tag{1.10}$$

so that the potential energy of the system is

$$\begin{aligned} U &= -\tfrac{1}{2}(M_{1x}H_{1x} + M_{1y}H_{1y} + M_{2x}H_{2x} + M_{2y}H_{2y}) \\ &= -\frac{M_1M_2}{4\pi\mu_0 r_{12}^3}(2\cos\theta_1\cos\theta_2 - \sin\theta_1\sin\theta_2). \end{aligned} \tag{1.11}$$

If the two dipoles have the same magnetic moment, $M_1 = M_2 = M$; and, in addition, if they are always parallel to each other, that is, if $\theta_1 = \theta_2 = \theta$, then (1.11) becomes

$$U = -\frac{3M^2}{4\pi\mu_0 r_{12}^3}(\cos^2\theta - \tfrac{1}{3}). \tag{1.12}$$

The potential energy (1.12) has its minimum value at $\theta = 0$, so that configuration (*a*) in Fig. 1.4 is stable, while configuration (*b*) is unstable. This interaction between the two dipoles is called the magnetic dipole-dipole interaction. In (1.10) we assumed that each magnet rotates only in the xy plane. In the general case the potential energy is given by

$$U = \frac{1}{4\pi\mu_0 r^3}\left[(\boldsymbol{M}_1\cdot\boldsymbol{M}_2) - \frac{3}{r^2}(\boldsymbol{M}_1\cdot\boldsymbol{r})(\boldsymbol{M}_2\cdot\boldsymbol{r})\right]. \tag{1.13}$$

1.2 Classification of Various Kinds of Magnetism

Substances which are magnetized, more or less, by a magnetic field are called magnetic substances. There are various kinds of magnetism, and each is characterized by its own magnetic structure. From this point of view, various kinds of magnetism will be classified in this section and their magnetic structures and magnetic properties will be described.

The magnetic moment per unit volume (1 m^3) of a magnetic substance is called the intensity of magnetization and is denoted by the vector $\boldsymbol{I}$. This vector points from the S pole to the N pole, which would appear if the portion in which the magnetization is being specified were isolated from the rest of the specimen. The unit of I is the weber per square meter (1 Wb/m^2 of $I = (1/4\pi) \times 10^4$ gauss $= 7.96 \times 10^2$ gauss). The magnetic induction or magnetic flux density $\boldsymbol{B}$ is also commonly used in engineering applications to describe the magnetization. The relationship between $\boldsymbol{B}$ and $\boldsymbol{I}$ is given by the formula

$$\boldsymbol{B} = \boldsymbol{I} + \mu_0 \boldsymbol{H}. \tag{1.14}$$

As seen from this formula, B is also measured in units of webers per square meter. It must be noted, however, that the relation corresponding to (1.14) is not rationalized (contains the factor 4π) in the cgs system, so that the conversion factor between mksa units and cgs units is different for B and I (1 Wb/m^2 of $B = 10^4$ gauss).

The relation between the intensity of magnetization I and the magnetic field H can be expressed by

$$I = \chi H, \tag{1.15}$$

where χ is the magnetic susceptibility. The unit of χ is the henry per meter, which is the same as that of μ_0; hence it is possible to measure χ in units of μ_0. The susceptibility thus measured is called a relative susceptibility and is usually denoted by $\bar{\chi}$, which is

$$\bar{\chi} = \frac{\chi}{\mu_0}. \tag{1.16}$$

$\bar{\chi}$ is a dimensionless quantity, and its value is 4π times larger than the susceptibility χ measured in the cgs system. Substituting for $\boldsymbol{I}$ of (1.14) the expression (1.15), we have

$$B = (\chi + \mu_0)H = \mu H, \tag{1.17}$$

where μ is the permeability the unit of which is also the henry per meter. Usually we use a relative permeability, which is

$$\bar{\mu} = \mu/\mu_0 = \bar{\chi} + 1. \tag{1.18}$$

The value of $\bar{\mu}$ is the same as that of the permeability measured in the cgs system.

The observed value of relative susceptibility ranges from 10^{-5} for very weak magnetism to 10^6 for very strong magnetism. In some cases it takes a negative value. Sometimes the relation between I and H is not linear, so that $\bar{\chi}$ depends on the intensity of magnetic field. We can interpret the type of behavior of $\bar{\chi}$ in terms of the magnetic structure of the material.

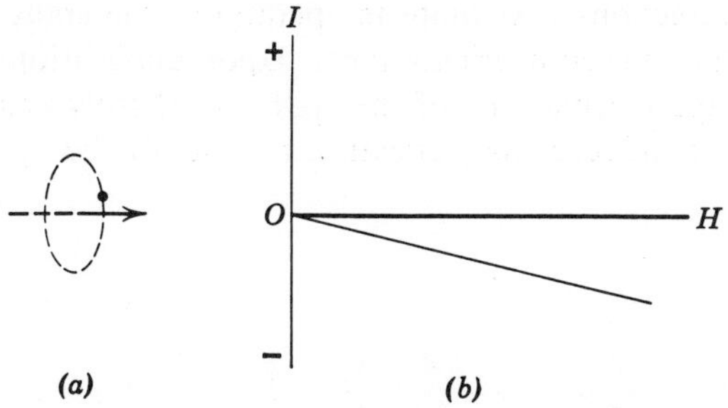

Fig. 1.5. Diamagnetism.

From this point of view, we can classify the various kinds of magnetism as follows.

Diamagnetism
Paramagnetism
Antiferromagnetism
Metamagnetism
Parasitic ferromagnetism
Ferrimagnetism
Ferromagnetism

There are two possible atomic origins of magnetism which lead to the magnetization of magnetic substances—orbital motion and spin of electrons—as will be explained in detail in Chapter 3. An atom which has a magnetic moment caused by spin or by orbital motion of electrons or by both is generally called a magnetic atom. Since the magnetic moments of the important magnetic atoms such as iron, cobalt, and nickel are caused mostly by spin motion of the electrons, we refer to the atomic magnetic moment simply as the "spin" hereafter.

Diamagnetism is a weak magnetism in which a magnetization is exhibited opposite to the direction of the applied field. The susceptibility is negative and the order of magnitude of $\bar{\chi}$ is usually about 10^{-5}. The origin of this magnetism is an orbital rotation of electrons about the nuclei induced electromagnetically by the application of an external field. As is well known from Lenz's law, the induced current produces a magnetic flux which opposes any change in the external field. This magnetism is so weak that, if there are some magnetic atoms in the substance, this is easily covered by the paramagnetism exhibited by these magnetic atoms.

In paramagnetism, the magnetization I is proportional to the magnetic field H. The order of magnitude of $\bar{\chi}$ is 10^{-3} to 10^{-5}. In most cases, paramagnetic substances contain magnetic atoms or ions whose spins are isolated from their magnetic environment and can more or less freely change their directions. At finite temperatures, the spins are thermally agitated and take random orientations. Upon application of a magnetic field, the average orientations of the spins are slightly changed so as to produce a weak induced magnetization parallel to the applied magnetic

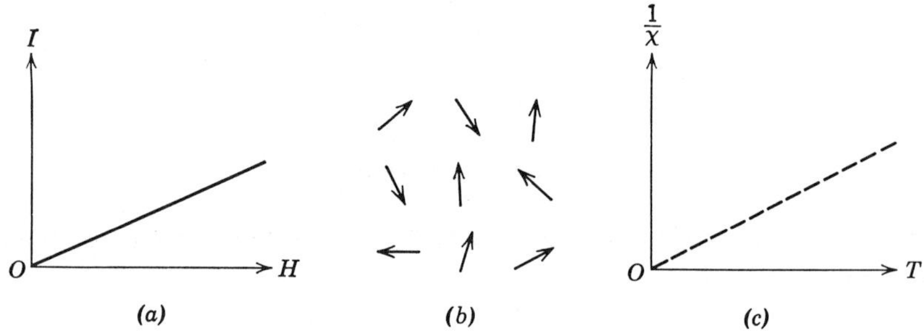

Fig. 1.6. Paramagnetism.

field. The susceptibility in this case is inversely proportional to the absolute temperature (the Curie law). Conduction electrons which form an energy band in metallic crystals also exhibit paramagnetism. Since, in this case, the excitation of minus spins to the plus spin band is opposed by an increase of kinetic energy of electrons irrespective of temperature, the susceptibility is independent of temperature (Pauli paramagnetism[1]).

Antiferromagnetism is a weak magnetism which is similar to paramagnetism in the sense of exhibiting a small positive susceptibility. The temperature dependence of the susceptibility of this magnetism is, however, characterized by the occurrence of a kink in the χ-T curve at the so-called Néel temperature. The reason for this is that below this temperature an antiparallel spin arrangement is established in which the plus and minus spins completely cancel each other (Fig. 1.7*a*): Since in such an antiferromagnetic arrangement of spins the tendency to be magnetized by the external field is opposed by a strong negative interaction acting between plus and minus spins, the susceptibility decreases with a decrease in temperature, contrary to the usual paramagnetic behavior. Above the Néel point the spin arrangement becomes random, so that the susceptibility now decreases with an increase of temperature.

Skipping the next two categories, we discuss the last two categories first, and then come back to metamagnetism and parasitic ferromagnetism later.

Ferrimagnetism is the term proposed by Néel[2] to describe the magnetism of ferrites. In these substances magnetic ions occupy two kinds of lattice sites, A and B, and spins on A sites point in the plus direction, while those on B sites point in the minus direction because of a strong negative interaction acting between the two spin systems on A and B. Since the number of magnetic ions and also the magnitude of spins of individual ions are

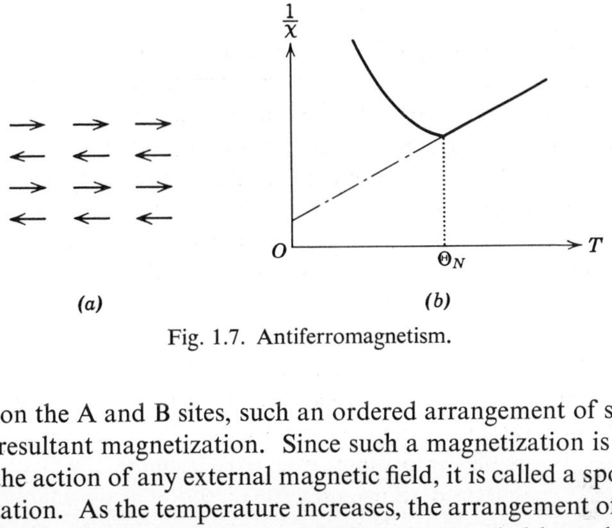

Fig. 1.7. Antiferromagnetism.

different on the A and B sites, such an ordered arrangement of spins gives rise to a resultant magnetization. Since such a magnetization is produced without the action of any external magnetic field, it is called a spontaneous magnetization. As the temperature increases, the arrangement of the spins is disturbed by thermal agitation, which is accompanied by a decrease of spontaneous magnetization. At a certain temperature, called the Curie point, the arrangement of the spins becomes completely random, and the spontaneous magnetization vanishes. There are several types of temperature dependence of the spontaneous magnetization, depending on the relative intensity of the interactions between A-A, B-B and A-B sites. One of the common types of temperature dependence for ferrites is shown in Fig. 1.8*b*. Above the Curie point, the substance exhibits paramagnetism, and the susceptibility decreases with increase of temperature. In the case of ferrimagnetism, the $1/\chi$ versus T plot is curved as shown in Fig. 1.8*b*, and the linear extrapolation of the high temperature portion usually intersects the temperature axis at its negative side. Ferrimagnetism is observed in various kinds of magnetic compounds, as will be discussed in detail in Section 5.3.

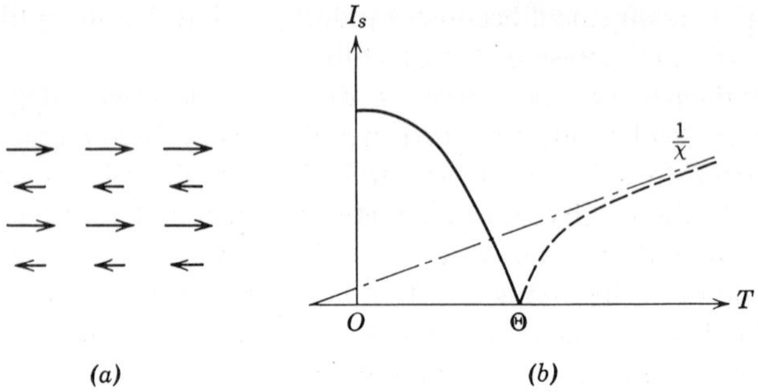

Fig. 1.8. Ferrimagnetism.

In the case of ferromagnetism, the spins are aligned parallel to one another as a result of a strong positive interaction acting between the neighboring spins (Fig. 1.9*a*). As the temperature increases, the arrangement of the spins is disturbed by thermal agitation, thus resulting in a temperature dependence of spontaneous magnetization as shown in Fig. 1.9*b*. Above the Curie point, the susceptibility obeys the Curie-Weiss law, which states that $1/\chi$ rises from zero at the Curie point and increases linearly with temperature. Ferromagnetism is exhibited mostly by metals and alloys, and by a few exceptional oxides such as CrO_2 (ref. 3) and EuO (ref. 4).

In spite of the presence of spontaneous magnetization, a block of ferromagnetic or ferrimagnetic substance is usually not spontaneously magnetized but exists rather in a demagnetized state. This is because the interior of the block is divided into many magnetic domains, each of which is spontaneously magnetized. Since only the direction of domain magnetization varies from domain to domain, the resultant magnetization can be changed from zero to the value of spontaneous magnetization. Actually, if an external field is applied, the apparent magnetization of the block is

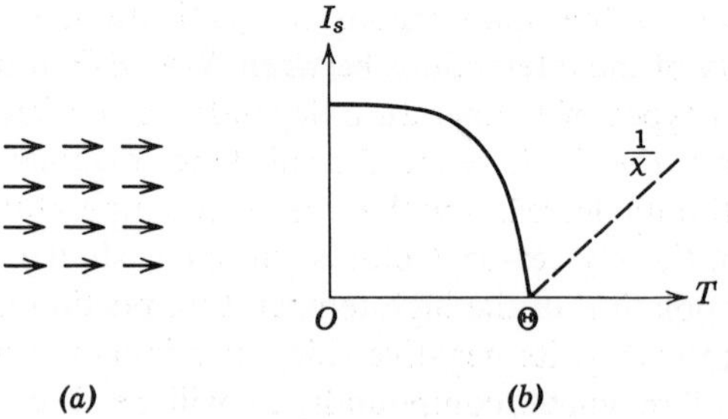

Fig. 1.9. Ferromagnetism.

changed as shown in Fig. 1.10 and finally reaches the saturation magnetization which is equal to the spontaneous magnetization. If the field is reduced, the magnetization is again decreased, but does not come back to the original value. Such an irreversible process of magnetization is called hysteresis. The presence of a saturation magnetization and a hysteresis is an important feature of ferro- and ferrimagnetism.

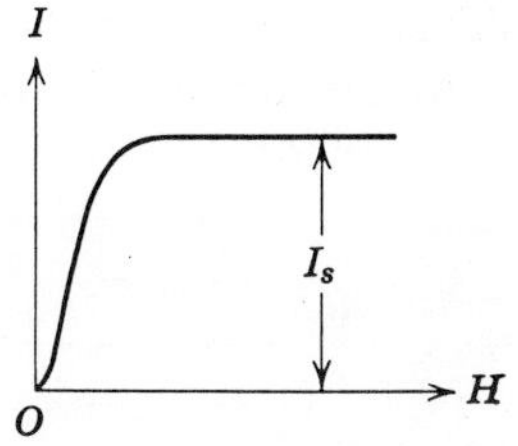

Fig. 1.10. Magnetization curve of a ferro- or ferrimagnetic substance.

Metamagnetism is the name given by Becquerel and van den Handel[5] to the phenomenon which is interpreted as a transition from antiferromagnetism to ferromagnetism and vice versa caused by the application of a strong field or by a change of temperature;[6,7] for example, $FeCl_2$ and $MnAu_2$ undergo transition by an application of a strong field. Such a reversal of spins from minus to plus directions should be distinguished from the flopping of the antiferromagnetic spin axis (cf. Section 5.1). Heavy rare earth metals such as Tb, Dy, Ho, Er, and Tm also exhibit a transition at some temperature from ferromagnetism to antiferromagnetism due to helical spin structures (cf. Section 20.2). Some nickel arsenide

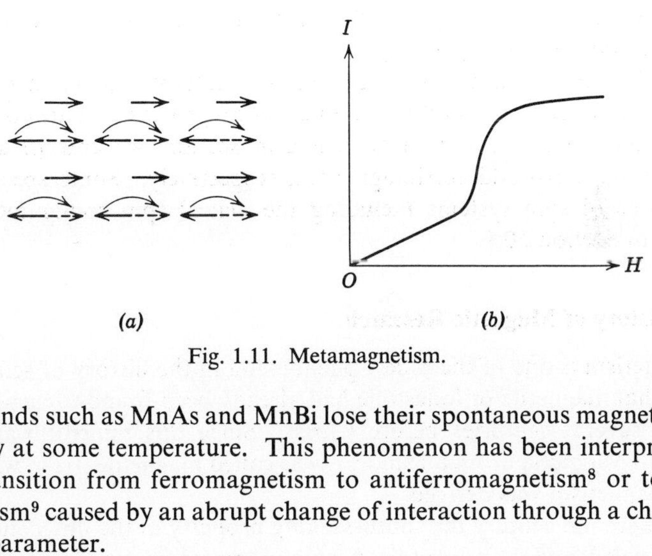

Fig. 1.11. Metamagnetism.

compounds such as MnAs and MnBi lose their spontaneous magnetization abruptly at some temperature. This phenomenon has been interpreted to be a transition from ferromagnetism to antiferromagnetism[8] or to paramagnetism[9] caused by an abrupt change of interaction through a change of lattice parameter.

Parasitic ferromagnetism is the name of a weak ferromagnetism which accompanies antiferromagnetism such as is observed for αFe_2O_3. The spontaneous magnetization of this form of magnetism disappears at the Néel point, where the antiferromagnetic arrangement of spins disappears.

There are several theories about the origin of this magnetism. One theory assumes the presence of a small amount of ferromagnetic impurities.[10] Another theory assumes an unbalanced distribution of plus and minus spins caused by some imperfection of the crystal or by the presence of antiferromagnetic domain boundaries.[11] On the other hand, Dzialoshinski[12] proposed the possibility that the directions of plus and minus spins deviate slightly from their original common axis, so that they give

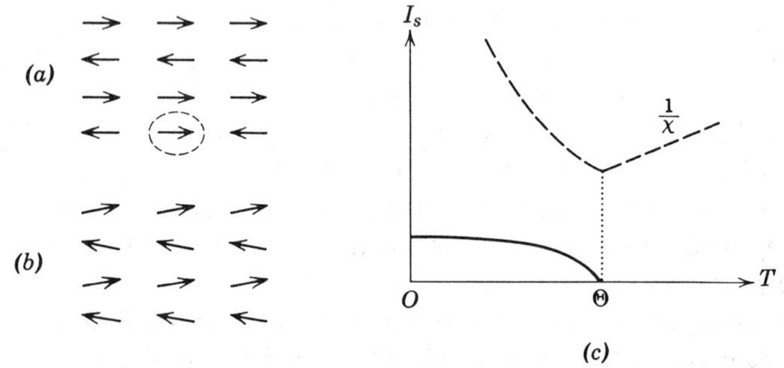

Fig. 1.12. Parasitic ferromagnetism.

rise to a feeble spontaneous magnetization perpendicular to the original common spin axis (cf. Fig. 1.12*b*). An actual mechanism of this anisotropy was worked out by Moriya.[13]

In this book, the main topics selected deal with ferro- and ferrimagnetism, their domain structures, and the magnetization processes. Paramagnetism and antiferromagnetism are also treated in Sections 4.1 and 5.1 as introductions to ferro- and ferrimagnetism, respectively. Some special configurations of spin systems including the helical spin configuration are treated in Section 20.1.

1.3 History of Magnetic Research

Magnetism is one of the oldest phenomena in the history of science. It is said that magnetite or lodestone had already been found to be a natural magnet several centuries before Christ. Since this mineral was found mostly in Magnesia of Asia Minor, it was called magnetite, from which the word magnetism was derived.

In the second century the south-seeking property of the magnetic needle was revealed and it was utilized as a compass in voyages. Scientific investigations were first made in the sixteenth century by W. Gilbert, who studied terrestrial magnetism, magnetic induction, and so on, and found that a magnet loses its magnetism at high temperature.

The most fruitful period in the study of electricity and magnetism came at the end of the eighteenth century and continued through the nineteenth century. The Coulomb law of magnetic interaction between two magnetic poles was discovered at the end of the eighteenth century. Magnetism due to electric currents was investigated by Oersted, Ampère, Biot, and Savart at the beginning of the nineteenth century. Arago tried to magnetize a magnetic substance by using an electric current. Discoveries of diamagnetism by Faraday, of magnetostriction—a deformation due to magnetization—by Joule, of the Curie law by P. Curie, of hysteresis by Ewing were all made during this period.

Ewing may have been the first person to study magnetic phenomena from the atomistic point of view. He tried to explain the phenomenon of hysteresis in terms of the magnetic interaction between molecular magnets. In this sense he was succeeded by Langevin and P. Weiss, who gave the correct interpretations of para- and ferromagnetism, respectively, from the atomistic standpoint.

The main current of magnetics research dates from the beginning of the twentieth century. A number of papers on magnetism began to appear at this time—at a rate of about 10 papers per year until 1920. Since then the development has been accelerated; 120 papers per year were recorded by 1939, and after the World War II this number rapidly increased to more than 500 papers per year.

There are so many varieties involved in this development that we can hardly describe the general history of magnetics research in this limited space. The history and development of investigations on individual subjects will be introduced in each section.

Problem

1.1. Describe various kinds of magnetism, and discuss their magnetic properties on the basis of their magnetic structures.

References

1.1. W. Pauli: *Z. Physik* **41,** 81 (1926).
1.2. L. Néel: *Ann. Physique* **3,** 137 (1948).
1.3. P. C. Guillaud, A. Michel, J. Benard, and M. Fallot: *Compt. rend.* **219,** 50 (1944).
1.4. B. T. Matthias, R. M. Bozorth, and J. H. Van Vleck: *Phys. Rev. Letters* **7,** 160 (1961).
1.5. J. Becquerel and J. van den Handel: *J. phys. radium* **10,** 10 (1939).
1.6. L. Néel: *Nuovo cimento Suppl.* **6,** No. 3, 942 (1957).

1.7. E. Vogt: *Z. angew. Physik.* **14,** 176 (1962).
1.8. C. Kittel: *Phys. Rev.* **120,** 335 (1960).
1.9. C. P. Bean and D. S. Rodbell: *Phys. Rev.* **126,** 104 (1962).
1.10. L. Néel: *Ann. Physique* **4,** 249 (1949).
1.11. Y. Y. Li: *Phys. Rev.* **101,** 1450 (1956).
1.12. I. Dzialoshinski: *J. Phys. Chem. Solids* **4,** 241 (1958).
1.13. T. Moriya: *Phys. Rev.* **120,** 91 (1960).

2

Magnetization of a Ferromagnetic Body

2.1 Magnetization Curve

One feature of ferromagnetic substances is that they exhibit a fairly complex change in magnetization upon the application of a magnetic field. This behavior can be described by a magnetization curve (Fig. 2.1).

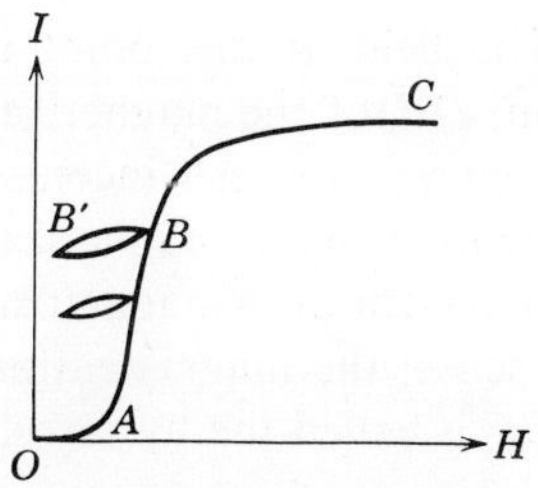

Fig. 2.1. Initial magnetization curve and minor loops.

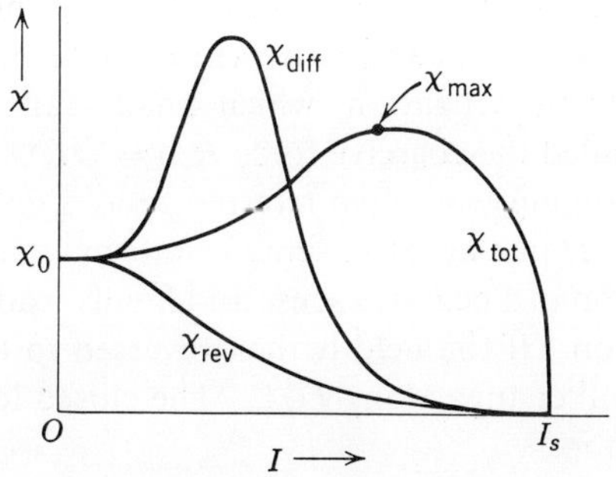

Fig. 2.2. Various kinds of magnetic susceptibilities as functions of the intensity of magnetization.

Starting from a demagnetized state ($I = H = 0$), the magnetization increases with an increase of the field along the curve $OABC$ and finally reaches the saturation magnetization which is normally denoted by I_s. In the region OA the process of magnetization is almost reversible; that is, the magnetization comes back to zero upon removal of the field. The inclination of the curve *OA* is called the initial susceptibility χ_a. Beyond this region the processes of magnetization are no longer reversible. If the field is decreased from its value at point *B*, the magnetization comes back, not along *BAO*, but along the minor loop *BB'*. The inclination of *BB'* is called the reversible susceptibility χ_{rev} or the incremental susceptibility. The slope of each portion of the initial magnetization curve *OABC* is called the differential susceptibility χ_{diff}, and the slope of the line which connects the origin *O* and each point on the initial magnetization curve is called the total susceptibility χ_{tot}. The maximum value of the total susceptibility, that is, the slope of the tangent line drawn from the origin to the initial magnetization curve, is called the maximum susceptibility χ_{max}; it is a good measure of the average inclination of the initial magnetization curve. Changes in χ_{rev}, χ_{diff}, and χ_{tot} along the initial magnetization curve are shown in Fig. 2.2. Starting from the value of χ_a, χ_{rev} decreases monotonically, while χ_{diff} has a sharp maximum, and χ_{tot} goes through its maximum value χ_{max} and drops off at $I = I_s$. The difference between χ_{diff} and χ_{rev} represents the susceptibility due to irreversible magnetization; it is called the irreversible susceptibility χ_{irr}; that is,

$$\chi_{\text{diff}} = \chi_{\text{rev}} + \chi_{\text{irr}}. \tag{2.1}$$

If the magnetic field is decreased from the saturated state *C* (Fig. 2.3), the magnetization *I* is gradually decreased along *CD*, not along *CBAO*, and at $H = 0$ it reaches the finite value I_r (= *OD*), which is called the residual magnetization or the remanence. Further increase of the magnetic field in a negative sense results in a continued decrease of the intensity of magnetization, which finally falls to zero. The field at this point is called the coercive force H_c (= *OE*). This portion, *DE*, of the magnetization curve is often referred to as a demagnetizing curve. Further increase of *H* in a negative sense results in an increase of the intensity of magnetization in a negative sense and finally leads to a negative saturation magnetization. If the field is then reversed to the positive sense, the magnetization will change along *FGC*. The closed loop *CDEFGC* is called the hysteresis loop.

Now we discuss the work necessary to magnetize a ferromagnetic substance. Suppose that the magnetization is increased from I to $I + \delta I$ under the action of a magnetic field H which is parallel to I. If we consider a cylindrical section of the magnetic substance whose length is l (parallel to I) and whose cross section is S, an increase of magnetization, δI, in the cylinder is attained by transporting the magnetic pole δIS through the distance l from the bottom to the top of the cylinder under the action of the force δISH. The work required for this transportation is $H\ \delta ISl$.

Since the volume of the cylinder is Sl, the work necessary to magnetize a unit volume of the magnetic substance is given by

$$\delta W = H\,\delta I. \tag{2.2}$$

Then the work required to magnetize a unit volume from $I = I_1$ to I_2 is expressed by

$$W = \int_{I_1}^{I_2} H\,dI. \tag{2.3}$$

For example, the work required to magnetize the volume from a demagnetized state to saturation, I_s, is given by (2.3) by putting $I_1 = 0$ and $I_2 = I_s$.

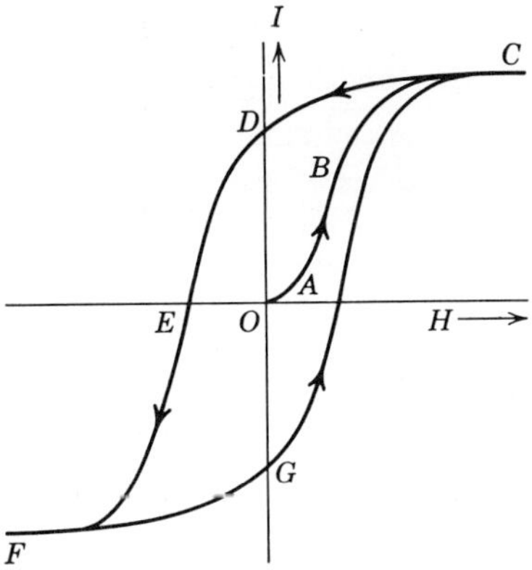

Fig. 2.3. Hysteresis loop.

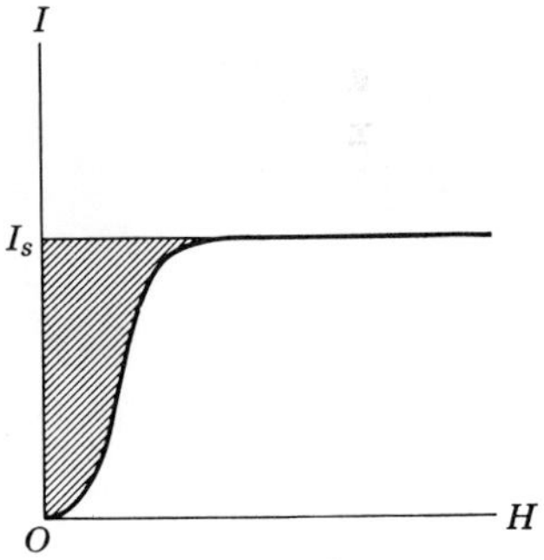

Fig. 2.4. Work required to saturate a unit volume of a ferromagnetic substance.

This is equal to the area surrounded by the ordinate axis, the line $I = I_s$, and the intial magnetization curve as shown in Fig. 2.4. The energy supplied by this work is partially stored as potential energy, and also partly dissipated as heat which is generated in the substance. During one cycle of the hysteresis loop the potential energy should return to its original value, so that the resultant work must be consumed as heat. This heat is called the hysteresis loss and is given by

$$W_h = \oint H\,dI, \tag{2.4}$$

which is equal to the area surrounded by the hysteresis loop.

For engineering applications, ferromagnetic substances can be classified into soft and hard magnetic materials. Soft magnetic materials are normally used for iron cores of transformers, motors, and generators, and for these purposes high permeability, low coercive force, and small hysteresis loss are required. On the other hand, hard magnetic materials are used as permanent magnets for various kinds of electric meters, loudspeakers, and

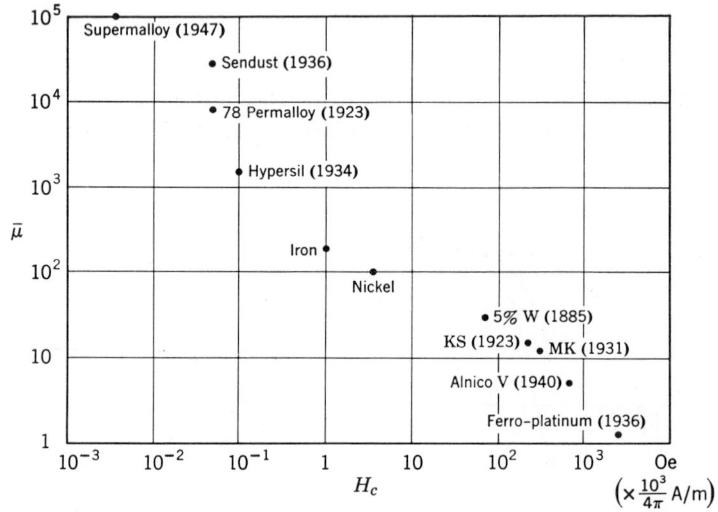

Fig. 2.5. Relative permeability and coercive force of various magnetic materials (after C. Kittel[G.6])†.

other apparatus for which high coercivity, high remanence, and large hysteresis loss are desirable. It is interesting that the main applications of ferromagnetic substances are thus separated into two fields which require almost opposite properties. As a result of the development of magnetic materials, the characteristics of existing magnetic materials range from extremely soft to extremely hard. Figure 2.5 shows the distribution of $\bar{\mu}$ and H_c for various kinds of magnetic materials; here we can see the history of the advances in magnetic materials.

In the beginning of the twentieth century, soft iron was used as a soft magnetic material. This substance has a hysteresis loop which is very wide compared to that of Permalloy, one of the typical soft magnetic materials, as shown in Fig. 2.6. Similarly a carbon steel which was used as a permanent magnet until the beginning of the twentieth century has a hysteresis loop

† The letter G preceding a reference number signifies that the General References, which appear at the end of the book, should be consulted.

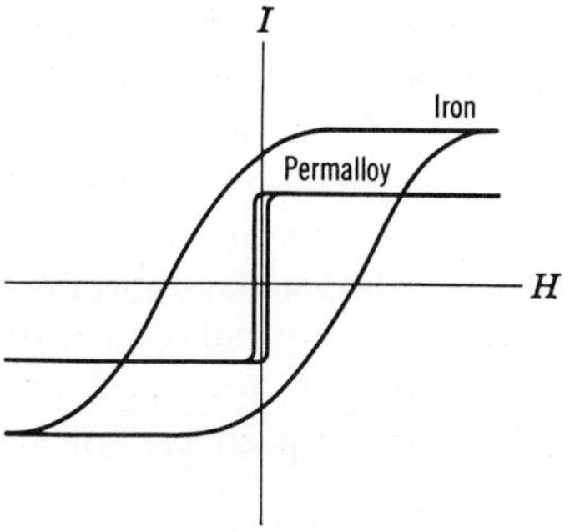

Fig. 2.6. Comparison of hysteresis curves of soft iron and Permalloy.

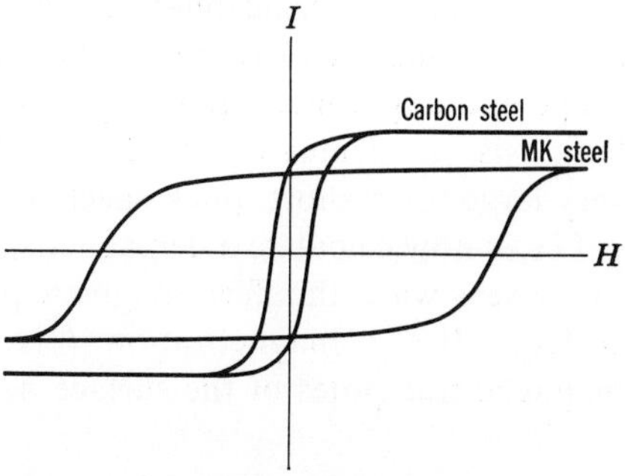

Fig. 2.7. Comparison of hysteresis curves of carbon steel and MK steel.

fairly narrow compared to that of MK magnet which is one of the typical permanent magnet materials (Fig. 2.7). Various hysteresis constants are listed in Table 2.1 for several typical magnetic materials.

Table 2.1. Magnetic Properties of Some Magnetic Materials (after Bozorth[G.10])

Material	I_s Wb/m^2	H_c A/m	H_c Oe	$\bar{\mu}_a$	$\bar{\mu}_{max}$
Mild steel	2.12	143	1.8	120	2,000
Oriented Si-steel	2.00	8	0.1	1,500	40,000
78 Permalloy	1.08	4	0.05	8,000	100,000
Supermalloy	0.79	0.16	0.002	100,000	1,000,000
Cobalt	1.79	797	10	70	250
Nickel	0.69	56	0.7	110	600
Alnico V	$I_r = 1.31$	50,900	640	—	—

2.2 Demagnetizing Factor

If a magnetic body of finite size is magnetized, free magnetic poles are induced on both its ends as shown in Fig. 2.8. These, in turn, give rise to a magnetic field in a direction opposite that of the magnetization. This field, called the demagnetizing field H_d, is proportional to the intensity of magnetization I. For regularly shaped specimens

$$H_d = \frac{NI}{\mu_0}, \tag{2.5}$$

where μ_0 is the permeability of vacuum as given in (1.2), and N is the demagnetizing factor (dimensionless quantity) which is a function of the shape of the specimen. If a long, thin specimen rod is magnetized along its long axis, the demagnetizing factor N is very small, whereas this factor is very large for a short, thick specimen.

Let us now calculate a demagnetizing factor for a simple case. Suppose that a very wide, thin ferromagnetic plate is magnetized normal to its wide surface. If the magnetization I is uniform throughout the specimen, magnetic free poles of the surface density $\pm I$ Wb/m² respectively will be

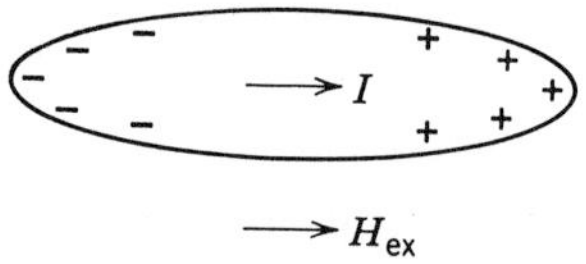

Fig. 2.8. Free pole distribution on the surface of a magnetized body.

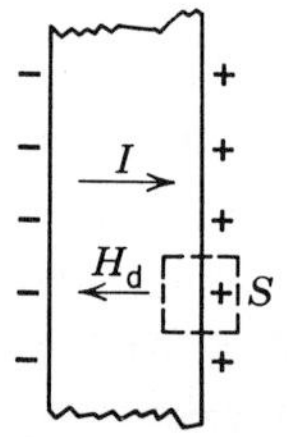

Fig. 2.9. Demagnetizing field due to the magnetization normal to the surface of a ferromagnetic plate.

induced on its surfaces. To calculate the demagnetizing factor we use Gauss' theorem, which states that the surface integral of the normal component of the field is given by the total free pole strength existent inside the enclosed region; that is,

$$\iint H_n \, dS = \frac{m}{\mu_0}. \tag{2.6}$$

We apply this theorem to a volume which includes the area S of the surface of the ferromagnetic plate (Fig. 2.9). The left side of (2.6) becomes $H_d S$, whereas the right side becomes IS/μ_0, and we have

$$H_d = \frac{I}{\mu_0}. \tag{2.7}$$

Comparing this result with (2.5), we have

$$N = 1. \tag{2.8}$$

If the same plate is magnetized parallel to its wide surface, free poles appear on its very narrow edges, and hence the demagnetizing field will be very weak. In the limit of an infinitely wide plate,

$$N = 0. \tag{2.9}$$

The demagnetizing factor is therefore dependent on the direction of magnetization.

In general cases the situation is not so simple. If a ferromagnetic body of irregular shape is magnetized, even a uniform distribution of magnetization gives rise to a non-uniform demagnetizing field, which, in turn, results in an irregular distribution of magnetization. Then the demagnetizing factor cannot be defined. It is only in the case of an ellipsoid that the demagnetizing field becomes uniform for a uniform distribution of magnetization. The calculated forms of demagnetizing factors of a general ellipsoid are fairly complex.[1,2] For a long, thin ellipsoid of rotation, for which the direction of magnetization is parallel to the long axis, the demagnetizing factor is given by

$$N = \frac{1}{k^2 - 1}\left[\frac{k}{\sqrt{k^2 - 1}} \log_e (k + \sqrt{k^2 - 1}) - 1\right], \tag{2.10}$$

where k is the dimension ratio, that is, the ratio of the length to the diameter. For $k \gg 1$, (2.10) becomes

$$N = \frac{1}{k^2} (\log_e 2k - 1). \tag{2.11}$$

For an oblate ellipsoid which is magnetized parallel to its circular plane,

$$N = \frac{1}{2}\left[\frac{k^2}{(k^2 - 1)^{3/2}} \sin^{-1} \frac{\sqrt{k^2 - 1}}{k} - \frac{1}{k^2 - 1}\right], \tag{2.12}$$

where k is the ratio of the diameter to the thickness. Numerical values of (2.10) and (2.12) are given in Table 2.2, together with the experimentally determined demagnetizing factor for a cylindrical specimen.

For a flat ellipsoid whose two long axes a and b are nearly equal to each other and very large compared to the short axis $c(a > b \gg c,)$ the demagnetizing factors for the directions of the a and b axes are approximately given by

$$N_a = \frac{\pi}{4}\frac{c}{a}\left[1 - \frac{1}{4}\frac{a - b}{a} - \frac{3}{16}\left(\frac{a - b}{a}\right)^2\right], \tag{2.13}$$

$$N_b = \frac{\pi}{4}\frac{c}{a}\left[1 + \frac{5}{4}\frac{a - b}{a} + \frac{21}{16}\left(\frac{a - b}{a}\right)^2\right]. \tag{2.14}$$

Although we have fairly complex expressions for the individual demagnetizing factors N_a, N_b, and N_c for the directions of the three principal axes of an ellipsoid, there is a simple relation between them:

$$N_a + N_b + N_c = 1. \tag{2.15}$$

Table 2.2. Demagnetizing Factors for Rods and Ellipsoids Magnetized Parallel to the Long Axis (after Bozorth[G.10])

Dimensional Ratio k	Rod	Prolate Ellipsoid	Oblate Ellipsoid
0	1.0	1.0	1.0
1	0.27	0.3333	0.3333
2	0.14	0.1735	0.2364
5	0.040	0.0558	0.1248
10	0.0172	0.0203	0.0696
20	0.00617	0.00675	0.0369
50	0.00129	0.00144	0.01472
100	0.00036	0.000430	0.00776
200	0.000090	0.000125	0.00390
500	0.000014	0.0000236	0.001567
1000	0.0000036	0.0000066	0.000784
2000	0.0000009	0.0000019	0.000392

From this relation we can easily derive the demagnetizing factor for simple cases. For instance, for a sphere, where $N_a = N_b = N_c$, we have

$$N = \tfrac{1}{3}. \tag{2.16}$$

For a very long cylinder, which is regarded as an ellipsoid with $a = b$ and $c = \infty$, we expect that $N_a = N_b$ and $N_c = 0$ so that

$$N_a = N_b = \tfrac{1}{2}. \tag{2.17}$$

For a flat plate, which is regarded as an ellipsoid with $a = b = \infty$, $N_a = N_b = 0$, so that

$$N_c = 1, \tag{2.18}$$

which coincides with (2.8).

If the intensity of magnetization is plotted as a function of external field for a specimen of finite shape, the magnetization curve becomes sheared as shown by the broken curve in Fig. 2.10. The reason is that the effective field H_{eff} inside the specimen is always less than the external field H_{ex} by the amount of the demagnetizing field; that is,

$$H_{\text{eff}} = H_{\text{ex}} - N\frac{I}{\mu_0}. \tag{2.19}$$

By using this relation we can correct the sheared magnetization curve to the normal one which is shown by the solid curve in Fig. 2.10.

As long as we are concerned with a ferromagnetic material, we normally cannot ignore the effect of demagnetizing fields. For instance, consider a

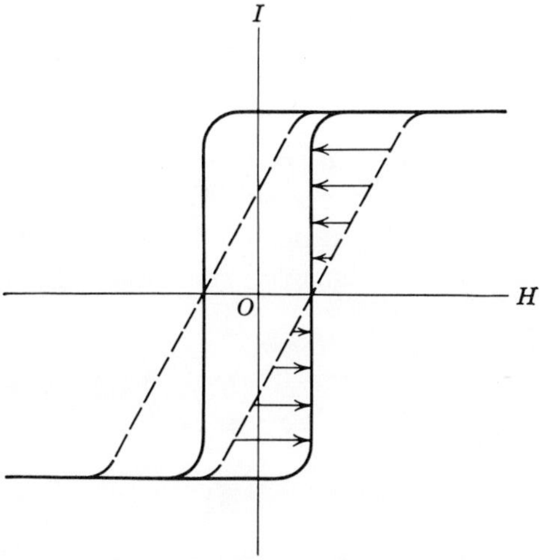

Fig. 2.10. Shearing correction of a magnetization curve.

sphere made of Permalloy. In order to magnetize it to its saturation magnetization, $I_s = 1.16$ Wb/m² (= 920 gauss), we must apply a magnetic field larger than its demagnetizing field. This is

$$H_d = N\frac{I_s}{\mu_0} = \frac{1}{3}\frac{1.16}{4\pi \times 10^{-7}} = 3.07 \times 10^5 \text{ A/m} \ (= 3860 \text{ Oe}), \quad (2.20)$$

which is about 10^5 times larger than its coercive force $H_c = 2$ A/m (= 0.025 Oe). Thus, when we use a soft magnetic material, we must be careful in designing the magnetic circuit. We shall discuss this point in the next section.

Finally let us discuss the magnetic field induced in an ellipsoidal hole in a magnetic body (Fig. 2.11). On the surface of the hole there appear

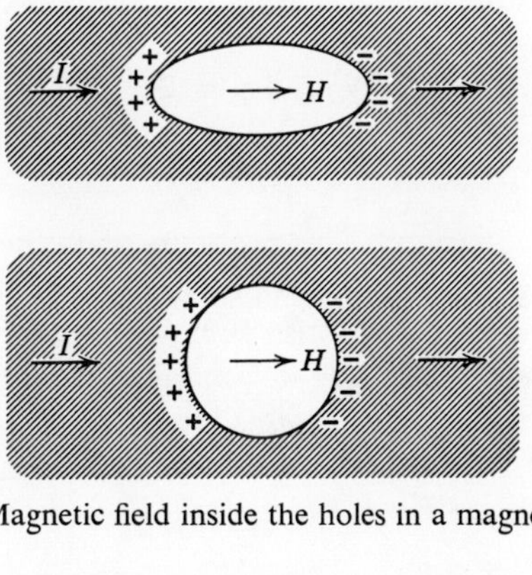

Fig. 2.11. Magnetic field inside the holes in a magnetized body.

magnetic free poles whose density is proportional to the intensity of magnetization of the body. The magnetic field induced by these free poles is parallel to the direction of magnetization and is expressed as

$$H_{\text{in}} = N \frac{I}{\mu_0}, \tag{2.21}$$

where N is the demagnetizing factor of the magnetic body the shape of which is the same as that of the hole. For instance, for a spherical hole,

$$H_{\text{in}} = \frac{I}{3\mu_0}. \tag{2.22}$$

This is usually referred to as the Lorentz field.

2.3 Magnetic Circuits

As mentioned in the preceding section, the demagnetizing field of a specimen of irregular shape is not uniform and a complicated distribution of magnetization results.

The same situation exists when we are confronted with solving a distribution of free electric charges in a space containing a dielectric substance of irregular shape. In the absence of true electric charge,

$$\text{div}\, \boldsymbol{D} = 0, \tag{2.23}$$

where $\boldsymbol{D}$ is the electric displacement or electric flux density. On the other hand,

$$\boldsymbol{D} = \epsilon \boldsymbol{E} = -\epsilon\, \text{grad}\, \phi, \tag{2.24}$$

where ϕ is the electric potential, so that (2.23) becomes

$$\text{div}\, (\epsilon\, \text{grad}\, \phi) = 0. \tag{2.25}$$

In a uniform dielectric substance we can put $\epsilon =$ const., so that (2.25) becomes

$$\text{div grad}\, \phi = 0,$$

or

$$\Delta \phi = 0, \tag{2.26}$$

which is Laplace's equation. The problem of finding a distribution of electric potential is reduced to the problem of solving the Laplace equation under the given boundary conditions. If, however, there are some dielectric substances distributed in an irregular manner, we must solve (2.25) for the given spatial distribution of ϵ; this is a fairly complex problem.

In magnetostatics, where there are some ferromagnetic substances of irregular shape, the situation is quite similar to the case just mentioned.

In magnetism, the relation

$$\operatorname{div} \boldsymbol{B} = 0 \tag{2.27}$$

is always valid. If we assume that the magnetization curve is given by a simple straight line without any hysteresis effect, we can simply write

$$\boldsymbol{B} = \mu \boldsymbol{H} = -\mu \operatorname{grad} \phi_m, \tag{2.28}$$

where ϕ_m is the magnetic potential. Then (2.27) becomes

$$\operatorname{div} (\mu \operatorname{grad} \phi_m) = 0, \tag{2.29}$$

which has exactly the same form as (2.25).

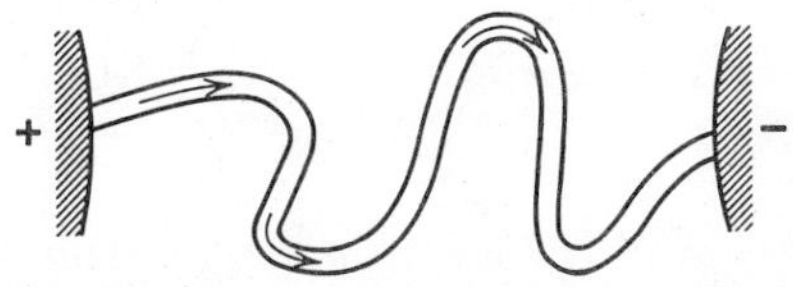

Fig. 2.12. Current distribution in a conductor of complex shape connecting two electrodes.

We also have a similar problem, that is, solving for the distribution of an electric current density in a conducting medium. For a steady current,

$$\operatorname{div} \boldsymbol{i} = 0. \tag{2.30}$$

Since

$$\boldsymbol{i} = \sigma \boldsymbol{E} = -\sigma \operatorname{grad} \phi, \tag{2.31}$$

where σ is the electric conductivity, (2.30) becomes

$$\operatorname{div} (\sigma \operatorname{grad} \phi) = 0, \tag{2.32}$$

which is also of the same form as (2.25) or (2.29).

If, however, the conductor is surrounded by insulators, the situation becomes very simple. For instance, if the conducting wire is placed in a vacuum as shown in Fig. 2.12, the electric current is concentrated into the wire no matter how complicated the shape of the wire. This is, however, not true for dielectric materials; here the $\boldsymbol{D}$ vector leaks more or less into the vacuum, because the dielectric constant of vacuum is finite ($\epsilon_0 = 8.85 \times 10^{-12}$ F/m). For ferromagnetic substances the situation is intermediate between the other two cases. Since the permeability of vacuum is finite ($\mu_0 = 4\pi \times 10^{-7}$ H/m), the situation is formally similar to the case in electrostatics. But actually the permeability of ferromagnetic substances is $10^3 \sim 10^5$ times larger than that of vacuum; hence the actual situation is quite similar to that of a steady current. In other words, if a magnetic substance forms a circuit, the magnetic induction $\boldsymbol{B}$ is almost fully confined inside the magnetic substance. Thus we can treat the magnetic circuit and the electric circuit similarly.

Figure 2.13 shows two typical magnetic circuits. One is magnetized by a permanent magnet, and the other is magnetized by a coil. Gray portions of the circuits are made from soft magnetic materials. The direction of magnetization should always be parallel to the circuit, because, if it is not parallel, free magnetic poles will appear on the surface, giving rise to a strong demagnetizing field that rotates the magnetization vector to a direction parallel to the circuit.

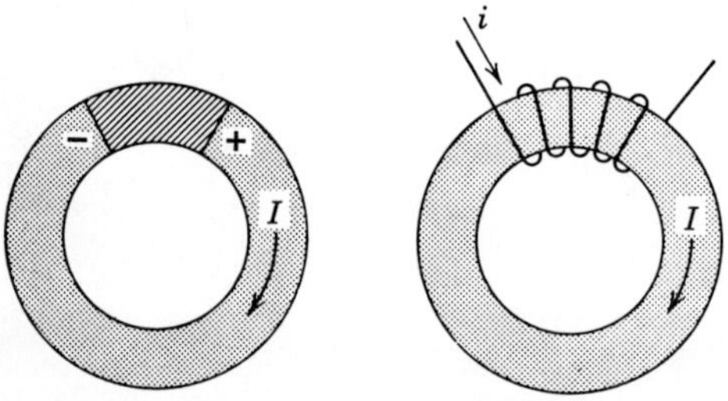

Fig. 2.13. Two kinds of magnetic circuits.

The flux density B in a magnetic circuit corresponds to the electric current density i of an electric circuit. The total flux of the circuit,

$$\Phi = \iint B_n \, dS, \tag{2.33}$$

corresponds to the total electric current. If B is uniform throughout the cross section of the magnetic circuit,

$$\Phi = BS. \tag{2.34}$$

The unit of magnetic flux is the weber (Wb).

When (2.29) and (2.32) are compared, the permeability μ is found to correspond to the conductivity σ. Thus, corresponding to the resistance of the electric circuit,

$$R = \int \frac{ds}{\sigma S}, \tag{2.35}$$

where S is the cross section of the conducting wire, and ds is a fragment of the circuit, we can define the magnetic resistance or reluctance as

$$R_m = \int \frac{ds}{\mu S}. \tag{2.36}$$

Corresponding to the electromotive force, we can define the magnetomotive force of the circuit as

$$V_m = \oint H_s \, ds, \tag{2.37}$$

where the integration should be carried out through one cycle of the circuit. When the magnetic circuit is magnetized by an electric current i, which flows in a coil of total N turns wound around the magnetic circuit, we have, from Ampère's theorem,

$$\oint H_s \, ds = Ni, \tag{2.38}$$

so that the magnetomotive force of the coil is given by

$$V_m = Ni. \tag{2.39}$$

The unit of magnetomotive force is the ampere-turn (AT) or simply the ampere (A). Using relations (2.28), (2.34), and (2.36), we have

$$V_m = Ni = \oint H_s \, ds = \oint \frac{B}{\mu} \, ds = \oint \frac{\Phi}{\mu S} \, ds = \Phi \oint \frac{ds}{\mu S} = \Phi R_m. \tag{2.40}$$

This relation corresponds to Kirchhoff's second law of the electric circuit; that is,

$$V = iR. \tag{2.41}$$

The calculation of the magnetomotive force for a permanent magnet is fairly complex. When a magnet supplies the magnetomotive force in a magnetic circuit, the direction of the magnetic field H_p inside the permanent magnet is always opposite to that of the magnetic flux B_p. The reason is that the demagnetizing field of the permanent magnet itself can be reduced in its value by attaching it to a magnetic circuit, but never changed in its direction unless another magnet or coil supplies an additional magnetomotive force. Thus the magnetic state of the permanent magnet is represented by a point, say P, in the second quadrant of the B-H space (Fig. 2.14). If the magnetic field is changed, the point P will move reversibly on the line RQ the slope of which, dB/dH, is given by the reversible permeability μ_r. If we translate the origin O to S, at which point the extrapolation of RQ intersects the abscissa, SR is regarded as a simple reversible magnetization curve. Thus a permanent magnet is equivalent to a magnetic material with permeability μ_r which has the ability to drive the magnetic flux against the hypothetical field OS $(= H_r)$. The magnetomotive force is given by integration of H_r through the length of the permanent magnet, so that

$$V_m = -H_r l_p = \frac{B_r l_p}{\mu_r}, \tag{2.42}$$

where l_p is the length of the permanent magnet and B_r is the remanent magnetic flux density (OR). It must be remarked, however, that the permanent magnet itself has a considerable reluctance given by

$$R_m' = \frac{l_p}{\mu_r S_p}. \tag{2.43}$$

Generally μ_r of a permanent magnet is very small, so that R_m' is usually fairly large, while the magnetomotive force (2.42) also has a fairly large

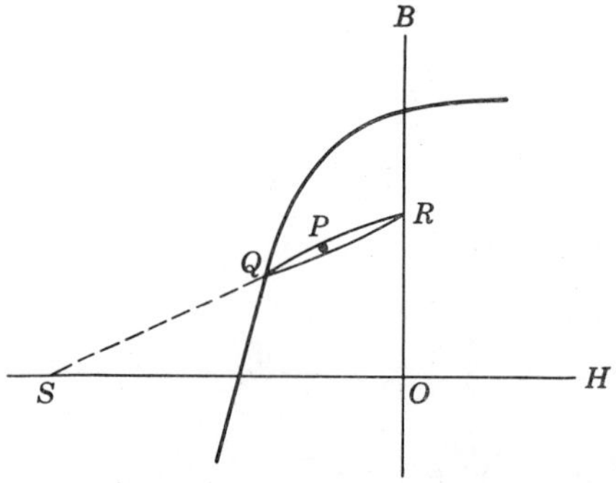

Fig. 2.14. Demagnetizing curve of a permanent magnet and the operating point P.

value. Thus a permanent magnet corresponds to a high voltage battery with a fairly large internal resistance which supplies a constant current independent of the output impedance. For a magnetic circuit with a permanent magnet,

$$\oint H_s\, ds = 0,$$

because of the absence of electric current. Then

$$-H_p l_p = \int_{\text{ex}} H_s\, ds, \tag{2.44}$$

where H_p is the field inside the permanent magnet and the integral is to be carried out along the part of the circuit excluding the permanent magnet. Thus

$$\begin{aligned} V_m &= \frac{B_r l_p}{\mu_r} = -H_r l_p = -\left(H_p - \frac{B}{\mu_r}\right) l_p = -H_p l_p + \frac{B l_p}{\mu_r} \\ &= \int_{\text{ex}} H_s\, ds + \Phi \frac{l_p}{\mu_r S_p} = \Phi\left(\int_{\text{ex}} \frac{ds}{\mu S} + \frac{l_p}{\mu_r S_p}\right) \\ &= \Phi(R_m + R_m'). \end{aligned} \tag{2.45}$$

This relation corresponds to Kirchhoff's second law.

If a magnetic circuit has a shunt or shunts, we have Kirchhoff's first law,

$$\Phi_1 + \Phi_2 + \cdots = 0, \tag{2.46}$$

where $\Phi_1, \Phi_2, \ldots$ are the fluxes which leave the shunt point through the branches, 1, 2, ...

By using Kirchhoff's first and second laws, we can solve any kind of magnetic circuit, as for an electric circuit. It must be noted, however, that a magnetic circuit has more or less leakage of the magnetic flux, just as an electric circuit dipped in an electrolytic solution has. Sometimes it happens that the leakage is so large that the actual flux is only a small fraction of the calculated value.

2.4 Measurement of Magnetization and Magnetic Field

In this section we survey various methods for the measurement of magnetization. From the physical principles, these methods are classified into (1) measurement of the force acting on the magnetic specimen, (2) measurement of the magnetic field produced by the specimen, and (3) measurement of the voltage or current induced by electromagnetic induction.

Measurement of the force acting on the specimen. When a magnetic specimen is placed in an inhomogeneous magnetic field, the specimen is acted on by the force given by (1.6), from which we can calculate the magnetic moment of the specimen. There are various methods by which the force can be measured. Most commonly used is the magnetic balance, in which one arm suspends the specimen between the pole pieces of an electromagnet while the other arm is balanced by weights or by a current-carrying coil which is placed in a radial magnetic field produced by a small electromagnet similar to that of a dynamic speaker (Fig. 2.15). In the latter the force can be measured by the current in the suspended coil. In the Sucksmith ring balance[3] a specimen is suspended from the bottom of an elastic metal ring the deformation of which, caused by the attraction of the specimen into the field, can be measured by an optical method through deflection of the mirrors attached to the ring (Fig. 2.16). These two methods are suitable for the measurement of weak magnetism only, for a ferromagnetic specimen will be attracted directly to the pole piece of the electromagnet by the gradient of the field or by the image force caused by the interaction between the magnetic moment of the specimen and its magnetic image induced in the pole piece.

A horizontal magnetic pendulum used by Weiss et al.,[4] shown in Fig. 2.17, has only one degree of freedom of motion for the specimen. This is in a direction perpendicular to the magnetic field, so that a fairly strong magnetism can be measured without being disturbed by the attraction of the specimen to the pole piece. Displacement of the arm due to the force

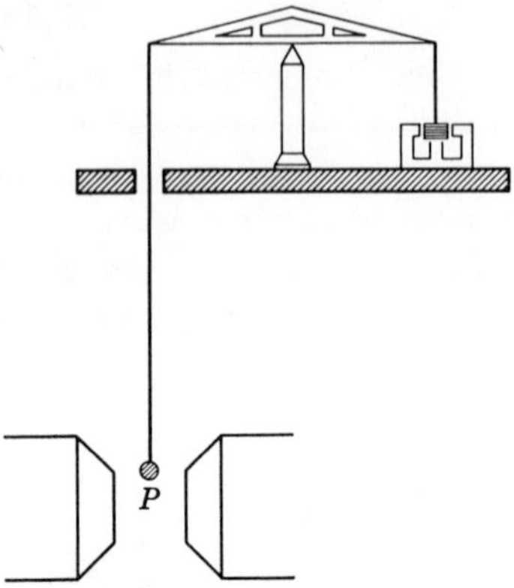

Fig. 2.15. Magnetic balance.

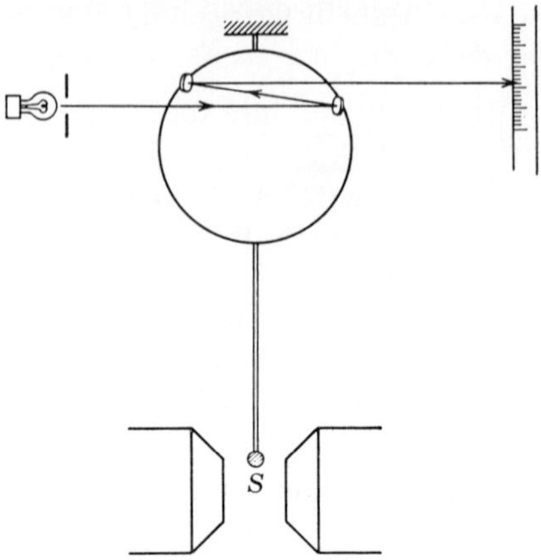

Fig. 2.16. Sucksmith's ring balance.

acting on the specimen is balanced by adjusting the current in a coil which encloses a piece of permanent magnet attached to the arm. The displacement of the arm is sensitively detected by the deflection of a mirror which is attached to the arm through a needle. The current in the coil is a direct measure of the magnetic moment of the specimen.

Bozorth and Williams[5] used a vertical magnetic pendulum suspended through a piece of elastic metal plate (Fig. 2.18). The specimen is attached

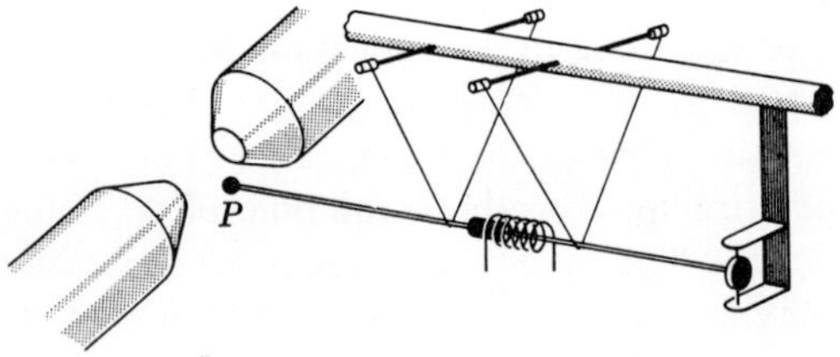

Fig. 2.17. Horizontal magnetic pendulum.

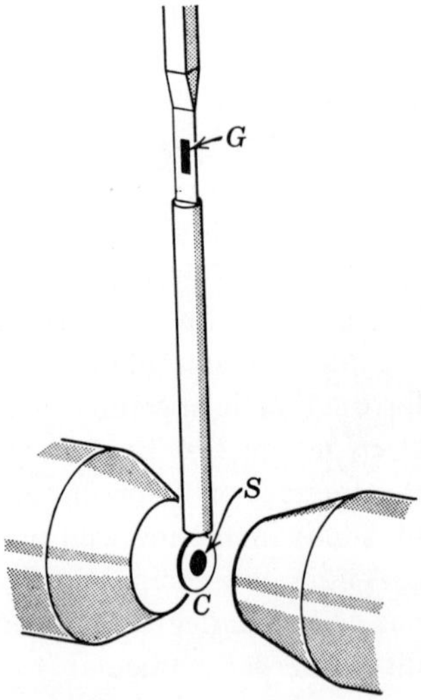

Fig. 2.18. Vertical magnetic pendulum.

to the bottom of the pendulum together with a coil which carries a current so as to produce a magnetic moment just compensating the magnetic moment of the specimen. The deflection of the pendulum is sensitively detected by a pair of strain gauges which are attached to both sides of the elastic metal plate. The specimen and the coil are placed in an inhomogeneous magnetic field produced by tapered pole pieces. The current in the coil also gives a direct measure of the magnetic moment of the specimen.

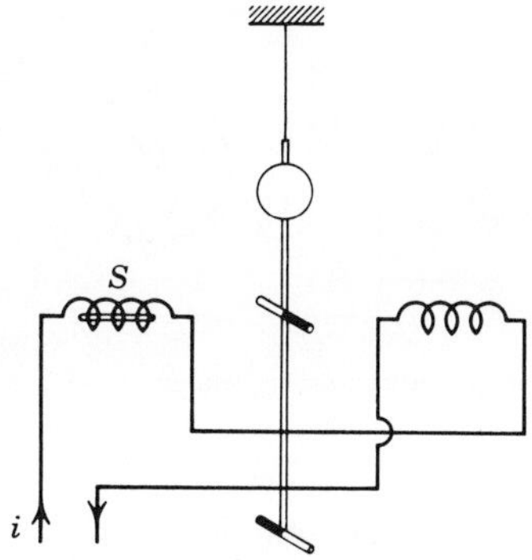

Fig. 2.19. Magnetometer.

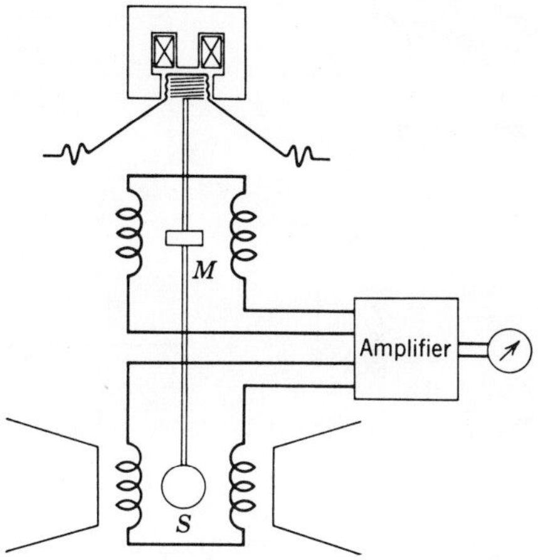

Fig. 2.20. Vibrating-specimen magnetometer.

Measurement of the magnetic field produced by the specimen. A classical method of this kind makes use of the magnetometer. This instrument consists of a pair of antiparallel magnetic needles and a pair of solenoids of the same size (Fig. 2.19). Since the two coils are so placed as to produce opposing fields at the position of the magnetic needles, only the field due to the magnetization of the specimen S is effective in causing a deflection of the needles which can be detected by optical means.

An advanced type of apparatus which belongs to this category is the vibrating-specimen magnetometer developed by S. Foner[6] (Fig. 2.20). The specimen is forced to vibrate in a vertical direction by a device similar to a dynamic speaker. The ac signal induced by the dipole field of the specimen in a pair of secondary coils placed on both sides of the specimen is amplified and compared with a signal produced by a standard magnet M, giving rise to an output signal which is exactly proportional to the magnetic moment of the specimen.

Measurement of voltage or current induced by electromagnetic induction. The ballistic galvanometer is the most common apparatus in measurements of this type (Fig. 2.21). If the magnetic flux inside the secondary coil is changed by changing the magnetization of the specimen (by increasing a magnetic field stepwise or by reversing a magnetic field), or by removing the specimen from the coil, a voltage $\partial\Phi/\partial t$ is induced in the coil The first deflection of the galvanometer is proportional to the electric charge which passes through it, which, in this instance, is proportional to the integrated value of $\partial\Phi/\partial t$ or to the flux change $\Delta\Phi$. The proportionality factor can be determined by a standard mutual inductance. The fluxmeter is a sort

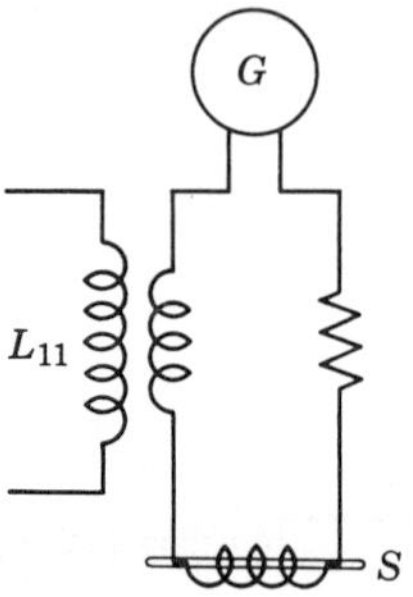

Fig. 2.21. Ballistic galvanometer circuit.

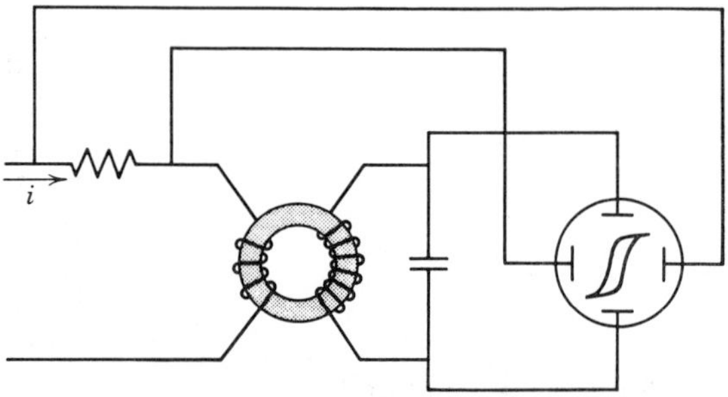

Fig. 2.22. Oscillograph hysteresis tracer.

of galvanometer whose coil is overdamped by electromagnetic means, so that the deflection can be easily read.

A conventional oscillograph hysteresis tracer is a device which integrates the voltage induced in the secondary coil by means of a capacitor as shown in Fig. 2.22. The driving field must be an ac field.

The most advanced type of fluxmeter is Cioffi's recording fluxmeter[7] (Fig. 2.23), in which the secondary coil of the specimen is connected to the secondary coil of the mutual inductance L_{11} through the galvanometer, the deflection of which is detected by a pair of photocells. The primary current of the mutual inductance L_{11} is automatically adjusted to make the deflection of the galvanometer practically zero. Under this condition the primary current of the mutual inductance is exactly proportional to the total flux change of the specimen, and the trace on the xy recorder is B as a function of H.

Measurement of magnetic field. The measurement of the magnetic field is also very important for the observation of the magnetization curve. For a solenoid we can easily calculate the coil constant, that is, the proportionality factor between H and i, if we know the number of turns and the diameter of each layer. For an electromagnet we can measure the intensity of a magnetic field by means of an induction method described above by using a small search coil. The magnetoresistance effect in bismuth and the

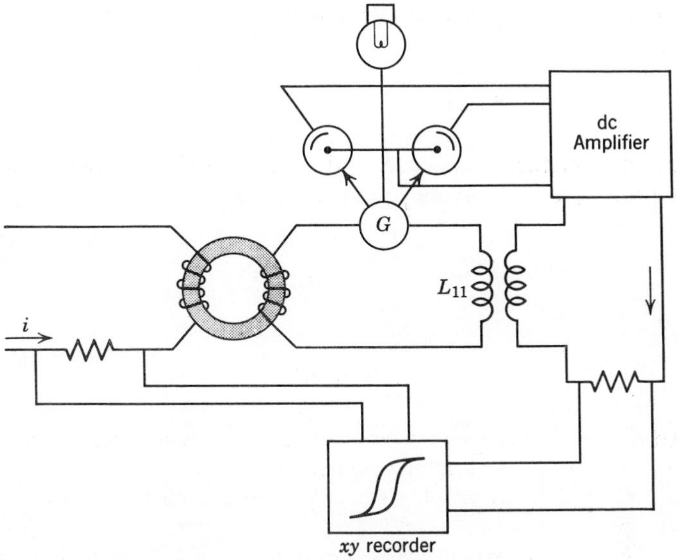

Fig. 2.23. Cioffi recording fluxmeter.

Hall effect in germanium can also be used as means of measuring a magnetic field.

The direct measurement of the effective field inside a ferromagnetic specimen is the most difficult problem. Since the tangential component of the magnetic field is continuous across the surface of the specimen, a small pick-up element which can be attached to the surface of the specimen can be used for a direct measurement of H_{eff}.

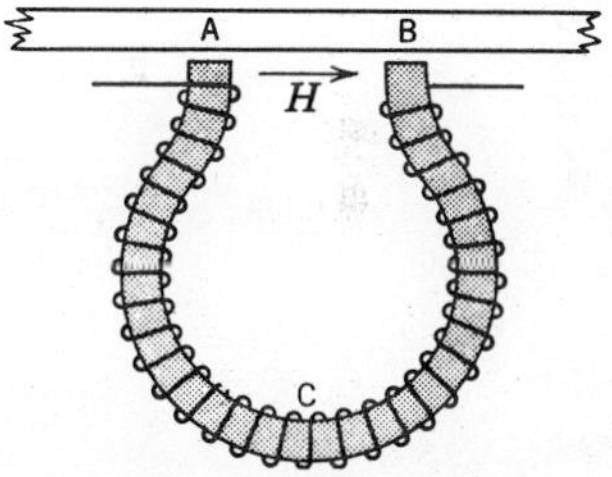

Fig. 2.24. Magnetic potentiometer.

Another ingenious device is the magnetic potentiometer,[8,9] which is a uniformly wound secondary coil of the shape shown in Fig. 2.24. It is easily shown that this coil is equivalent to the straight coil of length AB whose winding density and cross section are the same as those of the magnetic potentiometer. Since

$$\oint H_s \, ds = 0 \tag{2.47}$$

for the circuit ABCA, it follows, that

$$\int_{ACB} H_s \, ds = -\int_{AB} H_{eff} \, ds, \tag{2.48}$$

so that the total flux which goes through the magnetic potentiometer is given by

$$\Phi = \mu_0 \int_{\mathrm{ACB}} nSH_s\, ds = -nS\mu_0 H_{\mathrm{eff}} l, \tag{2.49}$$

where l is the distance between A and B and n is the number of turns per unit length of the coil. From (2.49), we have

$$H_{\mathrm{eff}} = -\frac{\Phi}{\mu_0 nSl}. \tag{2.50}$$

Problems

2.1. How strong a magnetic field is necessary to magnetize an iron sphere of diameter 1 mm to its saturation magnetization ($I_s = 2.12$ Wb/m^2)? Calculate the gradient of the magnetic field necessary to attract this sphere with a force of 1 g-wt.

2.2. Calculate the apparent relative permeability of a prolate ellipsoid whose dimension ratio is $e^4/2$ and whose relative permeability is $\bar{\mu}$.

2.3. Calculate the intensity of the magnetic field in the air gap of the magnetic circuit shown in the figure. Use the values $N = 200$, $i = 5$ A, $S_1 = 2.5 \times 10^{-3}$ m^2, $S_2 = 5 \times 10^{-4}$ m^2, $l_1 = 1$ m, $l_2 = 0.01$ m, $\bar{\mu} = 500$.

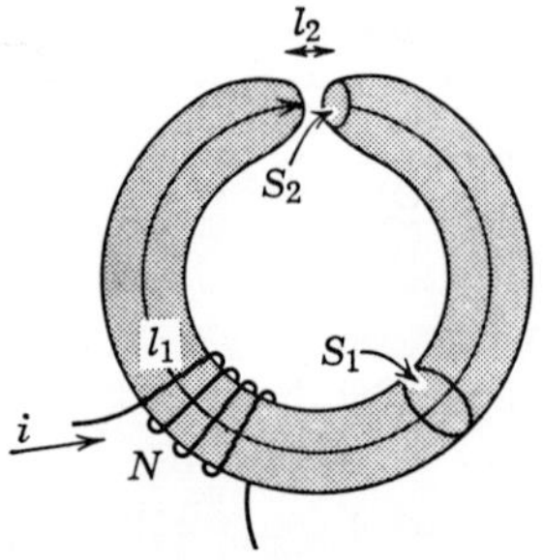

Problem 2.3.

References

2.1. J. A. Osborn: *Phys. Rev.* **67,** 351 (1945).
2.2. E. C. Stoner: *Phil. Mag.* [7] **36,** 803 (1945).
2.3. W. Sucksmith and R. R. Pearce: *Proc. Roy. Soc.* (*London*) **167A,** 189 (1938); W. Sucksmith: *ibid.*, **170A,** 551 (1939).

2.4. P. Weiss and G. Foëx: *J. Phys.* [5] **1,** 274 (1911).
2.5. R. M. Bozorth and H. J. Williams: *Phys. Rev.* **103,** 572 (1956).
2.6. S. Foner: *Rev. Sci. Instr.* **30,** 548 (1959).
2.7. P. P. Cioffi: *Rev, Sci. Instr.* **21,** 624 (1950).
2.8. A. P. Chattock: *Phil. Mag.* **24,** 94 (1887).
2.9. L. F. Bates: G.14.

part 2

INTRINSIC MAGNETIZATION

3

Origin of Magnetism

3.1 Orbital Motion and Spin

We learned in Chapter 1 that ferromagnetism is the result of an ordered alignment of the atomic magnetic moments. In this chapter we discuss the possible physical origins of the atomic moment and then describe two kinds of experiments, namely, the gyromagnetic experiment and ferromagnetic resonance, both of which serve to identify the source of the moment.

On the basis of an elementary knowledge of atomic structure, two possible origins may be imagined for the atomic magnetic moment. One of these is the orbital motion of the electron around the nucleus, and the other is a spin motion of the electron about its own axis. In order to distinguish between the two possibilities experimentally we can use the difference in the characteristic angular momentum-magnetic moment ratios of the two motions.

First we consider the simple model of an atom in which the electron moves in a circular orbit of radius r at an angular velocity ω. Since the electron makes $\omega/2\pi$ turns per second, its motion constitutes a current of $-e\omega/2\pi$ A where $-e$ is the electric charge of a single electron. The magnetic moment of a closed circuit of electric current i whose cross section is S m^2 is known from electromagnetic theory to be $\mu_0 iS$ Wb-m. Therefore the magnetic moment produced by the circular motion of the electron should be given by

$$M = -\mu_0 \frac{e\omega}{2\pi}(\pi r^2) = -\tfrac{1}{2}\mu_0 e\omega r^2. \tag{3.1}$$

Since the angular momentum of the circular motion is given by

$$P = m\omega r^2, \tag{3.2}$$

where m is the mass of the single electron, (3.1) may be written as

$$M = -\frac{\mu_0 e}{2m} P. \tag{3.3}$$

This motion of the electron is the famous problem which led to the quantum theory, and it is well known that we cannot give a complete description of the phenomenon on the basis of classical physics. It turns out that the electron may move in any orbit satisfying the condition that the angular momentum of the orbit be an integer multiple of $\hbar$, that is,

$$P = n\hbar, \tag{3.4}$$

where $\hbar$ denotes Planck's constant divided by 2π, that is,

$$\hbar = \frac{h}{2\pi} = \frac{6.626}{2\pi} \times 10^{-34} = 1.054 \times 10^{-34} \text{ J-sec.} \tag{3.5}$$

By use of relation (3.4), (3.3) becomes

$$M = -n\frac{\mu_0 \hbar e}{2m}. \tag{3.6}$$

This equation tells us that the magnetic moment of the orbital motion can change its value by the unit

$$M_B = \frac{\mu_0 \hbar e}{2m}, \tag{3.7}$$

which is the smallest possible magnetic moment and is usually called the Bohr magneton. By using the numerical values $\mu_0 = 4\pi \times 10^{-7}$, $e/m = 1.76 \times 10^{11}$ C/kg, and $\hbar = 1.054 \times 10^{-34}$ (J-sec), we have

$$M_B = 1.165 \times 10^{-29} \text{ Wb-m.} \tag{3.8}$$

Detailed information about the energy separations of various electronic states can be obtained from the spectroscopic study of atomic radiation. The concept of "spin" was first introduced to explain the multiplet structure of atomic spectra; it was thought to correspond to the spin motion of the electron about its own axis with angular momentum $\pm\hbar/2$. That this motion is accompanied by a magnetic moment was experimentally verified by the Zeeman effect. When the radiating atoms are placed in a magnetic field, a spectral line is split into two or three lines. From this fact we can infer that the relation between magnetic moment and angular momentum for spin motion is different from (3.3). Instead it is given by

$$M = -\frac{\mu_0 e}{m} P. \tag{3.9}$$

Comparing (3.3) and (3.9), we see that the proportionality factors for orbital and spin motion differ in the ratio 1:2. If we write the relation generally as

$$M = -g\frac{\mu_0 e}{2m} P, \tag{3.10}$$

it follows that $g = 1$ for orbital motion and $g = 2$ for spin. The factor g is called the "gyromagnetic ratio" or simply the "g factor."

Using values $\mu_0 = 4\pi \times 10^{-7}$, $e/m = 1.76 \times 10^{11}$ (C/kg), the proportionality factor in (3.10) is calculated as

$$\nu = g\frac{e\mu_0}{2m} = 1.105 \times 10^5\, g \quad \text{(m/A-sec)}, \tag{3.11}$$

which is referred to as gyromagnetic constant. Then (3.10) becomes

$$M = -\nu P. \tag{3.12}$$

Since the spin angular momentum changes by the unit $\hbar/2$, the unit of spin magnetic moment is also a Bohr magneton. In order to determine which mechanism is responsible for the observed ferromagnetism, we must therefore know the value of the g factor for the actual ferromagnetic substance. We have two kinds of experiments which give information about the g factor: the gyromagnetic experiment and ferromagnetic resonance.

3.2 Gyromagnetic Effect

The term gyromagnetic effect refers to such phenomena as the generation of angular momentum by the magnetization of a ferromagnetic body or the induction of magnetization by the rotation of the ferromagnetic substance. Since an atomic magnetic moment has angular momentum whether its source is the orbital or spin motion, its behavior should be similar to that of a top. It is well known that, when a spinning top is placed in a gravitational field, it precesses. This is the fundamental phenomenon of the gyromagnetic effect.

Before entering into the detailed description of the gyromagnetic effect, we discuss several important aspects of the behavior of a top. This will be useful for the understanding of the gyromagnetic effect itself as well as of the dynamical behavior of the magnetization to be discussed in Chapter 16.

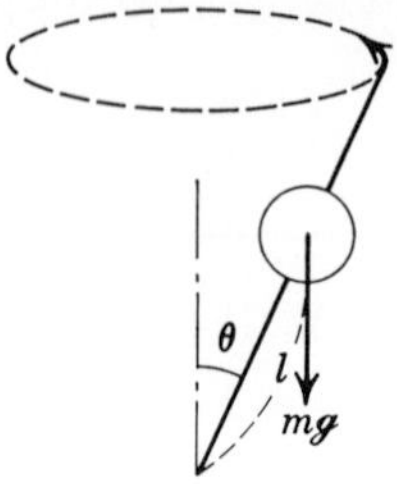

Fig. 3.1. Precession motion of a top in a gravitational field.

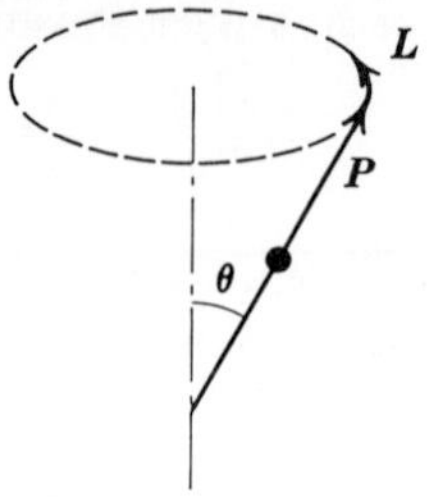

Fig. 3.2. Angular momentum vector and torque due to external force.

When a top is spinning in a field of gravitation, the moment of the couple which acts on it is given by

$$L = mgl \sin \theta \tag{3.13}$$

where m is the mass of the top, l the distance between the center of gravity and the bottom end of the axis, θ the angle of inclination of the axis from the vertical direction, and g the acceleration of gravity (cf. Fig. 3.1). For the magnetic moment M, which is in a magnetic field H, the moment of the couple is

$$L = -MH \sin \theta. \tag{3.14}$$

Since (3.13) and (3.14) have the same functional dependence on θ, we can express both by

$$L = L_0 \sin \theta. \tag{3.15}$$

Now, when the couple of forces $\boldsymbol{L}$ acts on a top having angular momentum $\boldsymbol{P}$, the vector $\boldsymbol{P}$ must change at the rate

$$\frac{d\boldsymbol{P}}{dt} = \boldsymbol{L}. \tag{3.16}$$

For a top which is placed in a gravitational field or a magnetic moment which is placed in a uniform magnetic field, the direction of the vector $\boldsymbol{L}$ is always perpendicular to the axis of the top or the magnetic moment and also to the direction of gravity or the magnetic field. The end point of the vector $\boldsymbol{P}$, therefore, moves along a circle whose radius is equal to $P \sin \theta$ as shown in Fig. 3.2. The angular velocity of the precession is, therefore,

$$\omega = \frac{L}{P \sin \theta} = \frac{L_0}{P}. \tag{3.17}$$

The value of ω thus obtained is seen to be independent of the angle θ; that is, the angular velocity of the precession is independent of the inclination of the top.

It should be emphasized that a spinning top, if free to precess under the action of an external force, will maintain its inclination with respect to the force. No matter how strong the force applied may be, the only effect of increasing the force on the freely precessing top is to make the precession faster. When, on the other hand, the top loses its freedom to precess, then, like a non-spinning body, it moves in the direction of the external force. For example, if a spinning gyroscope is allowed to move only in a vertical plane as shown in Fig. 3.3, it immediately drops to the lowest position and oscillates about this position independently of its spinning velocity. The reason is that the top first wants to start precessing about the vertical axis, but such a motion is opposed by the reaction from the horizontal axis. This reaction in turn induces precession about the horizontal axis. The final movement of the gyroscope is the same as the dropping motion of the non-spinning top. This behavior of the top is utilized in the recently discovered magnetic material called Ferroxplana which has excellent magnetic properties in the high frequency range. The magnetization in this material is confined to a certain crystallographic plane, being prevented from giving rise to a resonance rotation up to considerably high frequencies (cf. Section 16.2).

When a damping force acts on the precession motion, the top becomes partially obedient to the external force. For example, suppose that the spinning gyroscope is just stirring water by its precession motion as shown in Fig. 3.4. The reaction from the water should influence the motion of the gyroscope so as to make the rotation axis approach the vertical direction. The stirring motion decreases in amplitude gradually and finally fades away. Thus the top has the characteristic that it enjoys the sport of precession while it is free and does no useful work, but it becomes suddenly idle as soon as one intends it to do work. This is somewhat similar to the common nature of children.

In the situation shown in Fig. 3.4 the top does some work before it stops stirring the water; that is, the motion of the top gives up to the water a certain amount of angular momentum. According to the law of conservation of angular momentum, the momentum P_{ob} which the water obtained must be equal to the change in the angular momentum of the top. Thus

$$P_{ob} = P(\cos\theta - 1) \tag{3.18}$$

where P is the angular momentum of the top about its rotation axis, and θ is the angle between the top axis and the vertical direction. It may seem

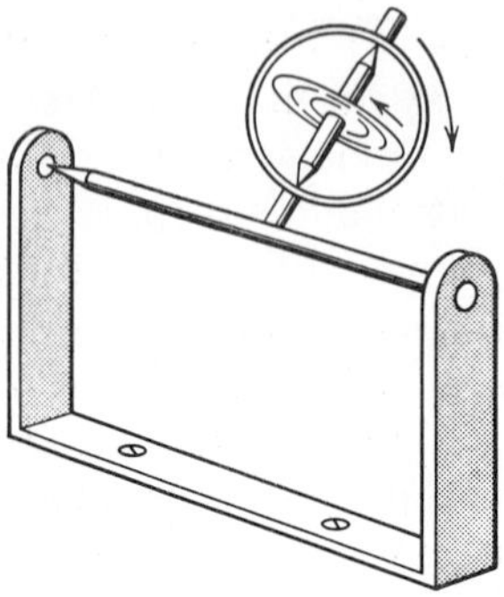

Fig. 3.3. Gyroscope restricted to vertical plane.

Fig. 3.4. Gyroscope's precession damped by water.

somewhat strange that P_{ob} does not depend on the damping characteristics of the water. It does not, however, because if the damping action of the water is weak the top will continue to stir the water so much longer, whereas if the damping action is strong the top will give up much more angular momentum in each revolution of stirring. It must also be noted that the work done in stirring the water is supplied not from the kinetic energy of the spinning top, but rather from the decrease in the potential energy of the top in the gravitational field.

It is worthwhile to add that the spinning top can exhibit inertia when it is suitably connected to an elastic device. Consider the top which is bound to a rotatable vertical axis by way of a spring as shown in Fig. 3.5. Since the couple of the force of gravity is counterbalanced by the elastic force of the spring, the top does not precess. If we now rotate the vertical axis counterclockwise as shown in the figure, this action induces a precession tending to make θ smaller. Then the couple of force due to the spring becomes smaller than that due to gravity, and the top begins to precess about the vertical axis. Since the sense of this motion is the same as that of our initial action, we can feel that this system has some inertia. If we rotate the system in the opposite direction, this action makes the angle θ larger and results in precession in the opposite direction. The inertia of the domain wall which is discussed in Chapter 16 comes from such a situation.

In a gyromagnetic experiment the g factor is measured by making use of the dynamical properties of the top described above. The first trial was made by Maxwell in 1861.[1] The procedure in his experiment is to rotate a permanent bar magnet about the vertical axis and to observe the resultant inclination of the magnet (Fig. 3.6). If the carrier of the magnetic moment has an angular momentum, the rotation should induce an inclination of the magnet in a manner similar to that described in connection with Fig. 3.5. The centrifugal reaction tends to make the bar magnet lie in the horizontal plane. Thus the gyromagnetic ratio should be determined from a knowledge of the magnetization of the bar and the balanced inclination of the magnet. Unfortunately this experiment has never succeeded. The main reason for this may be that the effect is too small compared to the mechanical disturbance which would arise from the slightest unbalance of the mechanical system.

The first successful gyromagnetic experiment was conducted by Barnett.[2] In his procedure a specimen rod is magnetized by high speed rotation of the specimen itself, and its magnetization is compared with that of another specimen rod which is magnetized by a magnetic field. When the specimen is rotated at angular velocity ω, the atomic magnetic moments should tend to align themselves with the rotation axis as in Fig. 3.5, inducing a weak magnetization parallel to the rotation axis. The inclination of the individual atomic moment can be determined by the condition (cf. 3.17)

$$L = P\omega \sin \theta, \tag{3.19}$$

where L is the couple of force which is supplied by the magnetic anisotropy

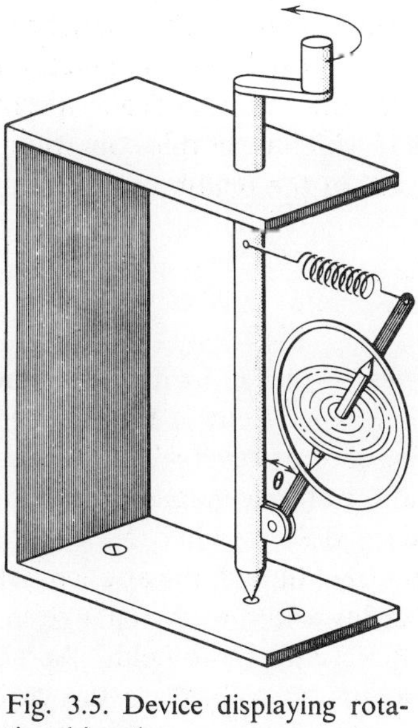

Fig. 3.5. Device displaying rotational inertia.

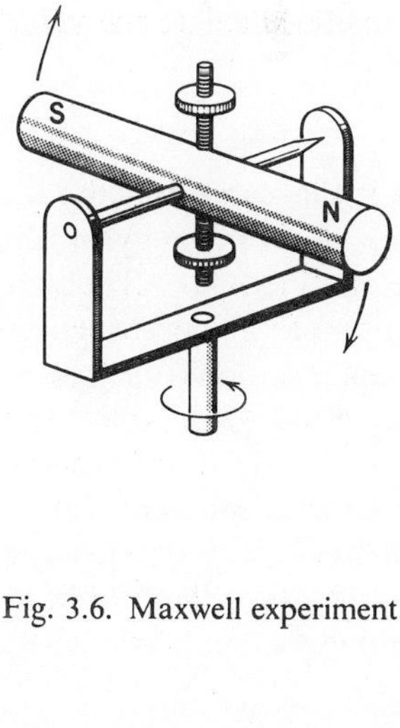

Fig. 3.6. Maxwell experiment.

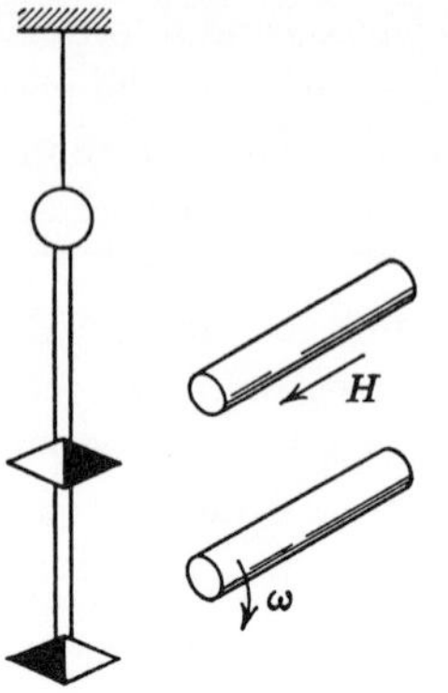

Fig. 3.7. Barnett method.

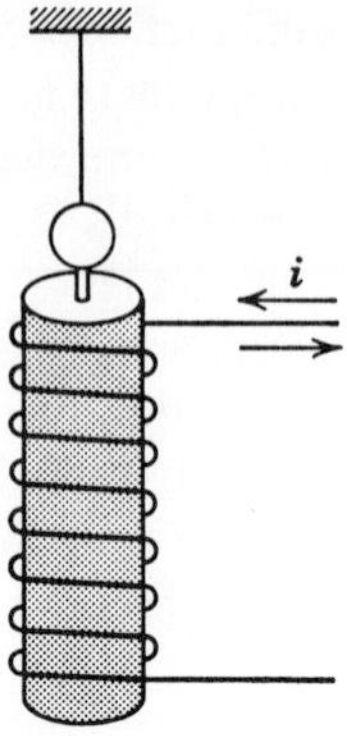

Fig. 3.8. Einstein-de Haas method.

of the substance. When the specimen is magnetized by the magnetic field, the inclination of the atomic magnetic moment must be given by

$$L = -MH \sin \theta. \tag{3.20}$$

If we adjust the magnetic field so as to satisfy the relation

$$P\omega = -MH, \tag{3.21}$$

the inclination of the individual atomic magnetic moments must be the same for the two cases, and thus the same intensity of magnetization results. We can easily find this condition by balancing a magnetometer which has two oppositely magnetized needles, one in front of each specimen (Fig. 3.7). Knowing the values of ω and H which meet this condition, we can easily calculate the value of M/P by use of the relation

$$\frac{M}{P} = -\frac{\omega}{H}. \tag{3.22}$$

This is the Barnett method.

The most refined gyromagnetic experiment is that utilizing the Einstein-de Haas method.[3] The procedure in this experiment is to observe the rotation of a freely suspended specimen induced by reversing the direction of magnetization (Fig. 3.8). The motion of the atomic magnetic moment in the material is similar to the motion of the top described in connection with Fig. 3.4. If the atomic moment is perfectly free, though the applied field is reversed, the moment should go on precessing without any change in the component of magnetization along the direction of the field. Actually, however, since the atomic magnetic moments are subject to the various damping actions of the lattice, they perform the helical motion shown in

Fig. 3.4, finally aligning themselves with the direction of the applied magnetic field. The change in the angular momentum of the crystal lattice which was transferred from the motion of the atomic moment is given by (3.18). On the other hand, this motion of the atomic moment results in a change in the vertical component of magnetic moment which is given by

$$M_{ob} = M(\cos\theta - 1). \tag{3.23}$$

Comparing (3.23) and (3.18), we have

$$\frac{M}{P} = \frac{M_{ob}}{P_{ob}}. \tag{3.24}$$

Since this relation is valid, independently of the inclination θ of the atomic moments, the ratio on the right side of (3.24) can be replaced by the ratio

Table 3.1. Summary of Results of Gyromagnetic Experiments (after Kittel[G.15])

Substance	g	ϵ
Iron	1.93	0.04
Cobalt	1.85	0.07
Nickel	1.84–1.92	0.04–0.08
Magnetite, Fe_3O_4	1.93	0.04
Heusler alloy, Cu_2MnAl	2.00	0.00
Permalloy, 78Ni, 22Fe	1.90	0.05
Supermalloy	1.91	0.05

of the total magnetization change to the total angular momentum change for the specimen.

There are a number of experimental techniques to measure the change of angular momentum, among which the simplest and most refined may be that used by Scott.[4] He used a very massive pendulum whose period of oscillation was about 26 sec. The ferromagnetic specimen was attached to it and the magnetic field was reversed in synchronism with the oscillations. From the observed change in the decay constant of the oscillation he calculated the change in angular momentum per single reversal of the magnetization.

These experiments yield the value of M/P, which we use to calculate the value of the g factor by the relation

$$g = -\frac{2m}{\mu_0 e}\frac{M}{P}. \tag{3.25}$$

The results are summarized in Table 3.1. We see in the table that the values of the g factor are close to 2 for almost all substances. This indicates that the origin of ferromagnetism is not an orbital motion but a spin motion of the electron. The fact that the g values are slightly less than 2 shows the presence of a small orbital contribution to the moment.† The factor ϵ is the ratio of the contribution of the orbit to that of the spin.

3.3 Ferromagnetic Resonance

We learned above that the angular velocity ω of precession of a top is independent of its inclination to the external force (cf. 3.17). For a magnetic moment in an applied field we have

$$\omega = -\frac{M}{P} H. \tag{3.26}$$

In order to maintain this precession motion for an actual atomic moment, energy must be fed into the system from without. Otherwise the precession must die out because of the damping action of the crystal lattice. One of the means available for doing this is to apply an alternating magnetic field normal to the axis of precession. To estimate the order of magnitude of ω for an actual magnetic moment, we substitute in (3.26) the value given by (3.12) for M, finding

$$\omega = \nu H. \tag{3.27}$$

If we assume the application of a fairly strong magnetic field, $H = 7.96 \times 10^5$ A/m (= 10,000 Oe), we have, with $g = 2$,

$$\omega = (1.106 \times 10^5) \times 2 \times (7.96 \times 10^5) = 1.76 \times 10^{11} \text{ Rad/sec}, \tag{3.28}$$

which corresponds to the frequency

$$f = \frac{\omega}{2\pi} = 2.8 \times 10^{10} \text{ sec}^{-1}. \tag{3.29}$$

The wavelength of the electromagnetic wave with this frequency is

$$\lambda = \frac{c}{f} = \frac{3 \times 10^{10}}{2.8 \times 10^{10}} \approx 1.1 \text{ cm}, \tag{3.30}$$

† For rare earth metals, the orbital magnetic moments remain almost completely unquenched (Chapter 20).

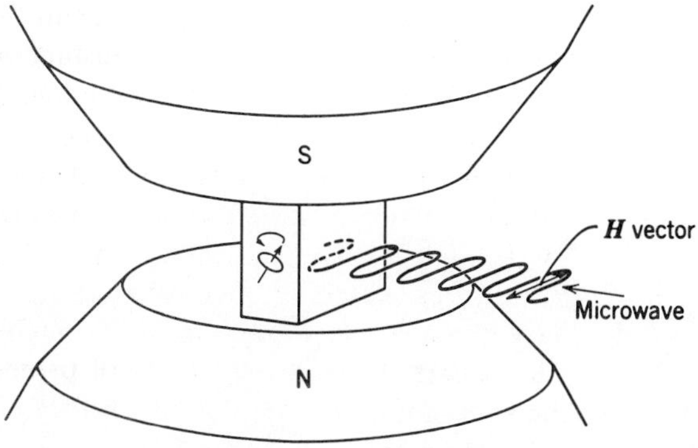

Fig. 3.9. Ferromagnetic resonance experiment.

which corresponds to the microwave region. The remarkable progress in microwave techniques during World War II made such an experiment possible. The first experiment was conducted by Griffiths[5] in 1946.

A schematic illustration of the ferromagnetic resonance experiment is given in Fig. 3.9. The atomic moment inside the ferromagnetic specimen which has been magnetized by an electromagnet is shown ready to begin its precession motion. When the microwave field with its magnetic vector pointing in a horizontal direction enters the ferromagnetic body, it induces the precession motion and is thereby absorbed. Since the precession frequency may be changed by changing the static magnetic field H (cf. 3.27), if we measure the absorption or the effective permeability as a

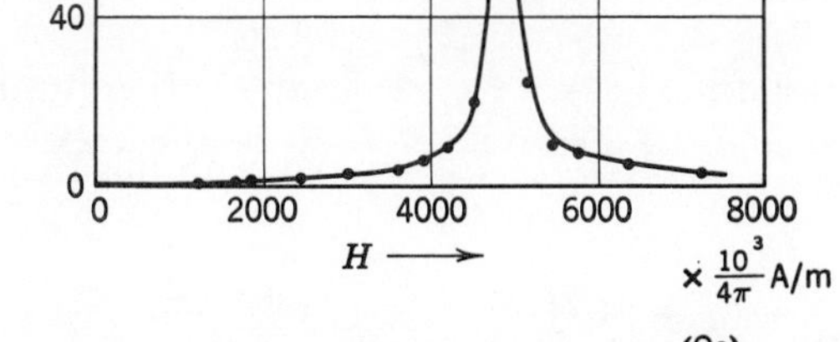

Fig. 3.10. Ferromagnetic resonance curve observed for Supermalloy using 24,000 Mc microwaves (after Yager and Bozorth[6]).

function of the static field we should observe the strongest absorption at that field where the precession frequency coincides with the frequency of the microwave. Figure 3.10 shows one of the typical ferromagnetic resonance curves observed for Supermalloy. From the value of H at which resonance occurs in a microwave field of frequency $\omega/2\pi$, we can calculate the value of g by use of (3.27) and (3.11). The g values thus obtained are listed in Table 3.2 for various materials. We see that almost all materials have g values which are close to 2, again indicating that a large part of the atomic magnetic moment comes from the spin motion.

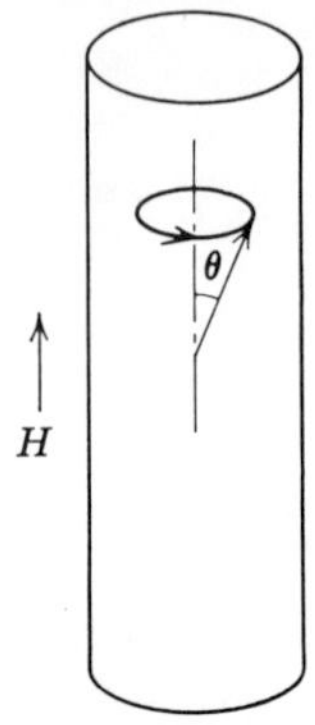

Fig. 3.11. Resonance in a cylindrical specimen.

We note here that equation (3.27) is valid only if there are no demagnetizing effects. The role of the demagnetizing fields in ferromagnetic resonance was first pointed out by Kittel.[7] Suppose, for example, that the specimen has the form of a long cylindrical rod (Fig. 3.11). When the magnetization I_s is inclined with respect to the cylinder axis by the angle θ, a demagnetizing field $(I_s \sin\theta)/2\mu_0$ is produced (cf. 2.17) which increases the couple of force acting on the atomic moment. Thus

$$\begin{aligned} L &= -MH\sin\theta - \frac{MI_s}{2\mu_0}\cos\theta\sin\theta \\ &\fallingdotseq -M\left(H + \frac{I_s}{2\mu_0}\right)\sin\theta \qquad (\theta \ll 1). \end{aligned} \tag{3.31}$$

The resonance frequency (3.27) becomes

$$\omega = \nu\left(H + \frac{I_s}{2\mu_0}\right). \tag{3.32}$$

If the specimen is in the shape of a plane, it can be shown that

$$\omega = \nu\left[H\left(H + \frac{I_s}{\mu_0}\right)\right]^{1/2}. \tag{3.33}$$

To obtain the result in (3.33), we consider the normal to the plane to be along the x axis and the direction of H to be the z axis. The x and y components of (3.16) are

$$\frac{dP_x}{dt} = L_x = M_y H \tag{3.34}$$

$$\frac{dP_y}{dt} = L_y = -M_x H - M\frac{I_x}{\mu_0} = -M_x\left(H + \frac{I_s}{\mu_0}\right). \tag{3.35}$$

By use of relation (3.12) we have from the preceding equations

$$\frac{d^2P_x}{dt^2} = \frac{dM_y}{dt} H = -\nu H \frac{dP_y}{dt} = -\nu^2 H \left(H + \frac{I_s}{\mu_0} \right) P_x. \qquad (3.36)$$

Solving this differential equation, we obtain for the angular frequency of precession the value (3.33).

Finally we discuss the deviation of the g values from the spin-only value of 2. We have already mentioned that the g values obtained from gyromagnetic experiments are slightly less than 2 is thought to be due to the

Table 3.2. Summary of Results of Ferromagnetic Resonance Experiments (after Kittel[G.15])

Substance	g	ϵ
Iron	2.12–2.17	0.06–0.09
Cobalt	2.22	0.11
Nickel	2.2	0.1
Magnetite, Fe_3O_4	2.2	0.1
Heusler alloy, Cu_2MnAl	2.01	0.005
Permalloy, 78Ni, 22Fe	2.07–2.14	0.04–0.07
Supermalloy	2.12–2.20	0.06–0.10

small contribution of the orbital magnetic moment. Let $(I_s)_{\text{spin}}$ denote the fraction of spontaneous magnetization which comes from spin and $(I_s)_{\text{orb}}$ denote the fraction which comes from orbital motion. Similarly let $(P_s)_{\text{spin}}$ and $(P_s)_{\text{orb}}$ denote the angular momentum due to spin and orbital motion, respectively. If we set

$$\frac{(I_s)_{\text{orb}}}{(I_s)_{\text{spin}}} = \epsilon, \qquad (\epsilon \ll 1) \qquad (3.37)$$

it follows that

$$\frac{(P_s)_{\text{orb}}}{(P_s)_{\text{spin}}} = 2\frac{(I_s)_{\text{orb}}}{(I_s)_{\text{spin}}} = 2\epsilon, \qquad (3.38)$$

so that the g value should be given, in this case, by

$$g = -\frac{2m}{\mu_0 e}\frac{(I_s)_{\text{orb}} + (I_s)_{\text{spin}}}{(P_s)_{\text{orb}} + (P_s)_{\text{spin}}} = -\frac{2m}{\mu_0 e}\frac{(I_s)_{\text{spin}}}{(P_s)_{\text{spin}}}\frac{1+\epsilon}{1+2\epsilon} \approx 2(1-\epsilon). \qquad (3.39)$$

The values of ϵ given in Table 3.1 are obtained from the experimental g values by using relation (3.39). We see that in most cases several per cent of the magnetization comes from the orbital motion.

On the other hand, the g values obtained from microwave resonance are all greater than 2, as may be seen in Table 3.2. If the atomic magnetic moment, which is comprised of spin and orbital contributions, were perfectly free of the influence of the surrounding lattice, the g values obtained should also be given by (3.39). Since, however, the orbital motion is quenched by the crystal field of the lattice, the angular momentum of the orbital motion does not take part in the precession. Therefore we have

$$g = \frac{2m}{\mu_0 e} \frac{(I_s)_{\text{orb}} + (I_s)_{\text{spin}}}{(P_s)_{\text{spin}}} = \frac{2m}{\mu_0 e} \frac{(I_s)_{\text{spin}}(1 + \epsilon)}{(P_s)_{\text{spin}}} = 2(1 + \epsilon). \quad (3.40)$$

This fact was first pointed out by Kittel.[8] He treated the problem on a quantum mechanical basis, but the reader may understand the situation in the following way: The orbital motion in this situation corresponds to a resonant system with a low Q value because of its strong interaction with the surrounding lattice. Although such an orbital motion does not contribute to the resonance because of this interaction, it still contributes to gyromagnetic effects, conserving angular momentum (the reader may recall the experiment depicted in Fig. 3.4, where P_{ob} is found to be independent of the damping action of the water). The values of ϵ obtained from resonance experiments by use of (3.40) are given in Table 3.2. It should be noted that these values correspond well to those of Table 3.1. We often identify the g values obtained from the gyromagnetic experiments and magnetic resonance, respectively, as the magnetomechanical factor and spectroscopic splitting factor. Measurements of magnetic anisotropy by means of magnetic resonance are discussed in Section 7.1. Various modes of magnetic resonances and the mechanism of line width are treated in Section 16.4.

3.4 Vector Model of Magnetic Atoms

In the preceding sections we have discussed the contributions of orbital and spin magnetic moments to ferro- and ferrimagnetism in ordinary magnetic substances. In these substances the orbital magnetic moment is almost quenched by the crystalline field which is produced by surrounding atoms. In the free state, however, the orbital moment remains unquenched. The energy states of orbits as well as those of spins have been fully studied through investigations of atomic spectra. Information on orbital and spin states from the area of spectroscopy is very valuable, because it serves not only for interpreting the spin states of magnetic atoms, but also for the explanation of the origin of magnetic anisotropy in terms of a partial orbital motion which plays an important role in this phenomenon. In rare earth metals the orbital moment remains almost completely unquenched, as we learn in Section 20.2; hence the total magnetic moment is composed of spin and orbital magnetic moments. In this section we study the spin and orbital states of free atoms and the way in which they are coupled to give rise to the total atomic magnetic moments.

The state of a free atom can be defined by four quantum numbers: (i) The total quantum number n with values 1, 2, 3, . . . determines the size of the orbit and defines the energy of the orbit. Strictly speaking, this energy is for one electron traveling about a nucleus as in a hydrogen atom. When the atom has more than one electron, the energy of the orbit should be somewhat modified through interactions with other electrons, as is discussed later. Groups of orbits with $n = 1, 2, 3, \ldots$ are refered to as the K, L, and M shells, respectively. (ii) The orbital angular momentum quantum number l is a measure of the angular momentum of the orbital motion of the electron. The value of the angular momentum is $l\hbar$ (cf. equation 3.5). The number l takes one of the values $0, 1, 2, 3, \ldots, (n - 1)$ depending on the shape of the orbit. The electrons with $l = 0, 1, 2, 3, 4, \ldots$ are referred to as $s, p, d, f, g, \ldots$ electrons, respectively. For example, the M shell ($n = 3$) has s, p, and d electrons. (iii) The spin momentum quantum number s is always $\frac{1}{2}$, which signifies that the magnitude of the associated angular momentum is $\hbar/2$. (iv) The total angular momentum quantum number j is the vector sum of $\boldsymbol{l}$ and $\boldsymbol{s}$ and thus has only half-integral values.

In the presence of a magnetic field the orientations of orbital and spin angular momenta are also quantized. The orbit with the angular momentum quantum number l can take $2l + 1$ directions with the projections of l upon the direction of the field $l_z = l, l - 1, l - 2, \ldots, 0, \ldots, -(l - 1), -l$. For example, the d electron ($l = 2$) can take five orientations with $l_z = 2, 1, 0, -1, -2$ (Fig. 3.12). It is not necessary to think about the

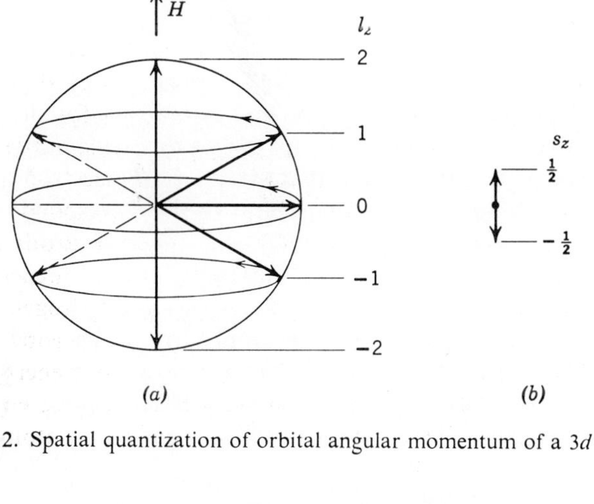

Fig. 3.12. Spatial quantization of orbital angular momentum of a $3d$ electron.

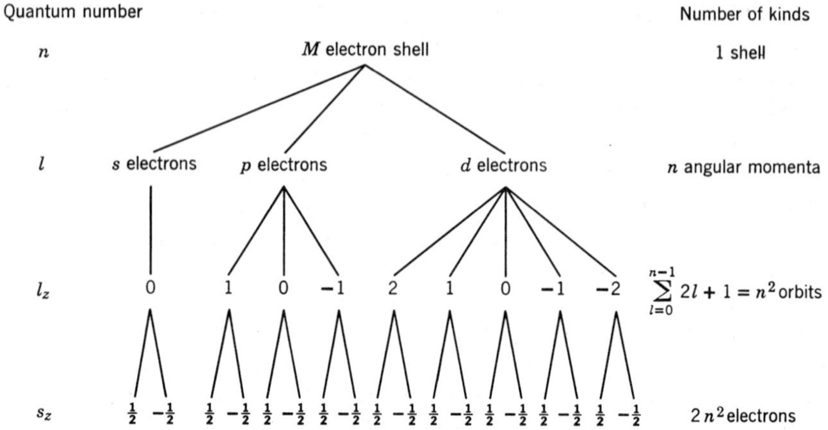

Fig. 3.13. Electronic states in the M electron shell.

azimuthal angle of the momentum vector with respect to the field direction, because the orbital plane is precessing about the axis of the external field. The spin can take only two directions, parallel and antiparallel to the field; hence the projection of the spin upon the field direction, s_z, is $\frac{1}{2}$ or $-\frac{1}{2}$.

Now, when the atom has a number of electrons, each electron must occupy a different orbit which is defined by four quantum numbers n, l, l_z, and s_z. In other words, each orbit defined by the four quantum numbers can accept only one electron. This rule is known as the Pauli principle.[9] Figure 3.13 shows the kinds of states of electrons in the M shell. As is illustrated in the figure, the maximum number of electrons which belong to the electron shell with principal quantum number n is given by $2n^2$. For instance, the maximum number of electrons is 2 for the K shell ($n = 1$), 8 for the L shell, 18 for the M shell, and so on.

As the atomic number Z increases, the number of electrons increases. These electrons fill up the states from the lowest energy state to higher energy ones. Generally speaking, the energy is lower for the state with smaller principal quantum number n as long as the shape of the orbit is of spherical symmetry. The reason is that the outermost electron of the atom with atomic number Z has the same energy as the corresponding electron of the hydrogen atom, because the $(Z - 1)$ inner electrons shield the nuclear charge $(Z - 1)e$; hence the outermost electron experiences the electric field of the unshielded nuclear charge e as in a hydrogen atom, and accordingly its energy is determined simply by the principal quantum number n. The inner electrons with smaller n have lower energy than the corresponding electron in the hydrogen atom because these electrons experience the less-shielded field of the nucleus. Thus, as the atomic number increases from $Z = 1$ (H) to $Z = 18$ (Ar) the electrons fill up in the order of $1s$, $2s$, $2p$, $3s$, and $3p$. Further increase of Z, however, results in an occupation of $4s$ levels, leaving vacant orbits in the $3d$ shell as shown in Table 3.3. The reason is that the shape of the $4s$ orbit in the Bohr atom model is so elliptic (Fig. 3.14) that the electron has a chance to approach the nucleus and to experience a strong electric field which lowers the energy of the orbit. In a series of elements from $Z = 19$ (K) to $Z = 29$ (Cu), the $3d$ levels are gradually filled up with an increase of Z. These elements are closely related to magnetism and are referred to as the transition elements.

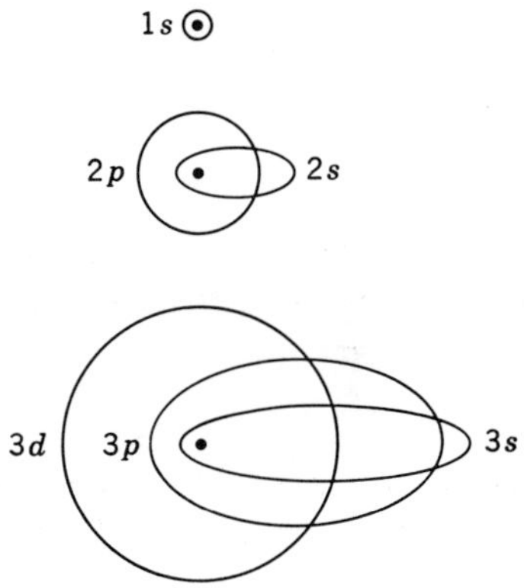

Fig. 3.14. Bohr atom model.

Now we discuss the spin and orbital states of the free atoms of transition elements. All the orbital states of $3d$ electrons have primarily the same energy, because they are all $n = 3$ and have the same orbital configuration. When, however, there are a number of electrons in the $3d$ shell, there should be strong interactions between the various angular momentum vectors: the spin vector $\boldsymbol{s}_i$ of the ith electron, the orbital vector $\boldsymbol{l}_i$ of the ith electron, $\boldsymbol{s}_j$ of the jth electron, and $\boldsymbol{l}_j$ of the jth electron. Among these interactions, it often happens that the interactions between $\boldsymbol{s}_i$ and $\boldsymbol{l}_i$ or those between $\boldsymbol{s}_i$ and $\boldsymbol{l}_j$ are small compared to the interactions between $\boldsymbol{s}_i$ and $\boldsymbol{s}_j$ and those between $\boldsymbol{l}_i$ and $\boldsymbol{l}_j$. Then the spin vectors $\boldsymbol{s}_i$ of individual electrons ($i = 1, 2, \ldots, n$) are coupled through the spin-spin interactions to give rise to the resultant spin vector:

$$\boldsymbol{S} = \sum_{i=1}^{n} \boldsymbol{s}_i. \tag{3.41}$$

Table 3.3. Electron Configuration of Transition Elements

Element	$1s$	$2s$	$2p$	$3s$	$3p$	$3d$	$4s$
22 Ti	2	2	6	2	6	2	2
23 V	2	2	6	2	6	3	2
24 Cr	2	2	6	2	6	5	1
25 Mn	2	2	6	2	6	5	2
26 Fe	2	2	6	2	6	6	2
27 Co	2	2	6	2	6	7	2
28 Ni	2	2	6	2	6	8	2
29 Cu	2	2	6	2	6	10	1
30 Zn	2	2	6	2	6	10	2

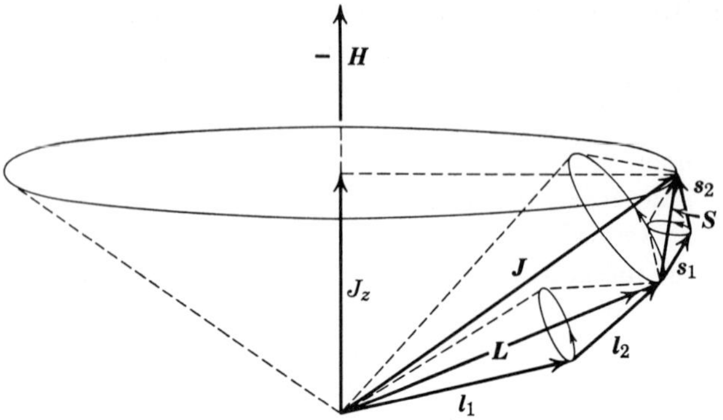

Fig. 3.15. Russell-Saunders coupling.

Similarly the orbital vectors $\boldsymbol{l}_i$ of the electrons ($i = 1, 2, \ldots, n$) combine to the resultant orbital vector:

$$\boldsymbol{L} = \sum_{i=1}^{n} \boldsymbol{l}_i. \tag{3.42}$$

The resultant $\boldsymbol{S}$ and $\boldsymbol{L}$ thus formed are rather loosely coupled through the spin-orbit interaction to form the resultant total angular momentum $\boldsymbol{J}$:

$$\boldsymbol{J} = \boldsymbol{L} + \boldsymbol{S}. \tag{3.43}$$

This sort of coupling is referred to as Russell-Saunders coupling[10] (Fig. 3.15), and it has been proved to be applicable to most magnetic atoms.

In order to determine the most stable spin and orbital configurations one may use Hund's rules[11] which state: (i) The spins s_i combine to give the maximum value of S consistent with the Pauli principle. For instance, $S = \frac{3}{2}$ for three $3d$ electrons, $S = \frac{5}{2}$ for five electrons, and $S = \frac{5}{2} - \frac{1}{2} = 2$ for six $3d$ electrons. (ii) The orbital vectors l_i combine to give the maximum value of L in conformity with the Pauli principle and condition (i). For instance, $L = 2 + 1 + 0 = 3$ for three $3d$ electrons, $L = 2 + 1 + 0 - 1 - 2 = 0$ for five electrons, and $L = 2 + 1 + 0 - 1 - 2 + 2 = 2$ for six electrons. (iii) The resultant L and S combine to form J, the value of which is $J = L - S$ if the shell is less than half occupied, and $J = L + S$ if the shell is more than half occupied. For instance, $J = 3 - \frac{3}{2} = \frac{3}{2}$ for three $3d$ electrons and $J = 2 + 2 = 4$ for six electrons.

The first and second rules are the consequence of electrostatic Coulomb interactions (repulsive) between electrons: rule (i) asserts that electrons are preferably distributed in different orbits; rule (ii) states that electrons should travel around the nucleus preferably in the same direction so as to avoid a chance to approach each other too closely.

Rule (iii) is mainly based on the spin-orbit interaction between spin $\boldsymbol{s}_i$ and the orbit $\boldsymbol{l}_i$ of the same electron. When the electron is traveling around the nucleus, the electron sees the nucleus as if the nucleus is traveling around the electron itself and feels the magnetic field caused by the circulating nuclear charge. As one may easily verify, the magnetic field thus caused is opposite in direction to the orbital magnetic moment and hence results in an antiparallel alignment of $\boldsymbol{s}_i$ and $\boldsymbol{l}_i$. If, therefore, there are less than half the full number of electrons the spin should be antiparallel to the resultant $\boldsymbol{L}$, whereas, if there are more than half the full number of electrons, only the electrons in the unfilled half shell are effective in spin-orbit interaction, because the orbital momenta in a full half shell combine to zero and are no longer effective in spin-orbit interaction. Since the spin of the electron in an unfilled half shell is opposite in direction to the resultant $\boldsymbol{S}$, whereas the orbit of the electrons is parallel to the resultant $\boldsymbol{L}$, the antiparallel coupling of $\boldsymbol{s}_i$ and $\boldsymbol{l}_i$ leads to the parallel alignment of $\boldsymbol{S}$ and $\boldsymbol{L}$. In general, the energy of the coupling between $\boldsymbol{L}$ and $\boldsymbol{S}$ is expressed in the form

$$w = \lambda \boldsymbol{L} \cdot \boldsymbol{S}, \tag{3.44}$$

where λ is the spin-orbit parameter, being positive for a less than half filled shell and negative for a more than half filled shell. The values of λ for various transition metal ions are given in Table 7.6.

The total magnetic moment of a free atom is closely related to the resultant total angular momentum $\boldsymbol{J}$. For orbital motion, comparison of (3.6) and (3.7) leads to the relationship between the orbital magnetic moment of the free atom, $\boldsymbol{M}_L$, and $\boldsymbol{L}$:

$$\boldsymbol{M}_L = -M_B \boldsymbol{L}. \tag{3.45}$$

whereas, because an electron with $s = \frac{1}{2}$ possesses $1M_B$ (not $\frac{1}{2}M_B$), the spin magnetic moment of the free atom, $\boldsymbol{M}_S$, is related to $\boldsymbol{S}$ by the relationship

$$\boldsymbol{M}_S = -2M_B \boldsymbol{S}. \tag{3.46}$$

Then the resultant magnetic moment is

$$\boldsymbol{M}_R = \boldsymbol{M}_L + \boldsymbol{M}_S = -M_B(\boldsymbol{L} + 2\boldsymbol{S}) \tag{3.47}$$

Figure 3.16 illustrates the composition of $\boldsymbol{L} + 2\boldsymbol{S}$ which is, in general, obviously not parallel to $\boldsymbol{J}$. Since $\boldsymbol{L}$ and $\boldsymbol{S}$ are precessing about $\boldsymbol{J}$ (Fig. 3.15), $\boldsymbol{L} + 2\boldsymbol{S}$ should also precess about $\boldsymbol{J}$. Then the total magnetic moment M is effectively parallel to J, and its magnitude is written as

$$\boldsymbol{M} = -gM_B \boldsymbol{J}, \tag{3.48}$$

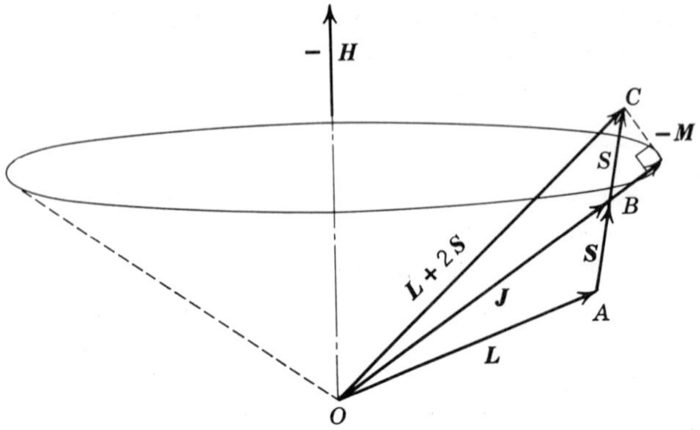

Fig. 3.16. Composition of atomic magnetic moment $\boldsymbol{M}$.

where g is the g factor as defined by (3.10). Comparing (3.47) and (3.48), we have

$$gJ = |\boldsymbol{L} + 2\boldsymbol{S}| \cos \angle BOC = J + S \cos \angle ABO. \tag{3.49}$$

In the $\triangle ABO$, there is a relationship

$$L^2 = J^2 + S^2 - 2JS \cos \angle ABO. \tag{3.50}$$

Using (3.50) in (3.49), we have

$$gJ = J + \frac{J^2 + S^2 - L^2}{2J}$$

or

$$g = 1 + \frac{J^2 + S^2 - L^2}{2J^2} \tag{3.51}$$

In an exact quantum mechanical treatment, the quantities of the form S^2, L^2, and J^2 should be replaced by $S(S + 1)$, $L(L + 1)$, and $J(J + 1)$. Then (3.51) should be

$$g = 1 + \frac{J(J + 1) + S(S + 1) - L(L + 1)}{2J(J + 1)} \tag{3.52}$$

This is the general form of the g factor which was first empirically derived by Landé[12] to describe the multiplet splitting of atomic spectra. When $S = 0$, $J = L$, and hence (3.52) gives $g = 1$; whereas, if $L = 0$, $J = S$, and hence (3.52) gives $g = 2$ in conformity with the discussion in Section 3.1.

When the atom is placed in a magnetic field, vector $\boldsymbol{J}$ precesses about the axis of H (cf. Fig. 3.16). As a result of spatial quantization, the component of J along the H axis, J_z, takes $2J + 1$ discrete values:

$$J_z = J, J - 1, J - 2, \ldots, 0, \ldots, -J + 2, -J + 1, -J \tag{3.53}$$

This fact is important in the calculation of magnetization (cf. Chapter 4).

The states of atoms described in this section are often expressed by spectroscopic notations such as $^2S_{1/2}$, $^3P_0$, $^4F_{41/2}$, and so on. In these expressions S, P, D, F, G, . . . signify that the resultant L is 0, 1, 2, 3, 4, . . . , respectively. A prefix to the capital letter represents $2S + 1$, and a suffix represents J. Electron configurations and spectroscopic ground terms are given in the Periodic Table in Appendix 6 for each element.

The atomic magnetic moments of 3*d* transition metal ions and rare earth elements are discussed in Sections 5.2 and 20.2, respectively. For further details the reader is referred to the books by Van Vleck,[13] Bates,[14] and Goodenough.[15]

Problems

3.1. Permalloy is magnetized to its saturation by a magnetic field of 100 A/m (= 1.26 Oe). Calculate the rotational speed of a long thin Permalloy rod around its axis which is necessary to magnetize it to saturation. Assume that $g = 2.00$.

3.2. How strong a magnetic field should be applied parallel to the long axis of a cylindrical Permalloy rod to attain the ferromagnetic resonance, using a microwave with a wavelength of 1 cm? Assume that $g = 2.00$ and $I_s = 1.1$ Wb/m^2.

3.3. A freely suspended iron block of volume 0.5 m^3 is magnetized to its saturation ($I_s = 2$ Wb/m^2), and then its magnetization is suddenly reversed. Assuming a g value of 2, find the change in angular momentum of the crystal lattice.

3.4. Knowing that Fe^{2+} ion has 6 electrons in the 3*d* shell, find S, L, J and describe the ground state by spectroscopic notation.

References

3.1. J. L. Maxwell: *Electricity and Magnetism*, §575.
3.2. S. J. Barnett: *Phys. Rev.* **6,** 171, 239 (1915).
3.3. A. Einstein and W. J. de Haas: *Verhandl. Deut. Physik. Ges.* **17,** 152 (1915); A. Einstein: *ibid.*, **18,** 173 (1916); W. J. de Haas: *ibid.*, **18,** 423 (1916).
3.4. G. G. Scott: *Phys. Rev.* **82,** 542 (1951).
3.5. J. H. E. Griffiths: *Nature* **158,** 670 (1946).
3.6. W. A. Yager and R. M. Bozorth: *Phys. Rev.* **72,** 80 (1947).
3.7. C. Kittel: *Phys. Rev.* **71,** 270 (1947); **73,** 155 (1948).
3.8. C. Kittel: *Phys. Rev.* **76,** 743 (1949).
3.9. W. Pauli: *Z. Physik.* **31,** 765 (1925).
3.10. H. N. Russell and F. A. Saunders: *Astrophys. J.* **61,** 38 (1925).
3.11. F. Hund: *Linienspektren und periodisches System der Elemente* (Julius Springer, Berlin, 1927).
3.12. A. Landé: *Z. Physik.* **15,** 189 (1923).
3.13. J. H. Van Vleck: G.1, p. 162.
3.14. L. F. Bates: G.14, p. 23.
3.15. J. B. Goodenough: G.39, p. 11.

4

Ferromagnetism

4.1 Langevin Theory of Paramagnetism

In this section we are concerned with the question of how the spin magnetic moments are aligned to produce ferromagnetism. By way of introduction we treat the Langevin theory of paramagnetism.

Let us suppose that each atomic magnetic moment in a paramagnetic system is provided by n spins which are parallel to one another. Then the atomic magnetic moment must be

$$M = nM_{\mathrm{B}}. \tag{4.1}$$

We also suppose that the atomic moments M of the system do not interact with each other and are therefore free to point in any direction. If a magnetic field H is applied to such a group of free moments, a couple of force $-MH \sin \theta$, where θ is the angle between the direction of H and that of the moment M, acts on each moment, tending to rotate it into the direction of H. This tendency is opposed by the thermal agitation if the temperature is finite. The potential energy of a moment M is given by

$$U_H = -MH \cos \theta. \qquad \text{(cf. 1.9)} \tag{4.2}$$

To obtain an order of magnitude for (4.2) we take $n = 1$, $\cos \theta = 1$, and $H = 10^6$ A/m (= 12,600 Oe). This field is easily obtained in laboratories but fairly strong. Then

$$\begin{aligned} |U_H| &\approx M_{\mathrm{B}}H \\ &= (1.17 \times 10^{-29}) \times 10^6 = 1.17 \times 10^{-23}\ \mathrm{J}. \end{aligned} \tag{4.3}$$

The energy of thermal motion is of the order of magnitude of kT (k is Boltzmann's constant) which is, at room temperature,

$$kT = 1.38 \times 10^{-23} \times 300 = 4.1 \times 10^{-21}\ \text{J}. \tag{4.4}$$

By comparison of (4.4) and (4.3) we find that the thermal energy is larger, by a factor of $10^2 \sim 10^3$, than the potential energy of the magnetic field. This means that the thermal agitation is sufficient to make the angular distribution of the atomic moments almost random, resulting in only a very small magnetization parallel to the magnetic field.

To calculate the magnetization we consider a system containing N atomic moments per unit volume. Let $n(\theta)\,d\theta$ denote the number of atomic moments in the unit volume which make an angle between θ and $\theta + d\theta$ with the direction of the field. This number must be proportional to the solid angle $2\pi \sin\theta\, d\theta$ and also to the Boltzmann factor $\exp(MH\cos\theta/kT)$. This factor describes the relative probability of an atomic moment to make an angle θ with H. Then we have

$$n(\theta)\,d\theta = 2\pi n_0 e^{(MH\cos\theta)/kT} \sin\theta\, d\theta, \tag{4.5}$$

where n_0 is a proportionality factor determined by the fact that the total density of atomic moments is N; that is,

$$\int_0^\pi n(\theta)\,d\theta = 2\pi n_0 \int_0^\pi e^{(MH\cos\theta)/kT} \sin\theta\, d\theta = N. \tag{4.6}$$

The intensity of magnetization is given by

$$I = \int_0^\pi M\cos\theta\, n(\theta)\,d\theta. \tag{4.7}$$

By use of (4.5) and (4.6), we have

$$I = NM\frac{\displaystyle\int_0^\pi n(\theta)\cos\theta\,d\theta}{\displaystyle\int_0^\pi n(\theta)\,d\theta} = NM\frac{\displaystyle\int_0^\pi e^{(MH\cos\theta)/kT}\cos\theta\sin\theta\,d\theta}{\displaystyle\int_0^\pi e^{(MH\cos\theta)/kT}\sin\theta\,d\theta}.$$

We set $MH/kT = \alpha$ and $\cos\theta = x$, and find

$$I = NM\frac{\displaystyle\int_1^{-1} e^{\alpha x} x\,dx}{\displaystyle\int_1^{-1} e^{\alpha x}\,dx} = NM\left(\frac{e^\alpha + e^{-\alpha}}{e^\alpha - e^{-\alpha}} - \frac{1}{\alpha}\right) = NM\left(\coth\alpha - \frac{1}{\alpha}\right). \tag{4.8}$$

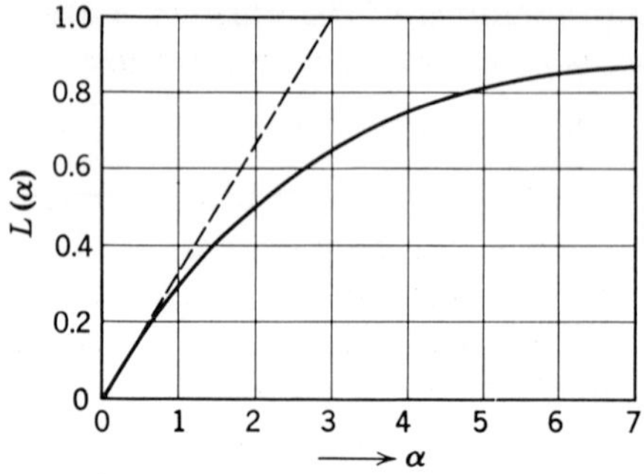

Fig. 4.1. Langevin function.

The function in parentheses is the Langevin function, $L(\alpha)$, which is plotted in Fig. 4.1. Inspection of the graph reveals that, as $\alpha \rightarrow \infty$, $L(\alpha)$ approaches 1. Thus, as $H \rightarrow \infty$, I approaches NM; that is, the moments tend to be perfectly aligned. In practice, however, we must use extremely high fields and very low temperatures to come near the saturation condition. Using the values given in (4.3) and (4.4), we find the order of magnitude of α at room temperature to be

$$\alpha = \frac{MH}{kT} = \frac{1.17 \times 10^{-23}}{4.1 \times 10^{-21}} = 2.8 \times 10^{-3}. \tag{4.9}$$

This result indicates that the actual magnetization is normally confined to values represented in the vicinity of the origin in the graph of Fig. 4.1. For $\alpha \ll 1$, the Langevin function may be expanded as

$$L(\alpha) = \frac{\alpha}{3} - \frac{\alpha^3}{45} + \cdots. \tag{4.10}$$

Neglecting the terms beyond the first one of (4.10), we have from (4.8) the result

$$I = \frac{NM^2}{3kT} H, \tag{4.11}$$

which yields the field independent susceptibility,

$$\chi = \frac{I}{H} = \frac{NM^2}{3kT}. \tag{4.12}$$

In this way we have derived the well-known Curie law, namely, the $1/T$ dependence of the susceptibility of a paramagnetic substance on the temperature.

In this calculation we made the assumption that each individual atomic moment can point in any arbitrary direction. For actual materials, this does not happen. Because of the spatial quantization of angular momentum the spin can take only discrete orientations (cf. Section 3.4). If an atom has gJ spins, where J is the total angular momentum number composed of spin and orbital angular momenta and g is the g factor (cf. equation (3.52)), its angular momentum along the z axis or the axis of quantization can take on the following $(2J + 1)$ values:

$$-J, -(J-1), \ldots, (J-1), J.$$

The value of the angular momentum along the z axis is denoted by the quantum number J_z, and the magnetic moment of the atom is given by gM_BJ_z. Since the maximum magnetic moment of the atom is

$$M = gM_BJ, \tag{4.13}$$

the magnetic moment of the atom is, in general, $(J_z/J)M$, and its potential energy in the magnetic field is $-(J_z/J)MH$. Now the total magnetization of the substance becomes

$$I = NM\frac{\sum_{J_z}\frac{J_z}{J}e^{J_zMH/JkT}}{\sum_{J_z}e^{J_zMH/JkT}}$$

$$= NM\left(\frac{2J+1}{2J}\coth\frac{2J+1}{2J}\alpha - \frac{1}{2J}\coth\frac{\alpha}{2J}\right). \tag{4.14}$$

The function in parentheses is the Brillouin function, $B_J(\alpha)$, which reduces to the Langevin function in the limit of very large J. Then we have

$$\frac{I}{NM_B} = gJB_J(\alpha). \tag{4.15}$$

In Fig. 4.2 are plotted the magnetization curves of several paramagnetic complex salts containing Cr^{3+}, Fe^{3+}, or Gd^{3+} ions, as measured at very low temperature. The shapes of the magnetization curves are well fitted by the theoretical curves calculated from (4.15), and the saturation values agree with those calculated from (4.13).

For $\alpha \ll 1$, $B_J(\alpha)$ may be expanded in the series

$$B_J(\alpha) = \frac{J+1}{3J}\alpha - \frac{[(J+1)^2 + J^2](J+1)}{90J^3}\alpha^3 + \cdots. \tag{4.16}$$

Then the Curie law is given by (4.12), where M must be replaced by

$$M_{\text{eff}} = gM_B\sqrt{J(J+1)}, \tag{4.17}$$

which we call the effective atomic moment, whereas (4.13) is referred to as the saturation moment. If $J \to \infty$, (4.17) becomes the same as (4.13), as expected.

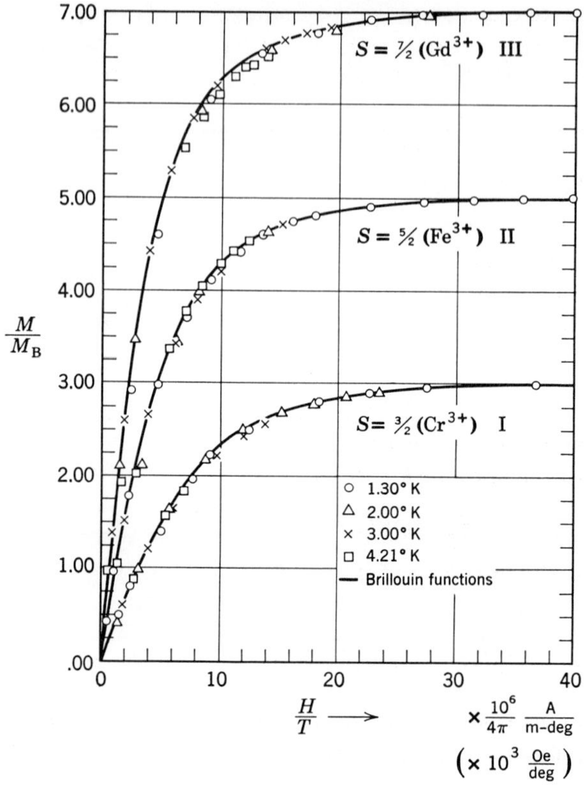

Fig. 4.2. Magnetization curves of paramagnetic salts: (I) potassium chromium alum, (II) ferric ammonium alum, and (III) gadolinium sulfate octahydrate (after W. E. Henry[1]).

4.2 Weiss Theory of Ferromagnetism

We have seen that it is very hard to magnetize an assembly of free atomic moments to any considerable extent at room temperature. The magnetic field necessary to reach saturation may be estimated by tentatively setting $\alpha = 7$ (cf. Fig. 4.1). We find

$$H = \frac{7kT}{M} \cong 2.5 \times 10^9 \text{ A/m} \quad (= 3.1 \times 10^7 \text{ Oe}). \tag{4.18}$$

On the other hand, the field required to magnetize a ferromagnetic substance to saturation is

$$H \sim 1 \text{ A/m} \quad (= 0.013 \text{ Oe}) \tag{4.19}$$

for Supermalloy, one of the typical soft magnetic materials, and

$$H \sim 5 \times 10^4 \text{ A/m} \quad (= 630 \text{ Oe}) \tag{4.20}$$

for Alnico, one of the typical permanent magnet materials. Most ferromagnetic materials can be saturated by a magnetic field whose intensity lies between these two values. Thus the fields required to magnetize ferromagnetic substances are considerably weaker than those required for paramagnetics. The explanation for the relative ease with which ferromagnetics may be magnetized is: The atomic magnetic moments in a ferromagnetic substance interact strongly with one another and tend to align themselves parallel to each other. The interaction is such as to correspond to an applied field of the order of magnitude of (4.18), and it results in a nearly perfect alignment of the spins in spite of the thermal agitation at room temperature. The effect of an externally applied field is merely to change the direction of the spontaneous magnetization.

The presence of a strong inner magnetic field was first postulated by P. Weiss.[2] He called it the molecular field and developed a fine theory of the temperature dependence of the saturation magnetization, which follows.

Again we assume that the atomic magnetic moment is composed of n spins, so that

$$M = nM_{\text{B}}. \tag{4.21}$$

We shall discuss the actual values of n for metals and alloys in Section 4.3 for ferrites in Section 5.2, and for various compounds in Section 5.3.

Each atomic moment is assumed to be acted on by the molecular field, which is proportional to the magnetization of its environment. We already know that, if we remove one atomic moment from a group of parallel moments as shown in Fig. 4.3, there is a magnetic field in this space which is parallel to the magnetization around it. If the space has a spherical shape, the field should be given by the Lorentz field (2.22). Thus we express the molecular field as

$$H_m = wI, \tag{4.22}$$

where w is the proportionality factor and is equal to $1/(3\mu_0)$ in the Lorentz field. As we shall see later, the actual value of w must be much larger than $1/(3\mu_0)$; hence we must attribute this field to some other origin.

Now, if a magnetic field H is applied parallel to the magnetization I of the system, an individual atomic moment has the potential energy

$$U_H = -M(H + wI)\cos\theta. \tag{4.23}$$

Since the probability for an atomic moment to have this energy is proportional to the Boltzmann factor, $e^{(-U_H/kT)}$, the average magnetization is given by

$$I = NM \frac{\int_0^\pi e^{[M(H+wI)\cos\theta]/kT} \cos\theta \sin\theta \, d\theta}{\int_0^\pi e^{[M(H+wI)\cos\theta]/kT} \sin\theta \, d\theta} = NML\left(M\frac{H + wI}{kT}\right). \tag{4.24}$$

The analytic solution of this equation for I as a function of T and H is very difficult, and it is best to seek a graphical solution. If we set the argument of the Langevin function equal to α, we have simultaneously

$$I = NML(\alpha) \tag{4.25}$$

from (4.24), and

$$I = \frac{\alpha kT}{Mw} - \frac{H}{w} \tag{4.26}$$

from the definition of α. In Fig. 4.4 we have a graphical representation of (4.25) and (4.26). Curve 1 represents the Langevin function, and curve 2, a straight line through the origin, represents (4.26) for $H = 0$. The magnetization I is given by the simultaneous solutions of (4.25) and (4.26), which are represented by the intersection points of the two curves. The actual stable solution is given by the point P, whereas the other intersection point O corresponds to an unstable solution. The latter point represents the state for which $I = 0$, the state in which there is a random distribution of the directions of the atomic moments. Since the molecular field wI is zero in this case, it is possible to maintain the random state. If, however, a parallel alignment should appear locally at any place in the lattice, a molecular field will be produced which, in turn, will promote further alignment of the atomic moments. This corresponds to the motion of the state point P' from O toward P on line 2. [It should be noted that, since (4.26) is the definition of α, point P' must stay on line 2. On the other hand, (4.25) gives the state which the system should attain in thermal equilibrium,

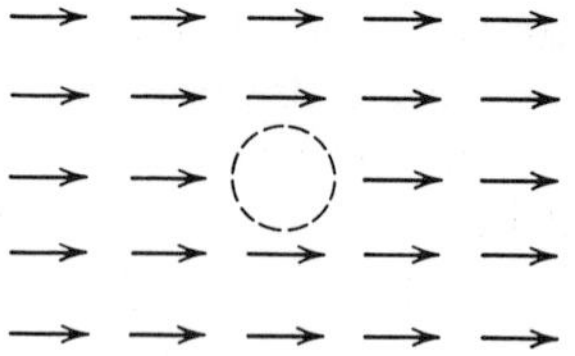

Fig. 4.3.

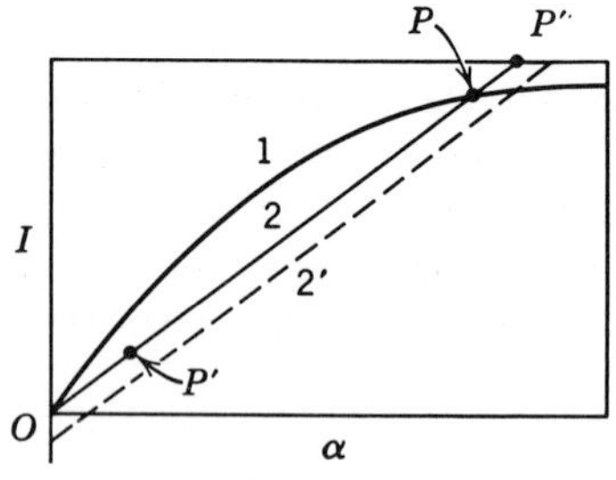

Fig. 4.4. Graphical representation of (4.25) and (4.26).

so that point P' always tends to move toward curve 1.] If the state point moves upward from P to , say, P'', the thermal equilibrium value of I is less than that of P'' and the system will return to state P. This means that point P represents the stable solution.

Now let us consider the temperature dependence of the magnetization. Since, at $T = 0$, the inclination of line 2 to the horizontal axis is infinitesimally small, the stable solution at that temperature occurs at $\alpha \to \infty$, and $L(\alpha) \to 1$. Then, according to (4.25), $I = NM$; this means that all the atomic moments are perfectly aligned parallel to one another. We put

$$I_0 = NM. \tag{4.27}$$

We shall refer to this quantity as the absolute saturation magnetization.

As the temperature increases, the inclination of line 2 also increases so that point P gradually moves down along curve 1, finally dropping rapidly to the origin at some finite temperature. At higher temperatures, point P remains at the origin, so that $I = 0$. The reader may imagine the change of I during this process following the temperature dependence of the magnetization shown in Fig. 1.9b. The Curie temperature Θ, at which I vanishes, can be obtained from the condition that the slope of line 2 (cf. 4.26)

$$\frac{\partial I}{\partial \alpha} = \frac{k\Theta}{Mw} \tag{4.28}$$

equals the initial slope of curve 1 (cf. 4.25)

$$\left(\frac{\partial I}{\partial \alpha}\right)_{\alpha=0} = NM\left(\frac{\partial L}{\partial \alpha}\right)_{\alpha=0} = \frac{NM}{3} \quad \text{(cf. 4.10).} \tag{4.29}$$

Thus we have

$$\Theta = \frac{NM^2 w}{3k}. \tag{4.30}$$

This formula indicates that the Curie temperature is an indicator of the magnitude of the molecular field factor w. If we take iron as an example, with $M = 2.2M_B$, $N = 8.54 \times 10^{28}$ m^{-3}, and $\Theta = 790°\text{C} = 1063°\text{K}$, we have

$$w = \frac{3k\Theta}{NM^2} = \frac{3(1.38 \times 10^{-23})(1063)}{(8.54 \times 10^{28})(2.2)^2(1.17 \times 10^{-29})^2} = 7.8 \times 10^8, \tag{4.31}$$

so that the molecular field is

$$\begin{aligned} H_m &= wI \\ &\approx wNM = (7.8 \times 10^8)(8.54 \times 10^{28}) \times (2.2) \times (1.17 \times 10^{-29}) \\ &= 1.7 \times 10^9 \text{ A/m} \quad (= 2.1 \times 10^7 \text{ Oe}). \end{aligned} \tag{4.32}$$

This value is too large to be explained by the Lorentz field, which is given by

$$\begin{aligned} H_L &= \frac{I}{3\mu_0} \\ &\approx \frac{NM}{3\mu_0} = \frac{(8.54 \times 10^{28})(2.2)(1.17 \times 10^{-29})}{3 \times 4\pi \times 10^{-7}} \\ &= 5.8 \times 10^5 \text{ A/m} \quad (= 7400 \text{ Oe}). \end{aligned} \tag{4.33}$$

The enormous size of the molecular field is discussed in Section 4.3.

We now discuss the susceptibility above the Curie temperature. As the external field is turned on, line 2 in Fig. 4.4 moves downward with constant slope (cf. 4.26). At the same time the intersection point P moves upward along curve 1, increasing the value of I. Let us suppose that a fairly strong field, $H = 10^6$ A/m ($\approx$ 12,600 Oe), is applied. The resultant motion of the line, as measured along the vertical axis, which is given by the second term of (4.26), is approximately

$$\frac{H}{w} = \frac{10^6}{7.8 \times 10^8} = 0.0013 \text{ Wb/m}^2. \tag{4.34}$$

This is much smaller than the value $I_0 \approx 2$ Wb/m². Thus at low temperatures the spontaneous magnetization can be changed but little by the external field. This is due to the fact that the external field is much smaller than the molecular field.

At temperatures just above the Curie temperature, however, where line 2 is almost parallel to the initial slope of curve 1, downward motion of line 2 results in a remarkable displacement of the intersection point. This means that the susceptibility is very large at, or just above, the Curie temperature.

Let us now calculate the susceptibility above the Curie temperature. Since the magnetization will be small compared to I_0, we may use, in (4.24), the first term in the series expansion for $L(\alpha)$ (4.10). Thus

$$I = \frac{NM^2}{3kT}(H + wI). \tag{4.35}$$

We solve this equation for I:

$$I = \frac{NM^2H}{3kT - NM^2w} = \frac{NM^2H}{3k(T - \Theta)} \qquad \text{(cf. 4.30)} \tag{4.36}$$

and therefore the susceptibility is

$$\chi = \frac{I}{H} = \frac{NM^2}{3k(T - \Theta)}. \tag{4.37}$$

We have thus derived the result that the susceptibility above the Curie temperature is inversely proportional to $T - \Theta$. This fact was discovered experimentally by Curie, and it is called the Curie-Weiss law. From a plot of $1/\chi$ vs T we can determine the Curie point Θ (using the T intercept) and the atomic moment M (from the slope $= 3k/NM^2$).

The foregoing calculation is based on the classical picture which assumes that the atomic magnetic moment is free to be oriented in any direction. Actually the spatial quantization must be invoked. Thus, for a total angular momentum number J, $L(\alpha)$ must be replaced, in (4.25), by $B_J(\alpha)$ (cf. 4.15), and some alteration in the shape of the I_s vs T curve results. Figure 4.5 shows I_s vs T curves for $J = \frac{1}{2}$, 1, and ∞. The experimental points for nickel and iron are much better fitted by the curves for $J = \frac{1}{2}$ or 1 than by $J = \infty$.

Referring to (4.13) and (4.16), we find that the Curie point for the total angular momentum number J is given by

$$\Theta = \frac{J(J+1)Ng^2M_B{}^2w}{3k}, \tag{4.38}$$

and the corresponding Curie-Weiss law becomes

$$\chi = \frac{J(J+1)Ng^2M_B{}^2}{3k(T-\Theta)}. \tag{4.39}$$

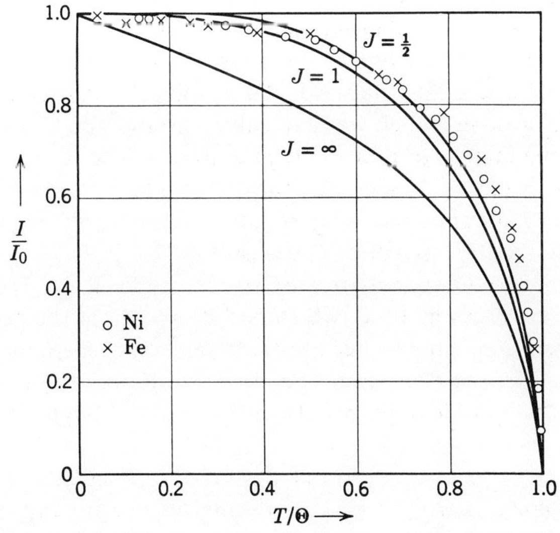

Fig. 4.5. Temperature dependence of spontaneous magnetization (after Becker and Doring[G.3]).

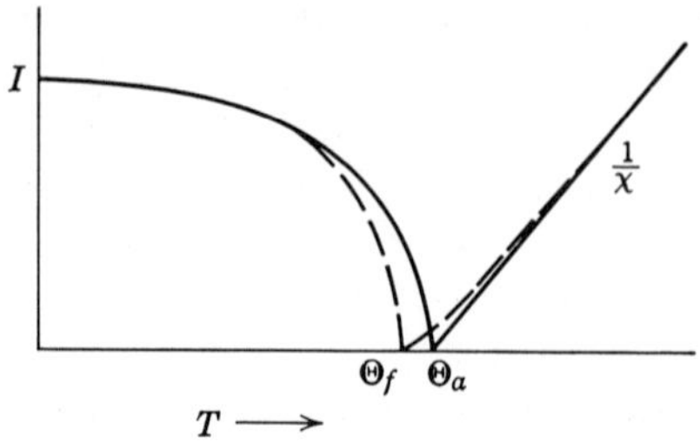

Fig. 4.6. Effect of cluster formation on the temperature dependence of spontaneous magnetization and the reciprocal of the susceptibility (broken line).

The susceptibility due to a change of the magnitude of the intrinsic magnetization below the Curie temperature is discussed in Section 13.3.

In the Weiss theory it is assumed that the molecular field is proportional to the average magnetization. It is more reasonable to assume that a local molecular field is produced by the surrounding spins and that the moments tend to lie parallel to the neighboring spins rather than to the average magnetization. The formation of parallel spin clusters in this way may tend to lower the Curie temperature because the direction of the cluster may more easily deviate from the direction of average magnetization. Above the Curie temperature the remaining clusters behave much as large paramagnetic molecules, with weaker interaction, and the susceptibility is less than expected from the Weiss theory. Thus the $1/\chi$ vs T curve should be shifted upward from the line predicted by the Weiss theory. With increasing temperature, the size of the clusters should decrease, so that the curve should approach the straight line of the Weiss theory at high temperature (Fig. 4.6). Two Curie points may now be defined for any ferromagnetic substance. One of these, which we shall call an asymptotic Curie point, is determined by the intersection of the temperature axis and the extrapolation of the $1/\chi$ vs T curve. It is usually somewhat higher than the temperature at which the spontaneous magnetization disappears; we shall call this temperature the ferromagnetic Curie point.

The formation of clusters was treated rigorously by P. R. Weiss.[3] He applied the Bethe-Peierls method, which had been used in the problem of short range ordering of superlattice alloys, to ferromagnetism and derived $1/\chi$ vs T curves displaying curvature above the Curie point, in agreement with our discussion. Some of his results will be used in Chapter 9 (cf. 9.2 to 9.4).

Another attack on the same problem is via the Bloch[4] theory of spin waves. The problem is treated by considering the correct expression for the exchange interaction (cf. Section 4.3) instead of the molecular field, and assuming that spin reversals, in small concentration, are freely propagated with various wavelengths through the crystal in which there is an almost perfect alignment of spins. Bloch found that the temperature dependence of spontaneous magnetization at low temperature is given by

$$I = I_0\left[1 - C\left(\frac{kT}{J}\right)^{3/2}\right], \tag{4.40}$$

where J is the exchange integral (cf. Section 4.3) and C is a constant which equals 0.1174 for the simple cubic lattice, 0.0587 for the body-centered cubic, and 0.0294 for the face-centered cubic lattice. The $T^{3/2}$ law of (4.40) agrees fairly well with experiment, but the proportionality factor does not.

It is interesting that the Bloch spin wave theory predicts that ferromagnetism should appear only for three-dimensional lattices, and not for one- and two-dimensional lattices. The same conclusion was reached by P. R. Weiss. On the other hand, the rigorous calculations made by Kramers and Wannier[5] and by Onsager[6] led to the conclusion that a two-dimensional lattice may be ferromagnetic. For further discussion the reader may refer to the reviews by Van Vleck[7] and by others.[8–10]

We close this section with a discussion of the internal energy accompanying spontaneous magnetization. The work done in increasing the magnetization from I to $I + \delta I$ under the action of the molecular field is

$$\delta E = -wI\,\delta I, \tag{4.41}$$

so that the internal energy of the state having spontaneous magnetization I is

$$E = \int \delta E = -\int_0^I wI\,dI = -\frac{w}{2}I^2. \tag{4.42}$$

The temperature dependence of I shown in Fig. 4.6 allows us to find the change in internal energy as a function of temperature, which is shown in Fig. 4.7. In Fig. 4.8 we have a plot of the specific heat as a function of

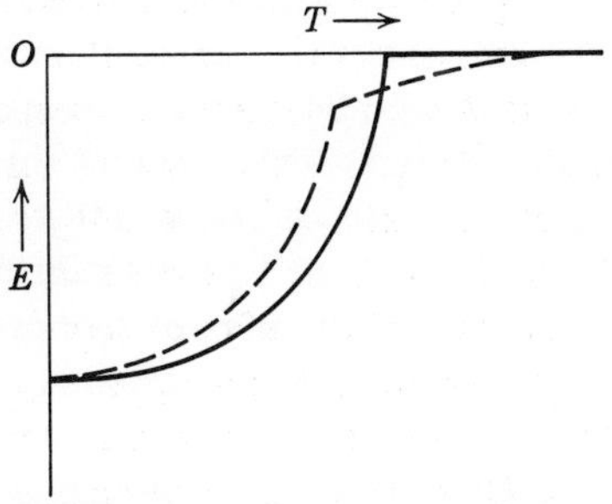

Fig. 4.7. Temperature dependence of the energy associated with the formation of spontaneous magnetization.

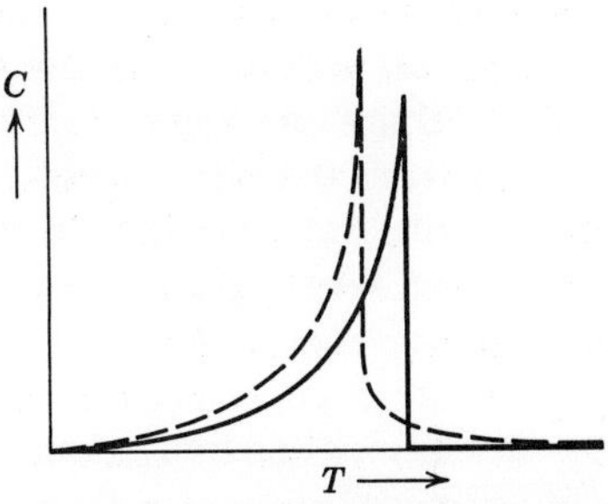

Fig. 4.8. Temperature dependence of the specific heat associated with the change of spontaneous magnetization.

temperature, given by

$$C_v = \frac{dE}{dT} = -\,wI\frac{dI}{dT}. \tag{4.43}$$

There is a sharp peak at the Curie point associated with the abrupt disintegration of the spontaneous magnetization. The broken curves in these two figures show the results when the formation of spin clusters is taken into account. The most striking evidence of spin clustering may be the tail of the specific heat peak which is observed experimentally.

4.3 Intrinsic Magnetization of Alloys

Although a large number of substances are in use as magnetic materials at the present time, only three pure metals exhibit ferromagnetism at room temperature. These metals are iron, cobalt, and nickel. Almost all magnetic alloys contain at least one of these ferromagnetic metals. By alloying these metals with other elements, we can prepare magnetic substances which have various magnetic properties. These properties are mainly determined by the magnetic anisotropy, magnetostriction, and secondary structures of the substances. We deal with these properties in Part 3. However, the intrinsic magnetization of an alloy is determined by its electronic structure.

We learned in Chapter 3 that magnetism has its origins in the spin and orbital magnetic moments in an unfilled electron shell. Actually each of the three ferromagnetic elements Fe, Co, and Ni has an unfilled 3*d* shell. The rare earth metals, most of which exhibit ferromagnetism at low temperatures, have an unfilled 4*f* shell. We deal with rare earth metals in Chapter 20 separately, and in this section we concentrate on the 3*d* transition metals.

In Fig. 4.9 is shown the average atomic magnetic moment of various alloys, which is calculated as the saturation magnetization at 0°K divided by the number of atoms in a unit volume, as a function of the average electron number per atom. As seen in this figure, most of the alloys are represented by points falling on a curve consisting of two straight lines: one of these lines rises from 0 Bohr magnetons at Cr at the rate of about $1M_B$ per electron, while the other falls from $2.5M_B$ at about 30 at % Co-Fe at the rate of about $-1M_B$ per electron. This curve is usually referred to as the Slater-Pauling curve. From it we see that ferromagnetism appears for average electron concentration ranging from 24 to 28.6. Since the argon shell (1*s* to 3*p*) is filled by 18 electrons, the number of 3*d* and 4*s* electrons in ferromagnetic alloys ranges from 6 to 10.6. If we assume that the number of conduction electrons is about 1.0 at Cr and 0.6 at Ni, the number of 3*d* electrons should be 5 to 10 in the range where ferromagnetism is realized.

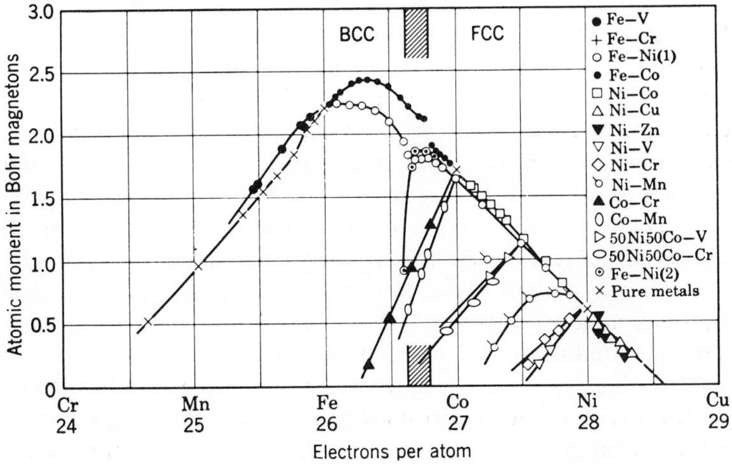

Fig. 4.9. Atomic magnetic moment of ferromagnetic metals and alloys as a function of the number of outer electrons per atom (Slater-Pauling curve) (after Bozorth[G.10]; Kono and Chikazumi[11] for NiCo-V and NiCo-Cr, and Kouvel and Wilson[12] for Fe-Ni(2)).

In order to explain the appearance of ferromagnetism, first of all we must identify the physical origin of the molecular field which gives rise to the parallel alignment of spins. We have shown by (4.32) that the molecular field which is obtained from the Curie point is too large to be explained by the Lorentz field. The accepted interpretation of the nature of the molecular field was first presented by Heisenberg[13] in 1928. According to his theory, the force which makes the spins line up is an exchange force of quantum mechanical nature. The potential energy between two atoms having spins $\boldsymbol{S}_i$ and $\boldsymbol{S}_j$ is given by

$$w_{ij} = -2J\boldsymbol{S}_i \cdot \boldsymbol{S}_j, \tag{4.44}$$

where J is the exchange integral. If J is positive, the energy is least when $\boldsymbol{S}_i$ is parallel to $\boldsymbol{S}_j$; if J is negative, the stable state is that in which $\boldsymbol{S}_i$ is antiparallel to $\boldsymbol{S}_j$.

Although it is impossible to understand the nature of the exchange force from the standpoint of classical physics, we can obtain a glimpse of it in the following way: As stated above, the Pauli principle permits an orbit to be occupied by one plus and one minus spin electron, while two electrons with the same kind of spin cannot approach one another closely. Thus the mean distance between two electrons should be different for parallel spins from that for antiparallel spins, and the Coulomb energy should be different in the two electrons. It may therefore be said that the exchange energy is substantially the electrostatic energy as it depends on the spin orientation through the Pauli principle.

There have been two points of view in interpreting spin configurations in ferromagnetic substances. One is based on a localized model in which the electrons responsible for ferromagnetism are regarded as localized at their respective atoms. Most ferrimagnetic oxides and compounds can be well explained in terms of this model, as will be discussed in Chapter 5. Rare earth metals are also good examples to be explained in terms of a localized model, because the electron spins responsible for the magnetism of these metals are confined to the deep inner $4f$ shells of individual atoms and the atomic moments interact with one another by an exchange interaction through conducting electrons (cf. Section 20.2).

The other point of view is an itinerant or collective electron model in which electrons responsible for ferromagnetism are thought of as wandering through the crystal lattice. Since the $3d$ shells of $3d$ transition metals are the most exposed except for the conducting $4s$ electrons, the $3d$ shells of individual atoms are thought to be nearly touching or overlapping with those of neighboring atoms. Then it would be natural to consider that the energy levels of $3d$ electrons are perturbed and spread to form an energy band.

The band theory of ferromagnetism was first proposed by Stoner[14,15] in 1933 and later by Slater[16] in 1936. On the basis of the knowledge of the density of states in the $3d$ shell of copper as calculated by Krutter,[17] Slater assumed that the density of states in nickel may be also very high at the top of the $3d$ band as it is for copper. Figure 4.10*a* shows the density of states in the $3d$ shell of nickel as a function of energy as calculated recently by Koster.[18] Figure 4.10*b* shows schematically the corresponding energy levels, each of which can be occupied by two electrons, one of plus spin and one of minus spin (Pauli principle). In order to have a net magnetic moment, therefore, it is necessary that some minus spin electrons be excited to higher energy levels and reverse the sign of their spins from minus to plus. Such an excitation should not require too much energy in the case of the $3d$ shell because of the high density of states. If, therefore, a positive exchange interaction is acting between $3d$ electrons, the number of plus spins should increase until they fill up half of the $3d$ shell, leaving vacant levels in the other half. Then the net magnetic moment will be proportional to the number of vacant levels in the $3d$ shell. In that case, if we add one electron to the atom, this addition should result in a decrease of 1 Bohr magneton per electron because the electron enters into a vacant minus spin level. In this way we can understand the $-45°$ inclination of the right half of the Slater-Pauling curve.

Recently neutron diffraction experiments have revealed a number of

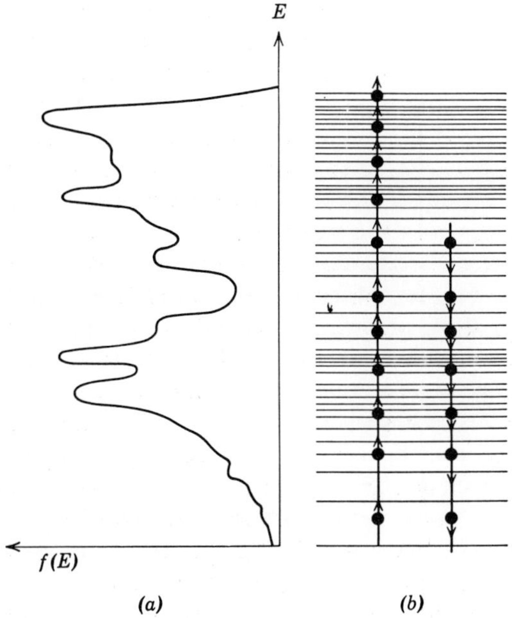

Fig. 4.10. State-density curve of $3d$ band of nickel (after Koster[15]) and the arrangement of spins in the band.

facts about the magnetic structures of metals and alloys. Several neutron experiments on nickel-rich iron-nickel alloys[19-21] have shown that atomic spins responsible for ferromagnetism are fairly well localized in each sort of atom: $2.6M_B$ for iron and $0.6M_B$ for nickel. These values are nearly constant over a wide range of composition. Friedel[22] explained this fact in the following way; When a solute atom with atomic number Z_s is introduced into the metallic matrix composed of the atoms with atomic number Z_m, the excess nuclear charge, $(Z_s - Z_m)\,|e| = \Delta Z\,|e|$, locally displaces the mobile electrons until the displaced charge exactly screens out the new nuclear charge. In transition metals such a localization of the excess electrons is thought to be particularly perfect because of the high density of states of the $3d$ shell. For iron-nickel alloys, $Z_s = 26$ for iron and $Z_m = 28$ for nickel; thus $\Delta Z = -2$, and hence the solute iron atom should have two more vacancies in the $3d$ shell than has a nickel atom in the matrix. It is, therefore, expected that an iron atom in iron-nickel alloys should have two more Bohr magnetons than a nickel atom, or $2.6M_B$, which is in good agreement with the experiment. In general, the average magnetic moment of the alloy with concentration c is expressed by $M_{av} = M_{matrix} - \Delta Z c M_B$, which is in accord with the right-hand half of the Slater-Pauling curve.

The alloys which contain vanadium or chromium show a great deviation from the main Slater-Pauling curve. A number of branches drawn from the main Slater-Pauling curve shown in Fig. 4.9 are those of vanadium or chromium alloys whose mother metals are cobalt, 50 at % Co-Ni, and nickel. Such a rapid diminution of the average magnetic moment with composition may be explained either by assuming an antiferromagnetic alignment of vanadium or chromium moment,[23] or by assuming the filling up of $3d$ vacancies with the outer electrons of vanadium or chromium atoms.[24–28] Neutron scattering experiments by Low and Collins[21] have revealed that the magnetic disturbance of vanadium or chromium atoms in Ni-Co or Ni-V alloys occurs not only at the solute atoms, but also on the neighboring nickel atoms. It was concluded from this experiment that most of the change in saturation magnetization is due to a decrease in the moments on nickel atoms in the vicinity of solute atoms rather than to changes in moment at the solute atoms, thus supporting the above-mentioned second possibility. Friedel[22] has considered that the energy states of vanadium or chromium atoms are strongly perturbed, because of their large negative nuclear charges ($\Delta Z < 0$), to such an extent that they pass up through the Fermi level and so empty their electrons into the vacancies of the $3d$ half-band with minus spin. By this transfer five electrons reverse their spins from plus to minus, so the average magnetic moment should be given by $M_{\text{av}} = M_{\text{matrix}} - (\Delta Z + 10)cM_{\text{B}}$. The slopes of the branches for vanadium or chromium alloys are fairly well expressed by this formula.

It should be noted that the saturation moment even as measured at 0°K drops off very sharply at an electron concentration of about 26.6 in a face-centered cubic region, as experimentally verified by Kondorsky and Sedov[29] and by Kouvel and Wilson.[12] It was observed for the 30 at % Ni-Fe alloy that the saturation magnetization recovers its large value as soon as the lattice transforms from face-centered cubic to body-centered cubic at low temperatures. It has been suggested[29,12] that the rapid decrease of saturation magnetization in this region may be related to some antiferromagnetic alignment of atomic moments.

The +45° slope of the left half of the Slater-Pauling curve has been less adroitly treated by the band theory. One possible explanation is that, if the high density of states portion at the top of the $3d$ band is able to contain 2.5 electrons, the plus spin band remains full until the minus spin band loses 2.5 electrons. Further loss of electrons would deplete the plus spin band because otherwise the Fermi surface of the minus spin band might drop to too low a level. The loss of plus spin electrons then results in a decrease of atomic magnetic moment. This model, however, does not necessarily include the realization of the +45° slope of the Slater-Pauling curve.

Zener[30] tried to explain this point in terms of antiferromagnetic alignment of two kinds of atomic moments, $+5M_B$ and $-1M_B$ for iron, on the two sublattices of the body-centered cubic lattice. On the basis of neutron diffraction or scattering experiments, however, Shull and others[31,32] concluded that, if there is some difference between the two sublattice moments, the difference should be less than $0.4M_B$ for iron provided magnetic ordering is of a long range nature, and it should be less than $0.6M_B$ if the ordering is of a short range nature. A number of theories[33-36] have been published on the ferromagnetism of metals and alloys, but none of them gives us a satisfactory understanding of the left half of the Slater-Pauling curve.

The presence of an antiferromagnetic alignment of atomic magnetic moments in metals and alloys has been verified by a number of experiments. Anomalous decreases of saturation magnetization resulting from lowering the temperature were observed by Kouvel, Graham, and Jacobs[37] for disordered nickel manganese alloys, and by Arrott and Sato[38] for iron-aluminum alloys. These phenomena have been interpreted in terms of antiferromagnetic alignment of atomic spins in these alloys. Neutron diffraction has revealed a number of antiferromagnetic structures[39] in chromium, α-manganese, γ-iron, Mn-Cr, Mn-Cu, and Au-Mn. Systematic investigations have been made on antiferromagnetic or helical spin configurations of heavy rare earth metals (cf. Section 20.2). This phenomenon has been well interpreted by Yosida and Watabe[40] in terms of the exchange interaction between localized $4f$ electrons and conducting electrons. For the $3d$ transition metals, however, the situation may not be so simple as for the rare earth metals because of the itinerant nature of $3d$ electrons.

Problems

4.1. Assuming that each molecule has a magnetic moment of 1 Bohr magneton, calculate the relative susceptibility of an ideal gas under atmospheric pressure at room temperature.

4.2. Nickel has a Curie point 358°C, melting point 1453°C, atomic weight 58.7, density 8.9 g/cc and can be assumed to have $J = \frac{1}{2}$ and $g = 2.00$. How strong an inhomogenous field is necessary to pull a drop of just melting nickel upwards by magnetic means?

4.3. The Slater-Pauling curve is composed of the two straight lines which connect the three points (24, 0), (26.1, 2.5), and (28.6, 0). Find the composition of the Fe-Cr, Co-Ni, and Ni-Fe alloys for which the average atomic magnetic moment is 1 Bohr magneton.

References

4.1. W. E. Henry: *Phys. Rev.* **88,** 559 (1952).
4.2. P. Weiss: *J. Phys.* **6,** 661 (1907).
4.3. P. R. Weiss: *Phys. Rev.* **74,** 1493 (1948).
4.4. F. Bloch: *Z. Physik* **61,** 206 (1930).
4.5. H. A. Kramers and G. H. Wannier: *Phys. Rev.* **60,** 252 (1941).
4.6. L. Onsager: *Phys. Rev.* **65,** 117 (1944).
4.7. J. H. Van Vleck: *Rev. Mod. Phys.* **17,** 27 (1945).
4.8. C. Kittel: *Nuovo Cimento Suppl.* **6,** 895 (1957).
4.9. J. Van Kranendonk and J. H. Van Vleck: *Rev. Mod. Phys.* **30,** 1 (1958).
4.10. J. H. Van Vleck: *J. Phys. Radium* **20,** 124 (1959).
4.11. Y. Kono and S. Chikazumi: *Kobayashi Riken Hokoku* **9,** 12 (1959).
4.12. J. S. Kouvel and R. H. Wilson: *J. Appl. Phys.* **32,** 435 (1961).
4.13. W. Heisenberg: *Z. Physik* **49,** 619 (1928).
4.14. E. C. Stoner: *Phil. Mag.* [7] **15,** 1080 (1933).
4.15. E. C. Stoner: *Repts. Progr. Phys.* **11,** 43 (1947).
4.16. J. C. Slater: *Phys. Rev.* **49,** 537, 981 (1936); **52,** 198 (1937).
4.17. H. Krutter: *Phys. Rev.* **48,** 664 (1935).
4.18. G. F. Koster: *Phys. Rev.* **98,** 901 (1955).
4.19. C. G. Shull and M. K. Wilkinson: *Phys. Rev.* **97,** 304 (1955).
4.20. M. F. Collins, R. V. Jones, and R. D. Lowde: *J. Phys. Soc. Japan* **17,** *Suppl.* B.III, 19 (1962).
4.21. G. G. E. Low and M. F. Collins: *J. Appl. Phys.* **34,** 1195 (1963).
4.22. J. Friedel: *Nuovo Cimento Suppl.* VII, 287 (1958); *J. Phys. et Rad.* **23,** 501, 692 (1962).
4.23. W. J. Carr: *Phys. Rev.* **85,** 590 (1952).
4.24. T. Hirone: *Sci. Rep. Tohoku Univ.* **27,** 101 (1938).
4.25. E. P. Wohlfarth: *Proc. Roy. Soc.* **195A,** 434 (1949).
4.26. J. Friedel: *Adv. Phys.* **3,** 446 (1954).
4.27. H. Watanabe: *Sci. Rep. Res. Inst. Tohoku Univ.* **6A,** 343 (1954); *J. Phys. Soc. Japan* **13,** 187 (1958).
4.28. G. F. Koster: *Stuttgart Colloq.* (1956).
4.29. E. I. Kondorsky and V. L. Sedov: *J. Appl. Phys.* **31S,** 331 (1960).
4.30. C. Zener: *Phys. Rev.* **81,** 446 (1951); **82,** 403 (1951); **83,** 299 (1951); **85,** 324 (1951).
4.31. C. G. Shull, E. O. Wollan, and W. C. Koehler: *Phys. Rev.* **84,** 912 (1951).
4.32. C. G. Shull and M. K. Wilkinson: *Rev. Mod. Phys.* **25,** 100 (1953).
4.33. L. Pauling: *Phys. Rev.* **54,** 899 (1938).
4.34. F. Bader: *Z. Naturforsch.* **8a,** 334 (1953); F. Bader, K. Ganzhorn, and U. Dehlinger: *Z. Physik* **137,** 190 (1954); refer also to (G.22).
4.35. N. F. Mott and K. W. H. Stevens: *Phil. Mag.* **8,** 1364 (1957).
4.36. J. B. Goodenough: *J. Appl. Phys.* **29,** 513 (1958).
4.37. J. S. Kouvel, C. D. Graham, Jr., and I. S. Jacobs: *J. Phys. Radium* **20,** 198 (1959).
4.38. A. Arrott and H. Sato: *Phys. Rev.* **114,** 1420 (1959); H. Sato and A. Arrott: *ibid.*, **114,** 1427 (1959).
4.39. Cf. refs. 21.51–57.
4.40. K. Yosida and A. Watabe: *Prog. Theor. Phys.* **28,** 361 (1962).

5

Ferrimagnetism

5.1 Theory of Antiferromagnetism

Recent progress in electronics, especially in high frequency techniques, has promoted the advancement of high frequency magnetic materials technology and finally has led to the tremendous development of oxide magnetic materials. One of the factors which made possible this rapid development was the systematic series of experimental investigations made by Snoek[1] and his collaborators. Another factor was the Néel[2] theory of ferrimagnetism. With this theory as the guiding principle, many interesting investigations have been made by researchers in various countries, especially in Holland. But, before going into the theory of ferrimagnetism, we present the theory of antiferromagnetism.

As stated in Chapter 1, by antiferromagnetism we mean an antiparallel alignment of the spins such as is shown in Fig. 1.7*a*. For example, MnO is the representative antiferromagnetic material which has the NaCl-type crystal structure. The Mn ions in this lattice form a face-centered cubic lattice, and their spins are aligned antiparallel to one another as shown in Fig. 5.1. This alignment of spins was determined by the neutron diffraction technique. As is well known, the neutron has no electric charge and accordingly is insensitive to the electric charge of the ions of the crystal lattice. On the other hand, it can be scattered by the magnetic moments of the spins and nuclei, because it has a magnetic moment itself. Therefore, when the magnetic moments of the ions are distributed in a regular manner, the scattered neutron beams interfere with one another, giving rise to

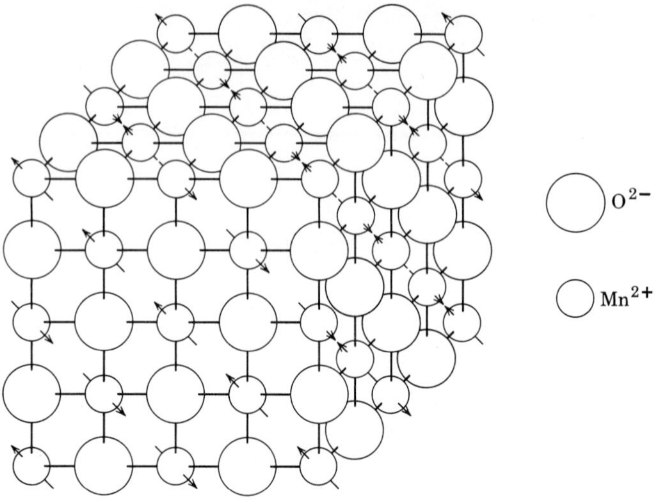

Fig. 5.1. Crystal and magnetic structure of MnO.

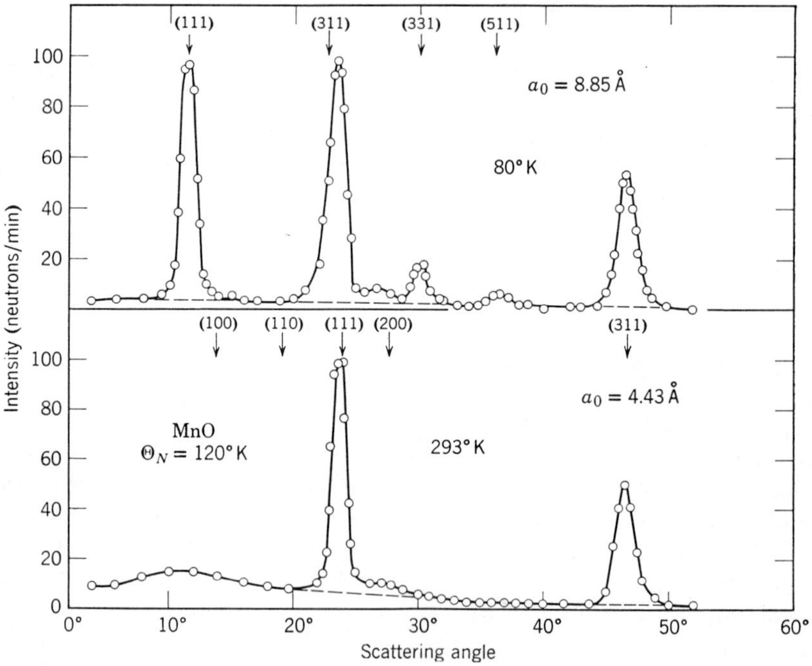

Fig. 5.2. Neutron diffraction lines from MnO crystal (above) and (below) its Néel temperature (after Shull and Smart[3]).

characteristic diffraction lines as in electron diffraction by charged ions. Figure 5.2 shows the neutron diffraction lines from an MnO crystal above and below the transition temperature as observed by Shull and Smart,[3] who discovered this phenomenon. Some of the lines disappear above the transition temperature where the regular alignment of spins is destroyed by the thermal agitation.

Since the magnetic ions in such an oxide are separated by oxygen ions, the direct exchange interaction between the magnetic ions is considered to be very weak. In spite of this fact, there seems to be a strong exchange interaction between magnetic ions, as indicated by the values of Néel temperature and asymptotic Curie temperature (5.9 and 5.15). This fact was explained first by Kramers[4] and more recently by P. W. Anderson[5] in terms of a superexchange interaction. The essential point in superexchange is that the spin moments of the metal ions on opposite sides of an oxygen ion interact with each other through the *p* orbit of the oxygen ion. Consider the system composed of two metal ions M_1 and M_2 separated by an oxygen ion O. The ground state of the oxygen ion should be doubly charged as O^{2-}, which, like neon, has the electronic configuration ($2s^2 2p^6$). In this state there are no spin couplings with the metal ions. There is, however, the possibility of having one of the two electrons of the O^{2-} ion excited and transferred to a neighboring metal ion (say M_1), in which the strong exchange interaction tends to direct the spin of the transferred electron in that direction such that the ion has a maximum spin magnetic moment (Hund's rule). In other words, if the M_1 ion with transferred electron has less than five 3*d* electrons, all the 3*d* electron spins tend to align themselves parallel to each other, while, if it has more than five electrons, the transferred electron must have its spin pointing antiparallel to the resultant magnetic moment of the ion. At the same time, the unpaired electron left in the *p* orbit of the O ion will be coupled with the other metal ion M_2, in which the transferred electron should interact with the M_2 electrons in the manner described above. Since, according to the Pauli principle, the two electrons which were in the *p* orbit of the O ion must have opposite spins, both metal ions should have antiparallel magnetic moments in order to fulfill Hund's rule for both ions. Such a superexchange interaction is expected to be strongest when M_1—O—M_2 lie along a straight line, because the *p* orbit of the O ion is stretched as shown in Fig. 5.3, and so it can overlap the orbits of both metal ions in this configuration. This is actually the case for the MnO crystal (the O ions are at the midpoint of the ⟨100⟩ bond between metal ions). If, however, M_1—O—M_2 makes an angle of, say, 90°, the superexchange is expected to be small. For further details of superexchange interaction, the reader may refer to the review by Nagamiya, Yosida, and Kubo.[6]

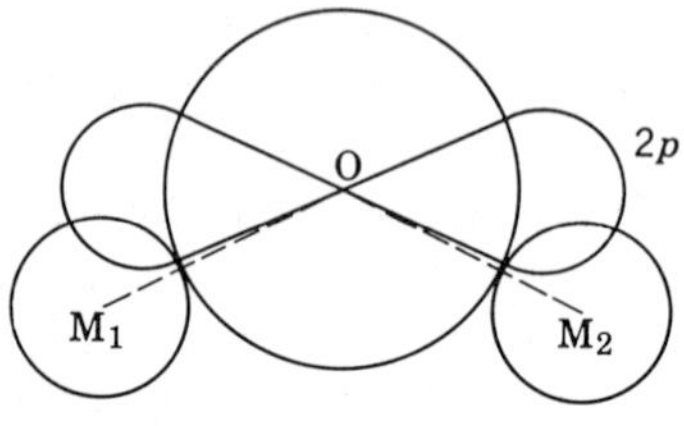

Fig. 5.3. Shape of $2p$ orbit of an oxygen ion.

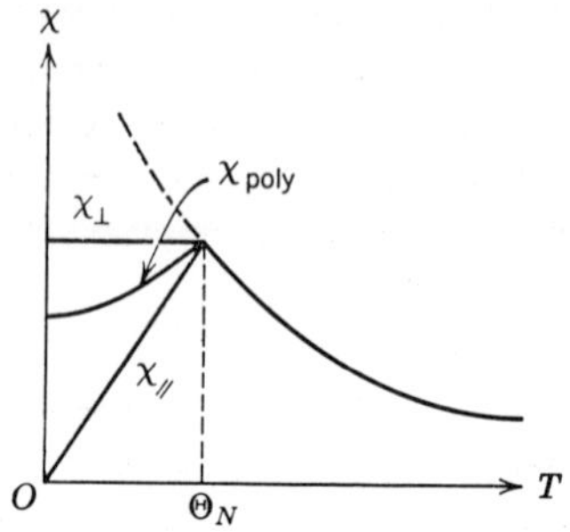

Fig. 5.4. Temperature dependence of the susceptibility of an antiferromagnetic substance.

Now, since an antiferromagnetic arrangement of spins has no net spontaneous magnetization, the substance does not show any kind of ferromagnetism. The tendency to maintain the antiferromagnetic arrangement of spins, however, opposes the magnetization due to the application of an external field and accordingly gives rise to a characteristic temperature dependence of the susceptibility. Thus, contrary to paramagnetism where the susceptibility increases with decrease of temperature, antiferromagnetic substances show a decrease of susceptibility with decrease of temperature below the transition temperature. This situation is represented by the curve denoted by $\chi_{\parallel}$ in Fig. 5.4; here the magnetic field is applied parallel to the spin axis. When the magnetic field is applied perpendicular to the spin axis, magnetization takes place by the rotation of each spin from the direction of the spin axis. Then the susceptibility becomes independent of the temperature as shown by the curve $\chi_{\perp}$ in Fig. 5.4. In a polycrystal the susceptibility is the average between the two cases as shown by the curve denoted by χ_{poly}. Above the transition temperature the susceptibility always decreases with an increase of temperature, independently of the direction of the applied magnetic field. Thus the susceptibility versus temperature curve shows a kink at the transition point. This fact is the feature of antiferromagnetism which distinguishes it from paramagnetism. We call this transition temperature the Néel temperature.

A theoretical treatment of antiferromagnetism was given first by Van Vleck[7] and later by Néel[2] and Anderson.[8] Let A denote the lattice sites which are to be occupied by plus spins, and B those by minus spins. As expected, the spins on A sites are under the influence of exchange forces due to the spins on B sites as well as to the other spins on A sites. As in the Weiss theory of ferromagnetism, the effect of exchange can be expressed by the molecular field. The molecular field which acts at an A site is given by

$$H_{mA} = w_{AA}I_A + w_{AB}I_B, \tag{5.1}$$

where I_A and I_B are the intensities of magnetization of the A and B sites, respectively, and the coefficient w_{AB} is always negative. Similarly, the molecular field at a B site is given by

$$H_{mB} = w_{BA}I_A + w_{BB}I_B. \tag{5.2}$$

Since the A and B sites are symmetrical in the case of antiferromagnetism, we can simply write

$$w_{AA} = w_{BB} = w_1, \quad w_{AB} = w_{BA} = w_2. \tag{5.3}$$

Furthermore, in the absence of an external field, the magnitudes of I_A and I_B must be equal to each other. Thus

$$I_A = -I_B. \tag{5.4}$$

Using relations (5.3) and (5.4), we can simplify (5.1) and (5.2). Thus

$$H_{mA} = (w_1 - w_2)I_A, \tag{5.5}$$

and

$$H_{mB} = (w_1 - w_2)I_B. \tag{5.6}$$

The thermal equilibrium values of I_A and I_B should be given by the following formulas, as they are for ferromagnetism (cf. 4.24):

$$I_A = \frac{NM}{2} L\left(\frac{M(w_1 - w_2)I_A}{kT}\right), \tag{5.7}$$

$$I_B = \frac{NM}{2} L\left(\frac{M(w_1 - w_2)I_B}{kT}\right), \tag{5.8}$$

where N is the number of magnetic atoms per unit volume and L is the Langevin function. Solving these equations, we have I_A and I_B as functions of the temperature, which should have the same functional form as that for the intrinsic magnetization of ferromagnetism. Namely, I_A and I_B decrease with increase of temperature and suddenly disappear at the Néel temperature, which is given by

$$\Theta_N = \frac{NM^2(w_1 - w_2)}{6k}. \tag{5.9}$$

When the substance is in an external field, I_A and I_B do not remain symmetrical, and we cannot use assumption (5.4). Let us suppose that the field H is applied in the plus direction. Then the thermal equilibrium values of I_A and I_B are given by

$$I_A = \frac{NM}{2} L\left(\frac{M(H + w_1 I_A + w_2 I_B)}{kT}\right), \tag{5.10}$$

and

$$I_B = \frac{NM}{2} L\left(\frac{M(H + w_2 I_A + w_1 I_B)}{kT}\right). \tag{5.11}$$

Differentiating these two equations, we have

$$\frac{\partial I_A}{\partial H} = \frac{NM^2}{2kT} L'(\alpha)\left(1 + w_1 \frac{\partial I_A}{\partial H} + w_2 \frac{\partial I_B}{\partial H}\right), \tag{5.12}$$

$$\frac{\partial I_B}{\partial H} = \frac{NM^2}{2kT} L'(\alpha)\left(1 + w_2 \frac{\partial I_A}{\partial H} + w_1 \frac{\partial I_B}{\partial H}\right). \tag{5.13}$$

Adding these equations and solving for $\partial I_A/\partial H + \partial I_B/\partial H$, we obtain

$$\chi = \frac{\partial I}{\partial H} = \frac{\partial I_A}{\partial H} + \frac{\partial I_B}{\partial H}$$

$$= \frac{(NM^2/kT)L'(\alpha)}{1 - (NM^2/2kT)L'(\alpha)(w_1 + w_2)}. \tag{5.14}$$

Putting

$$\Theta_a = \frac{NM^2(w_1 + w_2)}{6k} \tag{5.15}$$

and

$$C = \frac{NM^2}{3k}, \tag{5.16}$$

we can write (5.14) in the simpler form

$$\chi = \frac{3CL'(\alpha)}{T - 3L'(\alpha)\Theta_a}. \tag{5.17}$$

When the temperature is above the Néel temperature, so that the spin arrangement is perfectly random, it follows from (4.10) that

$$L'(\alpha) = \tfrac{1}{3}. \tag{5.18}$$

Then (5.17) becomes

$$\chi = \frac{C}{T - \Theta_a}. \tag{5.19}$$

This formula is quite similar to the Curie-Weiss law (4.37). If we plot $1/\chi$ as a function of T, we have a straight line which intersects the abscissa at $T = \Theta_a$. We learned in Chapter 4 that in ferromagnetism the asymptotic

Curie temperature Θ_a coincides in the first approximation with the ferromagnetic Curie temperature Θ_f. In antiferromagnetism, however, the asymptotic Curie temperature Θ_a is much lower than the Néel temperature Θ_N, as we see in (5.9) and (5.15). (It must be remembered that $w_2 < 0$.) Since the intersites interaction $|w_2|$ is usually larger than the intrasites interaction $|w_1|$, the asymptotic Curie temperature Θ_a is expected to be negative for most of the antiferromagnetic substances. That this is so is shown in Table 5.1 in which the constants of various antiferromagnetic substances are listed.

At the Néel temperature Θ_N, (5.19) becomes

$$\chi_{\max} = \frac{C}{\Theta_N - \Theta_a} = -\frac{1}{w_2}. \tag{5.20}$$

Below the Néel temperature, the magnetization on each site grows gradually with decrease of temperature, resulting in an increase of $L(\alpha)$ in (5.10) and (5.11). Since this produces a decrease of $L'(\alpha)$ from $\frac{1}{3}$, the susceptibility should decrease as expected from (5.17) and finally, for $T \to 0$, $\chi_\parallel$ approaches

$$\lim_{T\to 0} \chi_\parallel = \lim_{T\to 0} \frac{3CL''(\alpha)\,(\partial\alpha/\partial T)}{1 - 3\Theta_a L''(\alpha)\,(\partial\alpha/\partial T)} = 0. \tag{5.21}$$

This explains well the curve $\chi_\parallel$ in Fig. 5.4. The corresponding $1/\chi$ vs T curves are shown in Fig. 5.5.

When a magnetic field is applied perpendicular to the spin axis, χ does not become zero even at $T = 0$, because the substance can be magnetized by rotation of the magnetizations I_A and I_B from the original spin axis. Take the x axis parallel to the direction of the magnetic field H, and the

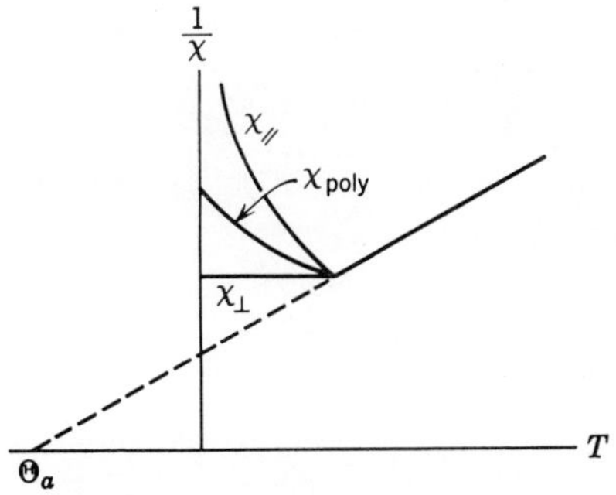

Fig. 5.5. Temperature dependence of the reciprocal susceptibility of an antiferromagnetic substance.

Table 5.1. Constants of Various Antiferromagnetic Substances (after Nagamiya, Yosida, and Kubo[6])

Substance	Crystal Type	Θ_N (from χ), °K	Θ_a, °K	C_{mole} in mks ($\times 10^{-11}$ Hm^2/mole · deg)	C_{mole} in cgs (cm^3/mole · deg)
MnO	NaCl	122	−610	6.94	4.40
FeO	NaCl	198	−570	9.83	6.24
CoO	NaCl	293	−280	4.82	3.055
NiO	NaCl	492 ~ 647			
CuO	monocl.	~453			
V_2O_3	Cr_2O_3	173			
V_2O_4	rutile	343	−720	0.87	0.55
Cr_2O_3	trig.	311	−1070	4.03	2.56
MnO_2	rutile	84			
α-Fe_2O_3	Cr_2O_3	~950	−2000	6.93	4.4
CrS	NiAs				
MnS	NaCl	165	−528	6.78	4.30
FeS	NiAs	613	−857	5.42	3.44
MnSe	NaCl	247†	−361	6.32	4.01
MnTe	NiAs	323	−690	7.25	4.59
FeTe	NiAs		−220	1.45	0.92
MnF_2	rutile	72	−113.2	6.43	4.08
FeF_2	rutile	79	−117	6.12	3.88
CoF_2	rutile	37.7†	−52.7	5.19	3.29
NiF_2	rutile	73.2†	−115.6	2.41	1.528
VCl_2			−565	3.35	2.13
VCl_3		30	−30.1	1.58	1.005
$CrCl_2$		40	−149	5.13	3.26
$FeCl_2$	$CdCl_2$	24	+48.0	5.62	3.56
$FeCl_3$	trig.		−11.5	6.42	4.07
$CoCl_2$	$CdCl_2$	25	+38.1	5.46	3.46
$NiCl_2$	$CdCl_2$	50	+68.2	2.14	1.36
$CuCl_2$		70	−109	0.885	0.536
$CuBr_2$		193			
CrSb	NiAs	~673	~ −1000		
MnAs	NiAs	399	+293	4.10	2.60
MnBi	NiAs	621 ~ 633†			
Cr	b.c.c.	~1673	−475		
α-Mn	complex	95†			
$MnCl_2 \cdot 4H_2O$		~1.68			
$MnBr_2 \cdot 4H_2O$		~2.2			
$CuCl_2 \cdot 2H_2O$		4.31†	−5	C_a = 0.703 C_b = 0.610	0.447 0.386
$Co(NH_4)_2(SO_4)_2 \cdot 6H_2O$	monocl.	0.084†			
$FeCO_3$	$NaNO_3$		−14	5.50	3.49
$FeCO_3 \cdot 2MgCO_3$					

† Determined from anomalous specific heat.

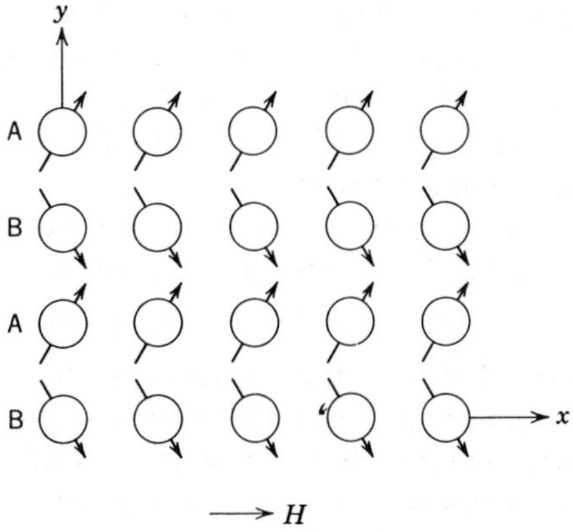

Fig. 5.6. Spin rotation in an antiferromagnetic substance.

y axis parallel to the original direction of I_A (cf. Fig. 5.6). Then the x and y components of the molecular field H_{mA} at A sites should be given by

$$\begin{aligned}(H_{mA})_x &= w_1 I_{Ax} + w_2 I_{Bx},\\ (H_{mA})_y &= w_1 I_{Ay} + w_2 I_{By}.\end{aligned} \tag{5.22}$$

Since the A sites and B sites are symmetrical, it follows that

$$I_{Bx} = I_{Ax}, \quad I_{By} = -I_{Ay}. \tag{5.23}$$

Then (5.22) becomes

$$\begin{aligned}(H_{mA})_x &= (w_1 + w_2) I_{Ax},\\ (H_{mA})_y &= (w_1 - w_2) I_{Ay}.\end{aligned} \tag{5.24}$$

When the external field H is applied parallel to the x axis, the spins should point parallel to the resultant field which is composed of the external field and the molecular field; thus

$$\frac{I_{Ax}}{I_{Ay}} = \frac{H_x}{H_y} = \frac{H + (H_{mA})_x}{(H_{mA})_y} = \frac{H + (w_1 + w_2) I_{Ax}}{(w_1 - w_2) I_{Ay}}. \tag{5.25}$$

Eliminating I_{Ay} from the preceding equation, we have

$$H + (w_1 + w_2) I_{Ax} = (w_1 - w_2) I_{Ax},$$

or

$$I_{Ax} = -\frac{H}{2w_2}, \tag{5.26}$$

so that the susceptibility at 0°K becomes

$$\chi_\perp = \frac{I_x}{H} = \frac{I_{Ax} + I_{Bx}}{H} = -\frac{1}{w_2}. \tag{5.27}$$

This value coincides with the value of χ at the Néel temperature (cf. 5.20), indicating that the susceptibility $\chi_\perp$ is independent of temperature from $T = 0°K$ to $T = \Theta_N$. This situation is shown in Figs. 5.4 and 5.5. In polycrystalline substances, which contain randomly-oriented crystallites, χ is given by the average between $\chi_\parallel$ and $\chi_\perp$.

We have seen that $\chi_\perp$ is always larger than $\chi_\parallel$ below the Néel temperature. Since the energy of magnetization is given by

$$E = -\tfrac{1}{2}\chi H^2, \qquad \text{(cf. 10.19)} \qquad (5.28)$$

this energy is always lower for the perpendicular field than for the parallel field. The whole spin arrangement, therefore, tends to rotate so as to make its spin axis perpendicular to the external field. Néel[9] predicted that such a flopping of the spin axis should occur when the field reaches some critical value which is determined by the magnitude of the magnetocrystalline anisotropy relative to the antiferromagnetic properties of the substance. This phenomenon was actually observed by Poulis et al.[10] for $CuCl_2{\cdot}2H_2O$ and by Jacobs[11] for MnF_2. We can interpret various interesting phenomena in antiferromagnetism in terms of this flopping of the spin axis; the reader is referred to the article by Nagamiya, Yosida, and Kubo[6] for further details.

5.2 Theory of Ferrimagnetism

As mentioned above, antiferromagnetics are not strongly magnetic because the magnetizations I_A and I_B of the two sublattices cancel each other. If, however, one of these magnetizations is stronger than the other (say $I_A > I_B$), it is to be expected that the difference between the two magnetizations will give rise to a strong magnetism. By reasoning along this line, Néel[2] treated the mechanism of the generation of intrinsic magnetization in ferrites theoretically and explained various characteristic temperature dependences of magnetization.

The term ferrite denotes a group of iron oxides which have the general formula $MO{\cdot}Fe_2O_3$, where M is a divalent metal ion such as Mn^{2+}, Fe^{2+}, Co^{2+}, Ni^{2+}, Cu^{2+}, Zn^{2+}, Mg^{2+}, or Cd^{2+}. The typical ferrite is magnetite, Fe_3O_4 (or $FeO{\cdot}Fe_2O_3$), which has been a well-known magnetic oxide since ancient times. By replacing the divalent iron in Fe_3O_4 by another divalent ion, we can produce various ferrites which have different intensities of intrinsic magnetization. Furthermore, just as we can get various magnetic alloys by mixing a number of metal elements, by mixing two or more kinds of M^{2+} ions we can obtain mixed ferrites, which show various interesting and useful magnetic properties. We shall be concerned with such properties of ferrites in Part 3 and thereafter. In this chapter we confine ourselves to the problem of the intrinsic magnetization of the ferrites.

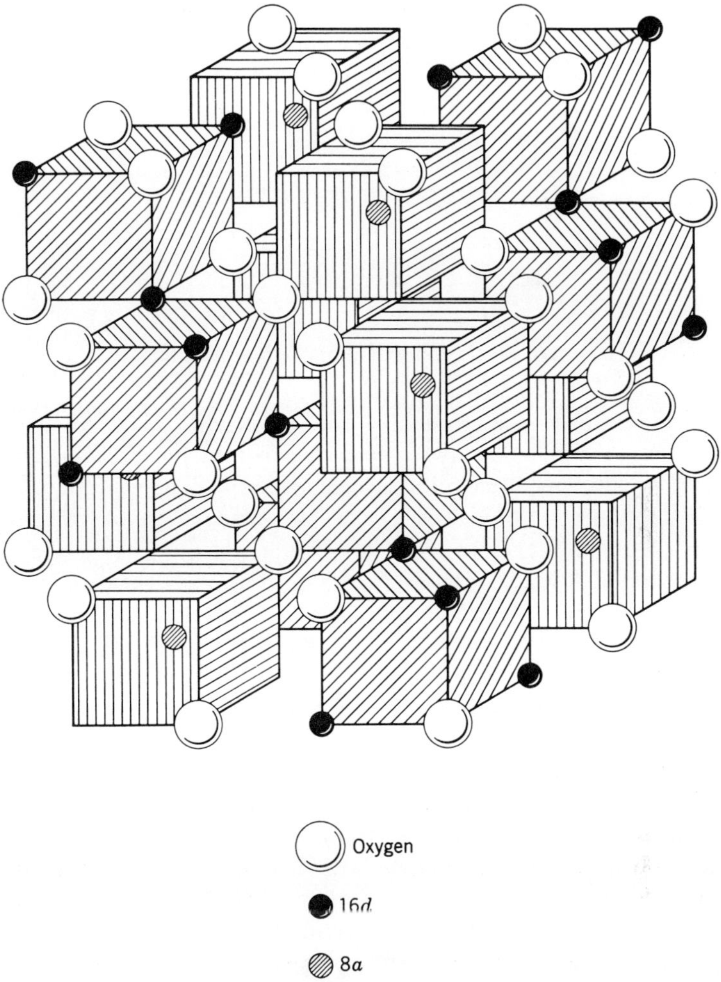

Fig. 5.7. Spinel structure.

Ferrites have the so-called spinel crystal structure which is shown in Fig. 5.7. The white circles in this figure represent the oxygen ions, and the black and hatched circles represent the metal ions. The radius of the oxygen ions is about 1.32 Å, which is much larger than that of metal ions (0.6 ~ 0.8 Å); hence the oxygen ions in this lattice touch each other and form a close-packed face-centered cubic lattice. In this oxygen lattice the metal ions take interstitial positions which can be classified into two groups: One is a group of lattice sites called tetrahedral sites or 8*a* sites, each of which is surrounded by four oxygens as shown by the hatched circles in

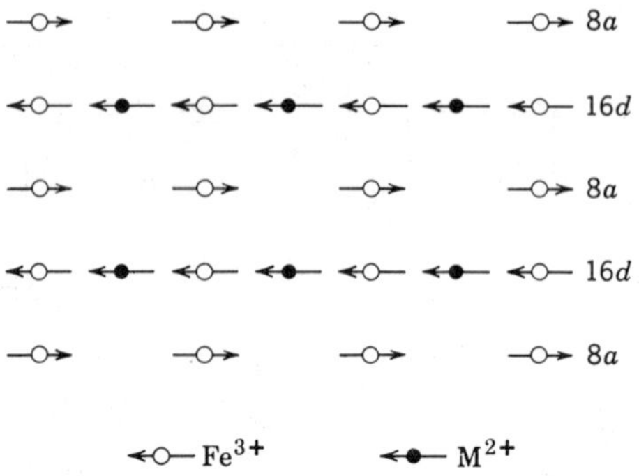

Fig. 5.8. Spin arrangement of inverse spinel ferrite.

the figure. The other is a group of sites called octahedral or 16*d* sites, each of which is surrounded by six oxygens as shown by the black circles. We shall, for brevity, hereafter call the former group A sites, and the latter B sites. From the point of view of valence, it seems reasonable to have M^{2+} ions on A sites and Fe^{3+} ions on B sites, because the number of oxygen ions which surround A and B sites are in the ratio of 2:3. This is actually the case for Zn ferrite. We call this structure a normal spinel. It turns out, however, that the ferromagnetic ferrites have half of the Fe^{3+} ions on A sites and the remaining half of the Fe^{3+} and all the M^{2+} ions on B sites. We call this crystal structure an inverse spinel. The factors which influence the distribution of metal ions over A and B sites have been considered to be the radii of the metal ions, the matching of the electronic configuration of the metal ions to the surrounding oxygen ions, and the electrostatic energy of the lattice (Madelung energy).

The intensity of intrinsic magnetization of ferrites can be explained by assuming the ion distribution of the inverse spinel type and also an antiparallel alignment of the spins on the A and B sites, of the character shown schematically in Fig. 5.8. Such an alignment of spins should be expected from the nature of the superexchange interaction; because the angle of A—O—B is about 125°, while that of A—O—A is about 80° and that of B—O—B 90°, the superexchange interaction (negative interaction) should be greatest between A and B sites (cf. Section 5.1). Now, as seen in Fig. 5.8, the same number of Fe^{3+} ions enter both sites, so that their magnetic moments should be cancelled. The net magnetization, therefore, should be due solely to the magnetic moments of M^{2+} ions which are on B sites.

Now, it is expected, on the basis of Hund's rule (cf. Section 3.4), that each transition metal ion has the maximum spin magnetic moment which can be attained by the alignment of its 3*d* spins within the restriction of the Pauli principle. Since Mn^{2+}, Fe^{2+}, Co^{2+}, Ni^{2+}, Cu^{2+}, and Zn^{2+} have

5, 6, 7, 8, 9, and 10 $3d$ electrons, respectively, their magnetic moments should be 5, 4, 3, 2, 1, and 0 Bohr magnetons (cf. the lower part of Fig. 5.9). Figure 5.9 shows the observed intrinsic magnetization of various ferrites, reduced to the magnetic moment per formula unit, as a function of the number of $3d$ electrons per M^{2+} ion. The solid line shows the theoretical magnetic moment which is to be expected when it is assumed that the net magnetization is due to the spin magnetic moment of the M^{2+} ions. It is seen that the experimental points are close to the theoretical line. The slight deviation of experimental points from the theoretical line has been explained in terms of the deviation of the distribution of metal ions from that of an ideal inverse spinel, the contribution of the orbital moment (hatched area), and angle formation by the ionic moments on the sites. In the case of $MnFe_2O_4$, the deviation cannot be explained in terms of ion distribution because both Mn^{2+} and Fe^{3+} have the same magnetic moment ($5M_B$). (Incidentally, it was confirmed by means of neutron diffraction by Hastings and Corliss[13] that 80% of the Mn^{2+} ions take the normal spinel positions, and the remaining 20% enter the inverse spinel positions.)

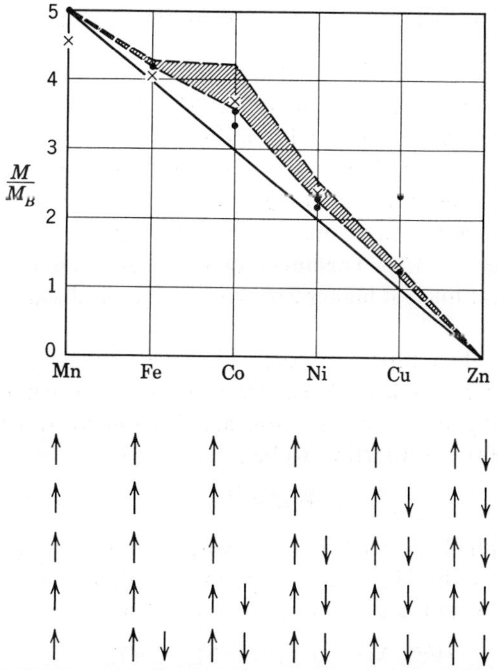

Fig. 5.9. Molecular absolute saturation moment of various simple ferrites as a function of the number of $3d$ electrons per M^{2+} ion. (The figures along the abscissa represent the spin alignment inside M^{2+} ions.) (After Neel[12]; cross points from G.32.)

Harrison et al.[14] explained this deviation by assuming the occurrence of Mn^{3+} and Fe^{2+} in B sites as the result of the transfer of an electron between Mn^{2+} and Fe^{3+}.

The molecular magnetic moment of a ferrite cannot, therefore, exceed $5M_B$ as long as it takes the inverse spinel structure. It is expected, however, that the molecular magnetic moment of a ferrite can be increased by adding some amount of normal spinel ferrite; for instance, $ZnFe_2O_4$. Then the Zn^{2+} ions should occupy the A sites and force the same number of Fe^{3+} ions from A sites to B sites. This should give rise to an additional magnetization of B sites. Thus, if we add a fraction x of $ZnO{\cdot}Fe_2O_3$ to the fraction $1 - x$ of

$$(\overrightarrow{\mathrm{Fe}^{3+}})\mathrm{O}\cdot(\overleftarrow{\mathrm{Fe}^{3+}}\overleftarrow{\mathrm{M}^{2+}})\mathrm{O}_3,$$

where the arrows denote the directions of the magnetic moments, the resultant structure is expressed by

$$(\overrightarrow{\mathrm{Fe}^{3+}_{1-x}}\mathrm{Zn}^{2+}_x)\mathrm{O}\cdot(\overleftarrow{\mathrm{Fe}^{3+}_{1+x}}\overleftarrow{\mathrm{M}^{2+}_{1-x}})\mathrm{O}_3.$$

The net magnetic moment of this molecule is calculated as

$$\begin{aligned} M &= [5(1 + x) + n(1 - x) - 5(1 - x)]M_B \\ &= [n + (10 - n)x]M_B, \end{aligned} \tag{5.29}$$

where n is the number of Bohr magnetons of M^{2+} moment. Thus M should increase from nM_B toward $10M_B$ with addition of Zn^{2+} ions. Guillaud[15] and Gorter[16] have actually found that it does increase. Figure 5.10 shows the dependence of the molecular magnetic moment of various mixed Zn ferrites on their composition, and we see the tendency of every curve toward the moment $10M_B$. The deviation of the actual magnetic moment from the theoretical lines at higher Zn concentration is discussed in Section 20.1.

Next we discuss the temperature dependence of the intrinsic magnetization of ferrites along the lines of Néel's[2] theory. The ferrite has the formula $MO{\cdot}Fe_2O_3$, and in general the Fe^{3+} ions are distributed over the A and B sites. We assume the proportion to be $\lambda:\mu$, where

$$\lambda + \mu = 1. \tag{5.30}$$

For a normal spinel, $\lambda = 0$, $\mu = 1$; whereas, for an inverse spinel, $\lambda = \mu = 0.5$. Since the M^{2+} ions occupy the remaining sites, the distribution of ions over A and B sites can be expressed by

$$(\mathrm{Fe}^{3+}_{2\lambda}\mathrm{M}^{2+}_{1-2\lambda})\mathrm{O}\cdot(\mathrm{Fe}^{3+}_{2\mu}\mathrm{M}^{2+}_{2-2\mu})\mathrm{O}_3.$$

Let us simplify the problem by assuming that Fe^{3+} ions carry magnetic moment M, whereas the M^{2+} ions have no magnetic moment at all. Since

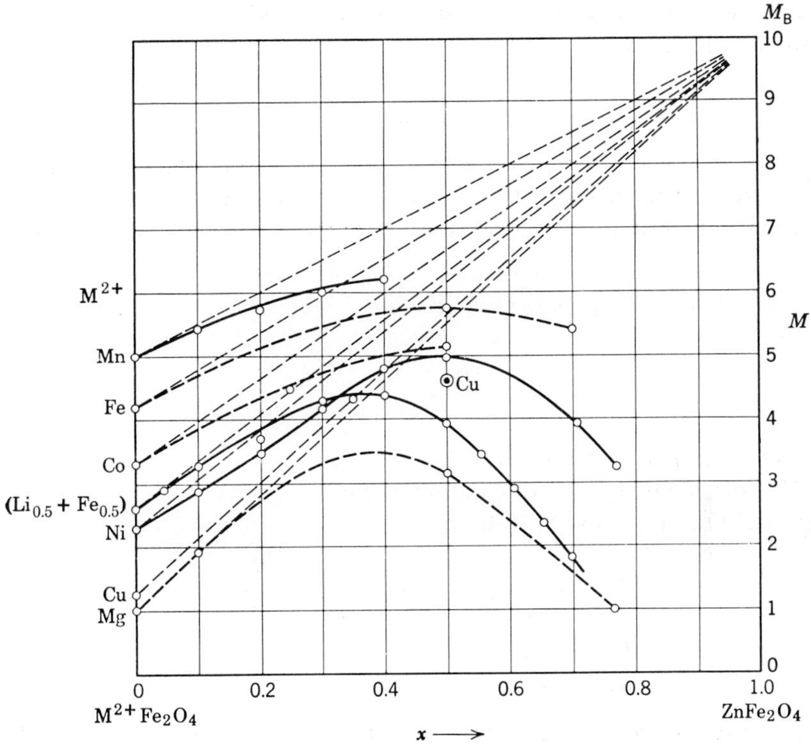

Fig. 5.10. Increase of molecular absolute saturation moment of various inverse ferrites by the addition of the Zn ferrite with a normal spinel structure (after Gorter[16])

the thermal agitation of the magnetic moment will be different in the different lattice sites because of the different intensity of the molecular field, the contribution of one magnetic ion to the net magnetization will be different for A and B sites. Let us designate these contributions by i_A and i_B for the A and B sites, respectively. Then the magnetization of each sublattice is given by

$$\begin{aligned} I_A &= N(2\lambda)i_A, \\ I_B &= N(2\mu)i_B, \end{aligned} \tag{5.31}$$

where N is the number of "molecules" of ferrite included in the unit volume of the spinel lattice. If we put

$$\begin{aligned} I_a &= 2Ni_A, \\ I_b &= 2Ni_B, \end{aligned} \tag{5.32}$$

we have the total magnetization,

$$I = I_A + I_B = \lambda I_a + \mu I_b. \tag{5.33}$$

Similarly to antiferromagnetism, the molecular fields at the A and B sites can be expressed by

$$\begin{aligned} H_{mA} &= w(\alpha\lambda I_a - \mu I_b), \\ H_{mB} &= w(\beta\mu I_b - \lambda I_a), \end{aligned} \tag{5.34}$$

where w is a positive coefficient and α and β are factors which express the intensity of intrasite interactions relative to the intersite interactions. The thermal equilibrium value of the magnetization of each sublattice is given by

$$\begin{aligned} I_a &= 2NML\left(\frac{Mw(\alpha\lambda I_a - \mu I_b) + MH}{kT}\right), \\ I_b &= 2NML\left(\frac{Mw(\beta\mu I_b - \lambda I_a) + MH}{kT}\right), \end{aligned} \tag{5.35}$$

where H is the intensity of the external field.

For the paramagnetism above the Curie temperature we can use the approximation $L'(\alpha) = \frac{1}{3}$, so that

$$\begin{aligned} \frac{\partial I_a}{\partial H} &= \frac{2NM^2}{3kT}\left[1 + w\left(\alpha\lambda \frac{\partial I_a}{\partial H} - \mu \frac{\partial I_b}{\partial H}\right)\right], \\ \frac{\partial I_b}{\partial H} &= \frac{2NM^2}{3kT}\left[1 + w\left(\beta\mu \frac{\partial I_b}{\partial H} - \lambda \frac{\partial I_a}{\partial H}\right)\right]. \end{aligned} \tag{5.36}$$

Solving for $\partial I_a/\partial H$ and $\partial I_b/\partial H$ in (5.36) and substituting in the formula

$$\frac{\partial I}{\partial H} = \lambda \frac{\partial I_a}{\partial H} + \mu \frac{\partial I_b}{\partial H},$$

we have

$$\frac{\partial I}{\partial H} = C \frac{T - Cw\lambda\mu(\alpha + \beta + 2)}{T^2 - Cw(\alpha\lambda + \beta\mu)T + C^2w^2\lambda\mu(\alpha\beta - 1)}, \tag{5.37}$$

where

$$C = \frac{2NM^2}{3k}. \tag{5.38}$$

Equation (5.37) may be written in the simplified form

$$\frac{1}{\chi} = \frac{T}{C} + \frac{1}{\chi_0} - \frac{\sigma}{T - \Theta}, \tag{5.39}$$

where

$$\begin{aligned} \frac{1}{\chi_0} &= w(2\lambda\mu - \lambda^2\alpha - \mu^2\beta), \\ \sigma &= Cw^2\lambda\mu[\lambda(\alpha + 1) - \mu(\beta + 1)]^2, \\ \Theta &= Cw\lambda\mu(\alpha + \beta + 2). \end{aligned} \tag{5.40}$$

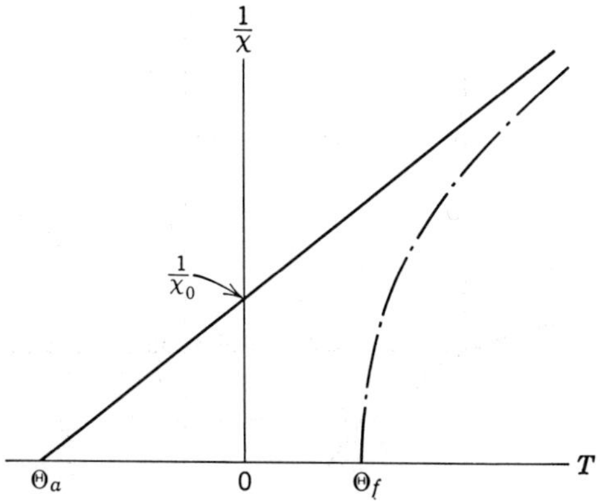

Fig. 5.11. The $1/\chi$ vs T curve of ferrimagnetism.

The first two terms on the right side of (5.39) represent the asymptotic line to which the $1/\chi - T$ curve should approach at high temperature, because the third term must vanish at $T \to \infty$ (cf. Fig. 5.11). The extrapolation of this asymptote intersects the abscissa at

$$\Theta_a = -\frac{C}{\chi_0}, \tag{5.41}$$

which shall be called the asymptotic Curie point. The third term, however, will increase rapidly as the temperature is decreased toward Θ, so that the $1/\chi - T$ curve deviates from the asymptotic line and finally drops off to zero at $T = \Theta_f$, before the temperature reaches Θ. Θ_f is the temperature at which $1/\chi$ becomes zero and at the same time a spontaneous magnetization appears, so it should be called a normal Curie point, or a ferrimagnetic Curie point. This temperature can be calculated by putting (5.39) equal to zero at $T = \Theta_f$; thus

$$\Theta_f{}^2 - Cw(\alpha\lambda + \beta\mu)\Theta_f + C^2w^2\lambda\mu(\alpha\beta - 1) = 0. \tag{5.42}$$

Solving this equation, we have

$$\Theta_f = \frac{Cw}{2}\left[\alpha\lambda + \beta\mu + \sqrt{(\alpha\lambda - \beta\mu)^2 + 4\lambda\mu}\right] \tag{5.43}$$

as the larger solution. If $\Theta_f < 0$, we have paramagnetism for the whole temperature range, whereas, if $\Theta_f > 0$, we have ferrimagnetism below this temperature. The spontaneous magnetization can be obtained by solving (5.35) with $H = 0$ for I_a and I_b and putting them into (5.33). The condition for obtaining ferrimagnetism is $\Theta_f > 0$, or, from (5.43),

$$\alpha\lambda + \beta\mu + \sqrt{(\alpha\lambda - \beta\mu)^2 + 4\lambda\mu} > 0$$

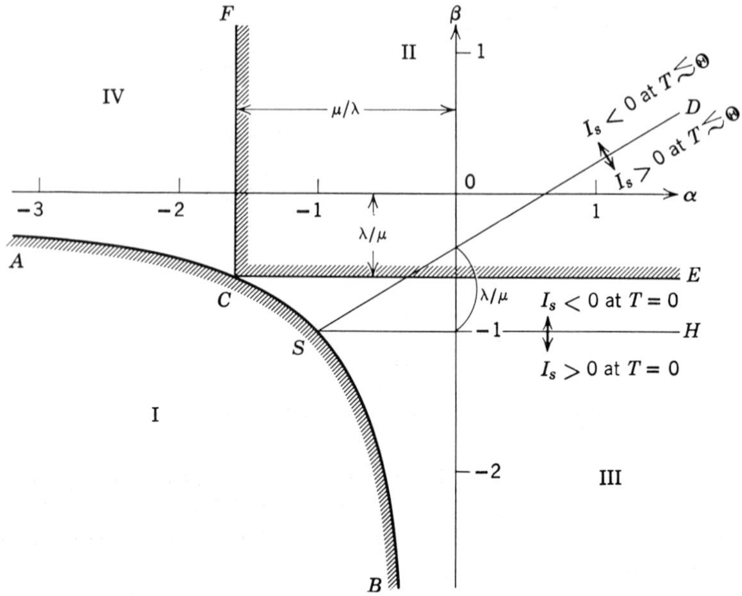

Fig. 5.12. Classification of various stable spin alignments in ferrite in the $\alpha\beta$ plane. (α and β measure the intensity of the intrasites interaction in A and B sites relative to the intersites interaction.) (After Néel.[2])

or, equivalently

$$\alpha > 0, \quad \beta > 0, \quad \text{and } \alpha\beta < 1 \quad \text{for negative } \alpha \text{ or } \beta. \tag{5.44}$$

This corresponds to the area above and to the right of the curve *ACB* ($\alpha\beta = 1$) in the $\alpha\beta$ plane (in Fig. 5.12). The area below and to the left of the curve *ACB* is the region of paramagnetism. The physical meaning of this is that sufficiently large negative values of α or β (intrasites interaction) destroy the ferrimagnetic alignment which is favored by the negative intersites interaction.

An assortment of temperature dependences, which is one of the features of ferrimagnetism, is to be expected from various combinations of α, β, λ, and μ. Let us first discuss the stable values of I_a and I_b at $T = 0$. The energy of spontaneous magnetization is given by

$$E = -\tfrac{1}{2}\lambda I_a H_{m\text{A}} - \tfrac{1}{2}\mu I_b H_{m\text{B}}, \tag{5.45}$$

similarly to ferromagnetism (cf. 4.41). Putting (5.34) in this equation, we have

$$E = -\tfrac{1}{2}w(\alpha\lambda^2 I_a^2 - 2\lambda\mu I_a I_b + \beta\mu^2 I_b^2). \tag{5.46}$$

Now we compare the energy for four cases.

I. Paramagnetism; that is, when

$$I_a = I_b = 0, \tag{5.47}$$

(5.46) gives

$$E = 0. \tag{5.48}$$

II. Both I_a and I_b fully saturated, that is, when

$$\begin{aligned} I_a &= 2NM, \\ I_b &= -2NM, \end{aligned} \tag{5.49}$$

(5.46) becomes

$$E = -2wN^2M^2(\alpha\lambda^2 + 2\lambda\mu + \beta\mu^2). \tag{5.50}$$

III. I_a is saturated, that is,

$$I_a = 2NM, \tag{5.51}$$

while I_b is determined so as to minimize the energy, or

$$\frac{\partial E}{\partial I_b} = 0,$$

from which we have

$$I_b = \frac{\lambda}{\beta\mu}\, 2NM. \tag{5.52}$$

The energy in this case is given by

$$E = -2wN^2M^2\lambda^2\left(\alpha - \frac{1}{\beta}\right). \tag{5.53}$$

IV. I_b saturated; that is,

$$I_b = -2NM, \tag{5.54}$$

while I_a takes the value

$$I_a = -\frac{\mu}{\alpha\lambda}\, 2NM \tag{5.55}$$

so as to minimize the energy. Equation (5.46) gives the energy as

$$E = -2wN^2M^2\mu^2\left(\beta - \frac{1}{\alpha}\right). \tag{5.56}$$

If condition (5.44) is not fulfilled, (5.50), (5.53) and (5.56) give positive values of E, so that case I (paramagnetism) is most stable in this region. This is consistent with the result given above on the basis of the sign of Θ_f. Next we assume that $\lambda < \mu$. In order to obtain case IV, it is necessary to have I_a less than $2NM$ or, by (5.55), $\alpha < -\mu/\lambda$. Actually we can show that the energy of case IV, as given by (5.56), has a lower value than that of (5.50); in other words, case IV is stable to the left side of the line FC,

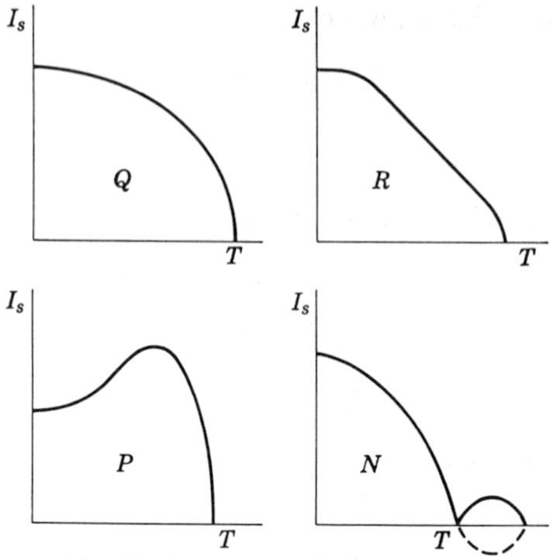

Fig. 5.13. Various types of temperature dependence of intrinsic magnetism of ferrites.

satisfying $\alpha = -\mu/\lambda$ (Fig. 12). Similarly, case III is stable in the region below the line *CE*, satisfying $\beta = -\lambda/\mu$. Case II is stable in the remaining region surrounded by the lines *FC* and *CE*.

It may be easily seen by comparison of (5.51) and (5.52) (cf. 5.33) that $I_A > |I_B|$ for $\beta < -1$, while $I_A < |I_B|$ for $\beta > -1$. In other words, the resultant magnetization is positive in the region below the line *SH*, and it is negative above this line. On the other hand, it is easy to verify that at the temperature just below the Curie point the net magnetization is positive when $\lambda(\alpha + 1) > \mu(\beta + 1)$, and that it is negative when $\lambda(\alpha + 1) < \mu(\beta + 1)$. The border line

$$\lambda(\alpha + 1) - \mu(\beta + 1) = 0 \tag{5.57}$$

is the line *SD* in the figure. In the region between the two lines *SD* and *SH*, therefore, it is expected that the magnetization changes its sign on the way from $T = 0$ to $T = \Theta$.

The temperature dependence of this spontaneous magnetization is shown as the *N*-type curve in Fig. 5.13. If the magnetization is measured under the application of an external field, as for the usual measurement, the magnetization disappears once at the temperature where it changes sign. This temperature is called a compensation point. This type of temperature dependence was first observed by Gorter[16] for Li-Cr ferrite (Fig. 5.14) and also by Bertaut and Pauthenet[17] for various kinds of ferrimagnetic iron garnets (cf. Fig. 5.21).

Even if the magnetization maintains its sign from $T = 0$ to $T = \Theta$, we nevertheless observe various temperature dependences, such as the *Q*, *R*,

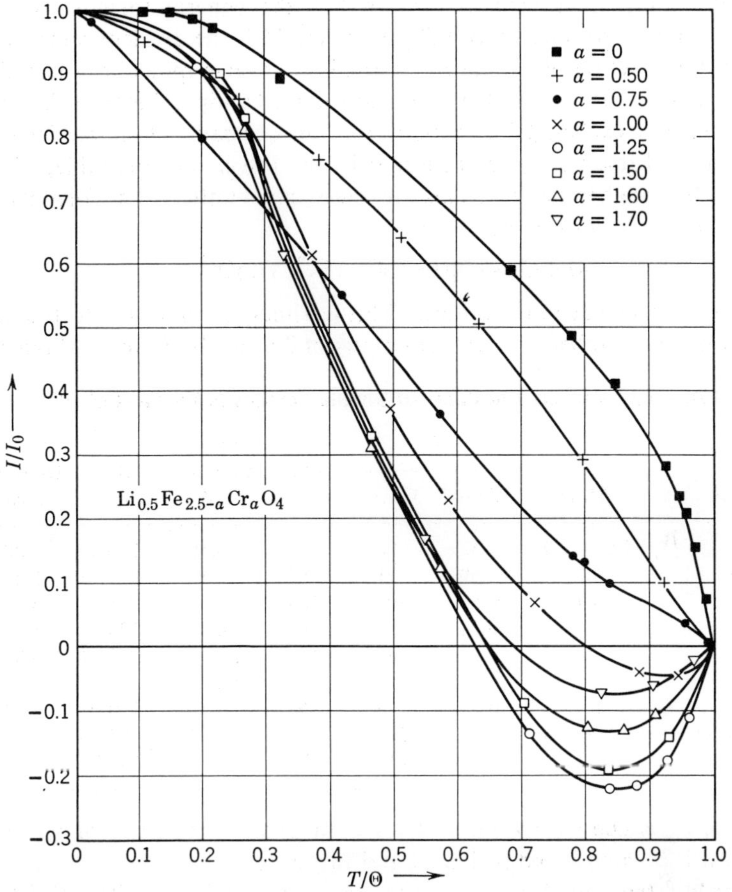

Fig. 5.14. Temperature dependence of intrinsic magnetization of Li-Cr ferrite (after E. W. Gorter[16]).

and P types shown in Fig. 5.13, for various combinations of α, β, λ, and μ. The Q type is the normal temperature dependence which we usually observe for metals and alloys. Almost all ferrites (Mn-Zn ferrite, Ni-Zn ferrite, and Co ferrite, for example) show the R-type temperature dependence. P-type temperature dependence was observed by Gorter[16] for some concentration ranges of Ni-Mn-Ti, Ni-Al, and Mn-Fe-Cr ferrites.

5.3 Ferrimagnetic Oxides and Compounds

In this section we survey various magnetic oxides and compounds and discuss their crystal structure and ferrimagnetism.

Ferrites. As mentioned in Section 5.2, the general formula of the ferrites can be expressed as $MO \cdot Fe_2O_3$, where M is the divalent metal ion. The crystal structure of a magnetic ferrite is an inverse spinel, in which the spin moments of the 8 *a* and 16 *d* sites are aligned antiparallel to each other by the negative exchange interaction. Various constants of simple ferrites are given in Table 5.2. γFe_2O_3, found in the last line of the table, is an oxide called maghemite, which also forms a spinel lattice. The formula of this oxide can be written as

$$4Fe_2O_3 \rightarrow 3[(Fe^{3+})O \cdot (Fe^{3+}_{5/3}V_{1/3})O_3],$$

where V is a vacancy in the lattice. This ferrimagnetism is due to the difference in the number of Fe^{3+} ions between 8 *a* and 16 *d* sites. Magnetic

Table 5.2. Various Constants of Simple Ferrites (after Gorter[G.21])

Substance	M/M_B	I_s Wb/m²	I_s Gauss	Θ_f, °C	Lattice Constant, Å	Mol. Wt.	Density, g/cc
$MnFe_2O_4$	4.6 ~ 5.0	0.52	408	300	8.50	230.63	5.00
$FeFe_2O_4$	4.1	0.60	471	585	8.39	231.55	5.24
$CoFe_2O_4$	3.7	0.50	392	520	8.38	234.64	5.29
$NiFe_2O_4$	2.3	0.34	267	585	8.34	234.41	5.38
$CuFe_2O_4$	1.3	0.17	133	455	8.64/8.24 (tetrag.)	239.24	5.38
$MgFe_2O_4$	1.1	0.14	110	440	8.36	200.02	4.52
$Li_{0.5}Fe_{2.5}O_4$	2.5 ~ 2.6	0.39	306	670	8.33	207.10	4.75
γFe_2O_3		0.52	417	675		159.70	

anisotropy, magnetostriction, and other magnetic properties of various ferrites will be mentioned in Part 3 and thereafter.

Corundum-type oxides. Hematite, αFe_2O_3, and ilmenite, $FeTiO_3$, belong to this category. These, together with magnetite, Fe_3O_4, and maghemite, γFe_2O_3, are the popular natural oxides. The oxides of this category crystallize in a lattice having rhombohedral symmetry, in which the metal ions occupy the various sites as shown in Fig. 5.15. Both hematite and ilmenite show parasitic ferromagnetism (cf. Section 1.2). The solid solutions of both oxides, however, show ferrimagnetism, the saturation magnetizations of which are shown in Fig. 5.16, together with those for the other $MTiO_3$-Fe_2O_3 series (M = Co, Ni, or Mn). The reason for the behavior observed is considered to be that the non-magnetic Ti^{4+} ions enter preferentially into the A sites while the magnetic M^{2+} ions enter into the B sites of the antiferromagnetic sublattices of αFe_2O_3, so that the saturation moment is determined by the magnetization due to M^{2+} ions in B sites.[19,20] Such an ordered arrangement of ions and alignment of

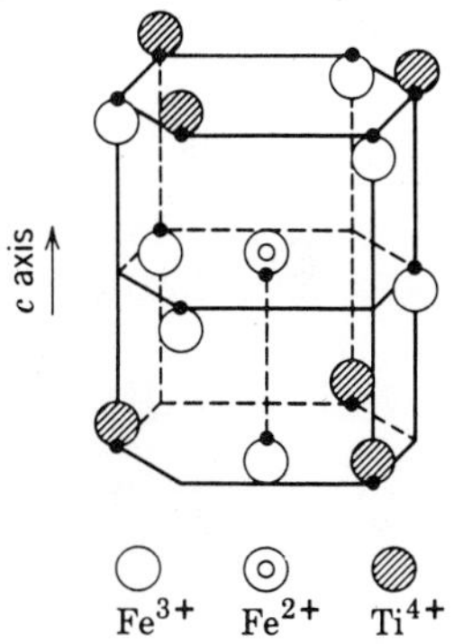

Fig. 5.15. Crystal structure of ilmenite.

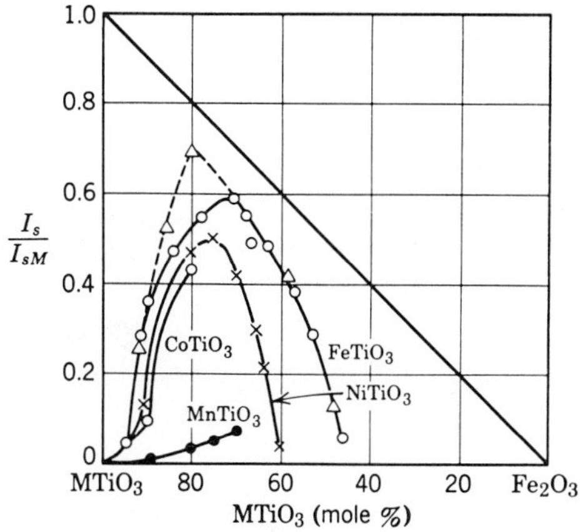

Fig. 5.16. Molecular saturation moment of $MTiO_3$-Fe_2O_3 (after Ishikawa[18]).

magnetic moments were confirmed by Shirane et al.[21] by means of neutron diffraction.

Magnetoplumbite structure oxide. Magnetoplumbite is the ferrimagnetic oxide the composition of which is given by $PbFe_{11}AlO_{19}$. The same type of oxide can be obtained[22] by mixing and sintering BaO, SrO, or PbO with Fe_2O_3. The general formula is $MO \cdot 6Fe_2O_3$ (so-called *M* type), where M represents the large divalent ions such as Ba^{2+}, Sr^{2+}, or Pb^{2+} (ionic radii: 1.43, 1.27, and 1.32 Å, respectively).

This type of oxide has a hexagonal structure, which is composed of an accumulation of several spinel ionic layers, separated by single ionic layers consisting of M^{2+}, O^{2-}, and Fe^{3+} ions (Fig. 5.17). Because of its low crystal symmetry, this type of oxide generally has a large magnetocrystalline anisotropy. Ba ferrite, which is used as a permanent magnetic material, belongs to this category.

There is also another group of hexagonal oxides[23] which are represented by the general formulas $BaO \cdot 2MO \cdot 8Fe_2O_3$ (*W* type), $2BaO \cdot 2MO \cdot 6Fe_2O_3$ (*Y* type), and $3BaO \cdot 2MO \cdot 12Fe_2O_3$ (*Z* type), where M denotes a divalent metal ion such as Mn^{2+} Fe^{2+}, Co^{2+}, Ni^{2+}, Zn^{2+}, or Mg^{2+}. They have a hexagonal crystal structure similar to the magnetoplumbite type. Some of them, however, have a negative anisotropy constant; this means that the basal plane is an easy plane of magnetization. Because of this type of anisotropy, some oxides such as Co_2Z and Mg_2Y exhibit excellent magnetic properties in the high frequency range. We discuss this matter in Chapter 16.

The saturation magnetization of these oxides is about the same as that of the ferrites, and their Curie points range from 400° to 500°C.

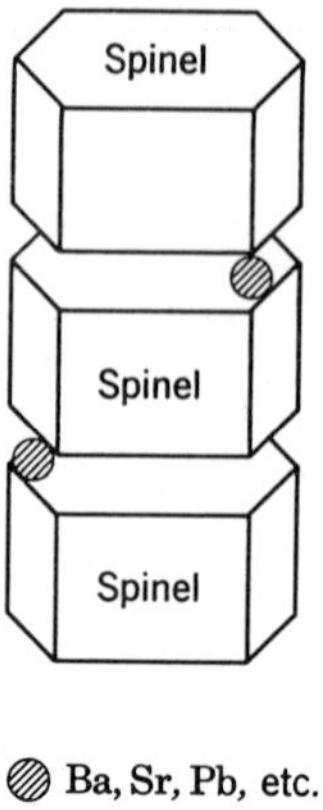

Fig. 5.17. Schematic representation of the magnetoplumbite-type crystal.

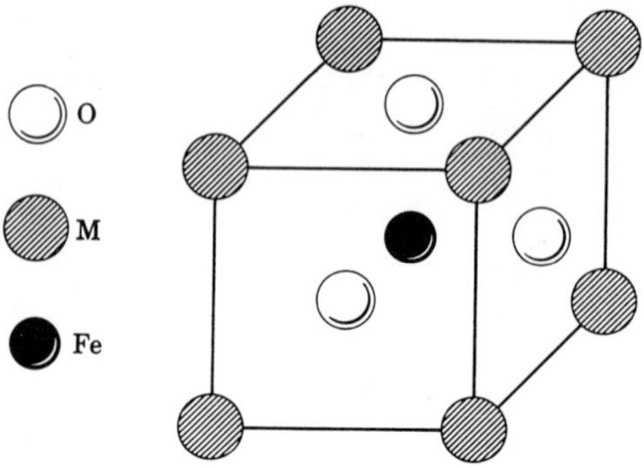

Fig. 5.18. Crystal structure of pervoskite-type magnetic oxides.

Perovskite-type oxides. Perovskite is the mineral having the composition $CaTiO_3$. By replacing Ti in this formula by Fe, we obtain a magnetic perovskite-type oxide of the formula $MFeO_3$, where M is a large metal ion such as La^{3+}, Ca^{2+}, Ba^{2+}, or Sr^{2+}. These oxides have cubic crystal structure as shown in Fig. 5.18. By replacing Fe in $MFeO_3$ by Mn or Co, we have also $MMnO_3$ and $MCoO_3$. The interesting thing about these oxides is that the ferromagnetism appears in the range of the solid solution[24] between $La^{3+}Mn^{3+}O_3$ and $Ca^{2}Mn^{4+}O_3$, $La^{3+}Mn^{3+}O_3$ and $Sr^{2+}Mn^{4+}O_3$, $La^{3+}Mn^{3+}O_3$ and $Ba^{2+}Mn^{4+}O_3$, and also $La^{3+}Co^{3+}O_3$ and $Sr^{2+}Co^{4+}O_3$, although the single oxides show only antiferromagnetism or weak parasitic ferromagnetism (Figs. 5.19 and 5.20). This fact can be explained by assuming the exchange interaction between Mn^{3+} and Mn^{4+} to be positive, and that between Mn^{3+} and Mn^{3+} and Mn^{4+} and Mn^{4+} to be negative. Zener[25] explained it in terms of a double exchange interaction. Goodenough[26] interpreted it in terms of a covalent and semicovalent exchange mechanism.

Ferrimagnetic garnets. The term garnet signifies a group of minerals which have the same type of composition as pyrope, $Mg_3Al_2(SiO_4)_3$. Replacing the Si by Fe,[27,28] we have the group of ferrimagnetic garnets $3M_2O_3{\cdot}5Fe_2O_3$, where M is a rare earth element such as Sm, Eu, Gd, Tb, Dy, Ho, Er, Tm, Yb, Lu, or Y.[17] It should be noted that all of these metal ions are trivalent.

These oxides have the most complex cubic crystal structure[29] the unit cell of which contains 160 atoms, 96 of them being oxygen ions occupying

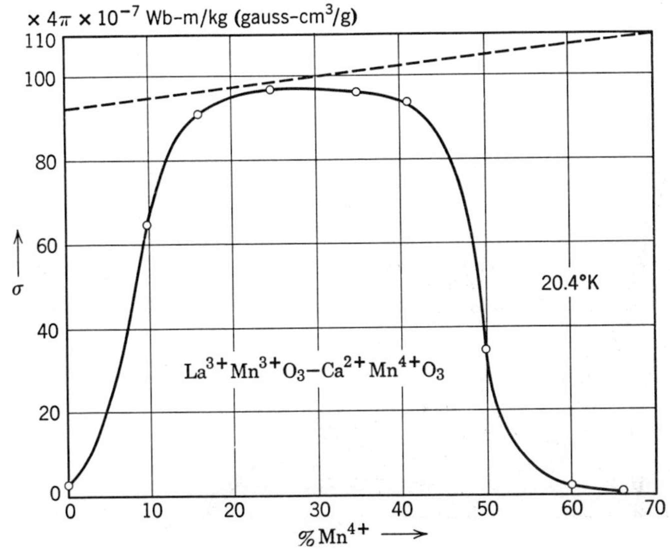

Fig. 5.19. Spontaneous magnetization in the $LaMnO_3$-$CaMnO_3$ series (after Jonker[24]). Solid line, experimental; broken line, calculated.

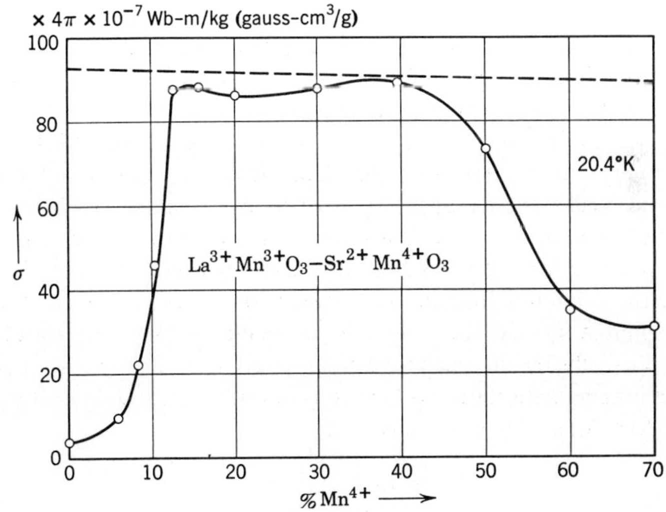

Fig. 5.20. Spontaneous magnetization in the $LaMnO_3$-$SrMnO_3$ series (after Jonker[24]). Solid line, experimental, broken line, calculated.

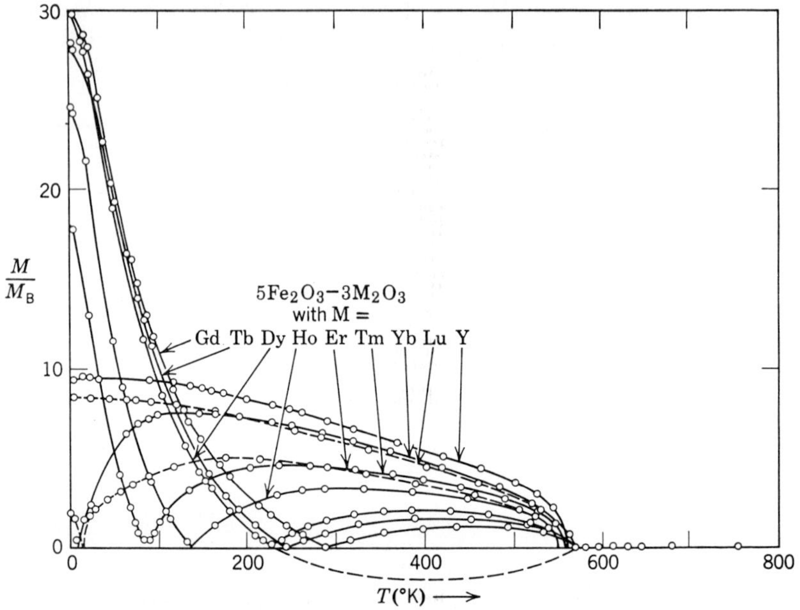

Fig. 5.21. Temperature dependence of the spontaneous magnetization of ferrimagnetic garnets (after Bertaut and Pauthenet[17]).

96 h sites, while 40 are the Fe^{3+} ions occupying 24 d and 16 a sites, and 24 are the M^{3+} ions occupying 24 c sites. The intrinsic magnetization of these oxides exhibits very interesting temperature dependence, as shown in Fig. 5.21. Most of them belong to the N type (cf. Fig. 5.13), and they also exhibit a very strong temperature dependence at low temperature. According to Néel,[30] these magnetic properties can be explained by assuming that the ferrimagnetism of this oxide is due to positive Fe^{3+} spins in 16 a sites, positive M^{3+} spins in 24 c sites, and negative Fe^{3+} spins in 24 d sites, with the exchange interaction between M^{3+} ions in 24 c sites being almost negligible so that the magnetic moments of M^{3+} ions should be aligned exclusively by the exchange interaction with Fe^{3+} ions. Let I_{Fe} denote the magnetization due to Fe^{3+} ions in both 16 a and 24 d sites, and I_M that of M^{3+} ions. Then we may write

$$I_{Fe} = I_{Fe}, \quad I_M = \chi(H + wI_{Fe}). \tag{5.58}$$

The resultant magnetization can be given by

$$I = I_{Fe} + I_M = I_{Fe} + \chi(H + wI_{Fe}) = I_{Fe}(1 + \chi w) + \chi H, \tag{5.59}$$

where the first term represents the spontaneous magnetization, and the second term shows the dependence on the external field. It is seen in this expression that the spontaneous magnetization vanishes when

$$I_{Fe} = 0, \tag{5.60}$$

which is attained at the Curie temperature Θ_f. It does also when

$$1 + \chi w = 0, \tag{5.61}$$

which condition should be satisfied at some temperature, because w is negative and also because χ should take a large value at low temperature and decrease with increase of temperature almost proportionally to $1/T$. This gives the compensation temperature Θ_c.

Table 5.3. Magnetic Properties of Ferrimagnetic Garnets, $3M_2O_3 \cdot 5Fe_2O_3$ (after Bertaut and Pauthenet[17])

M	Compensation Point, °K	Curie Point, °K	M/M_B	Lattice Constant, Å	Density, g/cc
Sm	—	560	9.3	12.52	6.235
Eu	—	570	5.0	12.52	6.276
Gd	290	564	30.3	12.48	6.436
Tb	246	568	31.4	12.45	6.533
Dy	220	563	32.5	12.41	6.653
Ho	136	567	27.5	12.38	6.760
Er	84	556	23.1	12.35	6.859
Tm	$4 \sim 20$	549	2.0	12.33	6.946
Yb	—	548	0	12.29	7.082
Lu	—	539	8.30	12.28	7.128
Y		560	9.41	12.38	5.169

The features of this type of oxide are high resistivity due to the lack of divalent ions, and also very low losses in the microwave region. Single crystals of this oxide are found to be transparent, so that the domain structure can be observed by purely optical means, making use of the Faraday rotation.[31]

The compensation temperature Θ_c, the Curie temperature Θ_f, the absolute saturation moment, the lattice constant, and the density are given in Table 5.3 for various ferrimagnetic garnets.

NiAs-Type compounds. The compounds between transition metals such as V, Cr, Mn, Fe, Co, and Ni and non-metal or semimetal elements such as O, S, Se, Te, P, As, Sb, and Bi usually show various interesting magnetic properties. Table 5.4 shows the crystal structures and the types of magnetism exhibited by these compounds. Most of them have the NiAs-type crystal structure, which is shown in Fig. 5.22. The magnetic moments of the metal ions, which point parallel to the *c* axis or to the basal plane, are aligned sometimes parallel (ferromagnetism) and sometimes antiparallel

Table 5.4. Crystal Type and Magnetism of NiAs-Type Magnetic Compounds (after Tsuya, Tsubokawa, Yuzuri[32] and Goodenough[G.39])

Crystal type: NC: NaCl type; NA: NiAs type; MP: MnP type

Magnetism: ferro: ferromagnetism; anti: antiferromagnetism; para: paramagnetism; dia: diamagnetism

Transition Metal	VIb group				Vb group			
	O	S	Se	Te	P	As	Sb	Bi
Ni	NC anti	NA anti	NA	NA dia	dia		NA dia	NA
Co	NC anti	NA anti	NA	NA anti	MP para		NA anti	
Fe	NC anti	NA anti	NA anti	NA anti	MP ferro	MP	NA	
Mn	NC anti	NC anti	NC anti	NA anti	MP ferro	NA ferro	NA ferro	NA ferro
Cr		NA anti	NA anti	NA ferro	MP	MP anti	NA anti	
V	NC anti	NA anti	NA anti	NA			NA anti	
Ti		NA para	NA para	NA para			NA para	

(antiferromagnetism) to each other.[34] Where there are deficiencies in metal ions, ferrimagnetism is expected to appear. It is a feature of this type of compound that ferro-, ferri-, and antiferromagnetism subsequently appear with a change of temperature and also with a change of composition.

The typical example of this type of compound is pyrrotite, $Fe_{0.875}S$, which shows ferrimagnetism. The magnetic properties of this mineral were first investigated by P. Weiss[35] in 1905. Then Haraldsen[36] made a detailed investigation of the crystal structure and also of the magnetic properties

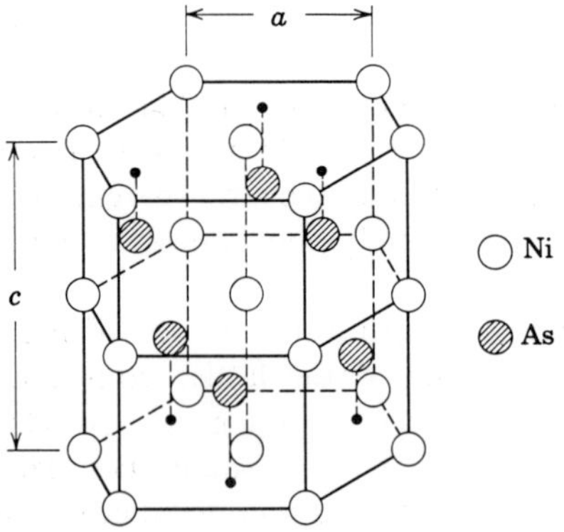

Fig. 5.22. Crystal structure of NiAs-type compounds (after Hirone[33]).

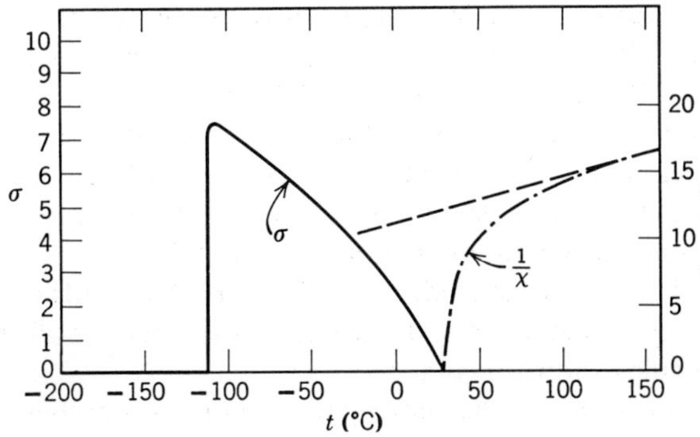

Fig. 5.23. Temperature dependence of the spontaneous magnetization and reciprocal of the susceptibility of $Cr_{1.17}S$ (after Yuzuri et al.[38]).

of the Fe-S series compounds; he found that they have NiAs-type crystal structure in the range of $Fe_{0.96}S$ to $Fe_{0.875}S$ and a spontaneous magnetization in the range $Fe_{0.90}S$ to $Fe_{0.875}S$. Bertaut[37] interpreted this ferrimagnetism by supposing that the vacancies in metal ion sites will form a superlattice, concentrating themselves on one of the antiferromagnetic sublattices, so that the difference in the number of Fe ions between the two sublattices gives rise to the spontaneous magnetization. Stoichiometric FeS shows antiferromagnetism due to the complete compensation of both sublattice magnetizations.

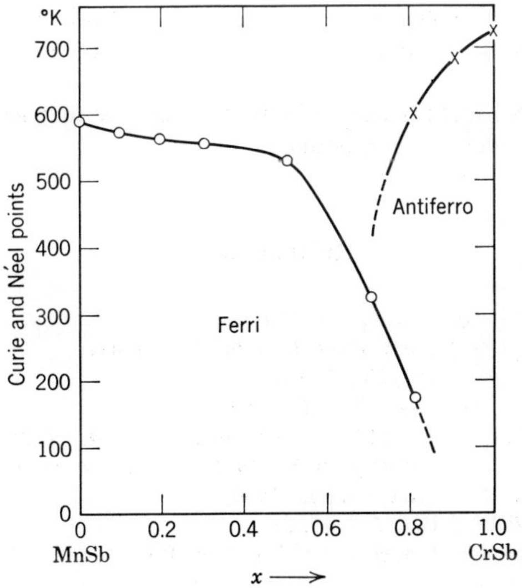

Fig. 5.24. Magnetic transition points of MnSb-CrSb series (after Hirone et al.[39]).

Another interesting example is the Cr-S system which was investigated by Yuzuri et al.[38] As shown in Fig. 5.23, $Cr_{1.17}S$ shows ferrimagnetism between $-114°C$ and its Curie point, 40°C, but the ferrimagnetism suddenly disappears below $-114°C$. One more example is the MnSb-CrSb system,[39] which shows ferrimagnetism on the side of MnSb and antiferromagnetism on the CrSb side, the Néel temperature of which is different from the ferrimagnetic Curie temperature (Fig. 5.24).

In Table 5.4 are listed the oxides NiO, CoO, FeO, and MnO, all of these having NaCl-type cubic crystal structure with the antiferromagnetic alignment of spins shown in Fig. 5.1.

The reader may find further information on the properties of magnetic compounds in the article by Nagamiya.[40]

Problems

5.1. Consider an antiferromagnetic substance which has a susceptibility χ_0 at its Néel point Θ_N. Assuming that the exchange interactions inside the A and B sites are negligibly small compared to the exchange interaction between A and B sites, find the values of the susceptibility which would be measured under the application of magnetic fields perpendicular to the spin axis at $T = 0$, $\Theta_N/2$, and $2\Theta_N$.

5.2. Consider a ferrimagnetic substance which has magnetic ions in A and B sites in the ratio of 3:2. Assuming that the exchange interactions within A and B sites are negligibly small compared to those between A and B sites, find the nature of the spin arrangement at 0°K and the ratio of the ferrimagnetic Curie point to the asymptotic Curie point, and examine the possibility of having a compensation point.

5.3. Describe and discuss the crystal structure and the magnetism of various kinds of magnetic oxides and compounds.

References

5.1. Ref. G.7.
5.2. L. Néel: *Ann. Physique* [12] **3,** 137 (1948).
5.3. C. G. Shull and J. S. Smart: *Phys. Rev.* **76,** 1256 (1949).
5.4. H. A. Kramers: *Physica* **1,** 182 (1934).
5.5. P. W. Anderson: *Phys. Rev.* **79,** 350 (1950).
5.6. T. Nagamiya, K. Yosida, and R. Kubo: *Advan. Phys.* **4,** 1 (1955).
5.7. J. H. Van Vleck: *J. Chem. Phys.* **9,** 85 (1941); *J. Phys. Radium* **12,** 262 (1951).
5.8. P. W. Anderson: *Phys. Rev.* **79,** 705 (1950).
5.9. L. Néel: *Ann. Phys.* [11] **5,** 232 (1936).
5.10. N. J. Poulis, Van der Handel, J. Ubbink, J. A. Poulis, and C. J. Gorter: *Phys. Rev.* **82,** 522 (1951).
5.11. I. S. Jacobs: *J. Appl. Phys.* **32,** 615 (1961).

5.12. L. Néel: *Proc. Phys. Soc.* (*London*) **65A,** 869 (1952).
5.13. J. M. Hastings and L. M. Corliss: *Phys. Rev.* **104,** 328 (1956).
5.14. F. W. Harrison, W. P. Osmond, and R. W. Teale: *Phys. Rev.* **106,** 865 (1957).
5.15. C. Guillaud and H. Creveaux: *Compt. Rend.* **230,** 1458 (1950); C. Guillaud and M. Sage: *ibid.*, **232,** 944 (1951).
5.16. E. W. Gorter: *Philips Res. Rept.* **9,** 295, 321, and 403 (1954).
5.17. F. Bertaut and R. Pauthenet: *Proc. I.E.E. Suppl.* **B104,** 261 (1957); R. Pauthenet: *J. Appl. Phys.* **29,** 253 (1958).
5.18. Y. Ishikawa: *Metal Phys.* (in Japanese) **6,** No l. 19 (1960).
5.19. Y. Ishikawa and S. Akimoto: *J. Phys. Soc. Japan* **12,** 1083 (1957).
5.20. R. M. Bozorth, E. W. Dorothy, and A. J. Williams: *Phys. Rev.* **108,** 157 (1957).
5.21. G. Shirane, S. J. Pickart, R. Nathans, and Y. Ishikawa: *Phys. Chem. Solids* **10,** 35 (1959).
5.22. J. J. Went, G. W. Rathenau, E. W. Gorter, and G. W. van Oosterhout: *Philips Tech. Rev.* **13,** 194 (1951/52).
5.23. G. H. Jonker, H. P. J. Wijn, and P. W. Brawn: *Philips Tech. Rev.* **18,** 145 (1956/57).
5.24. G. H. Jonker and J. H. van Santen: *Physica* **16,** 337 (1950); **19,** 120 (1953); G. H. Jonker: *ibid.*, **22,** 707 (1956).
5.25. C. Zener: *Phys. Rev.* **82,** 403 (1951).
5.26. J. B. Goodenough: *Phys. Rev.* **100,** 564 (1955).
5.27. F. Bertaut and F. Forrat: *Compt. Rend.* **242,** 382 (1956).
5.28. S. Geller and M. A. Gilleo: *Acta Cryst.* **10,** 239 (1957).
5.29. S. Geller and M. A. Gilleo: *Acta Cryst.* **10,** 787 (1957).
5.30. L. Néel: *Compt. Rend.* **239,** 8 (1954).
5.31. J. F. Dillon, Jr.: *J. Appl. Phys.* **29,** 1286 (1958).
5.32. N. Tsuya, I. Tsubokawa, and M. Yuzuri: *Metal Phys.* (in Japanese) **4,** 140 (1958).
5.33. T. Hirone: *Metal Phys.* **2,** No. 2, 59 (1956).
5.34. T. Hirone and K. Adachi: *J. Phys. Soc. Japan* **12,** 156 (1957).
5.35. P. Weiss: *J. Phys.* **4,** 469 (1905).
5.36. H. Haraldsen: *Z. Anorg. Chem.* **231,** 78 (1937); **246,** 169 (1941).
5.37. F. Bertaut: *Acta Cryst.* **6,** 557 (1953).
5.38. M. Yuzuri, T. Hirone, H. Watanabe, S. Nagasaki, and S. Maeda: *J. Phys. Soc. Japan* **12,** 385 (1957).
5.39. T. Hirone, S. Maeda, I. Tsubokawa, and N. Tsuya: *J. Phys. Soc. Japan* **11,** 1083 (1956).
5.40. T. Nagamiya: *J. Phys. Radium* **20,** 70 (1959).

part 3

MAGNETIC DOMAIN STRUCTURE

6

The Concept of Magnetic Domain

6.1 Discovery of Magnetic Domain

In Chapter 2 we learned of the mechanism by which spontaneous magnetization is generated in ferro- and ferrimagnetism. In this chapter we learn how the spontaneous magnetizations thus generated are distributed in the ferromagnetic body. We also discuss the factors which can influence the distribution of spontaneous magnetization, such as magnetic anisotropy, magnetostriction, structure of magnetic walls, and magnetostatic energy. These problems are common to ferro- and ferrimagnetism, and the reader should understand that every statement in this chapter is valid for both types of magnetism, unless there is a specific comment on this point.

Let us now consider a bar of ferromagnetic substance which is magnetized parallel to its length (Fig. 6.1). If we cut this bar into several pieces, each piece will act as a single magnet with N and S poles on its ends. This fact tells us that every part of the ferromagnetic substances contributes to the uniform magnetization. The internal magnetization, however, cannot be uniform down to the microscopic scale. If we were to observe the magnetization of the substance by means of an optical microscope with a magnification of about 100 times (this is possible, as will be mentioned later), we could actually look at many magnetic domains each of which is magnetized in a different direction. The magnetization inside each domain is made up of many atomic moments which are lined up by the action of the exchange force described in Part 2.

The presence of the ferromagnetic domain structure was first predicted by P. Weiss[1] in 1907 in his famous paper on the hypothesis of the molecular

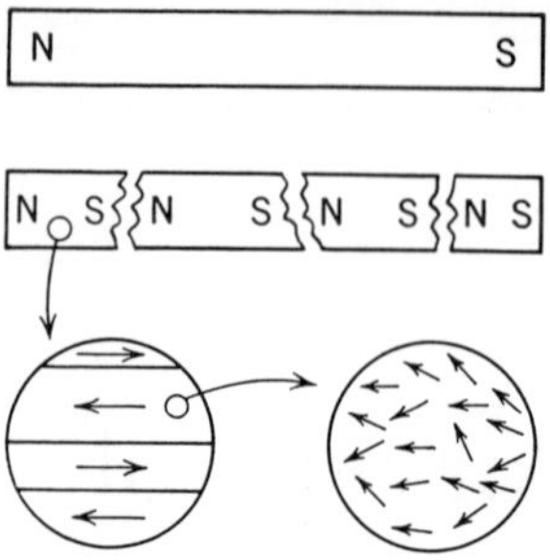

Fig. 6.1. Internal structure of a magnet.

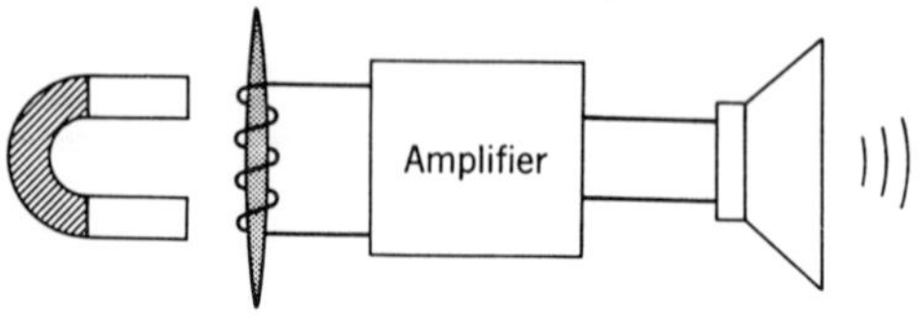

Fig. 6.2. Barkhausen effect.

field. The first experimental verification of the presence of ferromagnetic domains was indirect and was made by Barkhausen[2] in 1919. His experiment consists in amplifying the voltage induced in the secondary coil wound around the ferromagnetic specimen being magnetized, and observing the noise induced by the changing magnetization (Fig. 6.2). The presence of the characteristic noise proved that the magnetization process consists of many small discontinuous flux changes, which were thought, at that time, to correspond to the flopping of each small magnetic domain. It was estimated that each discontinuous magnetization event corresponds to the change in a domain volume of the order of 10^{-8} cc.

In 1926 Honda and Kaya[3] measured magnetization curves for single crystals of iron and found that the shape of the magnetization curve is dependent on the crystallographic orientation of the magnetization. There are several directions along which the magnetization takes place easily. These are called the directions of easy magnetization, or simply easy directions. For instance, [100], [010], and [001] are the directions of easy magnetization for iron. This means that the internal magnetization is stable when pointing parallel to one of these directions. The dependence of the internal energy on the direction of inner magnetization is called the magnetic anisotropy energy.

In 1931 Sixtus and Tonks[4] succeeded in obtaining a large magnetic domain by stretching a thin, long Permalloy wire. Generally a crystal of ferromagnetic substance is spontaneously deformed along the direction of internal magnetization. This is called magnetostriction. In the case of Permalloy, magnetostriction is positive; in other words, the crystal is spontaneously elongated along the direction of magnetization. When tension is applied along the Permalloy wire, the internal magnetization will point parallel or antiparallel to the axis of the wire, so as to produce an additional elongation in the direction of external tension (cf. Section 8.4).

After magnetizing the wire in one direction, Sixtus and Tonks applied a reverse field which was less than a certain critical value, and then they observed the motion of the domain boundary created by the starting coil C_2 (Fig. 6.3). They measured the velocity of its propagation by observing the time interval between the two signals induced in the search coils S_1 and S_2. This phenomenon is called a large Barkhausen effect.

In 1932 Bloch[5] showed theoretically that the boundary between domains is not sharp on an atomic scale but is spread over a certain thickness wherein the direction of spins changes gradually from one domain to the next. This layer is usually called a domain wall and is sometimes referred to as a Bloch wall. The calculated value of the energy stored in a domain wall was well in accord with the value from the experiment of Sixtus and Tonks.

Thus the problems of ferromagnetic domain structure had been solved successively, together with the problems of the magnetizing process. The reader may want to read the excellent survey of the development of the domain theory at this time in *Ferromagnetismus* written by Becker and Döring[G.3] in 1939. The concept of a ferromagnetic domain given in this book, however, must be corrected on at least the following two points: First, the size of the domain was considered to be as small as 10^{-2} mm in diameter. Secondly, the effect of magnetostatic energy was completely ignored in the consideration of domain structure.

The direct observation of ferromagnetic domain structure was attempted by Bitter[6] in 1931, and independently by Hamos and Thiessen[7] in 1932. These researchers applied a drop of ferromagnetic colloidal suspension, containing many fine ferromagnetic particles, to the polished surface of the ferromagnetic specimen and observed the image of the domain made by the ferromagnetic colloidal particles. Although some of the micrographs

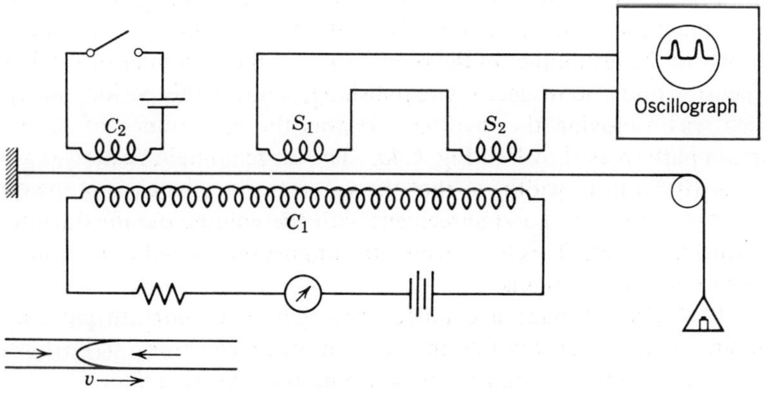

Fig. 6.3. Large Barkhausen effect.

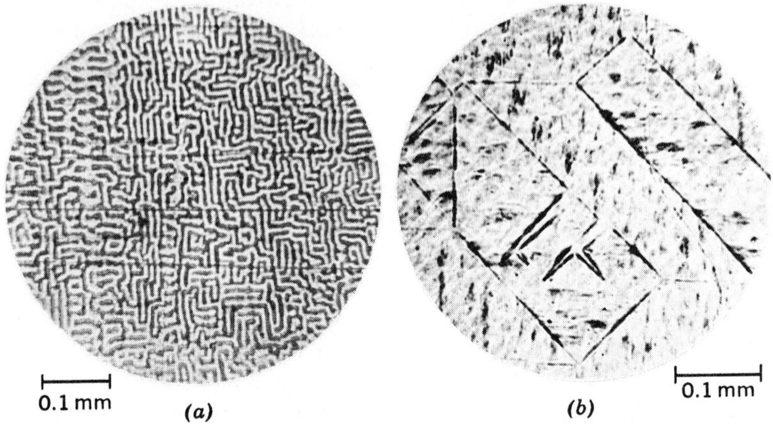

Fig. 6.4. Ferromagnetic domain pattern: (*a*) maze domain pattern; (*b*) true domain pattern observed after removal of 28 μ of the surface layer (after Chikazumi and Suzuki[9]).

in Bitter's paper revealed beautiful domain structures, he did not conclude that they were true images of domain structure, but simply explained them as being due to an inhomogeneity in ferromagnetic substances. The reason may be that the sizes of the observed domains were too large as compared to those expected on the basis of the then current concept of domain structure. Many investigations were made thereafter using the same technique, but no definite conclusion was drawn for about seventeen years after the discovery of this technique. The main reason may be that the surfaces of almost all single crystal specimens used were tilted somewhat from a principal crystallographic plane; hence the free magnetic poles appearing on the crystal surfaces prevented the appearance of a well-defined domain structure. Several improvements were made by Elmore[8] in the techniques of the preparation of the ferromagnetic colloidal suspension and also in electrolytic polishing. During this period, one thing which was annoying the investigators was the appearance of the maze domain pattern as shown in Fig. 6.4*a*. The maze domain pattern is about $10^{-2} \sim 10^{-3}$ mm in width, so that the volume of one block is of the order of 10^{-8} cc; this is in good agreement with the volume obtained from the Barkhausen effect. This fact seemed to support the "small block concept" of ferromagnetic domains.

In 1934 Kaya[10] made a detailed investigation of domain patterns of iron and nickel single crystals and showed the maze pattern to be dependent on the manner of grinding or polishing the crystal surface.

A theoretical treatment of ferromagnetic domain structure was first tried by Landau and Lifshitz[11] in 1935; they took into consideration the effect of magnetostatic energy. In 1944 Néel[12] also made a detailed calculation on a particular type of domain structure; he considered the magnetostatic energy, together with its magnetizing process.

Well-defined domain structures were finally observed by Williams, Bozorth, and Shockley[13] in 1949 on Si-Fe crystal which was cut precisely parallel to, or with a definite inclination to, the principal crystallographic plane. Fig. 6.4*b* shows the true domain structure observed by the Bitter technique on the electropolished surface. The reader can see that the domain structure thus observed is much bigger than the maze domain. These investigators showed that the observed domain structures are in good agreement with the theoretical predictions of Landau and Lifshitz and also those of Néel.

The structure of the maze domain was clarified thereafter; this domain turned out to be a sort of surface domain induced by the strain of grinding the surface[9,14] (cf. Section 9.3).

6.2 General Features of Magnetic Domains

In this section we discuss the fundamental character of domain structure and the various factors which influence its distribution in a ferromagnetic substance. Precise calculations of the size of ferromagnetic domains are treated in Chapter 11 after magnetic anisotropy, magnetostriction, domain wall, and magnetostatic energy have been discussed in Chapters 7, 8, 9, and 10, respectively.

First we discuss the domain structure of a uniform ferromagnetic crystal. Let us consider a spherical single crystal specimen. If this specimen were composed of a single domain, as shown in Fig. 6.5, the magnetic poles appearing on the surface would give rise to the strong demagnetizing field given by (2.16). In other words, such a system has a large magnetostatic energy. One way to avoid this is to make the inner magnetization rotate inside the sphere as shown in Fig. 6.6. Then there would be no magnetic poles and, accordingly, no magnetostatic energy being stored. Instead, however, the neighboring spins make some angle with one another, so that some amount of exchange energy must be stored. The choice between the two possibilities is essentially determined by comparison of the two kinds of energy, which are dependent on the shape and volume of the specimen as well as on the intensity of spontaneous magnetization and exchange interaction.

If the crystal has large magnetocrystalline anisotropy, the inner magnetization is forced to point parallel to a direction of easy magnetization. For instance, in iron crystal the easy directions are [100], [010], and [001],

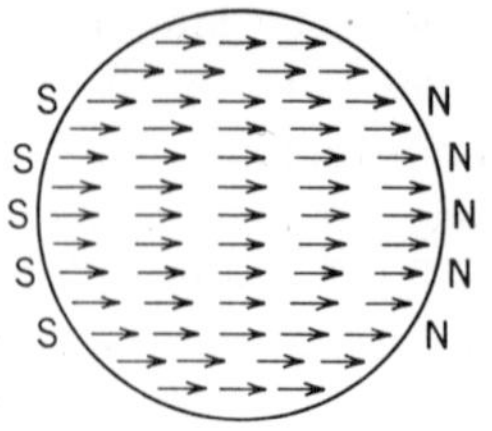

Fig. 6.5. Single domain structure.

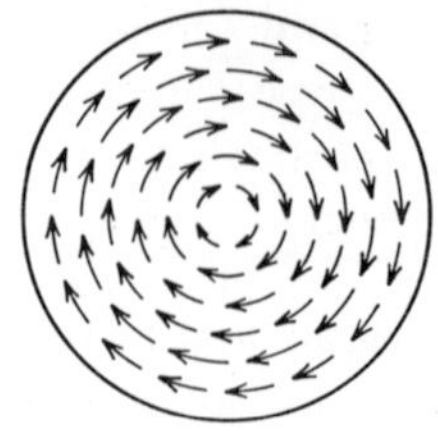

Fig. 6.6. Domain structure of a material with small crystal anisotropy.

so its domain structure might be as shown in Fig. 6.7. In such a case the energy should be stored mainly at the domain boundaries, where the spins rotate gradually from one direction to the other. In uniaxial anisotropy, such as that of cobalt metal, the inner magnetization must point either parallel or antiparallel to the easy direction. Then free poles with alternating signs will appear on the surface of the sphere (Fig. 6.8). The magnetostatic energy involved in this kind of distribution of the free poles is very small compared to that of Fig. 6.5, for the energy of the Coulomb interaction should be decreased by bringing free poles with opposite signs closer together. Roughly speaking, the magnetostatic energy should be decreased to $1/N$ of the original value when the specimen is divided into N domains. Since the total domain wall volume is increased in proportion to N, the actual size of the domain is determined by the counterbalance of wall energy and magnetostatic energy.

Magnetostriction also can influence the domain structure. If the substance of Fig. 6.7 has positive magnetostriction, each domain tends to expand parallel to and contract perpendicular to the direction of the domain magnetization (Fig. 6.9). Since the magnetostriction is a small strain ($\sim 10^{-5}$), such a deformation cannot break the crystal at the domain boundaries. In order to keep the domains in contact with each other,

Fig. 6.7. Domain structure of a material with large crystal anisotropy.

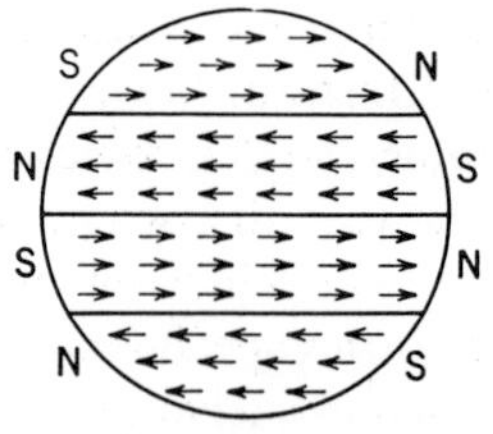

Fig. 6.8. Domain structure of a material with uniaxial anisotropy.

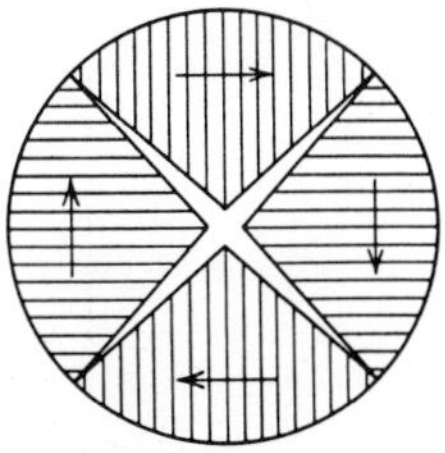

Fig. 6.9. Tendency of domain deformation for a material with positive magnetostriction.

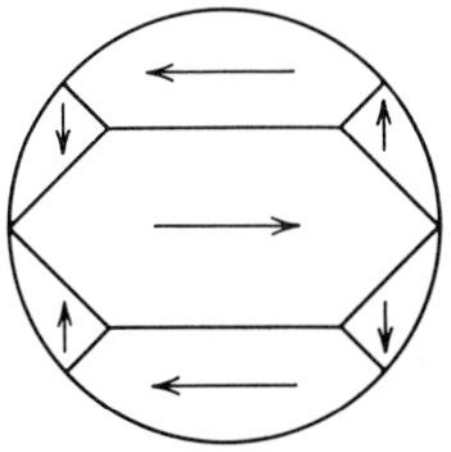

Fig. 6.10. Domain structure of a material with large magnetostriction.

however, a considerable amount of elastic energy must be stored in the substance. One possible way to avoid this is to increase the volume of the main domains which have their magnetizations parallel to a certain easy direction (Fig. 6.10). In this case the deformation of the whole volume will be determined exclusively by the magnetostriction of the main domains, and the elastic energy will be concentrated into the small flux-closure domains which are forced to strain to fit the deformation of the main domains. The sizes of the domains under these circumstances should be determined by the counterbalance between elastic energy and wall energy.

Under every condition mentioned above, the stable domain structure is determined so as to minimize the total energy. The size and shape of the stable domain thus determined are dependent on the size and shape of the specimen. This is the most striking difference between the new domain concept and the old one, in which the size of the domain was considered to be an inherent property of the magnetic substances.

In inhomogeneous magnetic materials, however, the size of the domain is determined exclusively by the inhomogeneity of the materials, independently of the shape of the specimen. The kinds of inhomogeneity which can affect the domain structure are voids, inclusions, precipitations, fluctuations in alloy compositions, internal stress, local directional order, crystal boundaries, etc. For instance, if a spherical inclusion or void exists in a ferromagnetic substance, the free poles which appear on its surface should make the magnetostatic energy quite large. In such a case the blade-shaped domain, as shown in Fig. 6.11*b*, would be induced so as to redistribute the free poles on the wider surface of the blade and thereby reduce the magnetostatic energy. Figure 6.12 is a micrograph of the actual blade-shaped domain observed on a (001) surface of 3% Si-Fe crystal. When a domain wall of one of the main domains approaches such an inclusion, the blade walls are attracted to the main wall, as shown in Fig.

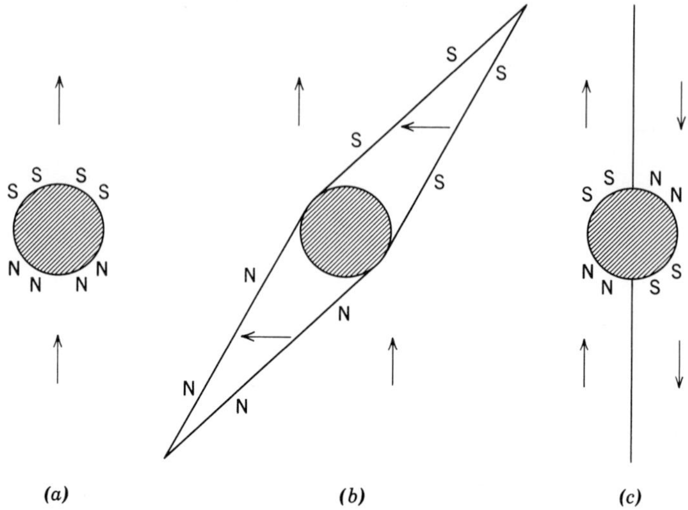

Fig. 6.11. Free pole distribution and the domain structure around a spherical inclusion.

6.13, so as to decrease the magnetostatic energy of the free pole distribution on the blade wall. Such a configuration of domains around an inclusion was first predicted by Néel[15] and first observed by Williams et al.[13] An inclusion which has no blade domain tends to attract a domain wall. In particular, when the wall intersects the inclusion as shown in Fig. 6.11c, the free poles of the inclusion are divided into four groups, reducing the

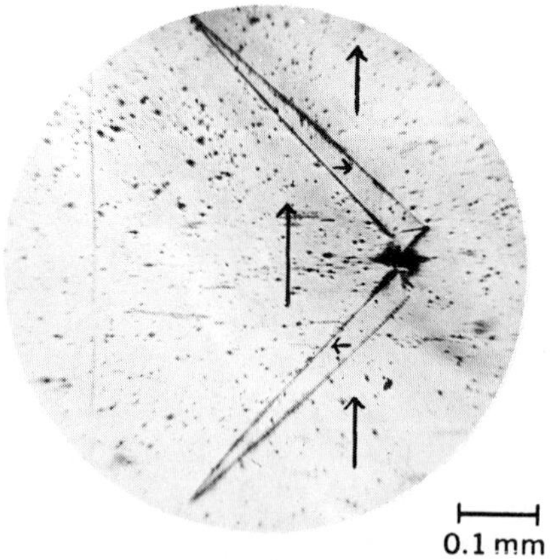

Fig. 6.12. Powder pattern of the blade-shaped domain around a void in an Si-Fe (001) surface.

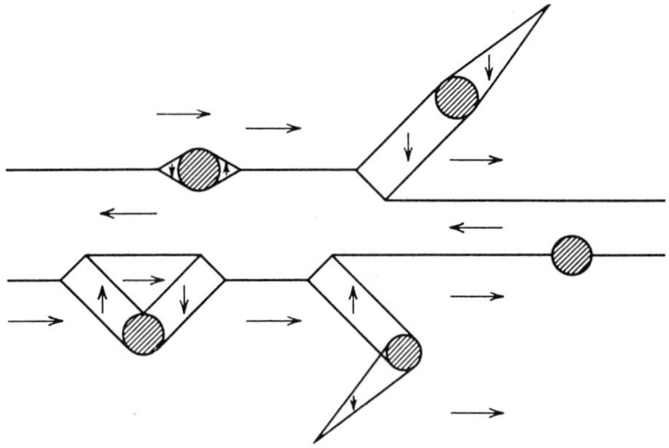

Fig. 6.13. Domain structure around inclusions.

magnetostatic energy. When many inclusions are distributed all over the material, the domain distribution shown in Fig. 6.13 may considerably reduce the magnetostatic energy of the system. The size and shape of the domains in such a case will be determined by the magnetostatic energy of the free poles around the inclusions and by the wall energy, independently of the size and shape of the specimen.

When the composition of a ferromagnetic alloy or a mixed ferrite differs from place to place (this holds more or less for actual materials), the intensity of spontaneous magnetization fluctuates from place to place because the spontaneous magnetization is usually a function of composition. In such a case there is a volume distribution of free poles the density of which is given by the divergence of the intensity of spontaneous magnetization. Since the sign of the free pole is changed by the reversal of the direction of domain magnetization, regions where there are high free pole densities should be divided into many domains so as to decrease the magnetostatic energy. Figure 6.14 shows the equi-composition surfaces of the material together with the expected domain structure.

When the internal stresses are distributed in a random manner in a magnetic material, a local anisotropy results, and its easy direction varies from place to place. The local directional ordering in a solid solution or a mixed ferrite (cf. Chapter 17) would also have the same effect as internal stress in producing local anisotropy. In such a case the spontaneous magnetization might change its direction from place to place (Fig. 6.15). A domain structure of this type should again be determined so as to minimize the total energy, which is, in these materials, composed of the local anisotropy energy and the exchange energy stored in the regions where the magnetization changes direction. The domain structure of Permalloy,

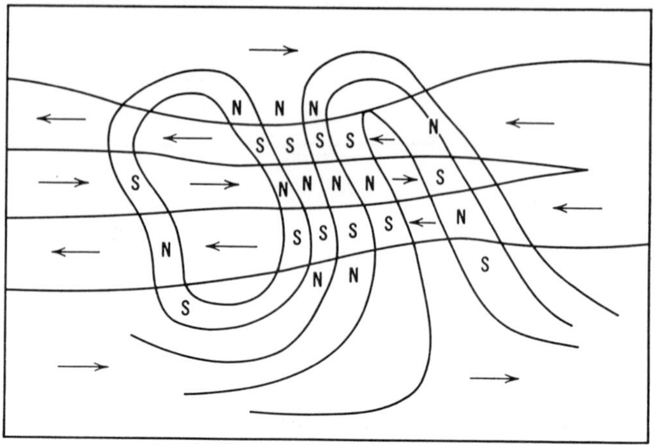

Fig. 6.14. Domain structure of a material in which the composition fluctuates from place to place.

which is a typical high permeability material, is thought to belong to this category.[16]

Next let us consider the domain structures of polycrystalline materials. Since each grain has its own easy directions, free poles are expected to appear on the crystal boundaries, unless the crystal boundary is oriented so as to include a bisector of the angle between the easy axes of the adjacent grains. The magnetostatic energy of such a free pole distribution is greatly decreased when the domain distribution in each grain is closely related to

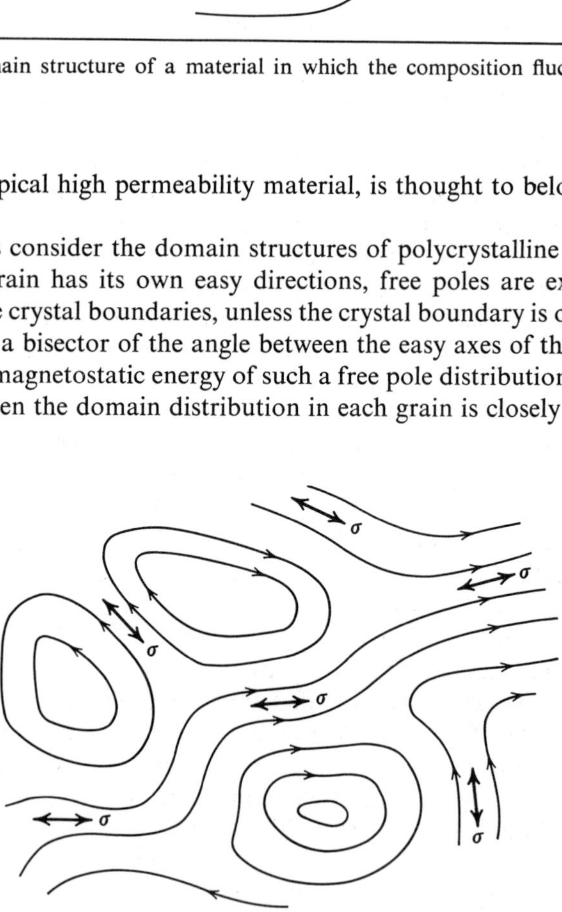

Fig. 6.15. Domain structure in a random distribution of internal stress.

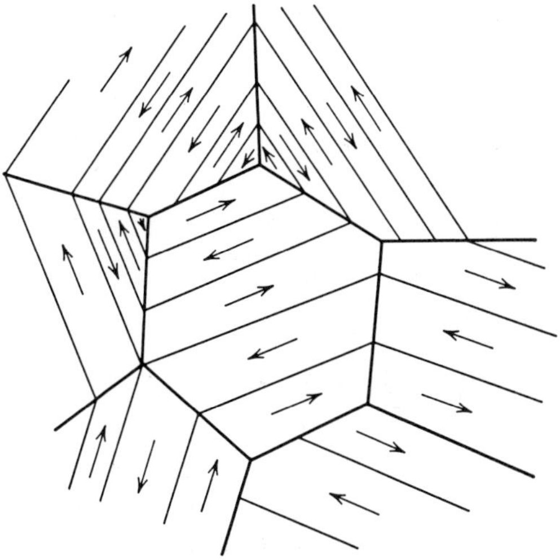

Fig. 6.16. Domain structure of a polycrystal.

that of the neighboring grain (Fig. 6.16). When the magnetostatic energy of the free poles which appear on the crystal boundaries is fairly large compared to that of the free poles appearing on the outside surface of the specimen, the domain structure should be determined by the grain size, independently of the outside shape of the specimen. When, however, the diameter of the grain is greater than the thickness of the specimen sheet, the domain structure becomes dependent on the sheet thickness.

6.3 Technical Magnetization

When a magnetic field is applied to a ferromagnetic substance, the domain structure changes in such a way as to increase the resultant magnetization parallel to the external field. We call this process technical magnetization. We study this phenomenon in detail in Part 4. In this section we merely outline the nature of the phenomenon.

The boundary between the domains—the domain wall—plays an important role in technical magnetization. Inside the domain wall the spins rotate gradually from one domain to the next (Fig. 6.17). When the external field is applied parallel to the magnetization of one domain, the spins inside both domains experience no torque resulting from the field simply because their directions are either parallel or antiparallel to the field (cf. 1.8). Since the spins inside the wall do make some angle with the field direction, they, under the action of a torque, start to rotate toward the field direction. As a result of the rotation of the spins inside the wall,

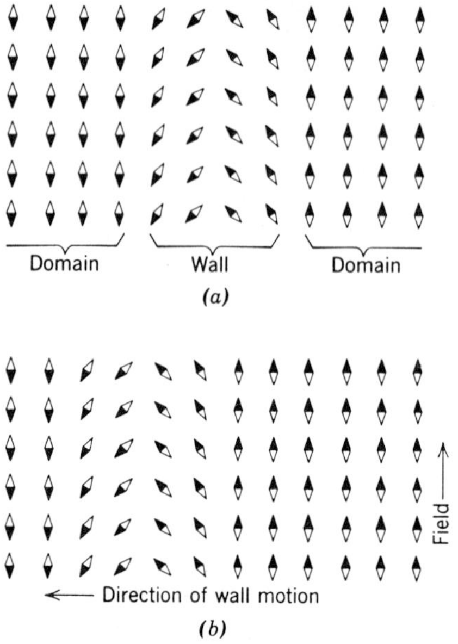

Fig. 6.17. Schematic explanation of domain wall displacement.

the center of the wall should be displaced as seen in Fig. 6.17*b*, resulting in an increase in the volume of the domain which has its magnetization parallel to the external field. We call this magnetization process domain wall displacement.

Let us now consider wall displacement in the domain structure shown in Fig. 6.18, which was actually observed on a Si-Fe (001) surface. When the field is applied parallel to one of the [100] directions as shown in Fig. 6.19*a*, the domains which have their magnetization parallel to the field should increase their volume at a sacrifice to the other domains and finally cover the whole volume of the specimen. Since there can be no further changes in the magnetization, this state corresponds to the saturation magnetization.

When the field is applied in a [110] direction (Fig. 6.19*b*), the two kinds of domains whose magnetization directions are closest to the field direction increase their volume and finally cover the whole volume of the specimen. If the field is increased further, the magnetizations in each domain rotate from the easy directions toward the field direction, and finally the specimen reaches saturation magnetization.

The magnetization curves corresponding to the processes mentioned above should be like those in Fig. 6.20. The curves in this figure are the actual magnetization curves measured for single crystals of iron. In [100]

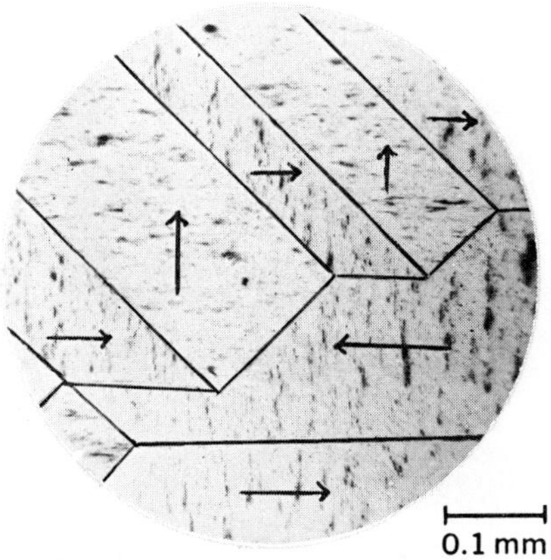

Fig. 6.18. Domain structure observed on the (001) surface of an Si-Fe crystal.

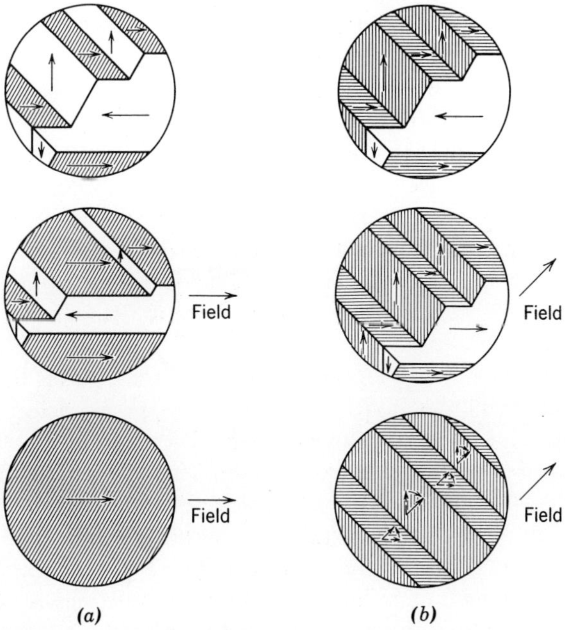

Fig. 6.19. Domain wall displacement and rotation magnetization: (*a*) magnetization in a [100], and (*b*) in a [110] direction.

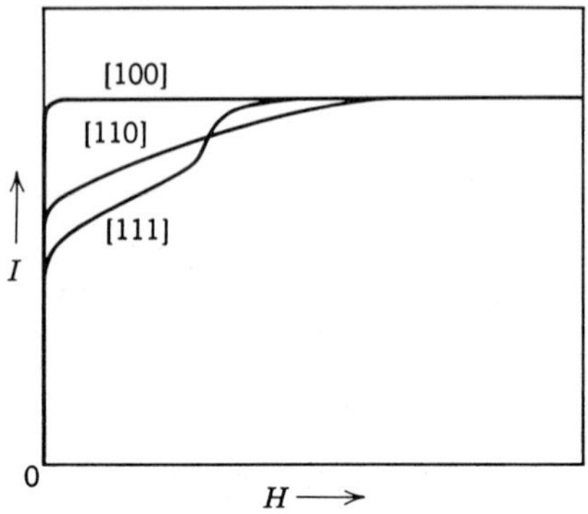

Fig. 6.20. Magnetization curves of single crystals with positive K_1.

magnetization, the entire process is performed by wall displacement which can be attained in a weak field, so that the magnetization curve reaches saturation in a weak field. On the contrary, [110] and [111] magnetizations require a farily large magnetic field to attain the saturation magnetization, because in most cases the rotation magnetization requires a fairly large amount of energy. We call this energy magnetic anisotropy energy.

In Section 6.2 we discussed domain structures by classifying magnetic substances into two groups, homogeneous and inhomogeneous. The magnetization process described above belongs to the category of homogeneous magnetic substances. In inhomogeneous magnetic materials, the displacement of a wall requires a fairly large magnetic field, so that the magnetization curve is bent from the initial stage of magnetization. This phenomenon is observed, more or less, for actual magnetic materials. The actual value of permeability is thus determined by the inhomogeneity of the materials. If there are too many irregularities, we cannot distinguish the two processes—wall displacement and rotation magnetization—on the magnetization curve.

Another feature of inhomogeneous magnetic materials is that the stable domain structure is not a unique function of the applied field. The domain structure which is attained by increasing the field to H is different from that attained by decreasing the field from a high value to H, so that the intensity of magnetization is different for both cases. This the reason why the magnetization of ferromagnetic substances shows hysteresis.

Homogeneity is thus the important factor which influences the permeability of ferromagnetic material. The situation, however, is not so simple if we want to have good magnetic properties in the high frequency range, where the velocity of the individual wall plays an important role in determining the actual value of permeability. We discuss this matter in Chapter 16.

References

6.1. P. Weiss: *J. Phys.* **6,** 661 (1907).
6.2. H. Barkhausen: *Phys. Z.* **20,** 401 (1919).
6.3. K. Honda and S. Kaya: *Sci. Rept. Tohoku Imp. Univ.* **15,** 721 (1926).
6.4. K. J. Sixtus and L. Tonks: *Phys. Rev.* **37,** 930 (1931); **39,** 357 (1932); **42,** 419 (1932); **43,** 70, 931 (1933).
6.5. F. Bloch: *Z. Physik.* **74,** 295 (1932).
6.6. F. Bitter: *Phys. Rev.* **38,** 1903 (1931); **41,** 507 (1932).
6.7. L. V. Hamos and P. A. Thiessen: *Z. Physik.* **71,** 442 (1932).
6.8. W. C. Elmore: *Phys. Rev.* **51,** 982 (1937); **53,** 757 (1938); **54,** 309 (1938); **62,** 486 (1942).
6.9. S. Chikazumi and K. Suzuki: *J. Phys. Soc. Japan* **10,** 523 (1955).
6.10. S. Kaya: *Z. Physik.* **89,** 796 (1934); **90,** 551 (1934); S. Kaya and J. Sekiya: *ibid.*, **96,** 53 (1935).
6.11. L. Landau and E. Lifshitz: *Physik. Z. Sowjetunion* **8,** 153 (1935); E. Lifshitz: *J. Phys. USSR* **8,** 337 (1944).
6.12. L. Néel: *J. Phys. Radium* **5,** 241, 265 (1944).
6.13. H. J. Williams, R. M. Bozorth, and W. Shockley: *Phys. Rev.* **75,** 155 (1949).
6.14. W. Stephan: *Exptl. Tech. Physik.* **4,** 153 (1956); **5,** 145 (1957).
6.15. L. Néel: *Cahiers Phys.* **25,** 21 (1944).
6.16. S. Chikazumi: *Phys. Rev.* **85,** 918 (1952).

7

Magnetic Anisotropy

7.1 Phenomenology of Magnetic Anisotropy

By magnetic anisotropy is meant the dependence of the internal energy on the direction of spontaneous magnetization. We call an energy term of this kind a magnetic anisotropy energy. Generally the magnetic anisotropy energy term possesses the crystal symmetry of the material, and we call it a crystal magnetic anisotropy or magnetocrystalline anisotropy. We can also produce magnetic anisotropy by applying mechanical stress to the material, as in the Sixtus-Tonks experiment. We call this magnetostrictive anisotropy. We can also control the magnetic anisotropy by heat treating the material in a magnetic field or by cold working it. We call this induced magnetic anisotropy. In this section we deal exclusively with crystal anisotropy. Magnetostrictive and induced anisotropies are treated in Section 8.4 and Chapter 17.

The simplest form of crystal anisotropy is uniaxial anisotropy. For example, hexagonal cobalt exhibits uniaxial anisotropy which makes the stable direction of internal magnetization (or easy direction) parallel to the c axis of the crystal at room temperature. As the internal magnetization rotates away from the c axis, the anisotropy energy increases with an increase of ϕ, the angle between the c axis and the internal magnetization, takes its maximum value at $\phi = 90°$, and then decreases to its original value at $\phi = 180°$. We can express this energy by expanding it in a series of powers of $\sin^2 \phi$:

$$E_a = K_{u1} \sin^2 \phi + K_{u2} \sin^4 \phi + \cdots . \tag{7.1}$$

Usually the first term is sufficient to express the actual anisotropy energy. The anisotropy is also dependent on the azimuthal angle about the *c* axis, but this term is as small as the third term of (7.1). For cobalt at room temperature,

$$K_{u1} = 4.1 \times 10^5 \text{ J/m}^3 \,(= 4.1 \times 10^6 \text{ ergs/cc}),$$
$$K_{u2} = 1.0 \times 10^5 \text{ J/m}^3 \,(= 1.0 \times 10^6 \text{ ergs/cc}). \tag{7.2}$$

Values of these constants for various uniaxial crystals are given in Table 7.4.

For cubic crystals such as iron and nickel the anisotropy energy can be expressed in terms of the direction cosines (α_1, α_2, α_3) of the internal magnetization with respect to the three cube edges. There are many equivalent directions in which the anisotropy energy has the same value, as shown by points A_1, A_2, B_1, B_2, C_1, and C_2 on an octant of the unit sphere in Fig. 7.1. Because of the high symmetry of the cubic crystal, the anisotropy energy can be expressed in a fairly simple way: We expand the anisotropy energy in a polynomial series in α_1, α_2, and α_3. Those terms which include odd powers of the α_i must vanish because a change in sign of any of the α_i should bring the magnetization vector to a direction which is equivalent to the original direction. The expression must also be invariant to the interchange of any two α_i's, so that the terms of the form $\alpha_i^l\alpha_j^m\alpha_k^n$ must have, for any given combination of l, m, n, the same coefficient for any interchange of i, j, and k. The first term, therefore, should have the form $\alpha_1^2 + \alpha_2^2 + \alpha_3^2$, which is always equal to 1. Next is the fourth order term which can be reduced to the form $\sum_{i>j} \alpha_i^2\alpha_j^2$ by the relation

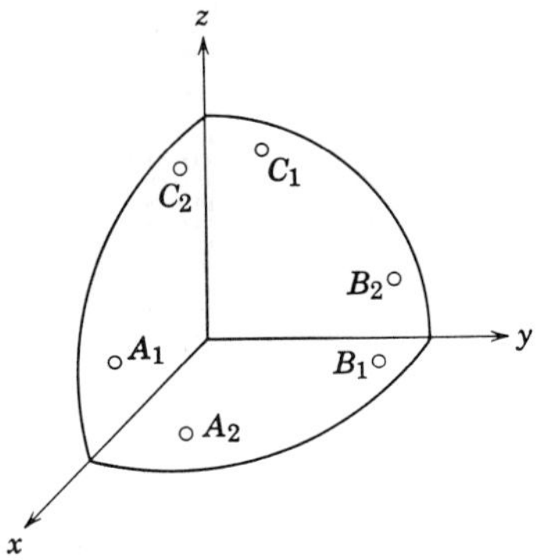

Fig. 7-1. Equivalent directions in a cubic crystal.

$$\alpha_1^4 + \alpha_2^4 + \alpha_3^4 = 1 - 2(\alpha_1^2\alpha_2^2 + \alpha_1^2\alpha_3^2 + \alpha_2^2\alpha_3^2). \tag{7.3}$$

Furthermore, by the relations

$$\alpha_1^6 + \alpha_2^6 + \alpha_3^6 = 1 - 3(\alpha_1^2\alpha_2^2 + \alpha_1^2\alpha_3^2 + \alpha_2^2\alpha_3^2) + 3\alpha_1^2\alpha_2^2\alpha_3^2, \tag{7.4}$$

$$\alpha_1^4\alpha_2^2 + \alpha_2^4\alpha_3^2 + \alpha_3^4\alpha_1^2 + \alpha_2^4\alpha_1^2 + \alpha_3^4\alpha_2^2 + \alpha_1^4\alpha_3^2 = 3(\alpha_1^2\alpha_2^2 + \alpha_2^2\alpha_3^2 + \alpha_3^2\alpha_1^2) - 9\alpha_1^2\alpha_2^2\alpha_3^2, \tag{7.5}$$

every sixth order term can be reduced to $\sum_{j>i} \alpha_i^2\alpha_j^2$ and $\alpha_1^2\alpha_2^2\alpha_3^2$. Thus we have the expression

$$E_a = K_1(\alpha_1^2\alpha_2^2 + \alpha_2^2\alpha_3^2 + \alpha_3^2\alpha_1^2) + K_2\alpha_1^2\alpha_2^2\alpha_3^2 + \cdots, \tag{7.6}$$

where K_1 and K_2 are the anisotropy constants. For iron at room temperature,[1]

$$\begin{aligned} K_1 &= 4.8 \times 10^4 \text{ J/m}^3 \quad (= 4.8 \times 10^5 \text{ ergs/cc}), \\ K_2 &= \pm 5 \times 10^3 \text{ J/m}^3 \quad (= \pm 5 \times 10^4 \text{ ergs/cc}), \end{aligned} \tag{7.7}$$

and, for nickel at room temperature,[2]

$$\begin{aligned} K_1 &= -4.5 \times 10^3 \text{ J/m}^3 \quad (= -4.5 \times 10^4 \text{ ergs/cc}), \\ K_2 &= 2.34 \times 10^3 \text{ J/m}^3 \quad (= 2.34 \times 10^4 \text{ ergs/cc}). \end{aligned} \tag{7.8}$$

When $K_1 > 0$ the first term of (7.6) takes on its minimum value at the [100], [010], and [001] directions, whereas when $K_1 < 0$ it does so at the [111],

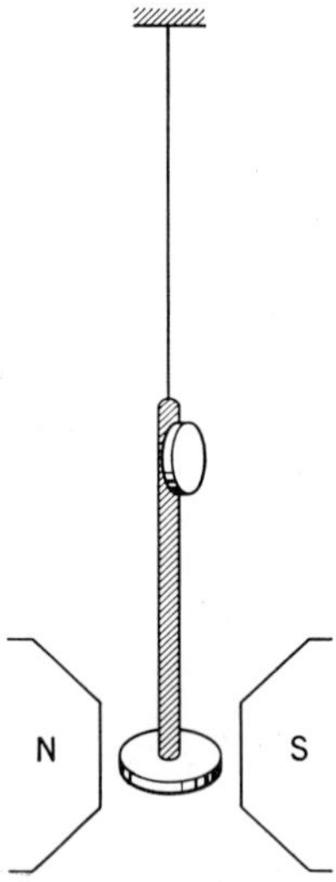

Fig. 7.2. Schematic diagram of a torque magnetometer.

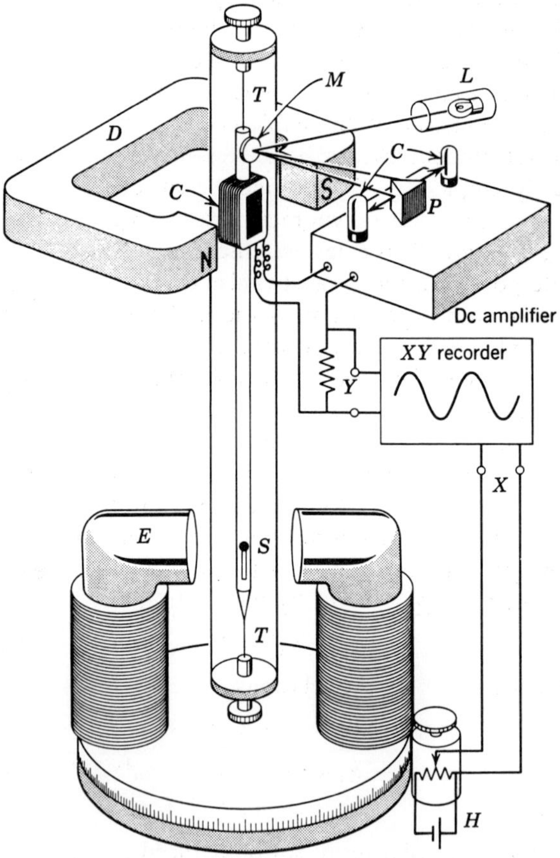

Fig. 7.3. Automatic torque magnetometer.

[$\bar{1}$11], [1$\bar{1}$1], and [11$\bar{1}$] directions. These, then, are the directions of easy magnetization.

There are various means for measuring the magnetic anisotropy. The torque magnetometer is an apparatus commonly used for this purpose. The principle of this apparatus is simply to suspend the specimen by a fine elastic string between the pole pieces of a rotatable electromagnet (Fig. 7.2). Usually a frictionless bearing is necessary to support the specimen holder and prevent the specimen from moving toward a pole piece. When a strong magnetic field is applied to the specimen, the internal magnetization is forced to line up with the field, and the specimen disk itself tends to rotate so as to make an easy direction approach the direction of magnetization. The torque exerted by the specimen can be measured by the angle of twist of the elastic string. If the magnet is rotated, the torque can be measured as a function of crystallographic direction of magnetization. We call this curve a magnetic torque curve, from which we can reproduce the magnetic anisotropy energy.

The most advanced type of torque magnetometer is shown in Fig. 7.3. The specimen holder is supported by a stretched wire the ends of which are supported at the top and bottom of the apparatus. When a torque is exerted on the specimen, the slight deflection of the mirror gives rise to an unbalanced current in a system of paired photocells. This current is amplified by the dc amplifier and supplied to the small coil which is attached at the upper part of the specimen holder. Since this coil is placed in a permanent magnet, the current gives rise to a torque which counterbalances the original torque so that the current in the coil is exactly proportional to the original torque. This current is recorded in the *XY* recorder as a function of a signal which is proportional to the angle of rotation of the electromagnet.[3,4]

The torque exerted by a unit volume of the specimen is

$$L = -\frac{\partial E_a}{\partial \phi}, \tag{7.9}$$

where ϕ is the angle made by the internal magnetization measured in the plane of the specimen disk. For magnetization confined to a (001) plane of a cubic crystal, we can write $\alpha_1 = \cos\phi$, $\alpha_2 = \sin\phi$, and $\alpha_3 = 0$, if ϕ is measured from the [100] direction. Then the first term of (7.6) becomes

$$E_a = K_1 \cos^2\phi \sin^2\phi = \tfrac{1}{4}K_1 \sin^2 2\phi. \tag{7.10}$$

From (7.9) we have

$$L = -\tfrac{1}{2}K_1 \sin 4\phi. \tag{7.11}$$

Figure 7.4 shows the actual torque curve observed for a (001) disk of 4% Si-Fe crystal the shape of which is well fitted by (7.11).

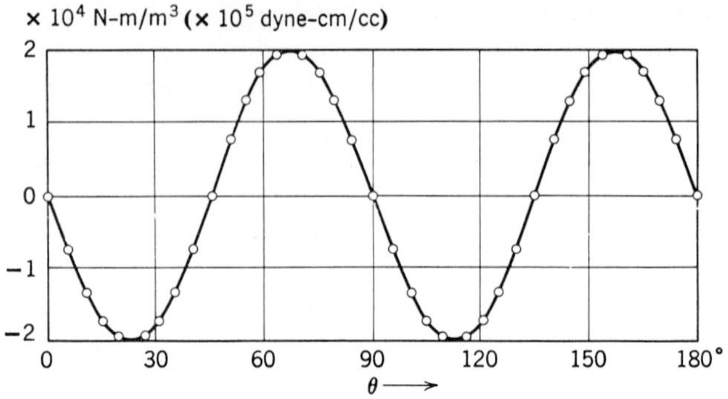

Fig. 7.4. Torque curve observed in the (001) plane of 4% Si-Fe crystal (Chikazumi and Iwata).

When the torque curve contains a lower order term, such as is caused by magnetostrictive anisotropy or induced anisotropy, we must determine the coefficient of each term by means of Fourier analysis. Suppose that the anisotropy is expressed by

$$E_a = -K_u \cos^2 (\phi - \phi_0') + \tfrac{1}{4} K_1 \sin^2 2(\phi - \phi_0), \tag{7.12}$$

where K_u is the uniaxial anisotropy constant, ϕ_0' the easy direction of the uniaxial anisotropy, and ϕ_0 the [100] direction. The corresponding torque becomes

$$L = A_2 \sin 2\phi + B_2 \cos 2\phi + A_4 \sin 4\phi + B_4 \cos 4\phi, \tag{7.13}$$

where

$$\begin{aligned} A_2 &= -K_u \cos 2\phi_0' \qquad & B_2 &= K_u \sin 2\phi_0', \\ A_4 &= -\tfrac{1}{2} K_1 \cos 4\phi_0, \qquad & B_4 &= \tfrac{1}{2} K_1 \sin 4\phi_0. \end{aligned} \tag{7.14}$$

If we can determine the values of the coefficients of each term in (7.13), we can calculate K_u, K_1, ϕ_0', and ϕ_0 from (7.14). In order to analyze the coefficients from the experimental curve we use the Fourier summations:

$$\begin{aligned} A_2 &= \frac{2}{n} \sum_{i=1}^{n} L \sin 2\phi_i, \\ B_2 &= \frac{2}{n} \sum_{i=1}^{n} L \cos 2\phi_i, \\ A_4 &= \frac{2}{n} \sum_{i=1}^{n} L \sin 4\phi_i, \\ B_4 &= \frac{2}{n} \sum_{i=1}^{n} L \cos 4\phi_i, \end{aligned} \tag{7.15}$$

where $\phi_i = (\pi/n)i$ and n is an integer. If the highest harmonics included in the torque curve are $\sin 2m\phi$ or $\cos 2m\phi$, it may be easily verified that the number of terms in the summation should be more than $2 + m$. For instance, if the torque curve contains no harmonics beyond sin 8ϕ or cos 8ϕ, we can perform the Fourier summation at every 30 degrees. The values of the harmonics necessary for this analysis are given in Table 7.1. The harmonics at every 22.5 degrees are listed in Table 7.2, and at every 15 degrees in Table 7.3.

Table 7.1. Coefficients for Fourier Analysis ($n = 6$)

θ	sin 2θ	cos 2θ	sin 4θ	cos 4θ
0°	0.000	1.000	0.000	1.000
30	0.866	0.500	0.866	−0.500
60	0.866	−0.500	−0.866	−0.500
90	0.000	−1.000	0.000	1.000
120	−0.866	−0.500	0.866	−0.500
150	−0.866	0.500	−0.866	−0.500

Table 7.2. Coefficients for Fourier Analysis ($n = 8$)

θ	sin 2θ	cos 2θ	sin 4θ	cos 4θ
0°	0.000	1.000	0.000	1.000
22.5	0.707	0.707	1.000	0.000
45.0	1.000	0.000	0.000	−1.000
67.5	0.707	−0.707	−1.000	0.000
90.0	0.000	−1.000	0.000	1.000
112.5	−0.707	−0.707	1.000	0.000
135.0	−1.000	0.000	0.000	−1.000
157.5	−0.707	0.707	−1.000	0.000

Table 7.3. Coefficients for Fourier Analysis ($n = 12$)

θ	sin 2θ	cos 2θ	sin 4θ	cos 4θ	sin 6θ	cos 6θ
0°	0.000	1.000	0.000	1.000	0.000	1.000
15	0.500	0.866	0.866	0.500	1.000	0.000
30	0.866	0.500	0.866	−0.500	0.000	−1.000
45	1.000	0.000	0.000	−1.000	−1.000	0.000
60	0.866	−0.500	−0.866	−0.500	0.000	1.000
75	0.500	−0.866	−0.866	0.500	1.000	0.000
90	0.000	−1.000	0.000	1.000	0.000	−1.000
105	−0.500	−0.866	0.866	0.500	−1.000	0.000
120	−0.866	−0.500	0.866	−0.500	0.000	1.000
135	−1.000	0.000	0.000	−1.000	1.000	0.000
150	−0.866	0.500	−0.866	−0.500	0.000	−1.000
165	−0.500	0.866	−0.866	0.500	−1.000	0.000

For a (110) surface the anisotropy energy is given by

$$E_a = -K_u \cos^2(\phi - \phi_0') + \frac{K_1}{32}[7 - 4\cos 2(\phi - \phi_0) - 3\cos 4(\phi - \phi_0)] + \frac{K_2}{128}[2 - \cos 2(\phi - \phi_0) - 2\cos 4(\phi - \phi_0) + \cos 6(\phi - \phi_0)], \tag{7.16}$$

so that the torque becomes

$$L = A_2 \sin 2\phi + B_2 \cos 2\phi + A_4 \sin 4\phi + B_4 \cos 4\phi + A_6 \sin 6\phi + B_6 \cos 6\phi \tag{7.17}$$

where

$$\begin{aligned}
A_2 &= -K_u \cos 2\phi_0' - \left(\frac{K_1}{4} + \frac{K_2}{64}\right)\cos 2\phi_0,\\
B_2 &= +K_u \sin 2\phi_0' + \left(\frac{K_1}{4} + \frac{K_2}{64}\right)\sin 2\phi_0,\\
A_4 &= -\left(\frac{3K_1}{8} + \frac{K_2}{16}\right)\cos 4\phi_0,\\
B_4 &= +\left(\frac{3K_1}{8} + \frac{K_2}{16}\right)\sin 4\phi_0,\\
A_6 &= +\frac{3K_2}{64}\cos 6\phi_0,\\
B_6 &= -\frac{3K_2}{64}\sin 6\phi_0.
\end{aligned} \tag{7.18}$$

In order to analyze A_6 and B_6, we can use $n = 12$ (the harmonics for this case are given in Table 7.3).

It must be noted that, when the intensity of the magnetic field is insufficient to align the internal magnetization in the field direction, the uniaxial anisotropy may give rise to a fourth order anisotropy. This phenomenon was first observed by Nesbitt, Williams, and Bozorth[5] who found that the fourth order torque of Fe_2NiAl decreases and changes sign with an increase of the field intensity. The same phenomenon was observed for Alnico 5 by Nesbitt and Williams[5] and for Co ferrite by Williams, Heidenreich, and Nesbitt.[6] The latter investigators found also that this phenomenon, which is called a torque reversal, does not occur for a material which does not respond to magnetic annealing.

The phenomenon of torque reversal can be explained as follows:[7] When a material which responds to magnetic annealing is cooled in zero magnetic field, the internal magnetization is stabilized in one of the easy directions. For a domain which is stabilized in the [100] direction, the energy when a field is applied in the $[\beta_1, \beta_2, \beta_3]$ direction is given by

$$E(x) = -K_u\alpha_1^2 - I_s H(\alpha_1\beta_1 + \alpha_2\beta_2 + \alpha_3\beta_3), \tag{7.19}$$

where $(\alpha_1, \alpha_2, \alpha_3)$ are the direction cosines of internal magnetization. The direction of the internal magnetization can be determined by minimizing (7.19) with respect to $(\alpha_1, \alpha_2, \alpha_2)$ under the restriction $\alpha_1^2 + \alpha_2^2 + \alpha_3^2 = 1$. Assuming that $I_s H \gg K_u$, we have

$$\begin{aligned} \alpha_1 &= \beta_1 + 2\frac{K_u}{I_s H}(\beta_1\beta_2^2 + \beta_1\beta_3^2), \\ \alpha_2 &= \beta_2 - 2\frac{K_u}{I_s H}\beta_2\beta_1^2, \\ \alpha_3 &= \beta_3 - 2\frac{K_u}{I_s H}\beta_3\beta_1^2. \end{aligned} \tag{7.20}$$

Putting (7.20) into (7.19), we have the equilibrium energy of the system as a function of the direction of the external field:

$$E(x) = -I_s H - K_u\beta_1^2 - \frac{4K_u^2}{I_s H}(\beta_1^2\beta_2^2 + \beta_1^2\beta_3^2). \tag{7.21}$$

Similar expressions can be obtained for y- and z-stabilized domains. Averaging over a uniform distribution, we finally have

$$E_a = -I_s H - \frac{K_u}{3} - \frac{8K_u^2}{3I_s H}(\beta_1^2\beta_2^2 + \beta_2^2\beta_3^2 + \beta_3^2\beta_1^2). \tag{7.22}$$

The last term has the same functional form as (7.6); hence, if this crystal has its own crystal anisotropy which may also be expressed approximately as $K_1(\beta_1^2\beta_2^2 + \beta_2^2\beta_3^2 + \beta_3^2\beta_1^2)$, it is expected that the resultant anisotropy $(K_1 - 8K_u^2/3I_s H)$ changes sign at $H = 8K_u^2/3I_s K_1$. The same effect may also be expected for a polycrystal of a material with uniaxial anisotropy whose crystallites are oriented with cubic symmetry.

The anisotropy constant can also be estimated from the magnetization curves of a single crystal. Let us consider the magnetization of a specimen which has positive K_1 in a magnetic field parallel to the [110] direction. As mentioned in Chapter 6 (cf. Fig. 6.19*b*), the domains remaining after the wall displacement is finished are reduced to [100] and [010] domains. Then the magnetization takes place by the rotation of domain magnetization. The direction of magnetization during this process can be determined by equating the torque $-I_s H \sin\theta$ exerted by the external field to the counterbalancing torque due to the crystal anisotropy, so that

$$-I_s H \sin\theta = -\frac{K_1}{2}\sin 4\phi \tag{7.23}$$

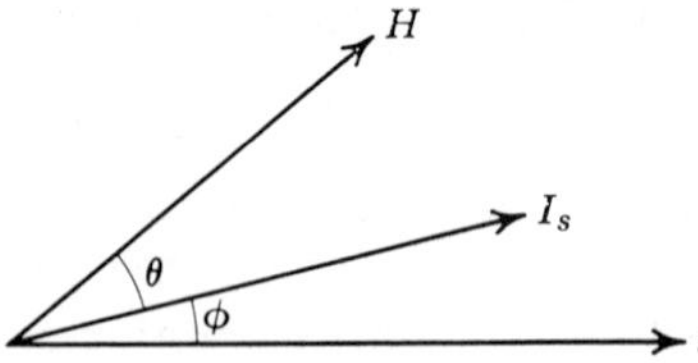

Fig. 7.5. Rotation of magnetization out of an easy direction.

where the torque given in (7.11) has been used since the rotation will take place in the (001) plane. From the relation $\phi = \pi/4 - \theta$ (Fig. 7.5) or $\sin 4\phi = \sin 4\theta = 4 \sin\theta \cos\theta \cos 2\theta$, (7.23) becomes

$$I_s H = 2K_1 \cos\theta \cos 2\theta, \tag{7.24}$$

and, if we put $\cos\theta = t$, it becomes

$$2t^3 - t - \frac{I_s H}{2K_1} = 0. \tag{7.25}$$

The magnetization is given by

$$I = I_s \cos\theta = I_s t. \tag{7.26}$$

In particular, the [110] curve is given by inserting the solution of (7.25) in (7.26). Figure 7.6 shows the magnetization curves of single crystals of iron first measured by Honda and Kaya,[8] together with the theoretical curves calculated by (7.26). We see that the theoretical curves reproduce

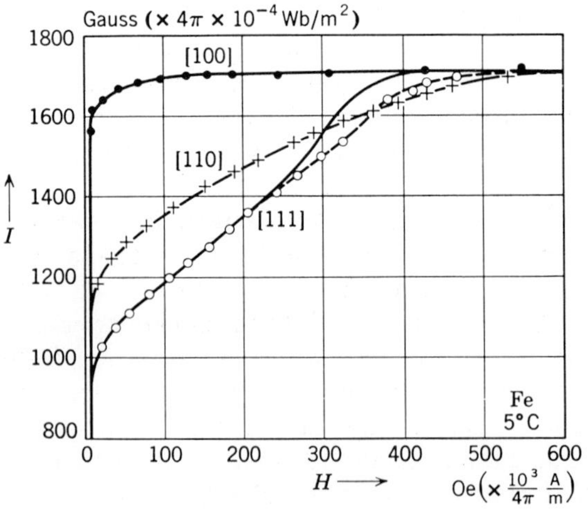

Fig. 7.6. Magnetization curve of iron single crystal measured by applying the magnetic field paralled to each of the three principal crystallographic directions (after Honda and Kaya[8]).

the experiments well. From the comparison of experiment and theory we can determine the value of K_1.

A more convenient method for determining the anisotropy constant from the magnetization curve is to calculate it from the area enclosed by the magnetization curve, the ordinate axis, and the line $I = I_s$; that is,

$$W = \int_{I_r}^{I_s} H\, dI, \tag{7.27}$$

which can be calculated for the [110] curve by using the relation (7.25):

$$\begin{aligned} W &= 2K_1 \int_{1/\sqrt{2}}^{1} (2t^3 - t)\, dt \\ &= 2K_1 \left| \frac{1}{2} t^4 - \frac{1}{2} t^2 \right|_{\frac{1}{\sqrt{2}}}^{1} \\ &= \frac{K_1}{4}. \end{aligned} \tag{7.28}$$

Since (7.27) represents the energy required to magnetize a unit volume of the specimen to its saturation state (cf. 2.3), this value can also be calculated from the increase of the anisotropy energy (7.10) in this process as

$$\begin{aligned} \Delta E_a &= E_a(\phi = \pi/4) - E_a(\phi = 0) \\ &= \frac{K_1}{4}. \end{aligned} \tag{7.29}$$

For uniaxial anisotropy, if we apply the magnetic field perpendicular to the easy axis, the magnetization curve becomes a straight line, which reaches saturation at $H_a = 2K_u/I_s$ (cf. Fig. 14.3). The area enclosed by the magnetization curve is, therefore given by

$$W = K_u. \tag{7.30}$$

If several kinds of anisotropy are mixed up in the material, it is very hard to analyze the individual anisotropies from the magnetization curve. It should also be noted that the area W is affected by the irreversible magnetization process, so that this method cannot be used for a material which has large hysteresis.

The magnetic anisotropy can also be measured by means of ferromagnetic resonance. As explained in Section 3.3 the resonance frequency depends on the external magnetic field, which exerts a torque on the precessing spin system. Since a magnetic anisotropy also causes a torque on a spin system if it points in other than easy directions, the resonance frequency is expected to be dependent on the magnetic anisotropy. For uniaxial anisotropy, which is generally expressed by the first term of (7.1), when the magnetization is nearly parallel to the easy direction, the anisotropy energy can be expressed by

$$E_a = K_{u1}\phi^2, \tag{7.31}$$

since $\sin\phi \approx \phi$ for $\phi \ll \pi$. This is equivalent to the presence of a magnetic field H_a parallel to the easy direction, because the energy is then expressed by

$$E = -I_s H_a \cos\phi \fallingdotseq \text{const.} + \tfrac{1}{2} I_s H_a \phi^2. \tag{7.32}$$

If we put

$$H_a = \frac{2K_{u1}}{I_s}, \tag{7.33}$$

we find that (7.31) and (7.32) are exactly identical. The field H_a is referred to as an anisotropy field. When a ferromagnetic resonance is observed by applying a magnetic field parallel to the easy direction, the resonance frequency should be given by

$$\omega = \nu(H + H_a), \tag{7.34}$$

instead of by (3.27). This means that the resonance occurs at an external field less by H_a than in the isotropic case.

For cubic anisotropy, we express the direction cosines by polar coordinates as

$$\begin{aligned} \alpha_1 &= \sin\theta\cos\phi \approx \theta\cos\phi \\ \alpha_2 &= \sin\theta\sin\phi \approx \theta\sin\phi \\ \alpha_3 &= \cos\theta \approx 1 - \tfrac{1}{2}\theta^2 \end{aligned} \tag{7.35}$$

for $\theta \ll \pi$, so that the first term of (7.6) becomes

$$\begin{aligned} E_a &= K_1[\theta^4 \sin^2\phi\cos^2\phi + (1 - \tfrac{1}{2}\theta^2)^2\theta^2] \\ &\approx K_1\theta^2. \end{aligned} \tag{7.36}$$

Comparing (7.36) and (7.32), we have the anisotropy field

$$H_a = \frac{2K_1}{I_s} \tag{7.37}$$

for $\langle 100 \rangle$ directions. When the magnetization is nearly parallel to the $\langle 111 \rangle$ direction, the anisotropy energy can be expressed by

$$E_a = \frac{K_1}{3} - \frac{2K_1}{3}\theta^2, \tag{7.38}$$

so that the anisotropy field is

$$H_a = -\frac{4K_1}{3I_s}. \tag{7.39}$$

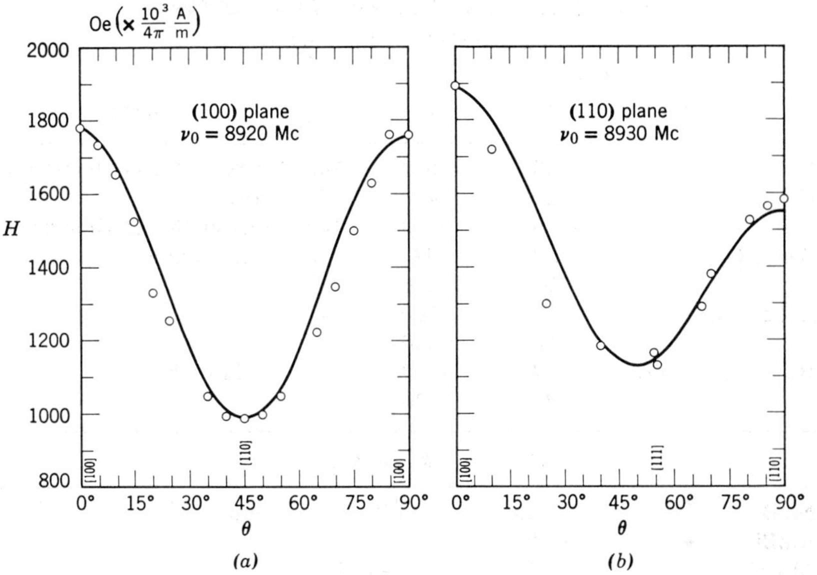

Fig. 7.7. Dependence of resonance field on the crystallographic direction of magnetization. Observed for (100) and (110) planes of magnetite at room temperature. (After L.R. Bickford.[9])

Therefore, when the field is rotated from $\langle 100 \rangle$ to $\langle 111 \rangle$, the shift of the resonance field is changed from (7.37) to (7.39) by the amount

$$\Delta H = \frac{2K_1}{I_s} - \left(-\frac{4K_1}{3I_s}\right) = \frac{10K_1}{3I_s}. \tag{7.40}$$

Figure 7.7 shows the dependence of the resonance field on the crystallographic direction of magnetization as observed by Bickford[9] for the (001) and (110) planes of single crystal magnetite. We can easily estimate the value of K_1 from these curves. This method has the advantage that it not only enables us to measure the anisotropy of a very small specimen but also offers information about the magnitude of local anisotropy. For instance, using the ferromagnetic resonance technique, Bickford observed the process of a change of uniaxial anisotropy occurring in the individual orthorhombic crystallites of magnetite just above the low temperature transition point (cf. Fig. 17.28).

The anisotropy constants determined for various substances are listed in Tables 7.4 and 7.5 and are also shown in Figs. 7.8, 7.9, 7.10, and 7.11 as functions of composition and temperature.

Table 7.4. Anisotropy Constants of Uniaxial Crystal
($\times 10^3$ J/m^3, $\times 10^4$ ergs/cc)

Substance	Temperature, °C	K_{u1}	K_{u2}	References
Co	20	530	100	Sucksmith and Thompson[10]
Co	20	430	120	Bozorth[11]
Co	−176	790	100	Sucksmith and Thompson[10]
$BaFe_{12}O_{19}$	20	320	0	Went et al.[12]
	−190	350	0	
$BaFe_{18}O_{27}$	20	300	—	
	−190	350	—	
$BaCo_{1.85}Fe_{16.2}O_{27}$	27	−400	—	Bickford[13]
	−173	−720	—	
	°C	$K_{u1} + K_{u2}$		
MnBi	20	120		Guillaud[14]
MnBi	−190	0		
Mn_2Sb	20	−0.25		
Mn_2Sb	−190	−1.3		
$Ba_2Zn_{1.5}Fe_{12.5}O_{22}$	20	−90[a]		Smit and Wijn[15]
$Ba_2Co_2Fe_{12}O_{22}$	20	−260[a]		
$Ba_3Co_2Fe_{24}O_{41}$	20	−180[a]		

[a] These values are for $K_{u1} + 2K_{u2}$.

Table 7.5. Anisotropy Constants of Various Ferrites
($\times 10^3$ J/m^3, $\times 10^4$ ergs/cc)

$Mn_xFe_yO_4$

x	y	K_1 at 17°C	K_1 at −183°C	References
0.1	2.9	−13	−6	Penoyer and Shafer[16]
0.4	2.6	−4.5	−17	
0.6	2.4	−0.5	−5	
0.8	2.2	−0.5	−3	
0.9	2.1	−2.0	−7.5	
1.0	2.0	−4.0	−20	
1.25	1.75	−2.5	−18	
1.55	1.45	−2.0	−15.5	
1.8	1.2	−1.5	−12.5	
1.9	1.1	−1.0	−12.0	
0	3.0	−14		

Table 7.5. Anisotropy Constants of Various Ferrites (Continued)

x	y	K_1 at 17°C	K_1 at −183°C	References
0.40	2.60	−5.0	−18	Miyata[17] (also refer to Funatogawa et al.,[18] Palmer,[19] Enz,[20] Pearson[21])
0.50	2.50	−1.6	−10.7	
0.68	2.32	0.6	−2.3	
0.75	2.25	0.9	0	
0.85	2.15	−0.5	−2.9	
0.95	2.05	−2.5	−12.3	
1.15	1.85	−3.3	−19.0	

$Mn_xZn_yFe_zO_4$

x	y	z	K_1 at 20°C	K_1 at −183°C	References
0.98	0.06	1.96	−4.0	−19.7	Ohta[22]
0.50	0.12	2.38	−0.2	−5.7	
0.28	0.19	2.53	−4.0	−19.5	
0.54	0.18	2.28	+0.4	−1.6	
0.59	0.20	2.21	+0.4	−4.5	
0.86	0.20	1.94	−1.0	−16.0	
1.01	0.21	1.78	−1.0	−15.2	
0.15	0.19	2.66	−7.6	−21.8	
0.66	0.30	2.04	−1.0	−12.0	
0.45	0.35	2.20	+0.2	−3.8	
0.58	0.38	2.04	−0.2	−6.0	
0.52	0.40	2.08	−0.1	−4.5	
0.57	0.40	2.03	−0.2	−12.0	
0.62	0.41	1.97	−0.2	−8.7	
0.31	0.39	2.30	−0.2	−8.2	
0.42	0.43	2.15	+0.2	−3.3	
0.49	0.45	2.06	+0.6	−3.7	
0.45	0.55	2.00	−0.38		Galt et al.[23]

$Mn_xNi_yFe_zO_4$

x	y	z	K_1 at 17°C	K_1 at −183°C	K_2 at −183°C	References
0.19	0.17	2.64	−5.9	−14.3	0	Miyata[17]
0.40	0.20	2.40	−3.9	−12.0	−6	
0.57	0.20	2.23	−2.8	−9.5	−8	
0.20	0.40	2.40	−4.5	−6.2	−8	
0.38	0.39	2.23	−4.4	−10.5	−3	
0.17	0.58	2.25	−4.7	−6.6	−9	

Table 7.5. Anisotropy Constants of Various Ferrites (Continued)

$Mn_xTi_yFe_zO_4$

x	y	z	K_1 at −268.8°C	K_2 at −268.8°C	References
1.0	0	2.0	−25	0	Smit et al.[24]
1.0	0.15	1.85	100	−230	
1.0	0.27	1.73	230	−440	
1.0	0.44	1.56	370	−520	

$Mn_xCo_yFe_zO_4$

x	y	z	K_1 at 17°C	K_1 at −73°C	References
0.98		1.99	−4.07	−10.53	Pearson[21,25]
0.99	0.009	1.98	−3.12	−4.68	
1.03	0.019	1.94	−2.22	+1.00	
0.98	0.038	1.97	−0.70	+11.22	
0.96	0.056	1.97	0.63	+21.00	
0.92	0.077	1.99	2.91	—	
0.91	0.094	1.97	3.73	+40.75	
0.747	0.245	1.99	17.1	—	
0.4	0.3	2.0	110		
0.891	0	2.093	−1.68	−4.85	
0.878	0.041	2.068	+1.59	+17.29	
0.819	0	2.135	−0.96	−3.55	
0.820	0.0358	2.135	+4.43	+22.40	
0.742	0	2.241	+0.49	−0.81	
0.722	0.0371	2.234	+9.36	+32.88	
0.679	0	2.291	+0.06	−1.72	
0.647	0.0372	2.31	+11.81	—	

$Co_xFe_yO_4$

x	y	K_1 at 27°C	K_1 at −143°C	References
0	3.0	−11.0	0	Bickford et al.[26]
0.01	2.99	0	69.2	
0.04	2.96	30.0	256	
		K_1 at 20°C	K_1 at −196°C	
0.8	2.2	290	440	Bozorth et al.[27]
1.1	1.9	180		
1.1	2.2	380	1750	Shenker[28]

Table 7.5. Anisotropy Constants of Various Ferrites (Continued)

$Co_xNi_yFe_zO_4$

x	y	z	K_1 at 20°C	K_1 at −196°C	References
0.002	0.7	2.2	−1.8	−12.5	
0.004	0.7	2.2	−1.0	−19.6	Shenker[28]
0.08	0.72	2.2	−4.3	−2.8	Bozorth et al.[27]

$Co_xZn_yFe_zO_4$

x	y	z	K_1 at 20°C	References
0.3	0.2	2.2	150	Bozorth et al.[27]
0.3	0.4	2.0	110	

$Ni_xFe_yO_4$

x	y	K_1 at 20°C	K_1 at −185°C	References
0	3.0	−11.8	—	Elbinger[29]
0.25	2.75	−7.2	+14.3	(also refer to Miyata,[17]
0.50	2.50	−3.65	+3.9	Galt et al.[30])
0.75	2.25	−4.8	−5.0	
1.0	2.0	−6.7	−10.6	

$Ni_xZn_yFe_zO_4$

x	y	z	K_1 at 20°C	K_1 at −185°C	References
0.3	0.45	2.25	−1.7	−16	Ohta[31]

$Cu_xFe_yO_4$

x	y	K_1 at 20°C	K_1 at −196°C	References
1	2	−6	−20.6	Okamura and Kojima[32]

$Zn_xFe_yO_4$

x	y	K_1 at 17°C	K_1 at −183°C	References
0.16	2.84	−9.0	−15.9	Miyata[17]
0.30	2.70	−5.0	−21.3	
0.48	2.52	−1.2	−15.3	
0.61	2.39	−0.7	−9.2	

Table 7.5. Anisotropy Constants of Various Ferrites (Continued)

$Li_xFe^{2+}_yFe^{3+}_zO_4$

x	y	z	K_1 at 17°C	References
0.013	0.908	2.058	−10.9	Miyata et al.[33]
0.034	0.880	2.069	−9.5	
0.069	0.805	2.107	−10.0	
0.115	0.823	2.079	−8.3	

$Mg_xFe_yO_4$

x	y	K_1 at 20°C	K_1 at −185°C	References
0	3	−11.8	—	Elbinger[29] (also refer to Rado et al.[34])
0.25	2.75	−7.75	+1.4	
0.50	2.50	−5.75	−6.0	
0.75	2.25	−4.8	−12.2	
1.00	2.00	−3.9	−15.0	

$Ga_xFe^{2+}_yFe^{3+}_zO_4$

x	y	z	K_1 at 17°C	K_1 at −73°C	References
0	0.969	2.021	−13.4	−14.4	Pearson[35]
0.077	0.956	1.953	−13.1	−14.9	
0.144	0.943	1.894	−11.4	−14.1	
0.274	0.938	1.768	−9.8	−13.1	
0.356	0.918	1.699	−8.7	−12.5	
0.442	0.929	1.606	−8.09	−12.18	Pearson[21]
0.628	0.980	1.385	−4.97	−9.78	
0.757	0.942	1.282	−3.69	−7.57	

$Al_xFe^{2+}_yFe^{3+}_zO_4$

x	y	z	K_1 at 17°C	K_1 at −73°C	References
0.105	0.946	1.931	−14.0	−15.4	Pearson[35]
0.209	0.984	1.802	−12.3	−15.6	

$Ti_xFe^{2+}_yFe^{+3}_zO_4$

x	y	z	K_1 at 27°C	K_1 at −73°C	References
0.00	1.00	2.00	−13.2	−13.5	Syono and Ishikawa[36]
0.04	1.04	1.92	−19.6	−20.6	
0.10	1.10	1.80	−22.4	−27.6	
0.18	1.18	1.64	−19.2	−25.1	
0.051	0.944	1.969	−10.1	−8.1	Miyata et al.[33]

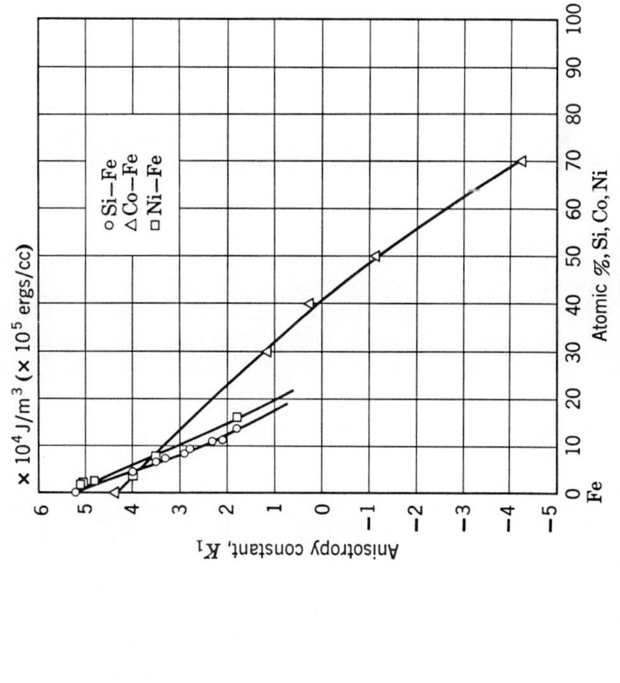

Fig. 7.8. Crystal anisotropy constants of Si-Fe (Tarasov[37]), Co-Fe (Shih[38]), and Ni-Fe (Tarasov[39]) alloys.

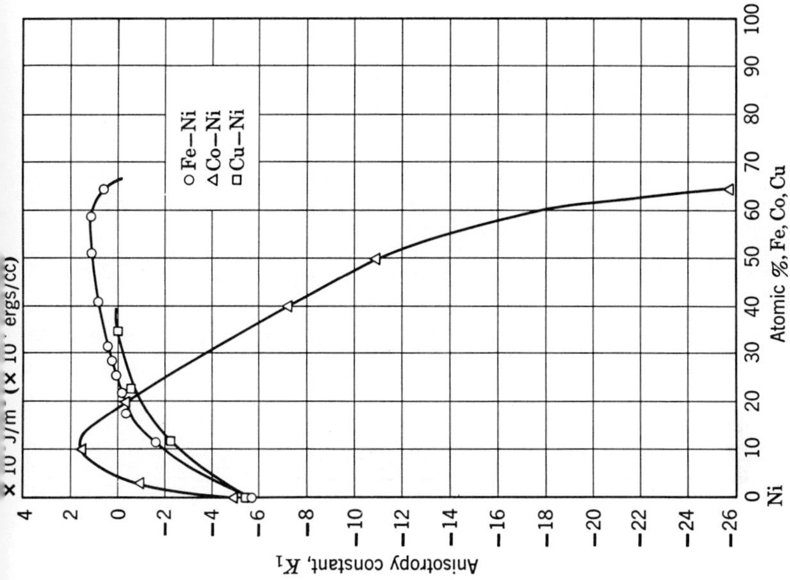

Fig. 7.9. Crystal anisotropy constants of Fe-Ni (Bozorth and Walker[40]), Co-Ni (Shih[41] and McKeehan[42]), and Cu-Ni (Bozorth[43]) alloys.

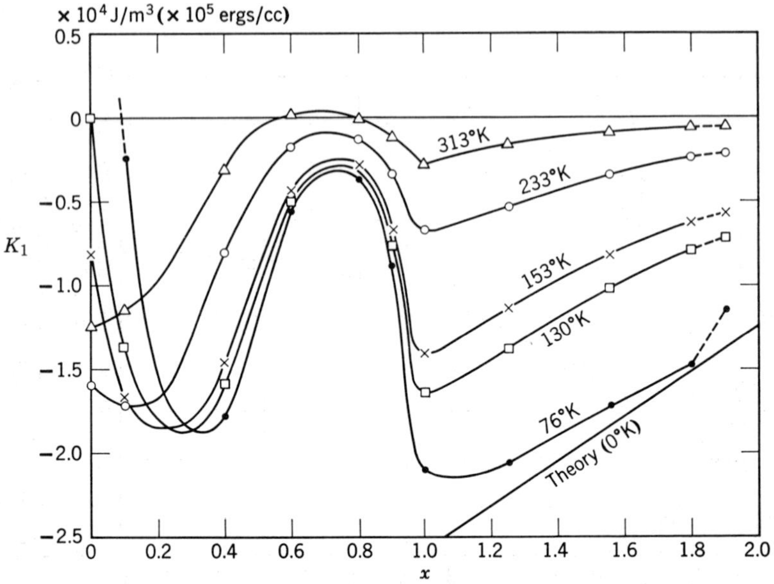

Fig. 7.10. Isothermal curves of K_1 vs x in $Mn_xFe_{3-x}O_4$ (after Penoyer and Shafer[16]).

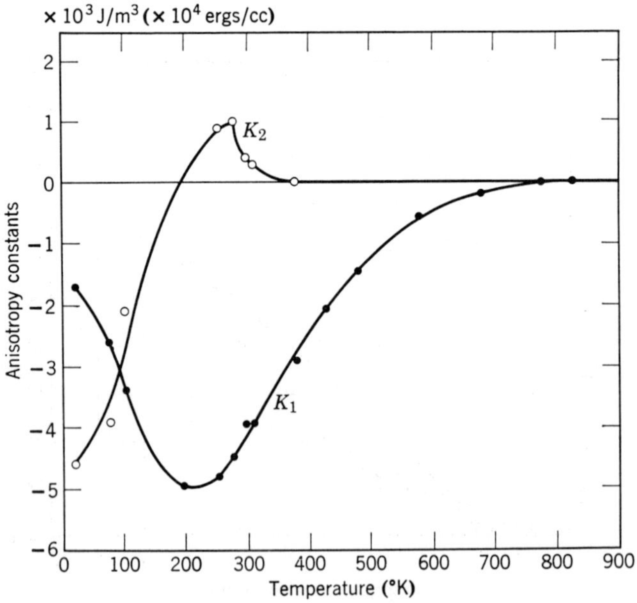

Fig. 7.11. K_1 and K_2 of NiCo ferrite ($Ni_{0.72}Fe_{0.20}Co_{0.08}Fe_2O_4$) as a function of temperature (after Shenker[44]).

7.2 Pair Model of Magnetic Anisotropy

Magnetic anisotropy describes the circumstance that the energy of a system changes with a rotation of the magnetization (Fig. 7.12). The attempt to interpret this phenomenon in terms of a change in energy of atomic pairs is called a pair model. This kind of theory was first developed by Van Vleck[45] in 1937. The most important interaction between the atomic magnetic moments is the exchange interaction. But this energy is dependent only on the angle between neighboring atomic moments, independently of their orientation relative to their bond direction, and so does not give rise to anisotropy. In order to explain magnetic anisotropy we may assume that the pair energy is dependent on the direction of the magnetic moment, ϕ, as measured from the bond direction (Fig. 7.13). In general, we can express the pair energy by expanding it in Legendre polynomials:

$$w(\cos\phi) = g + l(\cos^2\phi - \tfrac{1}{3}) + q(\cos^4\phi - \tfrac{6}{7}\cos^2\phi + \tfrac{3}{35}) + \cdots. \tag{7.41}$$

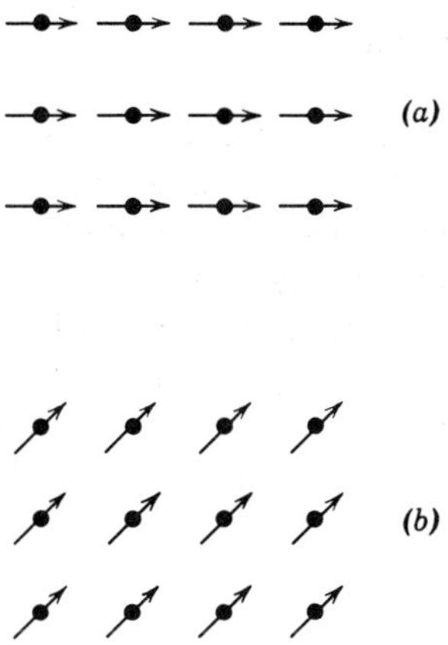

Fig. 7.12. Rotation of magnetization.

The first term is independent of ϕ; hence the exchange energy is to be included in this term. The second term is called the dipole-dipole interaction term because it has the same form as the magnetic interaction given by (1.12). If the pair energy were due exclusively to magnetic dipolar interaction, it should follow that

$$l = -\frac{3M^2}{4\pi\mu_0 r^3}, \tag{7.42}$$

and the other terms should vanish. The actual value of l can be evaluated from the uniaxial crystal anisotropy or from the values of magnetostriction constants (cf. Chapter 8). In most cases the estimated value is $10^2 \sim 10^3$ times larger than that given by (7.42). The origin of this strong interaction is believed to be the combined effect of spin-orbit interaction and exchange or Coulomb interaction between neighboring orbits. That is, if there are small amounts of orbital magnetic moment remaining unquenched by the crystalline field, a part of the orbit will rotate with a rotation of the spin magnetic moment because of a magnetic interaction between the two, and the rotation of the orbit will, in turn, change the overlap of the wave

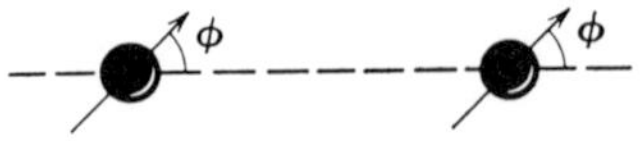

Fig. 7.13. A pair of magnetic atoms.

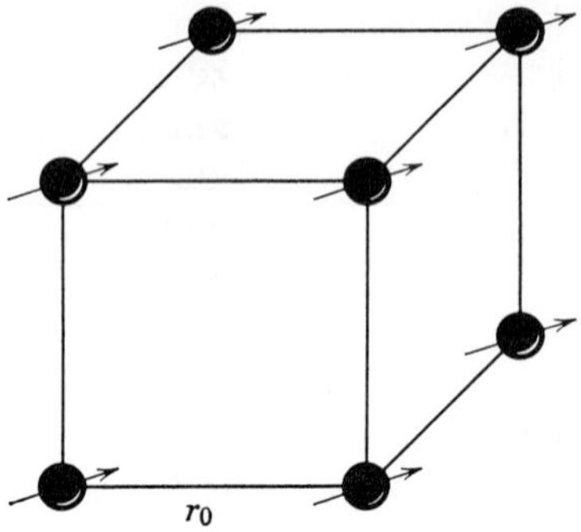

Fig. 7.14. A simple cubic lattice formed by magnetic atoms.

functions between the two atoms, giving rise to a change in the electrostatic or exchange energy. We call this type of interaction anisotropic exchange. Its various terms are called pseudodipolar, pseudoquadrupolar, etc., interactions.

Now let us discuss how an interaction like (7.41) between the atomic moments gives rise to the magnetic anisotropy. For simplicity we consider a simple cubic lattice formed by magnetic atoms (Fig. 7.14). The magnetic anisotropy energy can be obtained by adding up all the atomic pair energies thus

$$E_a = \sum_i w_i, \tag{7.43}$$

where i is an index attached to each of the atom pairs. Since the interaction is negligibly small between distant atoms, all we do is sum up the pair energies for the first nearest neighbor pairs, or at most for the first and second nearest neighbor pairs. In the present case, we confine ourselves to the first nearest neighbors. If we let (α_1, α_2, α_3) denote the direction cosines of the internal magnetization, cos ϕ in (7.41) can be replaced by α_1 for those atom pairs whose bond direction is parallel to the x axis, by α_2 for y pairs, and by α_3 for z pairs, and (7.43) becomes

$$\begin{aligned} E_a &= N[g + l(\alpha_1^2 - \tfrac{1}{3}) + q(\alpha_1^4 - \tfrac{6}{7}\alpha_1^2 + \tfrac{3}{35}) \\ &\quad + g + l(\alpha_2^2 - \tfrac{1}{3}) + q(\alpha_2^4 - \tfrac{6}{7}\alpha_2^2 + \tfrac{3}{35}) \\ &\quad + g + l(\alpha_3^2 - \tfrac{1}{3}) + q(\alpha_3^4 - \tfrac{6}{7}\alpha_3^2 + \tfrac{3}{35})] \\ &= Nq(\alpha_1^4 + \alpha_2^4 + \alpha_3^4) + \text{const.} \\ &= -2Nq(\alpha_1^2\alpha_2^2 + \alpha_2^2\alpha_3^2 + \alpha_3^2\alpha_1^2) + \text{const.}, \end{aligned} \tag{7.44}$$

where N is the number of atoms included in the unit volume. Comparing this with (7.6), we have

$$K_1 = -2Nq \qquad \text{(s.c. lattice).} \tag{7.45}$$

Similarly, for the body-centered cubic and face-centered cubic lattices we have

$$K_1 = \tfrac{16}{9}Nq \quad \text{(b.c.c. lattice)}, \tag{7.46}$$

$$K_1 = Nq \quad \text{(f.c.c. lattice)}, \tag{7.47}$$

by assuming only first nearest neighbor interactions.

It should be noted here that the dipolar term of (7.41) does not contribute to the interaction energy E_a, since we have postulated perfect parallelism of the atomic magnetic moments. The reason is that the dipole terms between atom pairs with different bond directions cancel out as long as their distribution maintains cubic symmetry. If, however, the crystal has a lower symmetry than the cubic crystal, as in a hexagonal crystal, the dipole-dipole interaction gives rise to magnetic anisotropy. For instance, the first term of (7.1) comes from this interaction. Generally speaking, the value of l is much larger than that of q, so that it is a general rule that the anisotropy is rather larger for a low symmetry crystal than for a cubic crystal. The reader may recognize this fact by comparing Tables 7.4 and Table 7.5.

Van Vleck[45] pointed out that the dipole-dipole interaction does give rise to cubic anisotropy since the perfect parallelism of the spin system is disturbed by the dipolar interaction itself. If, for example, $l < 0$ and all the spins are parallel, the dipole-dipole interaction gives rise to a large pair energy for $\phi = \pi/2$. In such a case, a more stable configuration of the spin pair will be an antiparallel alignment. Some of the spins will therefore take the antiparallel direction in an equilibrium state; the number doing so may depend on the direction of magnetization and thereby give rise to an anisotropy energy.

This picture is, however, conceivable only on the basis of quantum mechanical considerations. According to Van Vleck's calculation, the cubic anisotropy constants due to dipole-dipole interaction are

$$K_1 = \frac{9Nl^2}{4SMH_m}, \quad \text{(s.c. lattice)} \tag{7.48}$$

$$K_1 = -\frac{2Nl^2}{SMH_m}, \quad \text{(b.c.c. lattice)} \tag{7.49}$$

$$K_1 = -\frac{9Nl^2}{8SMH_m}, \quad \text{(f.c.c. lattice)} \tag{7.50}$$

at $T = 0°\text{K}$, where S is the total spin quantum number, M the atomic magnetic moment, and H_m the molecular field. In the classical picture, K_1 should vanish as expected from the preceding formulas by letting $S \to \infty$.

Since $NMH_m \approx 10^9$ J/m^3 and $Nl \approx 10^7$ J/m^3 (cf. 17.16), the order of magnitude of K_1 due to dipole-dipole interaction is found to be

$$K_1 \approx \frac{(Nl)^2}{NMH_m} \approx \frac{(10^7)^2}{10^9} \approx 10^5 \text{ J/m}^3 \tag{7.51}$$

which is sufficient to explain the magnitude of the observed anisotropy energy.

Judging from the origin of the anisotropy, it would be natural to suppose that the anisotropy constant decreases with an increase in the temperature and disappears at the Curie point. Actually this does occur, and the temperature dependence is more drastic than that of spontaneous magnetization. In order to treat this problem we must calculate the anisotropy energy for a thermally perturbed spin system. Zener[46] treated this problem in a simple way and explained the temperature dependence fairly well. He assumed that the pair energy is given by (7.41) even for the thermally perturbed spin system, since the neighboring spins maintain an approximately parallel alignment up to the Curie point, where, because of the strong exchange interaction, parallel spin clusters prevail in the spin system, as proved by the presence of the tailed part of the anomalous specific heat. Carr[47] applied this method to the calculation of the crystal anisotropy constant for iron, nickel, and cobalt.

Now let $(\alpha_1, \alpha_2, \alpha_3)$ denote the direction cosines of the average magnetization, and $(\beta_1, \beta_2, \beta_3)$ those of the local magnetization. Since we assume local parallelism in the spin system, the anisotropy energy should be given by the average of the local anisotropy energies, so that

$$E_a(T) = K_1(0)\langle \beta_1^2\beta_2^2 + \beta_2^2\beta_3^2 + \beta_3^2\beta_1^2 \rangle, \tag{7.52}$$

where $K_1(0)$ is the anisotropy constant at $T = 0°$K and $\langle\ \rangle$ denotes the averaging operation over-all possible orientations of local magnetization. Using the polar coordinates (θ, ϕ), where θ is the angle between the local spin and the average magnetization, and ϕ the azimuthal angle around the magnetization direction, we express (7.52) as

$$E_a(T) = K_1(0)\frac{\displaystyle\int_0^\pi \left[(1/2\pi)\int_0^{2\pi} (\beta_1^2\beta_2^2 + \beta_2^2\beta_3^2 + \beta_3^2\beta_1^2)\, d\phi\right] n(\theta)\, d\theta}{\displaystyle\int_0^\pi n(\theta)\, d\theta}, \tag{7.53}$$

where $n(\theta)\, d\theta$ is the number of spins which point in the solid angle between θ and $\theta + d\theta$. In this case we can use expression (4.5) for $n(\theta)$, since the distribution of spins should be determined mainly by the exchange interaction. Since

$$\frac{1}{2\pi}\int_0^{2\pi}(\beta_1^2\beta_2^2 + \beta_2^2\beta_3^2 + \beta_3^2\beta_1^2)\,d\phi$$

$$= \tfrac{1}{5}[1 - P_4(\cos\theta)] + P_4(\cos\theta)(\alpha_1^2\alpha_2^2 + \alpha_2^2\alpha_3^2 + \alpha_3^2\alpha_1^2), \quad (7.54)$$

where $P_4(\cos\theta)$ is the fourth order Legendre polynomial, (7.53) becomes

$$E_a(T) = K_1(0)\langle P_4(\cos\theta)\rangle(\alpha_1^2\alpha_2^2 + \alpha_2^2\alpha_3^2 + \alpha_3^2\alpha_1^2), \quad (7.55)$$

where

$$\langle P_4(\cos\theta)\rangle = \frac{\int_0^\pi P_4(\cos\theta)n(\theta)\,d\theta}{\int_0^\pi n(\theta)\,d\theta}, \quad (7.56)$$

which can be expressed in a polynomial series in $1/\alpha = kT/MwI$ (wI = molecular field) as

$$\langle P_4(\cos\theta)\rangle = 1 - \frac{10}{\alpha} + \frac{45}{\alpha^2} - \frac{105}{\alpha^3} + \cdots, \quad (7.57)$$

in the temperature range $T \lesssim 0.8\Theta$. On the other hand, in the same temperature range (4.25) gives

$$I(T) = I(0)\left(\coth\alpha - \frac{1}{\alpha}\right) \approx I(0)\left(1 - \frac{1}{\alpha}\right), \quad (7.58)$$

so that

$$\left[\frac{I(T)}{I(0)}\right]^{10} \approx 1 - \frac{10}{\alpha} + \frac{45}{\alpha^2} - \frac{120}{\alpha^3} + \cdots. \quad (7.59)$$

Comparing (7.59) and (7.57), we have

$$\left[\frac{K_1(T)}{K_1(0)}\right] = \left[\frac{I(T)}{I(0)}\right]^{10}. \quad (7.60)$$

This relation holds well for the temperature dependence of K_1 of iron as shown in Fig. 7.15. Carr also explained the temperature dependence of K_1

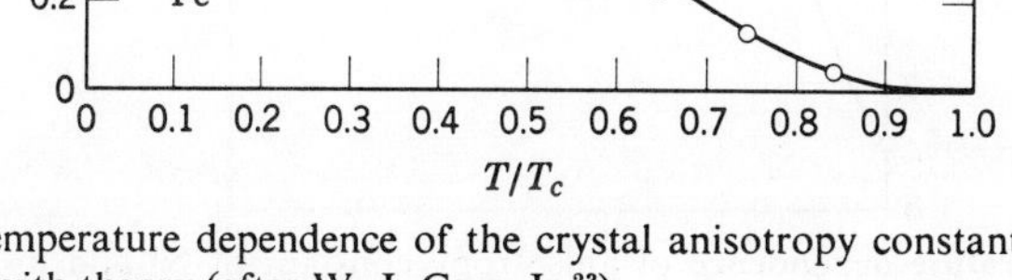

Fig. 7.15. Temperature dependence of the crystal anisotropy constant of iron and its comparison with theory (after W. J. Carr, Jr.[23]).

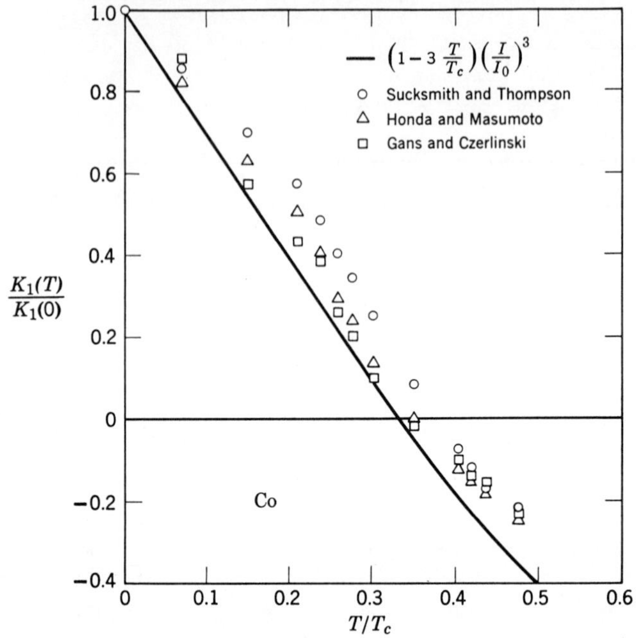

Fig. 7.16. Temperature dependence of the anisotropy constant of Co and its comparison with theory (after W. J. Carr, Jr.[47]).

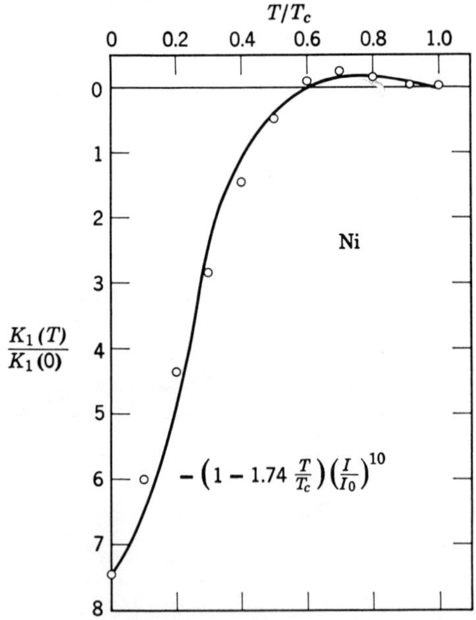

Fig. 7.17. Temperature dependence of anisotropy constant of Ni and its comparison with theory (after W. J. Carr, Jr.[47]).

for nickel and cobalt by taking into consideration the effect of thermal expansion of the crystal lattice (Figs. 7.16 and 17).

In contrast to Zener's theory, Van Vleck[45] obtained a much more gentle temperature dependence of magnetic anisotropy. Keffer[48] examined this point and showed that the Zener theory is valid at least at low temperature, while Van Vleck's theory is valid at high temperature.

7.3 One-Ion Model of Magnetic Anisotropy

As discussed in the preceding section, the orbital state of magnetic ions plays an important role in determining the magnetic anisotropy. In order to discuss this problem, we must first learn how the orbital state of a magnetic ion is influenced by a given crystalline field, and then we can see how the resultant orbital state gives rise to the magnetic anisotropy. We call such a treatment of the problem the "one-ion model." This model has been successful in interpreting the magnetic anisotropy of various antiferromagnetic and ferrimagnetic crystals.[49] The magnetic ions in ionic crystals are isolated by the surrounding anions, so that the situation is clearly suited to treatment by the one-ion model.

Now let us discuss the orbital state of a magnetic ion in a cubic ionic crystal. As discussed in Chapter 5, some of the magnetic ions are surrounded by four anions in a tetrahedral arrangement and some of them are surrounded by six ions in an octahedral array. These surrounding anions give rise to a strong electric field whose distribution has cubic symmetry about the magnetic ions. In free atomic states, every 3*d* electronic state has the same energy; in other words, their energy levels are degenerate. When, however, the atom is placed in a cubic field, the orbital states of 3*d* electrons are split into two groups: one is the triply degenerate $d\epsilon$ orbits the spatial distributions of which are expressed by xy, yz, or zx. The other is the doubly degenerate $d\gamma$ orbits whose distributions are expressed by $2z^2 - x^2 - y^2$ and $x^2 - y^2$. As shown in Fig. 7.18, $d\epsilon$ orbits extend to $\langle 110 \rangle$ directions, while $d\gamma$ orbits extend along the coordinate axes. In octahedral sites, the surrounding anions are found on the three coordinate axes, so that $d\gamma$ orbits, which extend toward the anions, have a much higher energy than $d\epsilon$ orbits because of the electrostatic repulsion between anions and *d* orbits. For tetrahedral sites the situation is just the reverse; that is, since no anions are found on the coordinate axes, $d\gamma$ orbits are more stable than $d\epsilon$ orbits (Fig. 7.19).

Next we discuss how the *d* electrons occupy the 3*d* energy levels. First we assume that the magnetic ion has only one 3*d* electron, as in Ti^{3+}. It will naturally occupy the lowest energy level. Since the three $d\epsilon$ levels

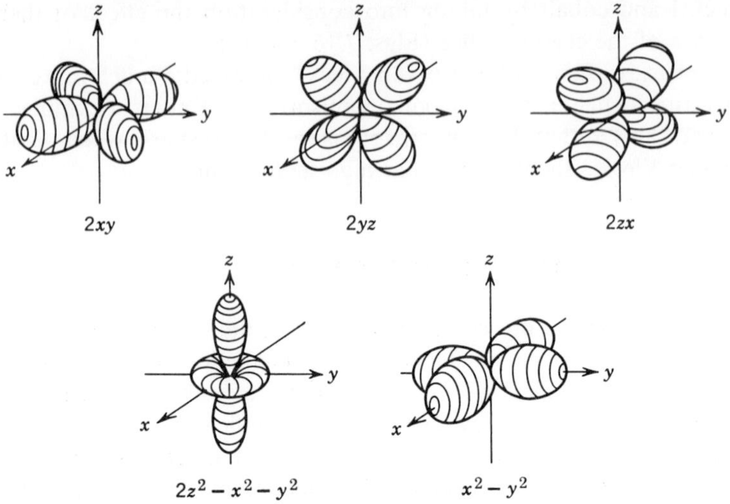

Fig. 7.18. Spatial distribution of $d\epsilon$ and $d\gamma$ orbits.

have the same energy, the lowest orbital state is triply degenerate (triplet). Such an orbital degeneracy plays an important role in determining the magnetic anisotropy, as will be discussed later. Now, if we have additional $3d$ electrons, they should occupy exclusively the plus spin levels (Hund's rule), because the exchange interaction between these $3d$ electrons is much larger than the energy separation between the $d\gamma$ and $d\epsilon$ levels. When the atom has two $3d$ electrons as in Ti^{2+} or V^{3+} ions, they will occupy two of the three plus spin $d\varepsilon$ levels. Since there are three ways of filling up two levels out of three, this ground state is also triply degenerate. In the case of $(3d)^3$, the three electrons occupy the three $d\epsilon$ levels, so that the ground state is not degenerate (singlet). As for $(3d)^4$, three electrons fill up the $d\epsilon$ levels and the remaining one occupies one of the two $d\gamma$ levels, so that the state is doubly degenerate (doublet). In the case of $(3d)^5$, all the electrons occupy plus spin levels, so that the ground state is a singlet. When there are more than five electrons, the first five fill up the plus spin levels and the remaining

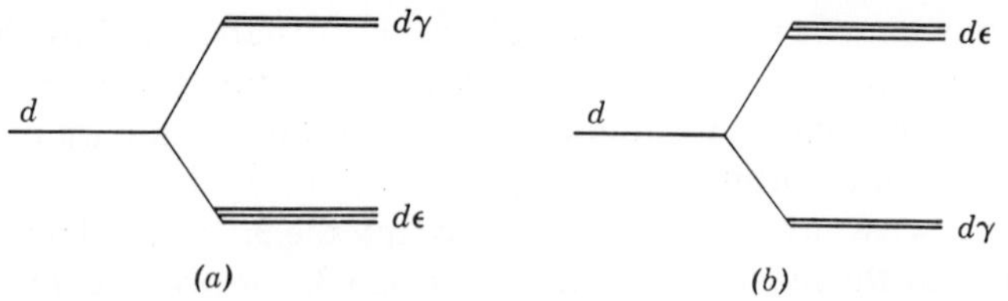

Fig. 7.19. Energy levels of $d\epsilon$ and $d\gamma$ electrons in (a) octahedral and (b) tetrahedral sites.

electrons occupy the minus spin levels in the same way as for the plus spin levels. The ground states and their degeneracies for various kinds of transition metal ions are given in Table 7.6 for octahedral and tetrahedral positions.

Now we discuss the relation between magnetic anisotropy and orbital degeneracy of the ground state. In a triplet, such as Fe^{2+} or Co^{2+} in an

Table 7.6. The Ground State and Degeneracy of Transition Metal Ions

			Octahedral Site		Tetrahedral Site			
Ion	No. of 3*d* Electrons	Ground State	State	Degeneracy	State	Degeneracy	λ, cm^{-1}	ρ, cm^{-1}
Ti^{3+}	1	^2D	$(d\epsilon)^1$	3	$(d\gamma)^1$	2	154	
Ti^{2+}	2	^3F	$(d\epsilon)^2$	3	$(d\gamma)^2$	1		0.24
V^{3+}	2	^3F	$(d\epsilon)^2$	3	$(d\gamma)^2$	1	104	0.24
V^{2+}	3	^4F	$(d\epsilon)^3$	1	$(d\gamma)^2(d\epsilon)^1$	3	55	0.4
Cr^{3+}	3	^4F	$(d\epsilon)^3$	1	$(d\gamma)^2(d\epsilon)^1$	3	87	0.44
Cr^{2+}	4	^5D	$(d\epsilon)^3(d\gamma)^1$	2	$(d\gamma)^2(d\epsilon)^2$	3	57	0.42
Mn^{3+}	4	^5D	$(d\epsilon)^3(d\gamma)^1$	2	$(d\gamma)^2(d\epsilon)^2$	3	85	0.8
Mn^{2+}	5	^6S	$(d\epsilon)^3(d\gamma)^2$	1	$(d\gamma)^2(d\epsilon)^3$	1		
Fe^{3+}	5	^6S	$(d\epsilon)^3(d\gamma)^2$	1	$(d\gamma)^2(d\epsilon)^3$	1		
Fe^{2+}	6	^5D	$(d\epsilon)^4(d\gamma)^2$	3	$(d\gamma)^3(d\epsilon)^3$	2	−100	0.95
Co^{2+}	7	^4F	$(d\epsilon)^5(d\gamma)^2$	3	$(d\gamma)^4(d\epsilon)^3$	1	−180	1.50
Ni^{2+}	8	^3F	$(d\epsilon)^6(d\gamma)^2$	1	$(d\gamma)^4(d\epsilon)^4$	3	−335	5.31
Cu^{2+}	9	^2D	$(d\epsilon)^6(d\gamma)^3$	2	$(d\gamma)^4(d\epsilon)^5$	3	−852	

octahedral site, the orbital configuration can be changed from one to the other without changing the energy, so that this state has an orbital angular momentum. Now let us consider the Fe^{2+} or Co^{2+} ions which occupy the octahedral sites of ferrites. As shown in Fig. 7.20, the octahedral site is surrounded by six neighboring oxygen ions, which give rise to the cubic field at the center. Besides these oxygen ions, we have six next nearest neighbor metal ions which belong to other octahedral sites. These metal ions give rise to a crystalline field whose distribution has trigonal symmetry around the [111] axis. In such a case the triply degenerate $d\epsilon$ level is split into a singlet and a doublet. The former has a wave function which extends along the trigonal ⟨111⟩ axis, while the latter functions spread into the plane perpendicular to the trigonal axis. As a result of the attraction from second neighboring cations, the singlet level should be more stable than the doublet level. As for the Fe^{2+} ion, the sixth electron should occupy the lowest singlet, so that the ground state is non-degenerate (Fig. 7.21*a*). On the other hand, the Co^{2+} ion has seven electrons, so that the last one should occupy the doublet (Fig. 7.21*b*). In such a case the orbit has the freedom to change its state in the plane which is normal to the

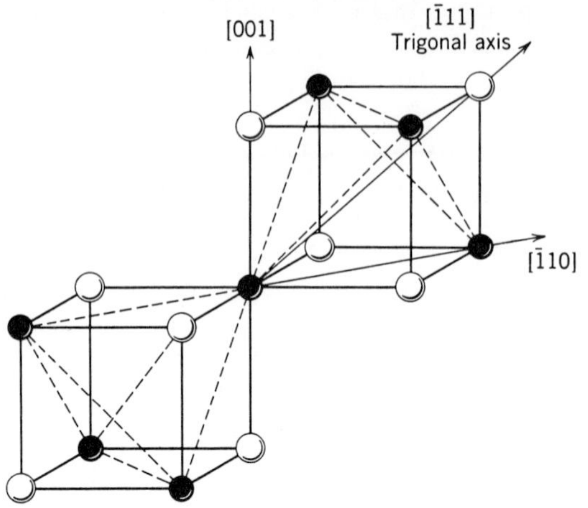

Fig. 7.20. Distribution of surrounding ions about the octahedral site of spinel structure. (Open circles represent oxygen ions, and solid ones represent cations.)

trigonal axis, so that it has an angular momentum parallel to the trigonal axis. Since this angular momentum is fixed in direction, it tends to align the spin magnetic moment parallel to the trigonal axis through the spin-orbit interaction. The energy of this interaction can be expressed as $-\lambda LS\,|\cos\theta|$, where λ is the spin-orbit parameter, L and S are the orbital- and spin-angular momentum and θ is the angle between the magnetization and the trigonal axis. (cf. 3.44). This model was first proposed by Slonczewski,[50] who explained the magnetic annealing effect in Co ferrite by this model (cf. Chapter 17). He also explained the temperature dependence of the anisotropy constant of cobalt-substituted magnetite. In the normal, or non-magnetically annealed, state, Co^{2+} ions should be distributed equally among the four kinds of octahedral sites each of which has its trigonal axis parallel to one of the four $\langle 111\rangle$ directions, so that the cubic

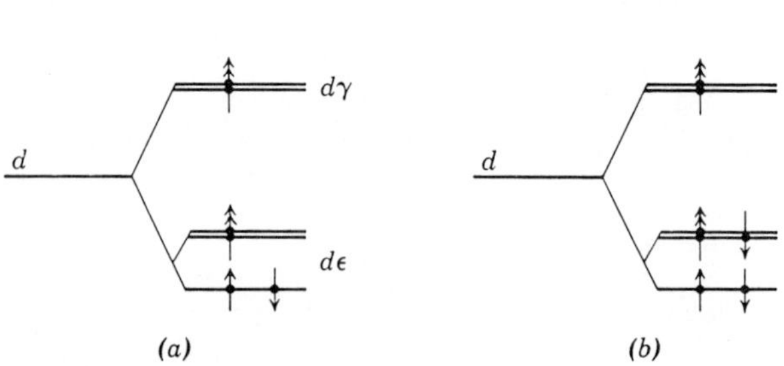

Fig. 7.21. Configuration of *d* electron for (*a*) Fe^{2+} and (*b*) Co^{2+} ions which are in an octahedral site.

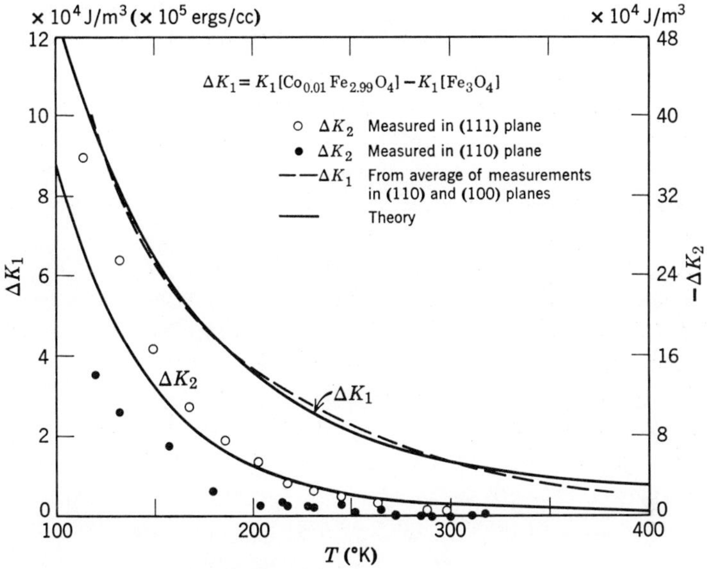

Fig. 7.22. Temperature dependence of the difference in anisotropy between $Co_{0.01}Fe_{2.99}O_4$ and Fe_3O_4 (after Slonczewski[50]).

anisotropy can be obtained by averaging the anisotropy energy

$$-\lambda LS\,|\cos\theta_i|$$

over four directions of trigonal axes. Figure 7.22 shows the temperature dependence of the additional anisotropy due to Co^{2+} ions, and we can see that the theory explains the experiment fairly well.

If the ground orbital state is non-degenerate, we cannot expect any orbital angular momentum so long as the atom stays in the ground state; in other words, the orbital angular momentum is quenched by the crystalline field. In such a case we cannot expect an anisotropy as large as that of the Co^{2+} ion in an octahedral site. There are, however, various sources which result in a fairly small magnetic anisotropy, which is nevertheless sufficiently large to account for the observed values. Yosida and Tachiki[51] calculated the various types of anisotropy and applied their results to the Mn, Fe, and Ni ferrites, which contain Mn^{2+}, Fe^{3+}, Fe^{2+}, and Ni^{2+} ions. First they found that magnetic dipole-dipole interaction (cf. Section 7.2) is too weak to account for the observed magnitude of anisotropy. The main source of anisotropy is thought to be the distortion of the 3*d* shell from spherical symmetry. In a distorted 3*d* shell the intra-atomic dipole-dipole interaction between the spin magnetic moments may depend on the direction of magnetization; this is similar to the dependence of the magnetostatic energy on the direction of magnetization in a fine elongated single domain particle (cf. 17.24), and it gives rise to the anisotropy energy. This mechanism was first mentioned by Pryce.[52] Table 7.6 contains the coefficient of intra-atomic dipole-dipole interaction, ρ of various free ions obtained by him from spectroscopic data. The anisotropy is also induced through spin-orbit interaction. That is, some amount of orbital angular momentum can be induced by spin-orbit interaction by exciting additional orbital states. In a distorted $3d$ shell this excitation is dependent on the direction of magnetization, giving rise to anisotropy. The spin-orbit parameter, λ, of various free ions are also given in Table 7.6. Yosida and Tachiki showed that the anisotropy due to these mechanisms should vanish for $S = \frac{1}{2}$, 1, and $\frac{3}{2}$, where S is the total spin angular momentum, and that anisotropy of this type cannot be expected for Ni^{2+} and Co^{2+} ions. Accordingly, the main source of the anisotropy of Mn, Fe, and Ni ferrites is considered to be the Mn^{2+}, Fe^{3+}, and Fe^{2+} ions. They made the exact calculation for the Fe^{2+} ion in an octahedral site and obtained the anisotropy energy per Fe^{2+} ion:

$$[\Delta K_1]^{B}_{Fe^{2+}} = -3.56 \times 10^{-2}\,\text{cm}^{-1}$$
$$= -7.10 \times 10^{-25}\,\text{J} \qquad (7.61)$$

at 0°K. Since the Ni^{2+} ion has no effect on the magnetic anisotropy in Ni ferrite, the difference in anisotropy energy between Fe and Ni ferrites must be explained by the anisotropy due to Fe^{2+} ions. Actually the value of (7.61) explains this difference well. Yosida and Tachiki also calculated the temperature dependence of the anisotropy constant for Mn ferrites and fitted the theory with experiment, as show in Fig. 7.23, by adjusting two

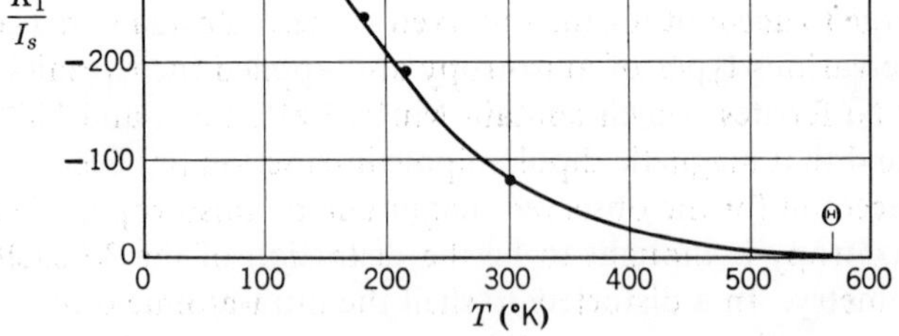

Fig. 7.23. Temperature dependence of K_1/I_s of Mn ferrite and its comparison with theory (a solid curve). Experimental points are due to Dillon, Geschwind and Jaccarino. (After Yosida and Tachiki[51].)

parameters. A similar theory was developed by W. P. Wolf,[53] who emphasized the importance of the local anisotropy of the individual ion because it may give rise to cubic anisotropy through a mechanism similar to that of torque reversal as mentioned in Section 7.1.

Problems

7.1. What kind of anisotropy can be expected in a (111) plane of cubic crystals? Assume $K_2 = 0$.

7.2. Calculate the anisotropy field for a cobalt crystal whose magnetization is close to the hexagonal axis. Use the value $I_s = 1.79$ Wb/m^2, $K_{u1} = 4.1 \times 10^5$ J/m^3.

7.3. Assuming that $q = 3 \times 10^{-25}$ J, calculate the cubic anisotropy constant of the simple magnetic metal which forms a body-centered cubic lattice with lattice constant 3 Å.

7.4. If the temperature dependence of the anisotropy constant is the same as that of the spontaneous magnetization, what should be the functional form of the anisotropy energy?

7.5. Discuss the reasons why a Co^{2+} ion can give rise to the large anisotropy energy when it is in an octahedral site.

References

7.1. C. D. Graham, Jr.: *Phys. Rev.* **112,** 1117 (1958).
7.2. H. Sato and B. S. Chandrasekhar: *Phys. Chem. Solids* **1,** 228 (1957).
7.3. R. F. Penoyer: G.25, p. 365.
7.4. R. F. Pearson: *J. Phys. Radium* **20,** 409 (1959).
7.5. E. A. Nesbitt, H. J. Williams, and R. M. Bozorth: *J. Appl. Phys.* **25,** 1014 (1954); E. A. Nesbitt and H. J. Williams: *ibid.*, **26,** 1217 (1955).
7.6. H. J. Williams, R. D. Heidenreich, and E. A. Nesbitt: *J. Appl. Phys.* **27,** 85 (1956)
7.7. S. Chikazumi: *J. Phys. Soc. Japan* **11,** 718 (1956).
7.8. K. Honda and S. Kaya: *Sci. Repts. Tohoku Imp. Univ.* **15,** 721 (1926).
7.9. L. R. Bickford, Jr.: *Phys. Rev.* **78,** 449 (1950).
7.10. W. Sucksmith and J. E. Thompson: *Proc. Roy. Soc. (London)* **A225,** 362 (1954).
7.11. R. M. Bozorth: *Phys. Rev.* **96,** 311 (1954).
7.12. J. J. Went, G. W. Rathenau, E. W. Gorter, and G. W. van Oosterhout: *Philips Tech. Rev.* **13,** 194 (1952).
7.13. L. R. Bickford, Jr.; *J. Phys. Soc. Japan Suppl.* B-I, 272 (1962).
7.14. C. Guillaud: Thesis, Strasbourg, pp. 1–129 (1943); *Cahiers Phys.* **3,** 15 (1943).
7.15. J. Smit and H. P. J. Wijn: G.32, p. 204.
7.16. R. F. Penoyer and M. W. Shafer: *J. Appl. Phys.* **30,** 315S (1959).
7.17. N. Miyata: *J. Phys. Soc. Japan* **16,** 1291 (1961).
7.18. Z. Funatogawa, N. Miyata, and S. Usami: *J. Phys. Soc. Japan* **14,** 1583 (1959).
7.19. W. Palmer: *J. Appl. Phys.* **33,** 1201S (1962).
7.20. U. Enz: *Physica* **24,** 609 (1958).

7.21. R. F. Pearson: *J. Appl. Phys.* **31,** 160S (1960).
7.22. K. Ohta: *J. Phys. Soc. Japan* **18,** 684 (1963).
7.23. J. K. Galt, W. A. Yager, J. P. Remeika, and F. R. Merritt: *Phys. Rev.* **81,** 470 (1951).
7.24. J. Smit, F. K. Lotgering, and R. P. Stapels: *J. Phys. Soc. Japan* **17,** Suppl. B-I, 268 (1962).
7.25. R. F. Pearson: *Proc. Phys. Soc.* **74,** 505 (1959).
7.26. L. R. Bickford, Jr., J. M. Brownlow, and R. F. Penoyer: *Proc. I.E.E.* **104*B*,** 238 (1956).
7.27. R. M. Bozorth, E. F. Tilden, and A. J. Williams: *Phys. Rev.* **99,** 1788 (1955).
7.28. H. Shenker: Thesis, Univ. of Maryland, 1955; Private communication.
7.29. G. Elbinger: *Naturwissenschaften* **48,** 498 (1961); *Z. Physik.* **14,** 273 (1962).
7.30. J. K. Galt, W. T. Matthias, and J. P. Remeika: *Phys. Rev.* **79,** 391 (1950).
7.31. K. Ohta: *Bull. Kobayasi Inst. Phys. Res.* **10,** 149 (1960).
7.32. T. Okamura and Y. Kojima: *Phys. Rev.* **86,** 1040 (1952).
7.33. N. Miyata, T. Kobayashi, K. Ito, and Z. Funatogawa: published at Annual Meeting of Phys. Soc. Japan, 1963.
7.34. G. T. Rado, V. J. Folen, and W. H. Emerson: *Proc. I.E.E.* **104*B*,** 198 (1956).
7.35. R. F. Pearson: G.31, p. 345.
7.36. Y. Syono and Y. Ishikawa: *J. Phys. Soc. Japan* **18,** 1230 (1963).
7.37. L. P. Tarasov: *Phys. Rev.* **56,** 1231 (1939).
7.38. J. W. Shih: *Phys. Rev.* **46,** 139 (1934).
7.39. L. P. Tarasov: *Phys. Rev.* **56.** 1245 (1939).
7.40. R. M. Bozorth and J. G. Walker: *Phys. Rev.* **89,** 624 (1953).
7.41. J. W. Shih: *Phys. Rev.* **50,** 376 (1936).
7.42. L. W. McKeehan: *Phys. Rev.* **51,** 136 (1937).
7.43. R. M. Bozorth: G.10, p. 573.
7.44. H. Shenker: *Phys. Rev.* **107,** 1246 (1957).
7.45. J. H. Van Vleck: *Phys. Rev.* **52,** 1178 (1937).
7.46. C. Zener: *Phys. Rev.* **96,** 1335 (1954).
7.47. W. J. Carr, Jr.: *J. Appl. Phys.* **29,** 436 (1958).
7.48. F. Keffer: *Phys. Rev.* **100,** 1692 (1955).
7.49. T. Nagamiya, K. Yosida, and R. Kubo: *Adv. in Phys.* **4,** 1 (1955).
7.50. J. C. Slonczewski: *Phys. Rev.* **110,** 1341 (1958).
7.51. K. Yosida and M. Tachiki: *Prog. Theor. Phys.* **17,** 331 (1957).
7.52. M. H. L. Pryce: *Phys. Rev.* **80,** 1107 (1950).
7.53. W. P. Wolf: *Phys. Rev.* **108,** 1152 (1957).

8

Magnetostriction

8.1 Phenomenology of Magnetostriction

Magnetostriction is that phenomenon wherein the shape of a ferromagnetic specimen changes during the process of magnetization. The deformation $\delta l/l$ due to magnetostriction is as small as $10^{-5} \sim 10^{-6}$. Such a deformation can be conveniently measured by means of a strain gauge technique the details of which are described later.

The strain due to magnetostriction changes with increase of magnetic field intensity as shown in Fig. 8.1 and finally reaches the saturation value λ. The reason is that the crystal lattice inside each domain is spontaneously deformed in the direction of domain magnetization and its strain axis rotates with a rotation of the domain magnetization, thus resulting in a deformation of the specimen as a whole (Fig. 8.2). In order to calculate

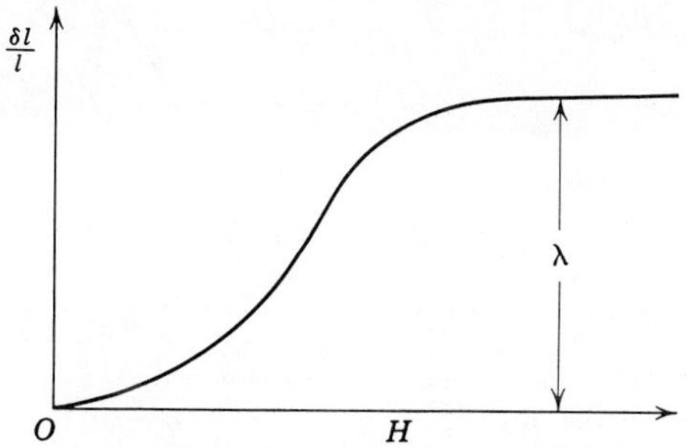

Fig. 8.1. Magnetostriction as a function of the field intensity.

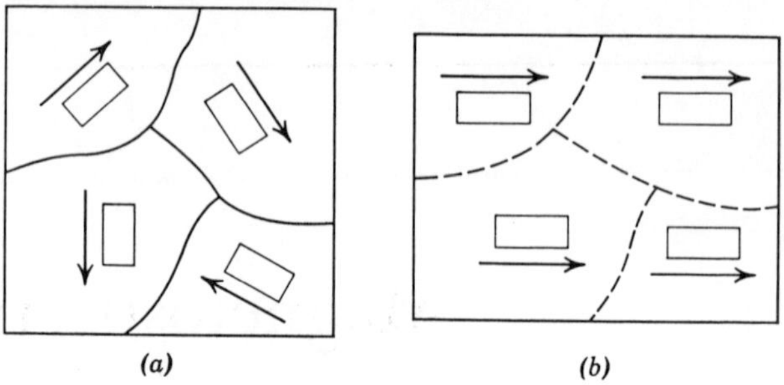

Fig. 8.2. Rotation of domain magnetization and accompanying rotation of the axis of spontaneous strain.

the dependence of the strain on the direction of magnetization we consider a ferromagnetic sphere, which is a real sphere only when it is demagnetized but is elongated by $\delta l/l = e$ in the direction of magnetization after it is magnetized (Fig. 8.3). The elongation of a diameter of the sphere along a direction which makes an angle ϕ with the direction of magnetization is given by

$$\frac{\delta l}{l} = e \cos^2 \phi. \tag{8.1}$$

When the domain magnetizations are distributed at random in a demagnetized state as shown in Fig. 8.2*a*, the average deformation is given by the average of (8.1); thus

$$\left(\frac{\delta l}{l}\right)_{\text{demag}} = \int_0^{\pi/2} e \cos^2 \phi \sin \phi \, d\phi = \frac{e}{3}. \tag{8.2}$$

Since, in the saturated state, as shown in Fig. 8.2*b*,

$$\left(\frac{\delta l}{l}\right)_{\text{sat}} = e, \tag{8.3}$$

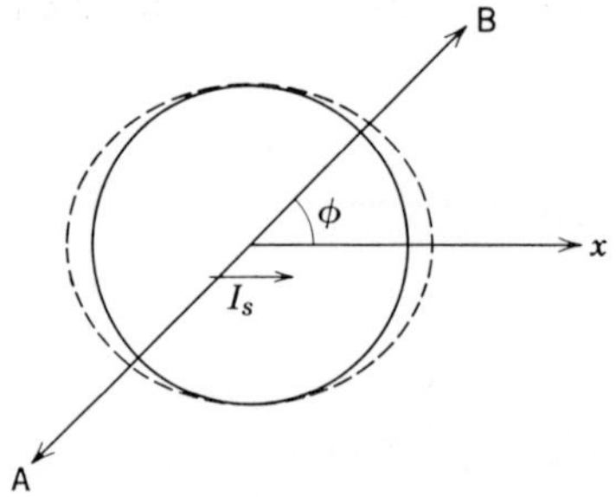

Fig. 8.3. Observation of elongation in a direction which makes an angle ϕ with the axis of spontaneous strain.

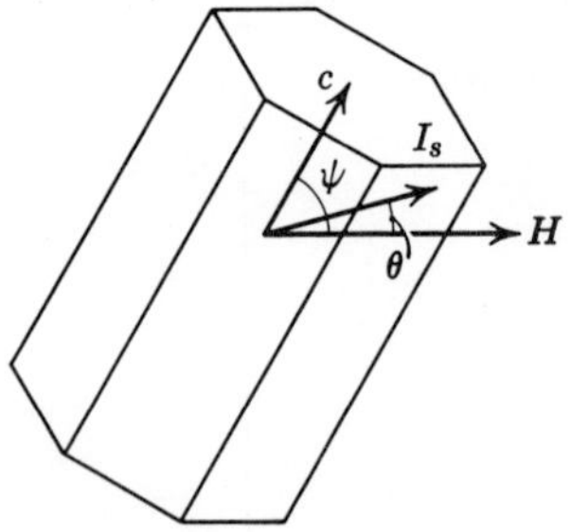

Fig. 8.4. Magnetization of a uniaxial crystal.

the saturation magnetostriction is given by

$$\lambda = \left(\frac{\delta l}{l}\right)_{\text{sat}} - \left(\frac{\delta l}{l}\right)_{\text{demag}} = \frac{2}{3} e. \tag{8.4}$$

Thus the spontaneous strain in a domain can be expressed in terms of λ as

$$e = \tfrac{3}{2}\lambda. \tag{8.5}$$

We assumed above that the magnitude of the spontaneous strain is independent of the crystallographic direction of magnetization. In this case we have to do with an isotropic magnetostriction.

Next we discuss how the magnetostrictive elongation changes as a function of intensity of magnetization. First we consider a ferromagnetic substance, such as cobalt, with a uniaxial anisotropy. If the magnetic field H makes an angle ψ with the easy axis (Fig. 8.4), the magnetization takes place by the displacement of 180° walls until the magnetization reaches the value $I_s \cos \psi$; during this process no magnetostriction can be observed. Then the domain magnetization rotates toward the direction of the applied field; during this process the elongation changes by the amount

$$\Delta\left(\frac{\delta l}{l}\right) = \frac{3}{2} \lambda(1 - \cos^2 \psi). \tag{8.6}$$

If H is parallel to the easy axis, that is, if $\psi = 0$, (8.6) gives $\Delta(\delta l/l) = 0$; in other words, there is no change in the length of the specimen from the demagnetized state to saturation. On the other hand, if H is perpendicular to the easy axis, that is, if $\psi = \pi/2$, (8.6) gives $\Delta(\delta l/l) = \frac{3}{2}\lambda$. Since, in this case magnetization takes place entirely by rotation, we put $I = I_s \cos \phi$ in (8.1); thus

$$\frac{\delta l}{l} = \frac{3}{2} \lambda \left(\frac{I}{I_s}\right)^2. \tag{8.7}$$

If ψ has some value between 0 and $\pi/2$, the magnetizations at which the displacement of 180° walls finish and the amount of change in the elongation are given by

$$\psi = 30^\circ \qquad I = \frac{\sqrt{3}}{2} I_s, \quad \Delta\left(\frac{\delta l}{l}\right) = \frac{1}{4}\left(\frac{3}{2}\lambda\right), \tag{8.8}$$

$$\psi = 45^\circ \qquad I = \frac{1}{\sqrt{2}} I_s, \quad \Delta\left(\frac{\delta l}{l}\right) = \frac{1}{2}\left(\frac{3}{2}\lambda\right), \tag{8.9}$$

$$\psi = 60^\circ \qquad I = \frac{1}{2} I_s, \quad \Delta\left(\frac{\delta l}{l}\right) = \frac{3}{4}\left(\frac{3}{2}\lambda\right). \tag{8.10}$$

The change in $\delta l/l$ is shown graphically in Fig. 8.5 as a function of magnetization for various values of ψ. For a polycrystal, if we assume that all

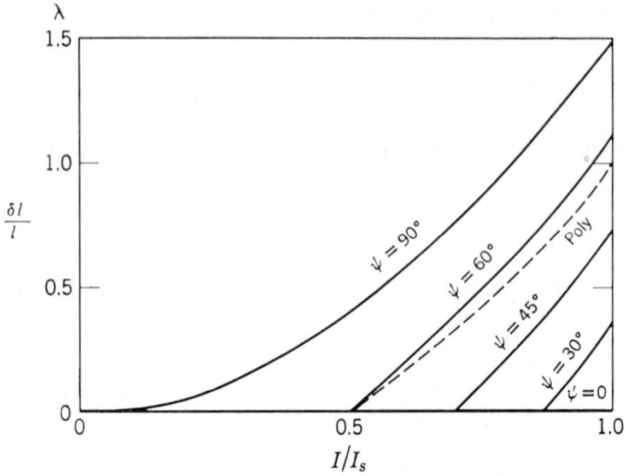

Fig. 8.5. Magnetostriction of a uniaxial ferromagnetic crystal as a function of the intensity of magnetization.

wall displacements are finished before the onset of rotation magnetization, we have, simply by averaging the above values,

$$\psi = \bar{\psi}, \quad I = \frac{I_s}{2}, \quad \Delta\left(\frac{\delta l}{l}\right) = \lambda. \tag{8.11}$$

The $\delta l/l$ vs I relation for this crystal is shown by the dotted curve in Fig. 8.5.

For a cubic crystal with $K_1 > 0$, the magnetization is along one of the $\langle 100 \rangle$ directions in each domain, so that the average elongation of the demagnetized specimen is given by $(\delta l/l)_{\text{demag}} = \lambda/2$, regardless of the direction of observation. If this crystal is magnetized to its saturation parallel to [100], $(\delta l/l)_{\text{sat}} = \frac{3}{2}\lambda$, so that

$$\Delta\left(\frac{\delta l}{l}\right) = \left(\frac{3}{2}\lambda\right) - \frac{\lambda}{2} = \lambda. \tag{8.12}$$

In this example, the entire magnetization takes place by the displacement of walls, among which only 90° walls are effective in giving rise to the elongation. Thus the $\delta l/l$ vs I/I_s curve depends on the ease of displacement of 90° walls relative to that of 180° walls. If 180° walls are very easily displaced, I is expected to increase to $I_s/3$ without changing the length of the specimen until the elongation due to 90° walls begins. Thus

$$\begin{aligned} \frac{\delta l}{l} &= 0 && \text{for } \frac{I}{I_s} \le \frac{1}{3}, \\ \frac{\delta l}{l} &= \frac{3}{2}\lambda\left(\frac{I}{I_s} - \frac{1}{3}\right) && \text{for } \frac{I}{I_s} \ge \frac{1}{3}. \end{aligned} \tag{8.13}$$

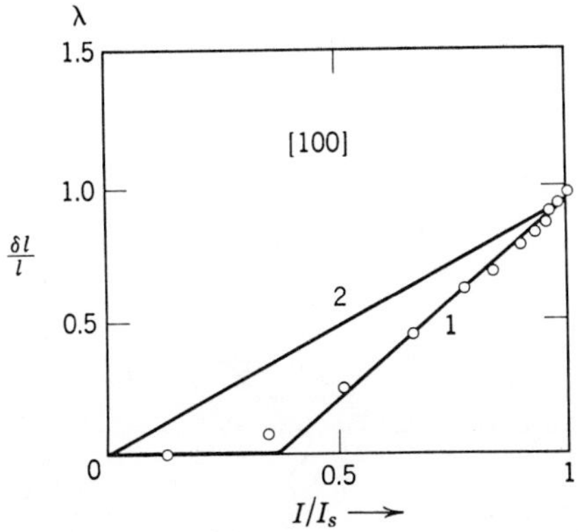

Fig. 8.6. Magnetostriction of an iron single crystal ⟨100⟩ bar as a function of the intensity of magnetization. The lines are drawn by assuming (1) preferential occurrence of 180° wall displacements before displacement of 90° walls; (2) simultaneous displacement of 180° and of 90° walls. Experimental points according to W. L. Webster[1].

On the other hand, if the displacements of 90° and 180° walls take place at the same time, the elongation should be given by

$$\frac{\delta l}{l} = \lambda \frac{I}{I_s}. \tag{8.14}$$

Both cases are shown in Fig. 8.6. The experimental points agree fairly well with graph 1 which represents (8.13).

If the same crystal is magnetized in the [111] direction, the domains are first reduced to [100], [010], and [001] domains by wall displacement. At the end of this stage $I = I_s/\sqrt{3} = 0.577 I_s$. Then domain magnetizations will rotate toward the direction of H; during this process $I = I_s \cos\theta$ and also $\delta l/l = (3/2)\lambda(\cos^2\theta - 1/3)$, where θ is the angle between I_s and H, so that

$$\begin{aligned} \frac{\delta l}{l} &= 0 && \text{for } \frac{I}{I_s} \leq \frac{1}{\sqrt{3}}, \\ \frac{\delta l}{l} &= \frac{3}{2}\lambda\left[\left(\frac{I}{I_s}\right)^2 - \frac{1}{3}\right] && \text{for } \frac{I}{I_s} \geq \frac{1}{\sqrt{3}}. \end{aligned} \tag{8.15}$$

This relation is shown graphically in Fig. 8.7. The experiment for iron reveals similar behavior, but the sign of the elongation is negative, just opposite that of [100] magnetization (cf. Fig. 8.8). Thus the sign of λ as well as its value depends on the crystallographic direction of magnetization. We call this anisotropic magnetostriction. For [100] magnetization the

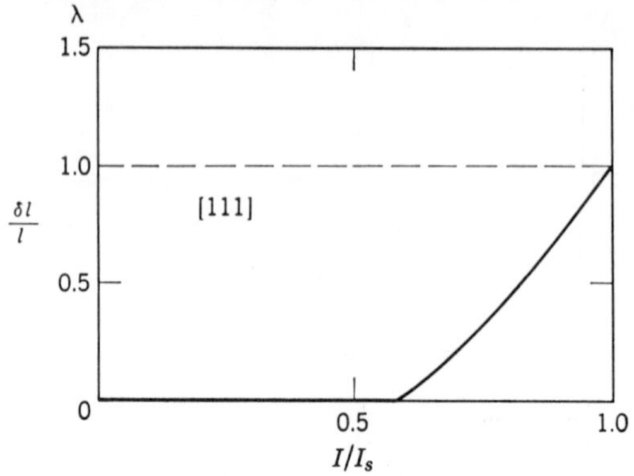

Fig. 8.7. Magnetostriction expected for ⟨111⟩ magnetization under the assumption of isotropic magnetostriction.

effect of anisotropic magnetostriction was not observed because I_s is always parallel to one of the ⟨100⟩ directions throughout the entire magnetization process.

Figure 8.9 shows the curve for [110] magnetization, where a slight positive elongation due to 90° wall displacement occurs in the early stages of magnetization, while a fairly large contraction is observed later during the rotation magnetization process.

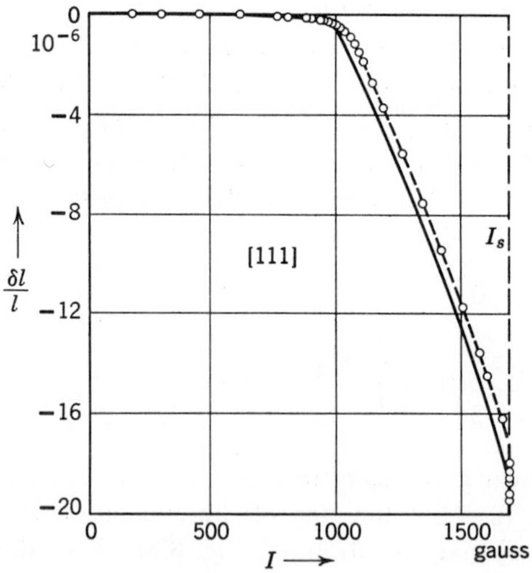

Fig. 8.8. Experimental magnetostriction data for iron for ⟨111⟩ magnetization (after Becker and Döring,[G.3] Kaya and Takaki[2]).

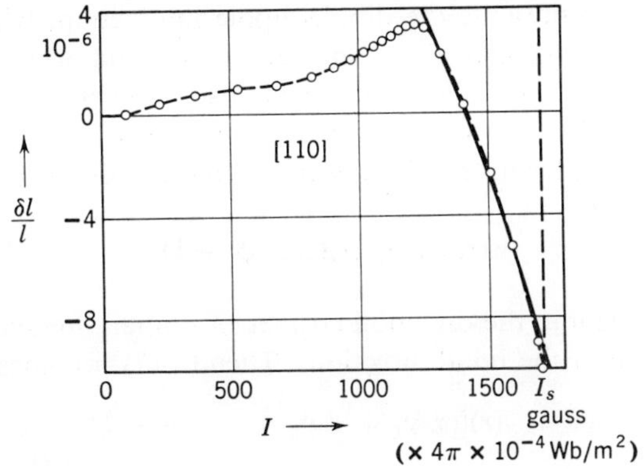

Fig. 8.9. Magnetostriction of iron for ⟨110⟩ magnetization (after Becker and Döring,[G.3] Kaya and Takaki[2]).

In order to discover the origin of such an anisotropic magnetostriction we must understand the actual mechanism of the generation of magnetostriction.

8.2 Mechanism of Magnetostriction

Magnetostriction originates in the interaction between the atomic magnetic moments, as in magnetic anisotropy. We discuss the origin of magnetostriction along the lines of Néel's[3] theory, which was developed in his paper on magnetic annealing and surface anisotropy. When the distance between the atomic magnetic moments is variable, the interaction energy (7.39) is expressed as

$$w(r, \cos\phi) = g(r) + l(r)(\cos^2\phi - \tfrac{1}{3}) + q(r)(\cos^4\phi - \tfrac{6}{7}\cos^2\phi + \tfrac{3}{35}) + \cdots, \quad (8.16)$$

where r is the interatomic distance (Fig. 8.10).

If the interaction energy is a function of r, the crystal lattice must be deformed upon the generation of a ferromagnetic moment, because such an interaction tends to change the bond length in a different way depending on the bond direction. The first term, $g(r)$, is the exchange interaction term; it is independent of the direction of magnetization. Thus the crystal deformation caused by the first term does not contribute to the usual magnetostriction, but it does play an important role in the volume magnetostriction, which is discussed in Section 8.3.

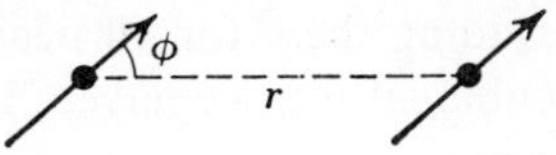

Fig. 8.10. Magnetic dipole pair.

The second term represents the dipole-dipole interaction, which depends on the direction of magnetization, and can be the main origin of the usual magnetostriction. The following terms also contribute to the usual magnetostriction, but normally their contributions are small compared to those of the second term. Neglecting these higher order terms, we express the pair energy as

$$w(r, \phi) = l(r)(\cos^2 \phi - \tfrac{1}{3}). \tag{8.17}$$

Let $(\alpha_1, \alpha_2, \alpha_3)$ denote the direction cosines of domain magnetization and $(\gamma_1, \gamma_2, \gamma_3)$ those of the bond direction. Then (8.17) becomes

$$w = l(r)[(\alpha_1\gamma_1 + \alpha_2\gamma_2 + \alpha_3\gamma_3)^2 - \tfrac{1}{3}]. \tag{8.18}$$

Now let us consider a deformed simple cubic lattice whose strain tensor components are given by e_{xx}, e_{yy}, e_{zz}, e_{xy}, e_{yz}, and e_{zx}. Under the crystal strain each pair changes its bond direction as well as its bond length. For instance, a spin pair with its bond direction parallel to the x axis has an energy in the unstrained state given by (8.18) with $\gamma_1 = 1$, $\gamma_2 = \gamma_3 = 0$; that is,

$$w_x = l(r_0)(\alpha_1^2 - \tfrac{1}{3}), \tag{8.19}$$

whereas, if the crystal strains, its bond length r_0 will be changed to $r_0(1 + e_{xx})$ and the direction cosines of the bond direction will be changed to $(\gamma_1 = 1, \gamma_2 = \tfrac{1}{2}e_{xy}, \gamma_3 = \tfrac{1}{2}e_{zx})$. Then the pair energy (8.18) will be changed by the amount

$$\Delta w_x = \left(\frac{\partial l}{\partial r}\right) r_0 e_{xx}(\alpha_1^2 - \tfrac{1}{3}) + l\alpha_1\alpha_2 e_{xy} + l\alpha_3\alpha_1 e_{zx}. \tag{8.20}$$

Similarly, for y and z pairs,

$$\Delta w_y = \left(\frac{\partial l}{\partial r}\right) r_0 e_{yy}(\alpha_2^2 - \tfrac{1}{3}) + l\alpha_2\alpha_3 e_{yz} + l\alpha_1\alpha_2 e_{xy}, \tag{8.21}$$

$$\Delta w_z = \left(\frac{\partial l}{\partial r}\right) r_0 e_{zz}(\alpha_3^2 - \tfrac{1}{3}) + l\alpha_3\alpha_1 e_{zx} + l\alpha_2\alpha_3 e_{yz}. \tag{8.22}$$

Adding these for all nearest neighbor pairs in a unit volume of a simple cubic lattice, we have

$$E_{\text{magel}} = B_1[e_{xx}(\alpha_1^2 - \tfrac{1}{3}) + e_{yy}(\alpha_2^2 - \tfrac{1}{3}) + e_{zz}(\alpha_3^2 - \tfrac{1}{3})] + B_2(e_{xy}\alpha_1\alpha_2 + e_{yz}\alpha_2\alpha_3 + e_{zx}\alpha_3\alpha_1), \tag{8.23}$$

where

$$B_1 = N\left(\frac{\partial l}{\partial r}\right) r_0, \qquad B_2 = 2Nl. \tag{8.24}$$

The energy thus expressed in terms of lattice strain and the direction of domain magnetization is called the magnetoelastic energy. Similar calculations for the body-centered cubic and face-centered lattices yield the same expression (8.23) with

$$B_1 = \tfrac{8}{3}Nl, \qquad B_2 = \frac{8}{9}N\left[l + \left(\frac{\partial l}{\partial r}\right)r_0\right], \tag{8.25}$$

for the body-centered cubic lattice and

$$B_1 = \tfrac{1}{2}N\left[6l + \left(\frac{\partial l}{\partial r}\right)r_0\right], \qquad B_2 = N\left[2l + \left(\frac{\partial l}{\partial r}\right)r_0\right]. \tag{8.26}$$

for the face-centered cubic lattice.

Since the magnetoelastic energy (8.23) is a linear function with respect to $e_{xx}, e_{yy}, \ldots, e_{zx}$, the crystal will be deformed without limit unless it is counterbalanced by the elastic energy which, for a cubic crystal, is given by

$$E_{\text{el}} = \tfrac{1}{2}c_{11}(e_{xx}^2 + e_{yy}^2 + e_{zz}^2) + \tfrac{1}{2}c_{44}(e_{xy}^2 + e_{yz}^2 + e_{zx}^2) + c_{12}(e_{yy}e_{zz} + e_{yy}e_{xx} + e_{zz}e_{xx}), \tag{8.27}$$

where c_{11}, c_{44}, and c_{12} are the elastic moduli. Since the elastic energy is a quadratic function of the strain of the crystal, it increases rapidly with increasing strain, so that equilibrium is attained at some finite strain. The condition for this is to minimize the total energy

$$E = E_{\text{magel}} + E_{\text{el}}. \tag{8.28}$$

That is,

$$\begin{aligned}
\frac{\partial E}{\partial e_{xx}} &= B_1(\alpha_1^2 - \tfrac{1}{3}) + c_{11}e_{xx} + c_{12}(e_{yy} + e_{zz}) = 0,\\
\frac{\partial E}{\partial e_{yy}} &= B_1(\alpha_2^2 - \tfrac{1}{3}) + c_{11}e_{yy} + c_{12}(e_{zz} + e_{xx}) = 0,\\
\frac{\partial E}{\partial e_{zz}} &= B_1(\alpha_3^2 - \tfrac{1}{3}) + c_{11}e_{zz} + c_{12}(e_{xx} + e_{yy}) = 0,\\
\frac{\partial E}{\partial e_{xy}} &= B_2\alpha_1\alpha_2 + c_{44}e_{xy} = 0,\\
\frac{\partial E}{\partial e_{yz}} &= B_2\alpha_2\alpha_3 + c_{44}e_{yz} = 0,\\
\frac{\partial E}{\partial e_{zx}} &= B_2\alpha_3\alpha_1 + c_{44}e_{zx} = 0.
\end{aligned} \tag{8.29}$$

Solving these equations, we have the equilibrium strain

$$
\begin{aligned}
e_{xx} &= \frac{-B_1}{c_{11} - c_{12}} (\alpha_1^2 - \tfrac{1}{3}), \\
e_{yy} &= \frac{-B_1}{c_{11} - c_{12}} (\alpha_2^2 - \tfrac{1}{3}), \\
e_{zz} &= \frac{-B_1}{c_{11} - c_{12}} (\alpha_3^2 - \tfrac{1}{3}), \\
e_{xy} &= \frac{-B_2}{c_{44}} \alpha_1\alpha_2, \\
e_{yz} &= \frac{-B_2}{c_{44}} \alpha_2\alpha_3, \\
e_{zx} &= \frac{-B_2}{c_{44}} \alpha_3\alpha_1.
\end{aligned} \tag{8.30}
$$

The elongation observed in the direction of $(\beta_1, \beta_2, \beta_3)$, which is given by

$$\frac{\delta l}{l} = e_{xx}\beta_1^2 + e_{yy}\beta_2^2 + e_{zz}\beta_3^2 + e_{xy}\beta_1\beta_2 + e_{yz}\beta_2\beta_3 + e_{zx}\beta_3\beta_1, \tag{8.31}$$

becomes

$$
\begin{aligned}
\frac{\delta l}{l} = \frac{-B_1}{c_{11} - c_{12}} &(\alpha_1^2\beta_1^2 + \alpha_2^2\beta_2^2 + \alpha_3^2\beta_3^2 - \tfrac{1}{3}) \\
&- \frac{B_2}{c_{44}} (\alpha_1\alpha_2\beta_1\beta_2 + \alpha_2\alpha_3\beta_2\beta_3 + \alpha_3\alpha_1\beta_3\beta_1).
\end{aligned} \tag{8.32}
$$

If the domain magnetization is along [100], the elongation in the same direction is given by putting $\alpha_1 = \beta_1 = 1$, $\alpha_2 = \alpha_3 = \beta_2 = \beta_3 = 0$ in (8.32); thus

$$\lambda_{100} = -\frac{2}{3} \frac{B_1}{c_{11} - c_{12}}. \tag{8.33}$$

Similarly, when the magnetization is along [111], the elongation is calculated to be

$$\lambda_{111} = -\frac{1}{3} \frac{B_2}{c_{44}}, \tag{8.34}$$

by putting $\alpha_i = \beta_i = 1/\sqrt{3}$ ($i = 1$, 2, and 3) in (8.32). By using λ_{100} and λ_{111}, (8.32) can be expressed as

$$
\begin{aligned}
\frac{\delta l}{l} = \tfrac{3}{2}\lambda_{100}&(\alpha_1^2\beta_1^2 + \alpha_2^2\beta_2^2 + \alpha_3^2\beta_3^2 - \tfrac{1}{3}) \\
&+ 3\lambda_{111}(\alpha_1\alpha_2\beta_1\beta_2 + \alpha_2\alpha_3\beta_2\beta_3 + \alpha_3\alpha_1\beta_3\beta_1).
\end{aligned} \tag{8.35}
$$

Thus the magnetostriction of a cubic crystal may be expressed in terms of λ_{100} and λ_{111}. The elongation in [110] is not independent of λ_{100} and λ_{111} but is related to them by

$$\lambda_{110} = \tfrac{1}{4}\lambda_{100} + \tfrac{3}{4}\lambda_{111} \tag{8.36}$$

as will be found by putting $\alpha_1 = \alpha_2 = \beta_1 = \beta_2 = 1/\sqrt{2}$, $\alpha_3 = \beta_3 = 0$ in (8.35). Putting (8.24), (8.25), and (8.26) into (8.33) and (8.34), we can express λ_{100} and λ_{111} in terms of the coefficient of the pair energy:

$$\left.\begin{aligned} \lambda_{100} &= -\frac{2}{3}\frac{N}{c_{11} - c_{12}}\left(\frac{\partial l}{\partial r}\right) r_0, \\ \lambda_{111} &= -\frac{4}{3}\frac{N}{c_{44}}\, l \qquad \text{for s.c.,} \end{aligned}\right\} \tag{8.37}$$

$$\left.\begin{aligned} \lambda_{100} &= -\frac{16}{9}\frac{N}{c_{11} - c_{12}}\, l, \\ \lambda_{111} &= -\frac{16}{27}\frac{N}{c_{44}}\left[l + \left(\frac{\partial l}{\partial r}\right) r_0\right] \qquad \text{for b.c.c.,} \end{aligned}\right\} \tag{8.38}$$

$$\left.\begin{aligned} \lambda_{100} &= -\frac{1}{3}\frac{N}{c_{11} - c_{12}}\left[6l + \left(\frac{\partial l}{\partial r}\right) r_0\right], \\ \lambda_{111} &= -\frac{2}{3}\frac{N}{c_{44}}\left[2l + \left(\frac{\partial l}{\partial r}\right) r_0\right] \qquad \text{for f.c.c.} \end{aligned}\right\} \tag{8.39}$$

If

$$\lambda_{100} = \lambda_{111} = \lambda, \tag{8.40}$$

(8.35) becomes

$$\frac{\delta l}{l} = \tfrac{3}{2}\lambda[(\alpha_1\beta_1 + \alpha_2\beta_2 + \alpha_3\beta_3)^2 - \tfrac{1}{3}] = \tfrac{3}{2}\lambda(\cos^2\theta - \tfrac{1}{3}), \tag{8.41}$$

where θ is the angle between the direction of domain magnetization and that of observation. The final form is the same as that of the isotropic magnetostriction given by (8.1). The condition for isotropic magnetostriction is thus expressed by equating the expressions for λ_{100} given by (8.37), (8.38), and (8.39) to those for λ_{111}. Since the coefficient of the pair energy and the elastic constants included in these expressions are entirely independent of each other, it is rather hard to discern any significant physical meaning for the isotropic magnetostriction.

For isotropic magnetostriction, that $\lambda = \lambda_{100} = \lambda_{111} = \lambda_{110}$ can be seen by putting (8.40) into (8.36). Figure 8.11 shows the dependence of magnetostriction constants on the alloy composition for iron-nickel alloys, where the isotropic magnetostriction is realized at 15% and at 40% Fe-Ni.

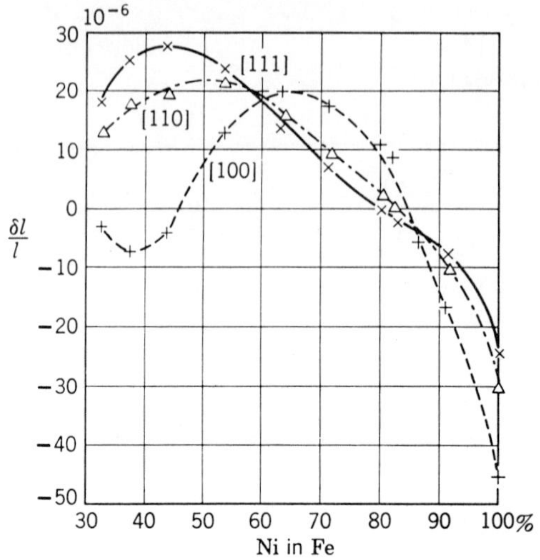

Fig. 8.11. Magnetostriction of iron-nickel alloy (Becker and Döring[G.3], F. Lichtenberger[4]).

For polycrystalline materials the longitudinal magnetostriction is calculated by averaging (8.35) for different crystal orientations by assuming $\alpha_i = \beta_i$ (i = 1, 2, and 3,); thus

$$\bar{\lambda} = \tfrac{2}{5}\lambda_{100} + \tfrac{3}{5}\lambda_{111}. \tag{8.42}$$

In the preceding discussion we considered only the dipole-dipole interaction term in (8.16). If we take into account the third term of (8.16), the expression for the magnetostriction of a cubic crystal becomes more precise than (8.35), as given below:

$$\begin{aligned}\frac{\delta l}{l} = {} & h_1(\alpha_1{}^2\beta_1{}^2 + \alpha_2{}^2\beta_2{}^2 + \alpha_3{}^2\beta_3{}^2 - \tfrac{1}{3}) \\ & + h_2(2\alpha_1\alpha_2\beta_1\beta_2 + 2\alpha_2\alpha_3\beta_2\beta_3 + 2\alpha_3\alpha_1\beta_3\beta_1) \\ & + h_4(\alpha_1{}^4\beta_1{}^2 + \alpha_2{}^4\beta_2{}^2 + \alpha_3{}^4\beta_3{}^2 + \tfrac{2}{3}s - \tfrac{1}{3}) \\ & + h_5(2\alpha_1\alpha_2\alpha_3{}^2\beta_1\beta_2 + 2\alpha_2\alpha_3\alpha_1{}^2\beta_2\beta_3 + 2\alpha_3\alpha_1\alpha_2{}^2\beta_3\beta_1) \\ & + h_3(s - \tfrac{1}{3}) \qquad \text{for } K_1 > 0 \\ & + h_3(s) \qquad\quad\ \ \text{for } K_1 < 0,\end{aligned} \tag{8.43}$$

where

$$s = \alpha_1^2\alpha_2^2 + \alpha_2^2\alpha_3^2 + \alpha_3^2\alpha_1^2. \tag{8.44}$$

In (8.43) h_1 and h_2 are related to λ_{100} and λ_{111} by $h_1 = \frac{3}{2}\lambda_{100}$ and $h_2 = \frac{3}{2}\lambda_{111}$.

Finally we discuss briefly the experimental procedures for determining magnetostriction constants. The most convenient and precise method is to use a single crystal disk which is cut parallel to $(1\bar{1}0)$, and to measure elongations along [001] and [110] during a revolution of the magnetization in the plane of the disk. The most convenient procedure for doing this is to attach two strain gauges, one parallel to [001] on the top surface, and the other parallel to [110] on the bottom; and place the disk in the gap of an electromagnet, and rotate the direction of magnetization in the crystal

Table 8.1. Elastic Moduli and Magnetostriction Constants of Various Metals

Substance	Elastic Moduli ($\times 10^{11}$ N/m^2)			Magnetostriction Constants ($\times 10^{-6}$)		Source
	c_{11}	c_{12}	c_{44}	λ_{100}	λ_{111}	
Fe	2.41	1.46	1.12	20.7	−21.2	Lee[5]
Ni	2.50	1.60	1.18	−45.9	−24.3	Lee[5]
40 at% Co-Fe	—	—	—	146.6	8.7	Azumi[6]
50 at% Co-Fe	—	—	—	119.3	41.3	Azumi[6]
70 at% Co-Fe	—	—	—	81.3	70.0	Azumi[6]

by rotating the electromagnet or by rotating the specimen disk. The elongation along [001] is given by

$$\frac{\delta l}{l} = \frac{3}{4}\lambda_{100} \cos 2\theta + \text{const.}, \tag{8.45}$$

as may easily be found by putting $\beta_1 = \beta_2 = 0$, $\beta_3 = 1$, $\alpha_1 = \alpha_2 = (1/\sqrt{2}) \sin\theta$, and $\alpha_3 = \cos\theta$ in (8.35). Similarly the elongation along [110] is given by

$$\frac{\delta l}{l} = -\frac{3}{8}(\lambda_{100} + \lambda_{111}) \cos 2\theta + \text{const.}, \tag{8.46}$$

by putting $\beta_1 = \beta_2 = 1/\sqrt{2}$ and $\beta_3 = 0$. From the amplitude of these sinusoidal curves we can calculate λ_{100} and λ_{111}. The values of λ_{100} and λ_{111} are listed for Fe, Ni, and Co-Fe alloys in Table 8.1, and shown graphically in Figs. 8.12 and 8.13 for various Fe and Ni alloys. The values are also given in Table 8.2 for various ferrites.

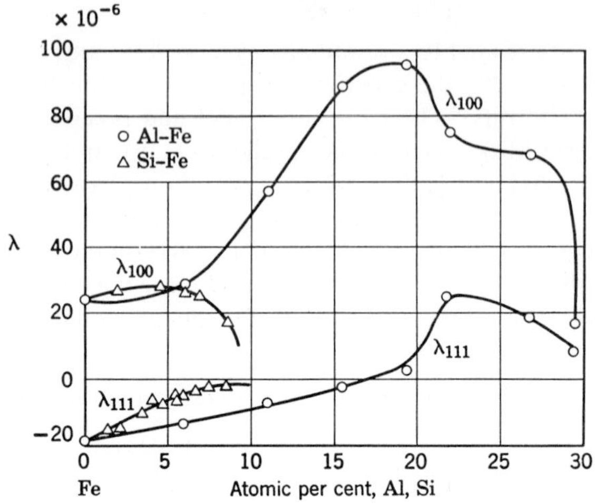

Fig. 8.12. Magnetostriction of various iron alloys (Al-Fe after Hall,[7] Si-Fe after Carr and Smoluchowski,[8] Tatsumoto,[9] Sturkin,[10] Takaki and Nakamura[11]).

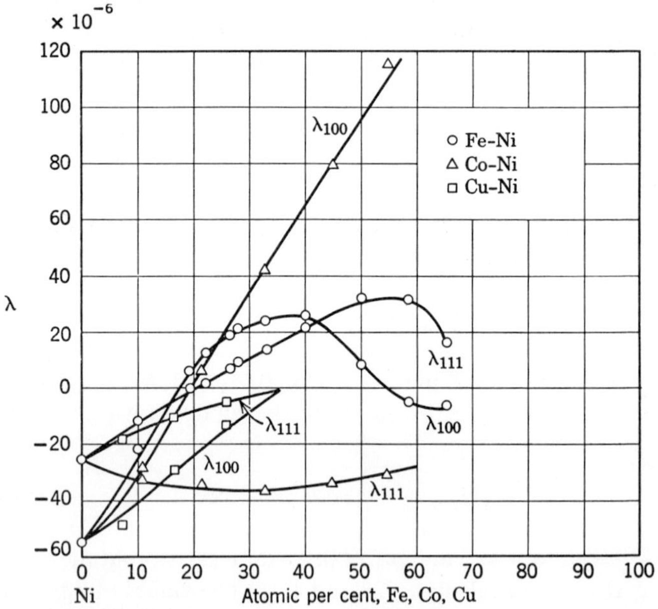

Fig. 8.13. Magnetostriction of various nickel alloys (Co-Ni after Bozorth and Walker,[12] Co-Ni and Cu-Ni after Yamamoto and Nakamichi.[13])

Table 8.2. Magnetostriction Constants of Various Ferrites ($\times 10^{-6}$)

$Mn_xFe_yO_4$

x	y	λ_{100} 20°C	λ_{100} −100°C	λ_{111} 20°C	λ_{111} −100°C	References
0	3	−20	−21	78	71	Bickford et al.[14]
0.10	2.90	−16	−22	75	71	Miyata and Funatogawa[15]
0.40	2.60	−6	−4	63	92	
0.60	2.40	−5	−2	45	73	
0.75	2.25	−8	−5	27	50	
0.85	2.15	−14	−15	16	32	
0.95	2.05	−22	−29	5	11	
1.05	1.95	−28	−37	4	10	

$Mn_xZn_yFe_zO_4$

x	y	z	λ_{100} at 20°C	λ_{111} at 20°C	References
0.6	0.1	2.1	−14	14	Bozorth et al.[16]

$Mn_xCo_yFe_zO_4$

x	y	z	λ_{100} at 20°C	λ_{111} at 20°C	References
0.4	0.3	2.0	−200	65	Bozorth et al.[16]

$Ni_xZn_yFe_zO_4$

x	y	z	λ_{100} at 20°C	λ_{111} at 20°C	References
0.8	0	2.2	−36	−4	Bozorth et al.[16]
0.3	0.45	2.25	−15	11	Ohta[17]

$Co_xZn_yFe_zO_4$

x	y	z	λ_{100} at 20°C	λ_{111} at 20°C	References
1.1	0	1.9	−250	—	Bozorth et al.[16]
0.8	0	2.2	−590	120	
0.3	0.2	2.2	−210	110	

Table 8.2 Magnetostriction Constants of Various Ferrites (Continued) ($\times 10^{-6}$)

$Ti_xFe_y^{2+}Fe_z^{3+}O_4$							
x	y	z	λ_{100}		λ_{111}		References
			20°C	−100°C	20°C	−100°C	
0.04	1.04	1.92	−6	−5	86	72	Syono and Ishikawa[18]
0.10	1.10	1.80	4	8	96	81	
0.18	1.18	1.64	46	92	109	112	

8.3 Volume Magnetostriction and Form Effect

Volume magnetostriction is that phenomenon in which the volume of the specimen is changed by magnetization. The fractional change of the volume is expressed in terms of strain tensor components as

$$\frac{\delta v}{v} = e_{xx} + e_{yy} + e_{zz}. \tag{8.47}$$

If we put the spontaneous strain given by (8.30) into this formula,

$$\frac{\delta v}{v} = 0. \tag{8.48}$$

That is, in the first approximation, a rotation of the magnetization does not give rise to any volume magnetostriction. It must be remarked, however, that the volume of the specimen had already been changed when the ferromagnetism was generated because of the first term of (8.16). The energy of the atom pair whose bond direction is $(\gamma_1, \gamma_3, \gamma_3)$ is changed by the lattice strain by the amount

$$\Delta w = \left(\frac{\partial g}{\partial r}\right) r_0 (e_{xx}\gamma_1^2 + e_{yy}\gamma_2^2 + e_{zz}\gamma_3^2 + e_{xy}\gamma_1\gamma_2 + e_{yz}\gamma_2\gamma_3 + e_{zx}\gamma_3\gamma_1). \tag{8.49}$$

For the simple cubic lattice, adding (8.49) for [100], [010], and [001] pairs, we have

$$E_{\text{volmag}} = N\left(\frac{\partial g}{\partial r}\right) r_0 (e_{xx} + e_{yy} + e_{zz}). \tag{8.50}$$

The conditions minimizing the total energy,

$$E = E_{\text{volmag}} + E_{\text{el}}, \tag{8.51}$$

are

$$\frac{\partial E}{\partial e_{xx}} = N\left(\frac{\partial g}{\partial r}\right)r_0 + c_{11}e_{xx} + c_{12}(e_{yy} + e_{zz}) = 0,$$
$$\frac{\partial E}{\partial e_{yy}} = N\left(\frac{\partial g}{\partial r}\right)r_0 + c_{11}e_{yy} + c_{12}(e_{zz} + e_{xx}) = 0, \qquad (8.52)$$
$$\frac{\partial E}{\partial e_{zz}} = N\left(\frac{\partial g}{\partial r}\right)r_0 + c_{11}e_{zz} + c_{12}(e_{xx} + e_{yy}) = 0,$$

from which we have

$$\frac{\delta v}{v} = e_{xx} + e_{yy} + e_{zz} = -\frac{3Nr_0}{c_{11} + 2c_{12}}\left(\frac{\partial g}{\partial r}\right). \qquad (8.53)$$

Similarly† we have

$$\frac{\delta v}{v} = -\frac{4Nr_0}{c_{11} + 2c_{12}}\left(\frac{\partial g}{\partial r}\right) \quad \text{for b.c.c. lattice,} \qquad (8.54)$$

and

$$\frac{\delta v}{v} = -\frac{6Nr_0}{c_{11} + 2c_{12}}\left(\frac{\partial g}{\partial r}\right) \quad \text{for f.c.c. lattice.} \qquad (8.55)$$

In order to observe this volume magnetostriction it is necessary to reduce the temperature of the specimen to below the ferromagnetic Curie point to establish ferromagnetism. Although we assume perfect parallelism for paired spins when we derive magnetic anisotropy and magnetostriction from (7.41) or (8.16), we do not necessarily assume parallelism in the present case because the g term does not include the direction cosines of the magnetization. Now we assume that g represents the average exchange energy for thermally agitated spin pairs. The internal energy of the spontaneous magnetization is then given by

$$E = \tfrac{1}{2}Nzg \qquad (8.56)$$

per unit volume of the specimen. Using this relation, together with the relation between the bulk modulus c and the elastic moduli,

$$c = \tfrac{1}{3}(c_{11} + 2c_{12}), \qquad (8.57)$$

we can rewrite (8.53), (8.54), and (8.55) as

$$\frac{\delta v}{v} = -\frac{1}{c}\frac{\partial E}{\partial \omega} = \frac{\partial E}{\partial p} \quad \text{where} \quad \omega = \frac{\delta v}{v} \qquad (8.58)$$

† It should be noted that $\partial g/\partial r$ depends not only on interatomic distance but also on crystal type, because in some cases a $d\epsilon$ orbit is effective for exchange coupling, while in other cases a $d\gamma$ orbit is (cf. Chapter 7).

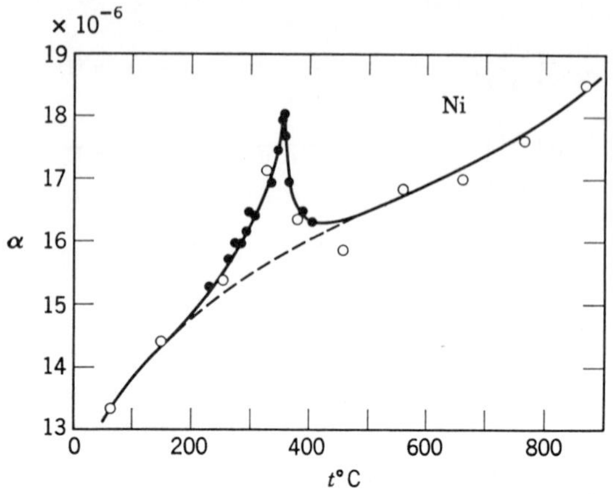

Fig. 8.14. Anomalous temperature dependence of thermal expansion coefficient of nickel [after Lee[5]; experimental points are according to Hidnert[19] (open circles) and Williams[20] (closed circles)].

for all crystal types. This relation can also be derived thermodynamically. The temperature coefficient of volume change, or the anomalous thermal expansion coefficient, is given by

$$\alpha = \frac{1}{3}\frac{\partial}{\partial T}\left(\frac{\delta v}{v}\right) = \frac{1}{3}\frac{\partial^2 E}{\partial T\,\partial p} = \frac{1}{3}\frac{\partial}{\partial p}\left(\frac{\partial E}{\partial T}\right) = \frac{1}{3}\frac{\partial C_p}{\partial p}, \tag{8.59}$$

where C_p is the anomalous specific heat. Figure 8.14 shows the temperature dependence of the thermal expansion coefficient which is similar to the temperature dependence of the anomalous specific heat (cf. Fig. 4.8).

It is also possible to change the value of g by applying a fairly strong field which is capable of increasing the alignment of the thermally agitated spin system. In the presence of a magnetic field the internal energy (4.42) becomes

$$E = -\tfrac{1}{2}wI_s^2 - I_s H, \tag{8.60}$$

so that from (8.58) the volume change becomes

$$\begin{aligned}\frac{\delta v}{v} &= -\frac{1}{c}\frac{\partial}{\partial H}\left(\frac{\partial E}{\partial \omega}\right)H \\ &= -\frac{1}{c}\frac{\partial}{\partial \omega}\left(\frac{\partial E}{\partial H}\right)H \\ &= \frac{1}{c}\frac{\partial I_s}{\partial \omega}H.\end{aligned} \tag{8.61}$$

This phenomenon is called forced magnetostriction. The anomalous thermal expansion and the forced magnetostriction mentioned above are substantially isotropic for cubic crystals since they originate in the exchange interaction.

Next we discuss the volume magnetostriction which is caused by the higher order terms of (8.16). As mentioned before, the second term of (8.16) gives rise to the usual magnetostriction, but not to any volume magnetostriction. The third term, which is the origin of crystal anisotropy, does give rise to a volume magnetostriction. For instance, solving for the spontaneous strain for a simple cubic lattice by taking the third term into account, we have

$$\frac{\delta v}{v} = e_{xx} + e_{yy} + e_{zz} = \frac{2N(\partial q/\partial r)r_0}{c_{11} + 2c_{12}}(\alpha_1^2\alpha_2^2 + \alpha_2^2\alpha_3^2 + \alpha_3^2\alpha_1^2 + \tfrac{2}{5}). \quad (8.62)$$

This volume magnetostriction is attributed to the rotation of the magnetization against the crystal anisotropy and is called the crystal effect. That is, by putting the anisotropy energy given by the first term of (7.6) into (8.58), we have

$$\frac{\delta v}{v} = -\frac{1}{c}\frac{\partial K_1}{\partial \omega}(\alpha_1^2\alpha_2^2 + \alpha_2^2\alpha_3^2 + \alpha_3^2\alpha_1^2) \quad (8.63)$$

from which we can easily derive (8.62) by use of (7.45). In volume magnetostriction, elongation does not depend on the crystallographic direction of observation. The h_3 term of (8.43) corresponds to the deformation due to the crystal effect.

The shape of the specimen can also influence the volume magnetostriction. Consider a specimen with demagnetizing factor N and volume $1 + \omega$. When it is magnetized to the intensity I, the magnetostatic energy is given by

$$U = \frac{1}{2}\frac{N}{\mu_0}\frac{M^2}{1 + \omega}, \quad (8.64)$$

where M is the magnetic moment of the specimen. Putting this into (8.58), we have

$$\frac{\delta v}{v} = \frac{NI^2}{2c\mu_0}, \quad (8.65)$$

since $M \approx I$. This phenomenon is called the form effect. When I reaches its saturation value I_s, this effect also reaches a saturation value given by

$$\left(\frac{\delta v}{v}\right)_s = \frac{NI_s^2}{2c\mu_0}. \quad (8.66)$$

The feature of this effect is that its value increases in proportion to I^2, so

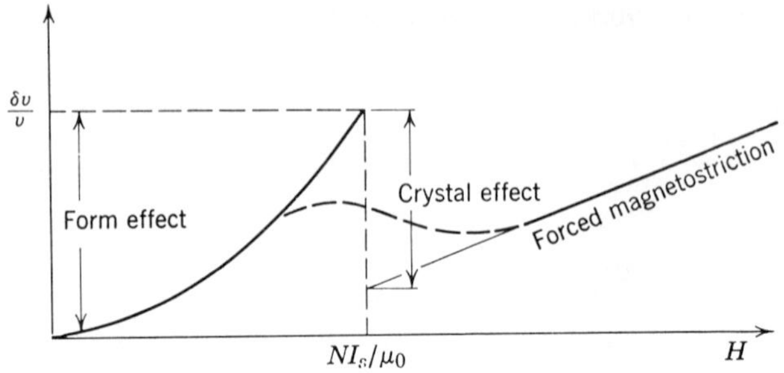

Fig. 8.15. Expected volume change as a function of the intensity of the applied field (Becker and Döring, ref. G.3, p. 295).

that if the magnetization curve is linear the volume change should be proportional to H^2.

To summarize these effects: the volume magnetostriction observed for a finite specimen shape is expected to be as shown in Fig. 8.15. First the volume should increase almost in proportion to H^2 owing to the form effect, and then, as soon as the rotation of magnetization starts, it begins to decrease because of the crystal effect.

The sign of this effect is expected to be negative always. To judge from the origin of the anisotropy, it should be reasonable to assume that $\partial |K_1|/\partial r < 0$, i.e., if $K_1 > 0$, $\partial K_1/\partial r < 0$; while, if $K_1 < 0$, $\partial K_1/\partial r > 0$. On the other hand, if $K_1 > 0$, the magnetization rotates from [100] to [111], so that the change in the function of α_i's in the parentheses of (8.63) is positive; if $K_1 < 0$, the magnetization rotates from [111] to [100], so that the change in the function is negative. In both cases, therefore, the volume change given by (8.63) becomes negative. Actually it has been found that the crystal effects observed for both iron and nickel are negative.

After the magnetization reaches saturation, the volume begins to increase again because of the forced magnetostriction. Generally speaking, the volume magnetostriction is a small effect, but it can give rise to a fairly large deformation for a very strong field, since the forced magnetostriction is proportional to H. Figure 8.16 shows two examples of volume magnetostriction, observed for specimens having different dimensional ratios.

It should be noted here that the form effect gives rise not only to volume magnetostriction but also to ordinary magnetostriction. Consider an ellipsoid of rotation with dimensional ratio k which is magnetized parallel to the rotation axis. Set the x axis parallel to the rotation axis. Then the demagnetizing factor N, expressed as a function of the strain, is

$$N = N_0 + \frac{dN}{dk}\left(\frac{\partial k}{\partial e_{xx}} e_{xx} + \frac{\partial k}{\partial e_{yy}} e_{yy} + \frac{\partial k}{\partial e_{zz}} e_{zz}\right). \tag{8.67}$$

Let l and d denote the length and width of the ellipsoid; then, since $k = l/d$,

$$\frac{\partial k}{\partial e_{xx}} = \frac{l}{d} = k, \quad \frac{\partial k}{\partial e_{yy}} = \frac{\partial k}{\partial e_{zz}} = -\frac{l}{2d} = -\frac{k}{2}. \tag{8.68}$$

For an elongated ellipsoid of rotation, from (2.11) we have

$$\frac{dN}{dk} = -\frac{1}{k^3}(2 \ln 2k - 3). \tag{8.69}$$

Putting (8.68) and (8.69) into (8.67), we obtain

$$N = N_0\{1 - a[e_{xx} - \tfrac{1}{2}(e_{yy} + e_{zz})]\}, \tag{8.70}$$

where

$$a = \frac{N'k}{N_0} = \frac{2 \ln 2k - 3}{\ln 2k - 1}. \tag{8.71}$$

If we neglect the volume magnetostriction, the magnetostatic energy of the system can be given by

$$E_N = \frac{1}{2\mu_0} NI^2 = \frac{1}{2\mu_0} N_0 I^2 [1 - a(e_{xx} - \tfrac{1}{2}(e_{yy} + e_{zz}))] \tag{8.72}$$

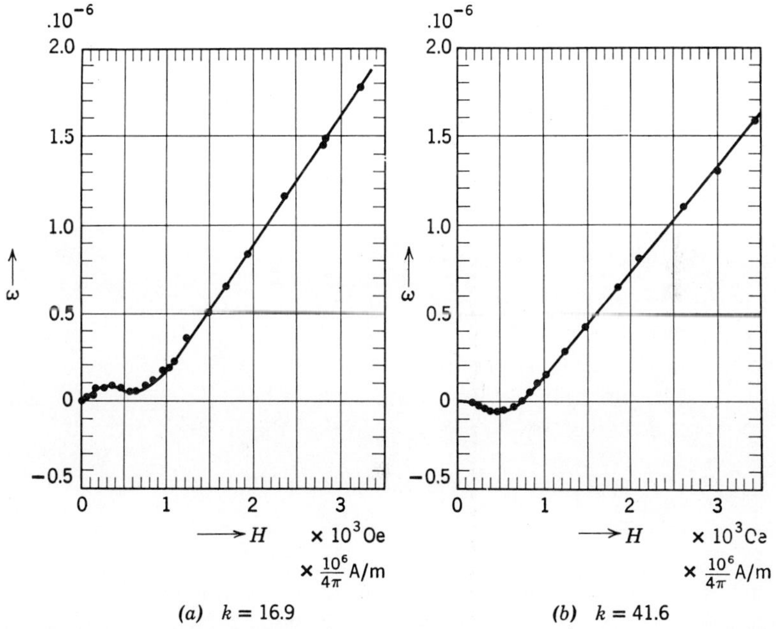

Fig. 8.16. Volume magnetostriction of ellipsoidal specimens of iron with different dimension ratios: (*a*) $k = 16.9$; (*b*) $k = 41.6$ (Becker and Döring,[G.3] p. 292; experimental points according to Kornetzki[21]).

It turns out from this formula that, in order to decrease the magnetostatic energy, e_{xx} should be increased and e_{yy} and e_{zz} must be decreased. The resultant strain can be determined by minimizing the total energy.

$$E = E_N + E_{\text{el}}. \tag{8.73}$$

That is,

$$\begin{aligned}
\frac{\partial E}{\partial e_{xx}} &= -\frac{a}{2\mu_0} N_0 I^2 + c_{11} e_{xx} + c_{12}(e_{yy} + e_{zz}) = 0, \\
\frac{\partial E}{\partial e_{yy}} &= \frac{a}{4\mu_0} N_0 I^2 + c_{11} e_{yy} + c_{12}(e_{zz} + e_{xx}) = 0, \\
\frac{\partial E}{\partial e_{zz}} &= \frac{a}{4\mu_0} N_0 I^2 + c_{11} e_{zz} + c_{12}(e_{xx} + e_{yy}) = 0.
\end{aligned} \tag{8.74}$$

Solving these equations, we have

$$\begin{aligned}
e_{xx} &= \frac{1}{2\mu_0} \frac{a N_0 I^2}{c_{11} - c_{12}}, \\
e_{yy} = e_{zz} &= -\frac{1}{4\mu_0} \frac{a N_0 I^2}{c_{11} - c_{12}},
\end{aligned} \tag{8.75}$$

where e_{xx} gives the longitudinal magnetostriction and e_{yy} and e_{zz} give the transverse magnetostriction. When N is fairly large, this effect cannot be neglected.

8.4 Inverse Magnetostrictive Effect

It is easy to imagine that a stress applied to a ferromagnetic body will affect the direction of domain magnetization through the magnetostriction. Suppose that a tension of intensity σ (N/m²) exists in a ferromagnetic body. Let the direction cosines of this tension be $(\gamma_1, \gamma_2, \gamma_3)$. The tensor components are given by $\sigma_{ij} = \sigma\gamma_i\gamma_j$, so that the strain is given by

$$\begin{aligned}
e_{xx} &= \sigma[s_{11}\gamma_1^2 + s_{12}(\gamma_2^2 + \gamma_3^2)] \cdots, \\
e_{xy} &= \sigma s_{44}\gamma_1\gamma_2 \cdots,
\end{aligned} \tag{8.76}$$

where s_{11}, s_{12}, and s_{44} are the elastic constants. Putting (8.76) into (8.23), we have

$$\begin{aligned}
E_\sigma = {} & B_1\sigma(s_{11} - s_{12})(\alpha_1^2\gamma_1^2 + \alpha_2^2\gamma_2^2 + \alpha_3^2\gamma_3^2 - \tfrac{1}{3}) \\
& + B_1\sigma s_{44}(\alpha_1\alpha_2\gamma_1\gamma_2 + \alpha_2\alpha_3\gamma_2\gamma_3 + \alpha_3\alpha_1\gamma_3\gamma_1).
\end{aligned} \tag{8.77}$$

Rewriting B_1 and B_2 in terms of λ_{100} and λ_{111} by relations (8.33) and (8.34), and also using the relations between c_{11}, c_{12}, and c_{44} and s_{11}, s_{12}, and s_{44}, that is,

$$c_{11} = \frac{s_{11} + s_{12}}{(s_{11} - s_{12})(s_{11} + 2s_{12})},$$

$$c_{12} = \frac{-s_{12}}{(s_{11} - s_{12})(s_{11} + 2s_{12})}, \tag{8.78}$$

$$c_{44} = \frac{1}{s_{44}},$$

we have

$$E = -\tfrac{3}{2}\lambda_{100}\sigma(\alpha_1{}^2\gamma_1{}^2 + \alpha_2{}^2\gamma_2{}^2 + \alpha_3{}^2\gamma_3{}^2) - 3\lambda_{111}\sigma(\alpha_1\alpha_2\gamma_1\gamma_2 + \alpha_2\alpha_3\gamma_2\gamma_3 + \alpha_3\alpha_1\gamma_3\gamma_1). \tag{8.79}$$

If the domain magnetization is parallel to $\langle 100 \rangle$, as it is for a material with positive K_1,

$$E_\sigma = -\tfrac{3}{2}\lambda_{100}\sigma\gamma_1{}^2 \tag{8.80}$$

for the [100] domains, and

$$E_\sigma = -\tfrac{3}{2}\lambda_{100}\sigma\gamma_2{}^2, \tag{8.81}$$

$$E_\sigma = -\tfrac{3}{2}\lambda_{100}\sigma\gamma_3{}^2 \tag{8.82}$$

for the [010] and [001] domains, respectively. The presence of such an energy difference between the domains is expected to give rise to a force on the 90° walls which separate these domains.

When $K_1 < 0$, (8.79) gives

$$E_\sigma = -\lambda_{111}\sigma(\gamma_1\gamma_2 + \gamma_2\gamma_3 + \gamma_3\gamma_1) \tag{8.83}$$

for the [111] domain. If we let ϕ denote the angle between [111] and the direction of the tension σ,

$$\cos\phi = \frac{1}{\sqrt{3}}(\gamma_1 + \gamma_2 + \gamma_3), \tag{8.84}$$

so that (8.83) becomes

$$E_\sigma = -\tfrac{3}{2}\lambda_{111}\sigma\cos^2\phi. \tag{8.85}$$

The same relation is valid for the other domains, provided ϕ denotes the angle between the easy direction and the direction of the tension σ. These relations will be used in Chapter 13, where the displacement of 90° walls is discussed.

It is assumed above that the direction of magnetization is fixed parallel to one of the easy directions. When the direction of magnetization is rotated from the easy direction, the energy is changed in a fairly complicated way, as shown by formula (8.79), unless σ is parallel to some crystallographic principal axis. In an isotropic magnetostriction this expression can be simplified by putting $\lambda_{100} = \lambda_{111} = \lambda$:

$$E_\sigma = -\tfrac{3}{2}\lambda\sigma(\alpha_1\gamma_1 + \alpha_2\gamma_2 + \alpha_3\gamma_3)^2 = -\tfrac{3}{2}\lambda\sigma\cos^2\phi, \qquad (8.86)$$

where ϕ is the angle between the tension σ and the direction of magnetization. Since this energy does not depend on the azimuthal angle about the axis of the tension, it gives a uniaxial anisotropy energy. This relation will be frequently used for simplicity in the discussion of the effect of stresses on the direction of domain magnetization.

Problems

8.1. Consider a cubic ferromagnetic crystal with a large magnetocrystalline anisotropy which first contains [100] and [$\bar{1}$00] domains in the demagnetized state, and then is magnetized parallel to [010]. Calculate the elongation measured in the [010] direction as a function of the intensity of magnetization in that direction.

8.2. When a single crystal sphere with magnetostriction constants h_1, h_2, and h_3 is magnetized to its saturation by a constant magnetic field which makes an angle θ with [100] in the plane of (001), how are the elongations parallel to [100] and [001] changed as a function of θ?

8.3. When a tension σ is applied parallel to [123] of a ferromagnetic crystal with positive K_1, how much energy can be stored in the x, y, and z domains? Assume $\lambda_{100}\sigma \ll K_1$.

References

8.1. W. L. Webster: *Proc. Roy. Soc. (London)* **A109,** 570 (1925).
8.2. S. Kaya and H. Takaki: *J. Fac. Sci. Hokkaido Univ.* **2,** 227 (1935).
8.3. L. Néel: *J. Phys. Radium* **15,** 227 (1954).
8.4. F. Lichtenberger: *Ann. Physik.* **10,** 45 (1932).
8.5. E. W. Lee: *Rept. Prog. Phys.* **18,** 184 (1955).
8.6. K. Azumi: *Unken Hokoku (Japan)* **4,** 1 (1955).
8.7. R. C. Hall: *J. Appl. Phys.* **28,** 707 (1957).
8.8. W. J. Carr, Jr., and R. Smoluchowski: *Phys. Rev.* **83,** 1236 (1951).
8.9. E. Tatsumoto: *J. Sci. Hiroshima Univ.* **A17,** 229 (1953).
8.10. D. A. Sturkin: *Compt. Rend. Acad. Sci. USSR* **58,** 581 (1947).
8.11. H. Takaki and Y. Nakamura: *J. Phys. Soc. Japan* **9,** 748 (1954).
8.12. R. M. Bozorth and J. G. Walker: *Phys. Rev.* **89,** 624 (1953).
8.13. M. Yamamoto and T. Nakamichi: *J. Phys. Soc. Japan* **2,** 228 (1958).
8.14. L. R. Bickford, Jr., J. Pappis, and J. L. Stull: *Phys. Rev.* **99,** 1210 (1955).

8.15. N. Miyata and Z. Funatogawa: *J. Phys. Soc. Japan* **17,** Suppl. B-I, 279 (1962).
8.16. R. M. Bozorth, E. F. Tilden and A. J. Williams: *Phys. Rev.* **99,** 1788 (1955).
8.17. K. Ohta: *Bull. Kobayasi Inst. Phys. Res.* **10,** 149 (1960).
8.18. Y. Syono and Y. Ishikawa: *J. Phys. Soc. Japan* **18,** 1231 (1963).
8.19. P. Hidnert: *J. Res. Natl. Bur. Standards* **5,** 1305 (1930).
8.20. C. Williams: *Phys. Rev.* **46,** 1011 (1934).
8.21. M. Kornetzki: *Z. Physik* **87,** 560 (1933).

9

Properties of Magnetic Domain Walls

9.1 Structure of Domain Wall

The structure of transition layers between the adjacent ferromagnetic domains was first investigated by F. Bloch[1] in 1932. This transition layer is called the magnetic domain wall; it is sometimes referred to as the Bloch wall.

First we discuss the reason why a domain wall has a finite thickness without abrupt change in the spin directions from one domain to the other. The origin of the force which lines up the spins is the exchange interaction of quantum mechanical nature. When the spin magnetic moments of the adjacent atoms i and j make an angle ϕ_{ij}, the exchange energy between the two moments is conveniently expressed in the form

$$w_{ij} = -2JS^2 \cos \phi_{ij}, \tag{9.1}$$

where J is the exchange integral and S the total spin quantum number of each atom. When $J > 0$, the lowest energy is attained when $\phi_{ij} = 0$ or the two spins are parallel to each other. As discussed in Section 4.2, the exchange interaction may be effectively expressed in terms of a molecular field whose coefficient w is related to the Curie temperature by (4.30). In the same way the exchange integral J can be related to Θ. According to the calculation by P. R. Weiss,[2]

$$J = 0.54k\Theta \quad \text{for s.c. lattice,} \quad S = \tfrac{1}{2}; \tag{9.2}$$

$$J = 0.34k\Theta \quad \text{for b.c.c. lattice,} \quad S = \tfrac{1}{2}; \tag{9.3}$$

$$J = 0.15k\Theta \quad \text{for b.c.c. lattice,} \quad S = 1. \tag{9.4}$$

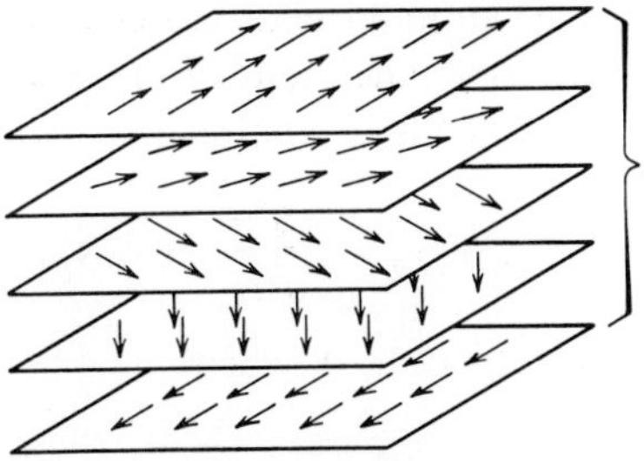

Fig. 9.1. Rotation of spins in the domain wall.

For iron, assuming $S = 1$,

$$J = 0.15 \times 1043 \times 1.38 \times 10^{-23} = 2.16 \times 10^{-21} \text{ J}. \tag{9.5}$$

When $\phi_{ij} \ll 1$, (9.1) is simplified by expanding $\cos \phi_{ij}$ as

$$w_{ij} = JS^2\phi_{ij}^2 + \text{const.} \tag{9.6}$$

The fact that w_{ij} is proportional to ϕ_{ij}^2 is the main reason why spins change their directions gradually inside the domain wall. Consider a 180° wall in which the spins rotate by 180° through N atomic layers (Fig. 9.1). For a rough estimation the rotation of spins is assumed to be uniform throughout the transition layers, so that $\phi_{ij} = \pi/N$. For a simple cubic lattice with lattice constant a the number of atoms per unit area of a (001) surface is $1/a^2$, so that the exchange energy stored per unit area of the transition layers is given by

$$\gamma_{\text{ex}} = \frac{N}{a^2} w_{ij} = \frac{JS^2\pi^2}{a^2 N}. \tag{9.7}$$

From this expression we see that the total exchange energy stored in the transition layers decreases with an increase of the number of transition layers, N. Thus the exchange energy tends to increase the wall thickness. If w_{ij} were proportional to ϕ_{ij}, (9.7) would be independent of N, and thus there should be no tendency to increase the wall thickness.

On the other hand, the rotation of the spins in the wall out of the direction of easy magnetization causes an increase in magnetocrystalline anisotropy energy. Roughly speaking, the deviation of spin directions from an easy direction increases the anisotropy energy by K per unit volume, where K is the anisotropy constant. Since the volume of the transition region is given by $(N/a^2) \times a^3 = Na$ per unit area of the wall surface, the anisotropy energy stored in this volume is given by

$$\gamma_a = KNa. \tag{9.8}$$

The actual thickness of the domain wall is determined by the counter-balance of the exchange energy which tends to increase the thickness, and the anisotropy energy which tends to diminish it. The total energy is

$$\gamma = \gamma_{ex} + \gamma_a = \frac{JS^2\pi^2}{a^2N} + KNa, \tag{9.9}$$

which is a minimum with respect to N when

$$\frac{\partial\gamma}{\partial N} = -\frac{JS^2\pi^2}{a^2N^2} + Ka = 0,$$

or

$$N = \sqrt{\frac{JS^2\pi^2}{Ka^3}}. \tag{9.10}$$

The thickness of the domain wall, δ, becomes

$$\delta = Na = \sqrt{\frac{JS^2\pi^2}{Ka}}. \tag{9.11}$$

For iron, using $J = 2.16 \times 10^{-21}$ (cf. 9.5), $S = 1$, $K = 4.2 \times 10^4$ and $a = 2.86 \times 10^{-10}$ m,

$$\delta = \sqrt{\frac{2.16 \times 10^{-21} \times 1 \times 3.14^2}{4.2 \times 10^4 \times 2.86 \times 10^{-10}}} = 4.2 \times 10^{-8}\ \text{m} \fallingdotseq 150\ \text{lattice const.} \tag{9.12}$$

The total energy of the wall is calculated by substituting N as given by (9.10) in (9.9). Thus

$$\gamma = \sqrt{\frac{JS^2\pi^2K}{a}} + \sqrt{\frac{JS^2\pi^2K}{a}} = 2\sqrt{\frac{JS^2\pi^2K}{a}}. \tag{9.13}$$

It is interesting to note that the exchange energy is equal to the anisotropy energy for the equilibrium spin arrangement, as shown by this equation.† For iron, we find

$$\gamma = 2\sqrt{\frac{2.16 \times 10^{-21} \times 1^2 \times 3.14^2 \times 4.2 \times 10^4}{2.86 \times 10^{-10}}}$$
$$= 1.1 \times 10^{-3}\ \text{J/m}^2\ (= 1.1\ \text{ergs/cm}^2) \tag{9.14}$$

We assumed above that the rotation of spins is uniform inside the domain

† When the total energy is given by a function of x containing two terms, one of which is a linear function of x and the other is inversely proportional to x, the two terms become equal to each other for the equilibrium value of x. In addition to the present example, we shall see another example pertaining to the size of domains in Section 11.2.

wall. This is, however, not true, because the exact nature of the rotation of the spins is also to be determined by the equilibrium condition.

Now let us set the z axis normal to the plane of the domain wall and its origin at the midpoint of the wall thickness. All the spins are assumed to rotate in the xy plane. The azimuthal angle of the spin orientation at $z = z$, as measured from the spin direction at $z = 0$, changes from $-\pi/2$ at $z = -\infty$ to $\pi/2$ at $z = \infty$ (see Fig. 9.2). Since the angle between neighboring spins is given by $(\partial\theta/\partial z)a$, the exchange energy of the spin pair is given by $JS^2a^2(\partial\theta/\partial z)^2$. The total exchange energy stored in the wall is, therefore,

$$\gamma_{\text{ex}} = \frac{JS^2}{a}\int_{-\infty}^{\infty}\left(\frac{\partial\theta}{\partial z}\right)^2 dz, \tag{9.15}$$

per unit area of the wall surface.

On the other hand, if we express the anisotropy energy as $g(\theta)$, the total anisotropy energy is given by

$$\gamma_a = \int_{-\infty}^{\infty} g(\theta)\, dz, \tag{9.16}$$

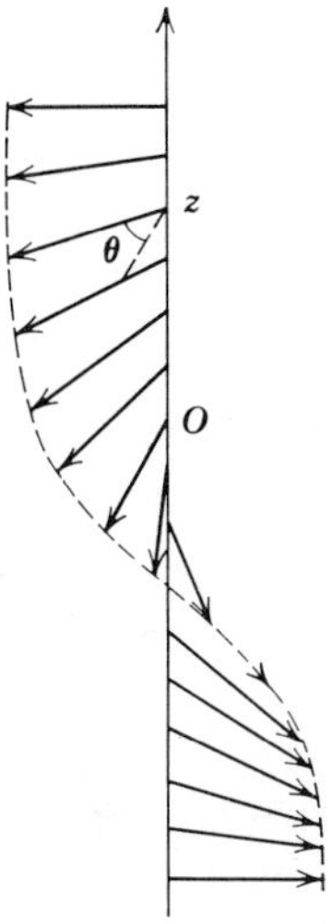

Fig. 9.2. Azimuthal angle of spin rotation.

per unit area of the wall surface, where $g(\theta)$ is defined to be zero at $z = \pm\infty$ or when the spin is parallel to the direction of easy magnetization.

Thus the total energy is given by

$$\gamma = \gamma_{\text{ex}} + \gamma_a = \int_{-\infty}^{\infty}\left[g(\theta) + A\left(\frac{\partial\theta}{\partial z}\right)^2\right] dz, \tag{9.17}$$

where A is the coefficient related to the exchange energy and is

$$A = \frac{JS^2}{a} \quad \text{for s.c. lattice,} \tag{9.18}$$

$$A = \frac{2JS^2}{a} \quad \text{for b.c.c. lattice,} \tag{9.19}$$

and

$$A = \frac{4JS^2}{a} \quad \text{for f.c.c. lattice.} \tag{9.20}$$

It should be noted here that the expressions for A in the formulas above are valid irrespective of the crystallographic orientation of the wall. For iron using $J = 2.16 \times 10^{-21}$, $S = 1$, and $a = 2.9 \times 10^{-10}$, we obtain

$$A = 1.49 \times 10^{-11}\ \text{J/m}. \tag{9.21}$$

Now the stable spin configuration can be obtained by minimizing the total energy (9.17) for a small variation of the spin arrangement inside the wall. When the angle θ of the spin at $z = z$ is changed by $\delta\theta$, the total energy of the wall is changed by

$$\delta\gamma = \int_{-\infty}^{\infty} \left[\frac{\partial g(\theta)}{\partial \theta}\, \delta\theta + 2A \left(\frac{\partial \theta}{\partial z}\right) \left(\frac{\partial\, \delta\theta}{\partial z}\right) \right] dz, \tag{9.22}$$

which should vanish for the stable spin arrangement. The second term of the integrand is treated by integrating by parts,

$$\begin{aligned} \int_{-\infty}^{\infty} 2A \left(\frac{\partial \theta}{\partial z}\right) \left(\frac{\partial\, \delta\theta}{\partial z}\right) dz &= \left| 2A \left(\frac{\partial \theta}{\partial z}\right) \delta\theta \right|_{-\infty}^{\infty} - \int_{-\infty}^{\infty} 2A \left(\frac{\partial^2 \theta}{\partial z^2}\right) \delta\theta\, dz \\ &= -\int_{-\infty}^{\infty} 2A \left(\frac{\partial^2 \theta}{\partial z^2}\right) \delta\theta\, dz, \end{aligned} \tag{9.23}$$

where the first term vanishes because $\delta\theta = 0$ at $z = \infty$ or $-\infty$. Then the vanishing of (9.22) requires that

$$\int_{-\infty}^{\infty} \left[\frac{\partial g(\theta)}{\partial \theta} - 2A \left(\frac{\partial^2 \theta}{\partial z^2}\right) \right] \delta\theta\, dz = 0. \tag{9.24}$$

In order that this condition be satisfied for any selection of $\delta\theta(z)$, the integrand must always be zero; that is,

$$\frac{\partial g(\theta)}{\partial \theta} - 2A \left(\frac{\partial^2 \theta}{\partial z^2}\right) = 0. \tag{9.25}$$

This equation is usually called the Euler equation of the variational problem. Multiplying by $\partial\theta/\partial z$ and integrating from $z = -\infty$ to $z = z$, we have

$$g(\theta) = A \left(\frac{\partial \theta}{\partial z}\right)^2, \tag{9.26}$$

from which

$$dz = \sqrt{A}\, \frac{d\theta}{\sqrt{g(\theta)}}. \tag{9.27}$$

On integrating, this becomes

$$z = \sqrt{A} \int_0^{\theta} \frac{d\theta}{\sqrt{g(\theta)}}. \tag{9.28}$$

We shall calculate this function for several examples later.

Equation (9.26) tells us that the anisotropy $g(\theta)$ is equal to the exchange energy $A(\partial\theta/\partial z)^2$ at any part of the wall. From this fact we can figure that the spin rotation is more rapid at any position where the spin takes on a higher anisotropy energy, and vice versa. Referring to (9.27), we obtain for the total surface energy of the wall given by (9.17)

$$\gamma = 2\sqrt{A}\int_{-\pi/2}^{\pi/2}\sqrt{g(\theta)}\,d\theta. \tag{9.29}$$

For uniaxial anisotropy, as produced by a magnetic annealing, roll magnetic anisotropy, or stress-induced anisotropy, or simply the uniaxial crystal anisotropy, $g(\theta)$ is expressed as

$$g(\theta) = K_u \cos^2\theta. \tag{9.30}$$

This expression is somewhat different from (7.1) or the first term of (13.1), because the origin of θ here is defined to correspond to the hard direction. Then (9.28) becomes

$$z = \sqrt{\frac{A}{K_u}}\int_0^{\theta}\frac{d\theta}{\cos\theta} = \sqrt{\frac{A}{K_u}}\log\tan\left(\frac{\theta}{2}+\frac{\pi}{4}\right). \tag{9.31}$$

This relation between z and θ is shown in Fig. 9.3. The most rapid change in θ can be seen at $z = 0$, where the anisotropy energy takes its maximum value. Since the rotation of spins becomes gradual on both sides of the wall, there are no sharp end surfaces of the wall. For convenience, however, we define the thickness of the wall by assuming that the rate of the rotation at $z = 0$; that is,

$$\left(\frac{\partial z}{\partial\theta}\right)_{z=0} = \frac{1}{2}\sqrt{\frac{A}{K_u}}\left[\frac{\sec^2(\theta/2+\pi/4)}{\tan(\theta/2+\pi/4)}\right]_{\theta=0} = \sqrt{\frac{A}{K_u}} \tag{9.32}$$

holds throughout the wall. Then the thickness of the wall becomes

$$\delta = \pi\left(\frac{\partial z}{\partial\theta}\right)_{z=0} = \pi\sqrt{\frac{A}{K_u}}. \tag{9.33}$$

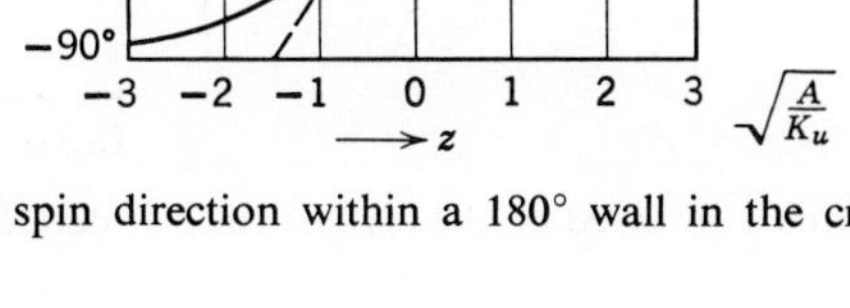

Fig. 9.3. Variation of spin direction within a 180° wall in the crystal with uniaxial anisotropy.

This result agrees well with the rough estimate (9.11). The surface energy of the wall is calculated by applying (9.30) to (9.29):

$$\gamma = 2\sqrt{AK_u}\int_{-\pi/2}^{\pi/2} \cos\theta \, d\theta = 4\sqrt{AK_u}, \tag{9.34}$$

which is also nearly equal to the rough estimate (9.13).

9.2 180° Domain Wall

Domain walls can be classified into 180° walls, in which the spins rotate by 180° from one domain to the other, and 90° walls, in which the spins rotate by 90° or so. Consider iron for instance, where there are six possible stable orientations of magnetization: [100], [$\bar{1}$00], [010], [0$\bar{1}$0], [001], and [00$\bar{1}$]. The domain boundaries between [100] and [$\bar{1}$00] domains form 180° walls, and those between [100] and [010] domains form 90° walls. For a substance with negative K_1, the easy directions are ⟨111⟩ directions, between which there are 180°, 71°, and 109° walls. Figure 9.4 is the domain pattern observed on a (110) crystal surface of perfectly ordered Permalloy, where we see three kinds of domain walls. We normally refer to 71° and 109° walls as 90° walls, because they are all distinguished from the 180° wall with respect to the stress-sensitive property. In this section we discuss in detail the properties of 180° walls.

First let us consider the shape of a 180° wall between the [100] and [$\bar{1}$00]

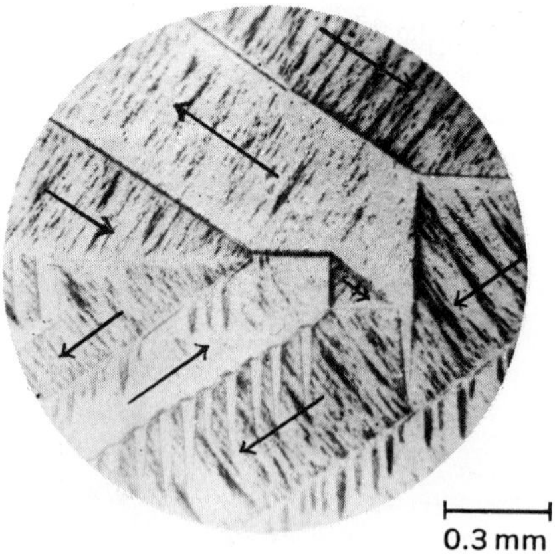

Fig. 9.4. Domain pattern observed on the (110) surface of perfectly ordered Permalloy (Chikazumi[3]).

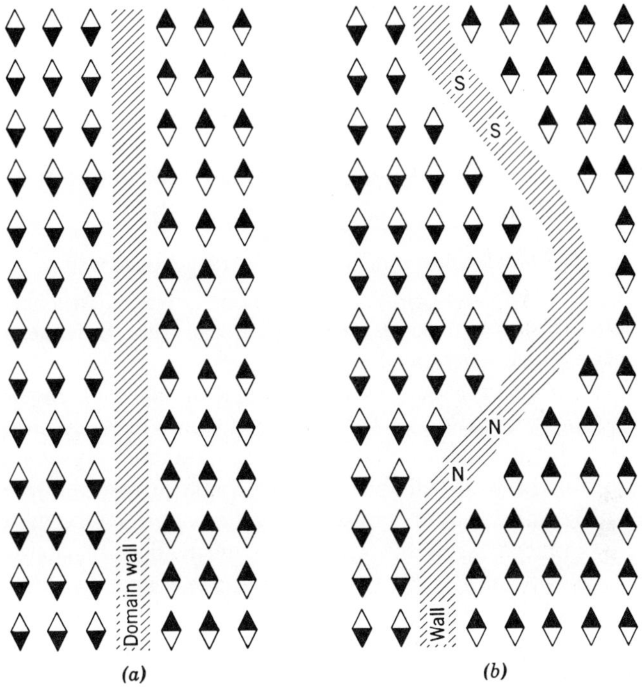

Fig. 9.5. Diagrams showing the appearance of magnetic free poles at the curved portion of a domain wall.

domains. As one looks at this domain wall on the (001) crystal surface, the cross-sectional view of the wall is expected to be a straight line as shown in Fig. 9.5*a*, because, if the wall is bent as shown in Fig. 9.5*b*, strong magnetic poles are induced at the bent portion of the wall, giving rise to a fairly large demagnetizing field at the swollen portion of the domain. Therefore, if there are no obstacles to sustain the swollen shape of the wall, this portion should be straightened by the action of the demagnetizing field.

When one looks at the domain wall from a direction parallel to the domain magnetization, the cross-sectional view of the wall is expected to be winding as shown in Fig. 9.6, because such winding does not result in the appearance of any magnetic free poles on the wall surface. Such winding, however, does increase the area of the wall surfaces and accordingly the total surface energy, so the wall tends to decrease the area of its surface unless there is some reason to sustain the windings of the wall. The possible mechanisms which sustain the wall shape result from the presence of inclusions or voids, irregular distribution of the internal stress

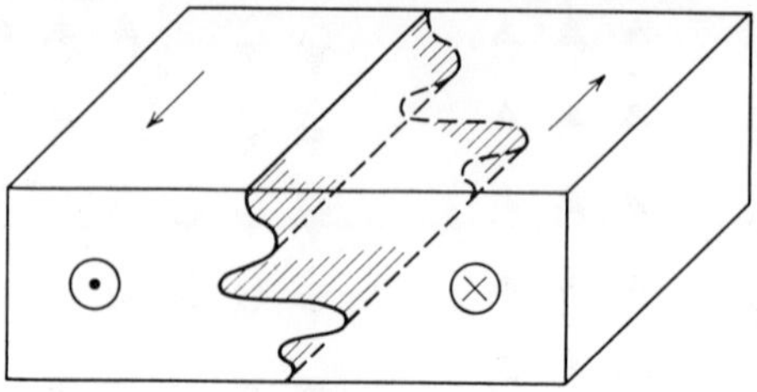

Fig. 9.6. Possible configuration of a 180° domain wall.

and alloy composition, and the dependence of the wall energy on its crystallographic orientation.

Now in order to consider the dependence of the wall energy on the crystallographic orientation of the wall, let us imagine a cylindrical 180° wall whose axis is parallel to x (Fig. 9.7). Let $\boldsymbol{n}$ denote the normal to the local portion of the wall, which makes the angle ϕ from the y axis. The spins inside the wall should rotate about the axis $\boldsymbol{n}$, since free poles would otherwise be induced inside the wall. (The illustration of rotation shown in Fig. 6.17 is schematic.) Across the wall the magnetization direction changes from $+x$ to $-x$.

The magnetocrystalline anisotropy energy density is

$$E_a = K_1(\alpha_1^2\alpha_2^2 + \alpha_2^2\alpha_3^2 + \alpha_3^2\alpha_1^2), \tag{9.35}$$

as already shown in (7.6). Now, if we set the z' axis parallel to $\boldsymbol{n}$ and x' parallel to x, the direction cosines of the magnetization vector $(\alpha_1, \alpha_2, \alpha_3)$ with respect to the old coordinate axes can be related to those in the new

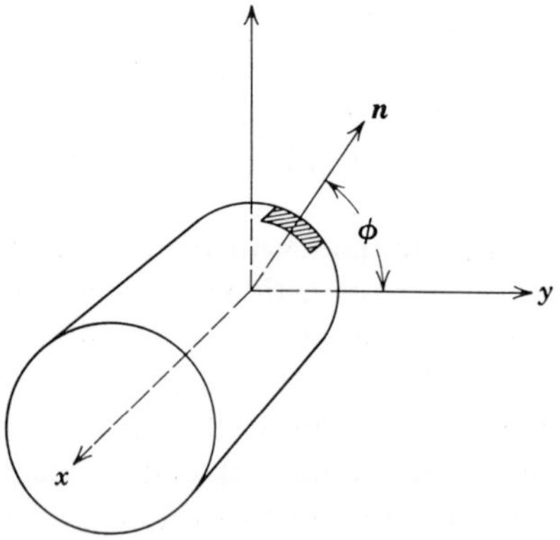

Fig. 9.7. A cylindrical domain wall and the angle ϕ indicating the direction of a normal to the wall surface.

coordinate system, $(\alpha_1', \alpha_2', \alpha_3')$, by

$$\begin{aligned} \alpha_1 &= \alpha_1', \\ \alpha_2 &= \alpha_2' \sin\phi + \alpha_3' \cos\phi, \\ \alpha_3 &= -\alpha_2' \cos\phi + \alpha_3' \sin\phi. \end{aligned} \tag{9.36}$$

Since $\alpha_3' = 0$ throughout the wall, (9.35) becomes, through use of (9.36),

$$E_a = K_1(\alpha_1'^2 + \alpha_2'^2 \sin^2\phi \cos^2\phi)\alpha_2'^2, \tag{9.37}$$

within the wall. If we define the azimuthal angle θ as shown in Fig. 9.2, it follows that

$$\begin{aligned} \alpha_1' &= \sin\theta, \\ \alpha_2' &= \cos\theta, \end{aligned} \tag{9.38}$$

so that

$$g(\theta) = K_1(\sin^2\theta + \cos^2\theta \sin^2\phi \cos^2\phi)\cos^2\theta. \tag{9.39}$$

Using this expression, (9.29) becomes

$$\gamma = 2\sqrt{AK_1}\int_{-\pi/2}^{\pi/2} \sqrt{\sin^2\theta + \cos^2\theta \sin^2\phi \cos^2\phi}\,\cos\theta\, d\theta \tag{9.40}$$

Setting $\sin\phi\cos\phi = s$, $\sin\theta = t$ or $\cos\theta\, d\theta = dt$, we have

$$\begin{aligned} \gamma &= 2\sqrt{AK_1}\int_{-1}^{+1} \sqrt{(1-s^2)t^2 + s^2}\, dt \\ &= \sqrt{AK_1}\left| t\sqrt{(1-s^2)t^2 + s^2} + \frac{s^2}{\sqrt{1-s^2}} \log\left(t + \sqrt{t^2 + \frac{s^2}{1-s^2}}\right)\right|_{-1}^{1} \\ &= 2\sqrt{AK_1}\left(1 + \frac{s^2}{\sqrt{1-s^2}} \log\frac{1+\sqrt{1-s^2}}{s}\right). \end{aligned} \tag{9.41}$$

If the plane of the wall is parallel to the xy plane ($\phi = \pi/2$) or to the xz plane ($\phi = 0$), it follows that $s = 0$; hence the wall energy becomes

$$\gamma_{100} = 2\sqrt{AK_1}. \tag{9.42}$$

The dependence of γ on the orientation of the wall, ϕ, is calculated by (9.41) and is graphically shown in Fig. 9.8. As seen in this graph, the energy of the wall takes the maximum value

$$\gamma_{110} = 2.88\sqrt{AK_1} \tag{9.43}$$

at $\phi = 45°$ or when the plane of the wall is parallel to (011). For iron, using the values of A and K_1 given by (9.21) and (7.7), we obtain

$$\gamma_{100} = 1.6 \times 10^{-3}\ \mathrm{J/m^2}\quad (= 1.6\ \mathrm{ergs/cm^2}), \tag{9.44}$$

$$\gamma_{110} = 2.4 \times 10^{-3}\ \mathrm{J/m^2}\quad (= 2.4\ \mathrm{ergs/cm^2}). \tag{9.45}$$

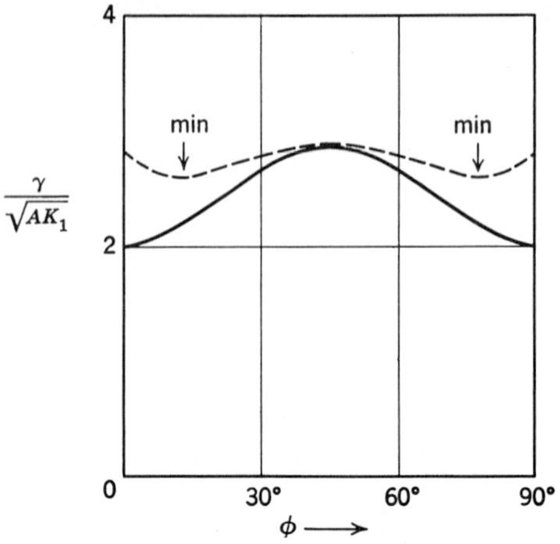

Fig. 9.8. Dependence of the surface energy of a 180° wall on its crystallographic orientation.

Since the wall energy is thus lower for the (100) plane than for the (110) plane, the wall tends to rotate its plane parallel to the (100) plane. For a single crystal plate whose wide plane is parallel to $(01\bar{1})$, therefore, the most stable orientation of the wall is not perpendicular to the wide plane, but will tilt by some angle in order to approach the (100) or (010) plane (cf. Fig. 9.9). Such a tilt of the wall, however, increases the area of the wall, so that the total energy will increase again with a tilt of the wall. The total energy of the wall is given by

$$U = ld \frac{\gamma(\phi)}{\sin(\phi - 45°)}, \tag{9.46}$$

where l is the length of the crystal measured along the direction of the domain magnetization and d is the thickness of the crystal. $U/(ld)$ is plotted as shown by the dotted curve in Fig. 9.8 as a function of ϕ, where we find two minima, at $\phi = 13°$ and 77°. It is expected, therefore, that the actual shape of the wall is zigzag, each segment making an angle of $\pm 32°$ from

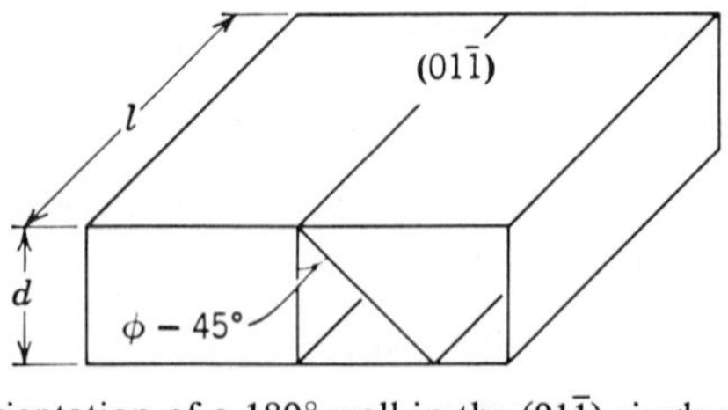

Fig. 9.9. Orientation of a 180° wall in the $(01\bar{1})$ single crystal plate.

the plane perpendicular to the crystal surface (Fig. 9.10). It is almost impossible to observe such a cross-sectional view of the 180° wall on the side surface of the crystal, because on this surface there must be a complicated domain structure induced to avoid the appearance of strong magnetic free poles. Graham and Neurath[4] confirmed such a tilt of the wall by observation of the same wall on the top and the bottom surfaces. They showed that the tilt angle was in good agreement with the calculation.

The spin structure of the 180° wall is calculated from (9.28), by using (9.39), as

$$\begin{aligned} z &= \sqrt{\frac{A}{K_1}} \int_0^\theta \frac{d\theta}{\sqrt{\sin^2\theta + \cos^2\theta \sin^2\phi \cos^2\phi}\cos\theta} \\ &= \sqrt{\frac{A}{K_1}} \int_0^{\sin\theta} \frac{dt}{\sqrt{(1-s^2)t^2 + s^2(1-t^2)}} \qquad \begin{pmatrix} s = \sin\phi\cos\phi \\ t = \sin\theta \end{pmatrix} \\ &= \frac{1}{2}\sqrt{\frac{A}{K_1}} \left| \log\left(\frac{\sqrt{(1-s^2)t^2+s^2}+t}{\sqrt{(1-s^2)t^2+s^2}-t}\right)\right|_0^{\sin\theta} \\ &= \frac{1}{2}\sqrt{\frac{A}{K_1}} \log\frac{\sqrt{(1-s^2)\sin\theta + s^2}+\sin\theta}{\sqrt{(1-s^2)\sin\theta+s^2}-\sin\theta}. \end{aligned} \tag{9.47}$$

The relation between θ and z is shown graphically in Fig. 9.11 for $\phi = 2.87°$ and 45°. When $\phi = 45°$, the rotation of spins is monotonic, as for the uniaxial anisotropy treated in Section 9.1; whereas, when $\phi = 2.87°$, the spin rotation is composed of two stages of abrupt rotation separated by a gradual rotation at its center. The reason is that the spin passes the vicinity of [001], another easy direction, where the anisotropy energy $g(\theta)$ is nearly equal to zero; hence the exchange energy should also be very small. C. Kittel[5] pointed out, however, that even when $\phi = 0$, a 180° wall would not be separated into two 90° walls, because the intermediate domain between the two 90° walls could not perform a free magnetostrictive elongation under the constriction of the domains on both sides; thus it must store a fairly large elastic energy, $\frac{1}{2}(\frac{3}{2}\lambda_{100})^2(c_{11} - c_{12})$. The

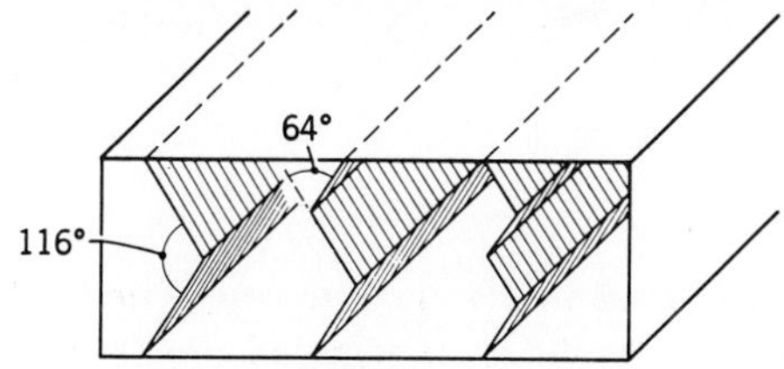

Fig. 9.10. Configuration of 180° walls in a (10$\bar{1}$) single crystal plate.

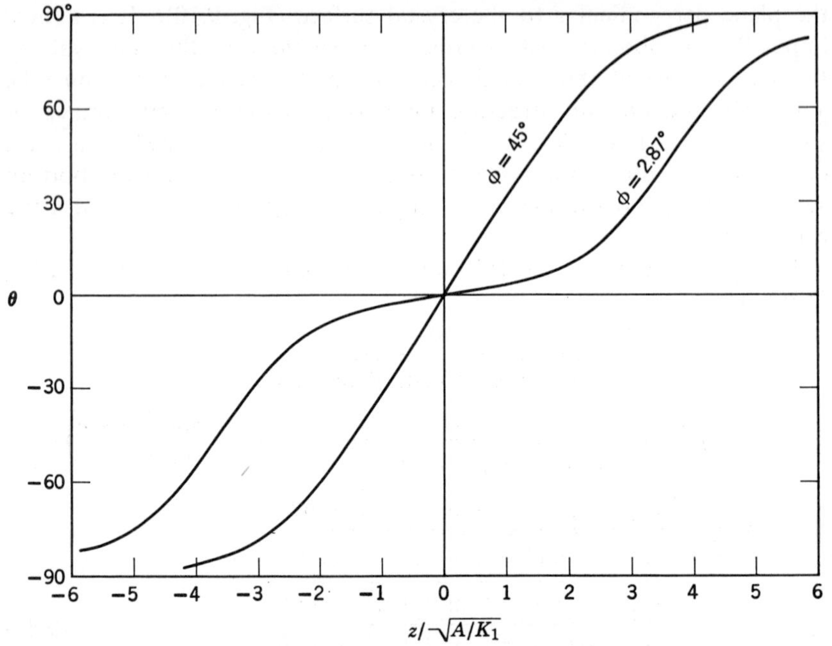

Fig. 9.11. Variation of spin direction within the 180° walls which makes the angles $\phi = 45°$ and $\phi = 2.87°$ with the (001) plane.

separation of two 90° walls, therefore, should increase the elastic energy, thus making such a configuration unstable.

This circumstance depends, of course, on the magnitude of the magnetostriction constants of the material. For a material with small magnetostriction a 180° wall is expected to split into two 90° walls. Such a double wall behaves in the same way as a single 180° wall for the field applied parallel to the direction of magnetization on either side of the double wall. When, however, the field is applied parallel to the magnetization of the intermediate domain, the double wall may be separated into two independent 90° walls, thus generating a new domain which is magnetized perpendicular to the original domains. This description presents one of the possible mechanisms of nucleation of 90° walls in the process of demagnetization.

9.3 90° Domain Walls

In this section we discuss in detail some properties of 90° walls.

The orientation of 90° walls is determined under the restriction of the continuity of the normal component of domain magnetization, as it is for

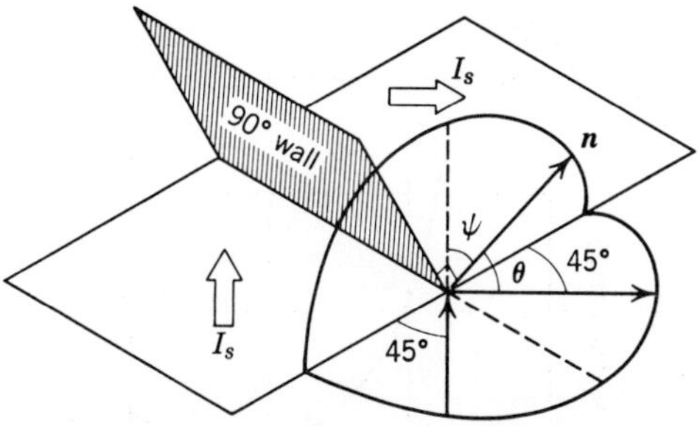

Fig. 9.12. Orientation of a 90° wall which has no magnetic free poles on its surface.

180° walls. It is easily verified that the plane of the wall which satisfies this condition must contain the bisector of the angle between the two directions of domain magnetizations on both sides of the wall. The 90° wall has, therefore, the freedom to change its orientation about this bisector (Fig. 9.12). Let the angle between the normal $\boldsymbol{n}$ to the 90° wall and the normal to the plane which contains the two directions of domain magnetizations be ψ, and the angle between $\boldsymbol{n}$ and the direction of magnetization be θ. θ must be maintained constant throughout the domain wall in order to satisfy the condition of continuity of magnetization. ψ and θ are related by the geometrical relation

$$\sin \psi = \sqrt{2} \cos \theta. \tag{9.48}$$

Now let us investigate the dependence of the wall energy on the angle ψ. In order to make it comprehensible, let us set the plane of the 90° wall parallel to the xy plane as shown in Fig. 9.13. The azimuthal angle of the

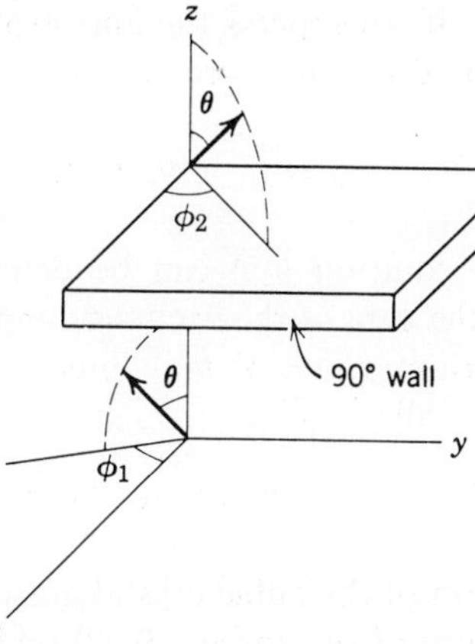

Fig. 9.13. Orientations of spins above and below a 90° wall.

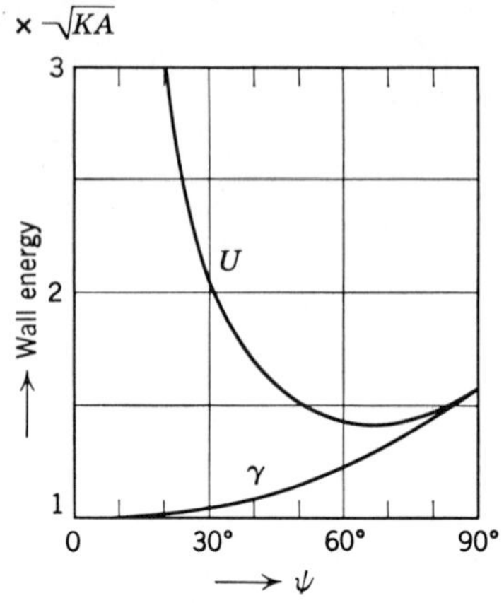

Fig. 9.14. Dependence of the surface energy of a 90° wall on its crystallographic orientation.

spin orientation changes from ϕ_1 to ϕ_2, both of which are connected to θ by

$$\sin\theta = \frac{1}{\sqrt{2}\sin\phi_1} = \frac{1}{\sqrt{2}\sin\phi_2} \tag{9.49}$$

provided $|\phi_1| = |\phi_2|$.

Since the angle between the adjacent spins is given by $\sin\theta(\partial\phi/\partial z)a$, where a is the spacing between the neighboring atomic layers, the exchange energy per unit area of the wall is given by

$$\gamma_{ex} = A\sin^2\theta\int_{-\infty}^{\infty}\left(\frac{\partial\phi}{\partial z}\right)^2 dz. \tag{9.50}$$

On the other hand, if we express the anisotropy energy by $g(\theta, \phi)$, the anisotropy energy stored in a unit area of the wall is given by

$$\gamma_a = \int_{-\infty}^{\infty} g(\theta, \phi)\, dz. \tag{9.51}$$

The details of the rotation $\theta(z)$ can be determined by a variational method to minimize the sum of the exchange and the anisotropy energies. By using the equilibrium value of $\theta(z)$ thus determined, we obtain the surface energy of the wall,

$$\gamma = 2\sqrt{A}\sin^2\theta\int_{\phi_1}^{\phi_2}\sqrt{g(\theta, \phi)}\, d\phi. \tag{9.52}$$

If we adopt the K_1 term of the cubic crystal anisotropy energy for $g(\theta, \phi)$, we have the dependence of the energy (9.52) on the wall orientation ψ as shown by the curve in Fig. 9.14. It is seen that the wall energy takes its minimum value at $\psi = 0$, where the wall is parallel to (001). The reason for this is considered to be that the total azimuthal rotation of the spin throughout the wall is only 90° here, while it is 180° for $\psi = 90°$. Thus it is concluded that a 90° wall tends to make itself parallel to one of the cubic surfaces.

Let us discuss briefly the actual shape of 90° walls which are induced by a mechanical scratch made on the (100) surface of a silicon iron single crystal. It was found that the mechanical stress induced by the scratch results in the appearance of many fine domains which have their magnetizations pointed normal to the crystal surfaces (Fig. 9.15). These domains are connected by closure domains at the crystal surface to prevent the appearance of free magnetic poles. The 90° walls between the fundamental domains and the closure domains tend to rotate their segments toward an orientation parallel to (001) (xy plane in Fig. 9.15) in order to attain the minimum surface energy. Since in this case 90° walls must stretch along the x axis, each segment must be out of the (001) plane to decrease the total area of the wall. The equilibrium shape of the wall becomes, therefore, a zigzag as shown in Fig. 9.15. The energy of the zigzag wall is

$$U = \frac{\gamma}{\sin \psi}, \tag{9.53}$$

per unit area of the basal plane of the wall. U is graphically shown in Fig. 9.14 as a function of the angle ψ. As seen in this graph, the energy takes the minimum value at $\psi = 62°$. It is concluded, therefore, that the most stable configuration of 90° walls is a zigzag shape composed of the two kinds of segments which make the angle $\psi = \pm 62°$ with the xy plane. When we observe this wall on the (100) crystal surface (yz plane in Fig. 9.15), the angle between the adjacent segments of the zigzag line (zigzag angle) is 106°.

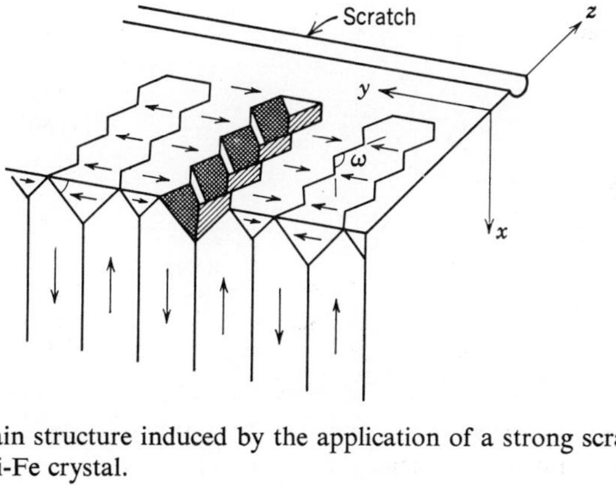

Fig. 9.15. Domain structure induced by the application of a strong scratch on a (100) surface of 3% Si-Fe crystal.

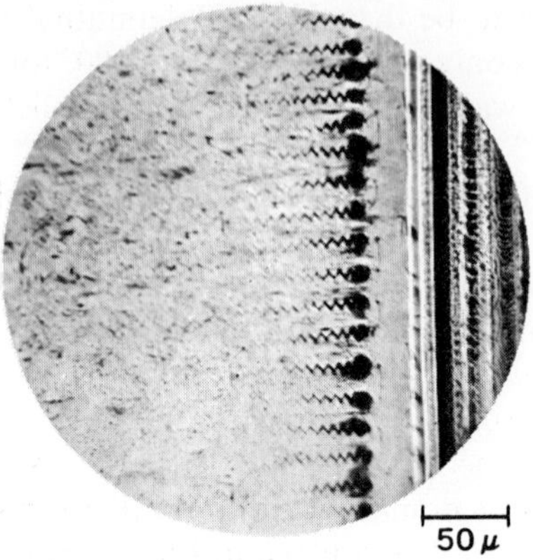

Fig. 9.16. Zigzag 90° walls produced by a scratch made on a (100) surface of 3% Si-Fe crystal (Chikazumi and Suzuki[6]).

Figure 9.16 shows the actual domain pattern observed on a (100) plane of 3% Si-Fe crystal; the zigzag 90° wall is induced by a mechanical scratch made on the crystal surface. It was observed that the zigzag angle ω is approximately 106° at places distant from the scratch, whereas it becomes smaller in the vicinity of the scratch as shown in Fig. 9.17. The reason for this is considered to be as follows: Because of a strong stress prevailing at the vicinity of the scratch, the depth of closure domains becomes smaller; hence some amount of free poles will appear on the 90°

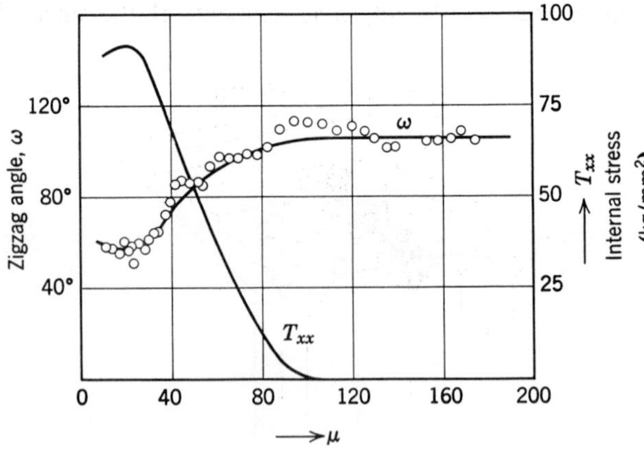

Fig. 9.17. Variation of the zigzag angle ω and the calculated internal stress T_{xx} with distance from the scratch (Chikazumi and Suzuki[6]).

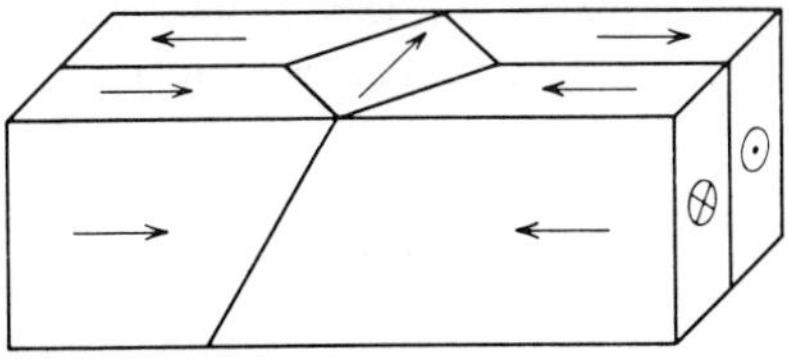

Fig. 9.18. Orientation of the 90° wall observed in an iron whisker.

wall owing to the discontinuity of the normal component of magnetization. It should be noted that the total amount of free poles thus induced is independent of the zigzag angle, while its magnetostatic energy can be extensively decreased by reducing the zigzag angle because of the dilution of the free poles over wider surfaces. Thus the stable zigzag angle can be calculated for a given intensity of internal stress. Figure 9.17 shows the value of internal stress estimated from the observed zigzag angle. From this experiment it was estimated that the maximum internal stress is as large as 90 kg/mm^2† in the vicinity of the scratch.

This experiment[6] was performed for the purpose of clarifying the nature of the maze-like domain pattern which is normally observed on mechanically polished crystal surfaces (cf. Fig. 6.4*a*). It was found by careful observation that the domain boundaries in this pattern, which are very thick and vague, are composed of fine zigzag lines. A similar observation was reported by Stephan.[7,8]

Direct observation of the crystallographic orientation of the 90° wall was made by De Blois and Graham[9] for iron whisker crystals which were grown by reducing ferrous bromide by hydrogen gas at about 700°C. The domain pattern observed on this crystal revealed that the 90° wall is inclined from the normal to the crystal surface as shown in Fig. 9.18.

Problems

9.1. Assuming that a crystal with the isotropic magnetostriction constant λ and with negligibly small crystal anisotropy energy is under the action of the uniform tension σ, formulate the surface energy of a 180° wall which is present in this crystal.

9.2. Calculate the surface energy of 180° walls which are parallel to the (100) and (110) planes in the crystal with $K_1 = 0$ and $K_2 > 0$.

† In the original paper[6] the μ^* correction was neglected in the estimation of magnetostatic energy. With this correction being taken into consideration, the estimated value of the internal stress was reduced to about one-fourth of the original value.

9.3. Calculate the surface energy of a 180° wall which is parallel to (100) of the b.c.c. crystal with S (spin) $= \frac{1}{2}$, a (lattice constant) $= 3$ Å, Θ (Curie point) $=$ 527°C, and $K_1 = 6 \times 10^4$ J/m^3.

References

9.1. F. Bloch: *Z. Physik.* **74,** 295 (1932).
9.2. P. R. Weiss: *Phys. Rev.* **74,** 1493 (1948).
9.3. S. Chikazumi: *Phys. Rev.* **85,** 918 (1952).
9.4. C. D. Graham, Jr., and P. W. Neurath: *J. Appl. Phys.* **28,** 888 (1957).
9.5. C. Kittel: G.6, p. 563.
9.6. S. Chikazumi and K. Suzuki: *J. Phys. Soc. Japan* **10,** 523 (1955).
9.7. W. Stephan: *Exptl. Tech. Physik.* **4,** 153 (1956).
9.8. W. Stephan: *Exptl. Tech. Physik.* **5,** 145 (1957).
9.9. R. W. De Blois and C. D. Graham, Jr.: *J. Appl. Phys.* **29,** 526 (1958).

10

Magnetostatic Energy

10.1 Fundamental Formulas

The correct concept of magnetic domain structures was first obtained by Landau and Lifshitz[1] and by Néel[2] when they took the magnetostatic energy into consideration in the calculation of the size of magnetic domains. This calculation was experimentally verified by Williams, Bozorth, and Shockley[3] in 1949.

The magnetostatic energy is due to the Coulomb interaction between magnetic free poles. Consider a system consisting of several permanent magnets (Fig. 10.1). Let the intensities of the free poles be $m_1, m_2, \ldots, m_i, \ldots, m_n$, and the magnetic potential at the position of each free pole be $\phi_1, \phi_2, \ldots, \phi_i, \ldots, \phi_n$. Then the potential energy of the system is

$$U = \frac{1}{2} \sum_{i=1}^{n} m_i \phi_i. \tag{10.1}$$

Since ϕ_i is the potential due to the free poles other than m_i, it is expressed as

$$\phi_i = \sum_{j \neq i} \frac{m_j}{4\pi\mu_0 r_{ij}}, \tag{10.2}$$

where r_{ij} is the distance between the ith and the jth free poles. If the free poles are distributed over space, the energy is expressed by the volume integral

$$U = \tfrac{1}{2} \iiint \rho_m \phi \, dv, \qquad (10.3$$

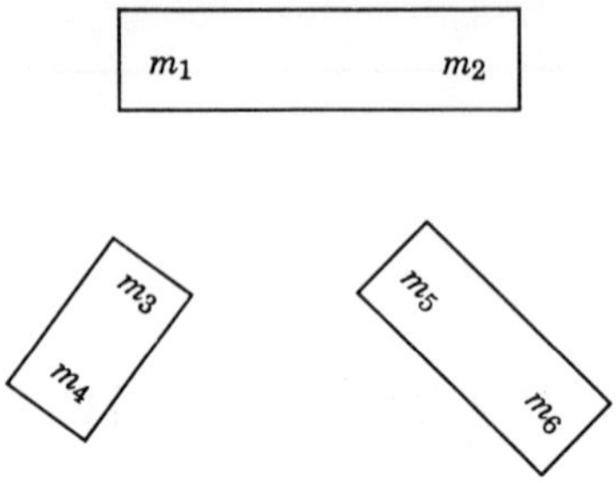

Fig. 10.1. Assembly of permanent magnets.

where the potential ϕ is calculated by

$$\phi = \iiint \frac{\rho_m}{4\pi\mu_0 r} dv. \tag{10.4}$$

It is also calculated by solving the Poisson equation,

$$\Delta\phi = -\frac{\rho_m}{\mu_0}, \tag{10.5}$$

(under the given boundary conditions), or the Laplace equation,

$$\Delta\phi = 0, \tag{10.6}$$

at points where there are no magnetic free poles.

The free pole density ρ_m induced in a ferromagnetic medium is expressed in terms of the magnetization as

$$\operatorname{div} \boldsymbol{I}_s = -\rho_m. \tag{10.7}$$

If the substance is homogeneous, the magnitude of $\boldsymbol{I}_s$ should be constant, so that div $\boldsymbol{I}_s$ is expressed in terms of direction cosines $(\alpha_1, \alpha_2, \alpha_3)$ of $\boldsymbol{I}_s$ as

$$\frac{\partial\alpha_1}{\partial x} + \frac{\partial\alpha_2}{\partial y} + \frac{\partial\alpha_3}{\partial z} = -\frac{\rho_m}{I_s} \tag{10.8}$$

or

$$\operatorname{div} \boldsymbol{\alpha} = -\frac{\rho_m}{I_s}, \tag{10.9}$$

where $\boldsymbol{\alpha}$ is the unit vector which is parallel to $\boldsymbol{I}_s$.

For example, for a sphere which is radially magnetized as shown in Fig. 10.2, (10.9) becomes

$$\frac{1}{r^2}\frac{\partial}{\partial r}(r^2) = -\frac{\rho_m}{I_s}. \tag{10.10}$$

Solving this equation, we have

$$\rho_m = -\frac{2I_s}{r}. \tag{10.11}$$

Then (10.5) becomes

$$\frac{1}{r^2}\frac{\partial}{\partial r}\left(r^2\frac{\partial\phi}{\partial r}\right) = \frac{2I_s}{\mu_0 r}. \tag{10.12}$$

Solving (10.12), we obtain

$$\phi = \frac{I_s}{\mu_0}(r - R). \tag{10.13}$$

The magnetostatic energy of the system is then

$$U = \frac{4\pi I_s^{\,2}}{\mu_0}\int_0^R (r^2 - rR)\,dr = \frac{I_s^{\,2}}{\mu_0}\frac{2\pi}{3}R^3. \tag{10.14}$$

The magnetostatic energy can be expressed in terms of I and H instead of ρ_m and ϕ. Let the length of the kth magnet be l_k and the magnetic poles at its ends by m_k and $-m_k$. Then the magnetic moment is

$$M_k = m_k l_k \tag{10.15}$$

On the other hand, the difference in the magnetic potential between the positions of the positive and the negative magnetic poles is

$$\phi_+ - \phi_- = -H_{\|} l_k, \tag{10.16}$$

where $H_{\|}$ is the component of the magnetic field parallel to the magnetic moment M_k. Then

$$m_k\phi_+ - m_k\phi_- = -\left(\frac{M_k}{l_k}\right) \times (H_{\|} l_k)$$

$$= -(\boldsymbol{M}_k \cdot \boldsymbol{H}_k), \tag{10.17}$$

and (10.1) becomes

$$U = -\frac{1}{2}\sum_{k=1}^{n}(\boldsymbol{M}_k \cdot \boldsymbol{H}_k), \tag{10.18}$$

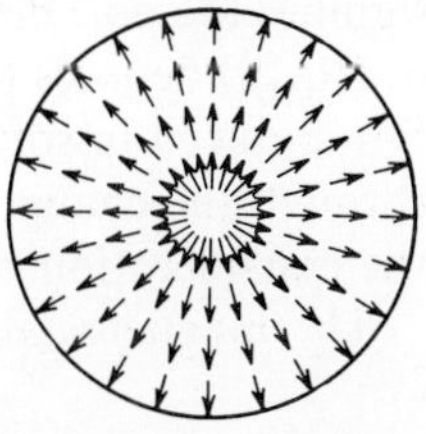

Fig. 10.2. A ferromagnetic sphere magnetized radially.

where n is the total number of magnets. When the magnetization $\boldsymbol{I}$ is distributed over the space, the magnetostatic energy is

$$U = -\tfrac{1}{2}\iiint(\boldsymbol{I} \cdot \boldsymbol{H})\,dv. \tag{10.19}$$

By using this equation, we can calculate the magnetostatic energy of the sphere shown in Fig. 10.2. Since the intensity of the magnetic field inside the sphere is calculated to be

$$H_r = -\frac{\partial \phi}{\partial r} = -\frac{I_s}{\mu_0}, \tag{10.20}$$

and it is zero outside the sphere, the magnetostatic energy can be calculated from (10.19) as

$$U = \frac{1}{2}\frac{I_s^{\,2}}{\mu_0}\frac{4\pi R^2}{3} = \frac{I_s^{\,2}}{\mu_0}\frac{2\pi}{3}R^3, \tag{10.21}$$

which agrees exactly with (10.14).

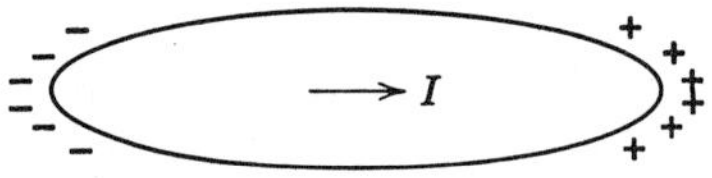

Fig. 10.3. A magnetized ferromagnetic body with a finite size.

Let us calculate the magnetostatic energy for another example. Consider a magnetic body with demagnetizing factor N magnetized to the intensity I (Fig. 10.3). Since the internal field is

$$H = -\frac{N}{\mu_0} I, \tag{10.22}$$

the magnetostatic energy is expressed as

$$U = \frac{1}{2\mu_0} N I^2 v, \tag{10.23}$$

where v is the volume of the magnetic body.

It should be noted that (10.3) and (10.19) can also be used for the calculation of magnetostatic energy, even when some soft magnetic materials are present with the permanent magnets. In this case ρ_m in (10.3) signifies the pole density produced only by permanent magnetization, and I in (10.19) denotes the permanent magnetization. The effect of magnetization induced in soft magnetic materials comes into (10.3) and (10.19) only through the change in ϕ or H, respectively. We discuss this point later in the consideration of the μ^* correction.

The magnetostatic energy can be also expressed in terms of B and H:

$$U = \tfrac{1}{2}\iiint (\boldsymbol{B}\cdot\boldsymbol{H})\, dv. \tag{10.24}$$

The integration should be carried out over the space where magnetic field $\boldsymbol{H}$ is present. It should be noted that $\boldsymbol{B}$ in the integrand is calculated by excluding the permanent magnetization†. When there are no soft magnetic

† Equation (10.24) was derived from (10.3) by using the relation $\rho_m = \operatorname{div} \boldsymbol{B}$. If the permanent magnetization is included in $\boldsymbol{B}$, div $\boldsymbol{B}$ must always be zero.

materials, $\boldsymbol{B}$ should equal $\mu_0 \boldsymbol{H}$, so that (10.24) becomes

$$U = \frac{\mu_0}{2} \iiint H^2 \, dv. \tag{10.25}$$

If there are some soft magnetic materials, (10.24) becomes

$$U = \frac{\mu}{2} \iiint_{\substack{\text{magnetic} \\ \text{material}}} H^2 \, dv + \frac{\mu_0}{2} \iiint_{\substack{\text{other} \\ \text{space}}} H^2 \, dv. \tag{10.26}$$

Considering again the example shown in Fig. 10.2, the magnetic field inside the sphere is given by (10.20) and $H = 0$ outside the sphere; hence (10.25) gives

$$U = \frac{I_s^2}{\mu_0} \frac{2\pi}{3} R^3, \tag{10.27}$$

which is again in agreement with (10.14) and (10.21).

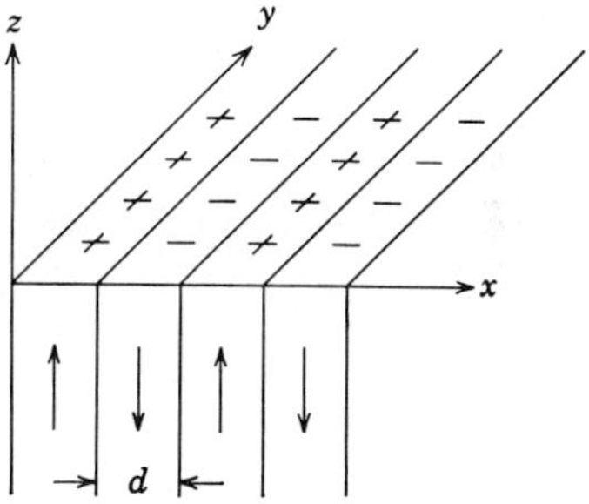

Fig. 10.4. Surface distribution of the magnetic free poles at the end surface of laminated domains.

10.2 Surface Distribution of Magnetic Poles

In this section we calculate the magnetostatic energy of actual domain structures. Let us consider the domain structure shown in Fig. 10.4. Set the crystal surface parallel to the xy plane at $z = 0$. The crystal is divided into many domain sheets which are parallel to the yz plane, with magnetizations parallel to $\pm z$. It is assumed for simplicity that the crystal is infinitely large in the direction of $\pm x$, $\pm y$, and $-z$. The surface density of free poles, ω, which appears on the crystal surface is expressed as

$$\omega = \begin{cases} I_s & 2kd < x < (2k+1)d \\ -I_s & (2k+1)d < x < 2(k+1)d, \end{cases} \tag{10.28}$$

where d is the width of the domains and k is an integer.

The first task we must perform is the calculation of the magnetic potential ϕ at the crystal surface, because we can then calculate the magnetostatic energy immediately by using (10.3). The magnetic potential must satisfy the Laplace equation,

$$\frac{\partial^2 \phi}{\partial x^2} + \frac{\partial^2 \phi}{\partial z^2} = 0, \tag{10.29}$$

where we have at $z = 0$ the boundary condition

$$\left(\frac{\partial \phi}{\partial z}\right)_{z=+0} - \left(\frac{\partial \phi}{\partial z}\right)_{z=-0} = -\frac{\omega}{\mu_0}. \tag{10.30}$$

Since the magnetic poles exist in the plane $z = 0$, ϕ is expected to be symmetrical about this plane; hence

$$\left(\frac{\partial \phi}{\partial z}\right)_{z=+0} = -\left(\frac{\partial \phi}{\partial z}\right)_{z=-0}. \tag{10.31}$$

Then (10.30) becomes

$$\left(\frac{\partial \phi}{\partial z}\right)_{z=+0} = \frac{\omega}{2\mu_0}. \tag{10.32}$$

The solution of the Laplace equation (10.29) must be in the form

$$\phi = \sum_{n=1}^{\infty} A_n \sin n\left(\frac{\pi}{d}\right)x \cdot e^{n(\pi/d)z}. \qquad (n\text{: odd integer}). \tag{10.33}$$

We recognize easily that (10.33) is a possible solution, because each term in the summation of (10.33) satisfies (10.29). The coefficient A_n of each term must be determined so as to satisfy the boundary condition (10.32). On substituting ϕ and ω given by (10.33) and (10.28), we have

$$\frac{\pi}{d}\sum_{n=1}^{\infty} nA_n \sin n\left(\frac{\pi}{d}\right)x = \begin{cases} \dfrac{I_s}{2\mu_0} & \text{for } 2kd < x < (2k+1)d \\[2ex] -\dfrac{I_s}{2\mu_0} & \text{for } (2k+1)d < x < 2(k+1)d. \end{cases}$$

Multiplying both sides of the preceding equation by $\sin n(\pi/d)x$ and integrating from $2kd$ to $2(k+1)d$, we have

$$\begin{aligned} A_n &= \frac{I_s}{2\pi\mu_0 n}\left[\int_{2kd}^{(2k+1)d} \sin n\left(\frac{\pi}{d}\right)x\,dx - \int_{(2k+1)d}^{2(k+1)d} \sin n\left(\frac{\pi}{d}\right)x\,dx\right] \\ &= \frac{2I_s d}{\pi^2\mu_0 n^2} \qquad (n\text{: odd}). \end{aligned} \tag{10.34}$$

Then the magnetic potential at $z = 0$ becomes

$$\phi_{z=0} = \frac{2I_s d}{\pi^2\mu_0}\sum_{n=1}^{\infty}\frac{1}{n^2}\sin n\left(\frac{\pi}{d}\right)x, \tag{10.35}$$

so that the magnetostatic energy per unit area of the crystal surface

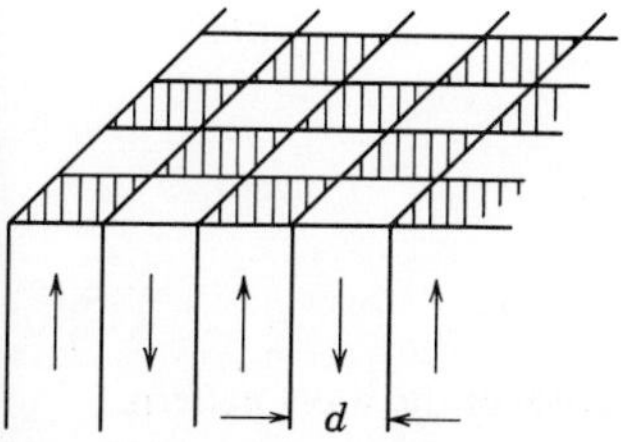

Fig. 10.5. Surface distribution of magnetic free poles of the checkerboard pattern.

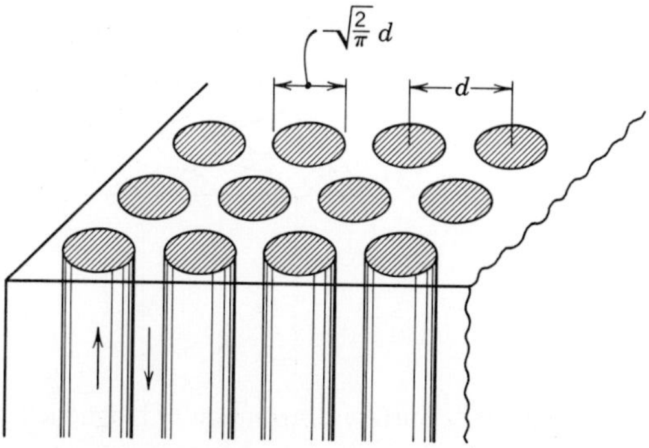

Fig. 10.6. Surface distribution of magnetic free poles of the circular pattern.

becomes

$$\begin{aligned}\epsilon_m &= \frac{I_s^2 d}{\pi^2 \mu_0} \sum_{n=1}^{\infty} \frac{1}{n^2 d} \int_0^d \sin n\left(\frac{\pi}{d}\right) x\, dx \\ &= \frac{2I_s^2 d}{\pi^3 \mu_0} \sum_{n\,:\,\mathrm{odd}}^{\infty} \frac{1}{n^3} \\ &= \frac{2I_s^2 d}{\pi^3 \mu_0} \times 1.0517 \\ &= 5.40 \times 10^4 I_s^2 d \;(= 0.852 I_s^2 d \text{ in cgs}).\end{aligned} \tag{10.36}$$

Kittel[4] also calculated the magnetostatic energy of the free pole distribution of the checkerboard pattern and that of the circular pattern as shown in Figs. 10.5 and 10.6 and obtained

$$\epsilon_m = 3.36 \times 10^4 I_s^2 d \quad (= 0.53 I_s^2 d \text{ in cgs}) \tag{10.37}$$

for the checkerboard pattern, and

$$\epsilon_m = 2.37 \times 10^4 I_s^2 d \quad (= 0.374 I_s^2 d \text{ in cgs}) \tag{10.38}$$

for the circular pattern.

It should be kept in mind that the magnetostatic energy is always proportional to the width d of the domain. The numerical coefficient is smaller for the checkerboard and the circular patterns than that of coplanar strips. The reason is that the fine mixture of N and S free poles is a natural consequence of the attractive force between N and S poles. Goodenough[5] calculated the magnetostatic energy of the wavy distribution of free poles as shown in Fig. 10.7 and showed the numerical coefficient is closer to that of the checkerboard pattern rather than that of the

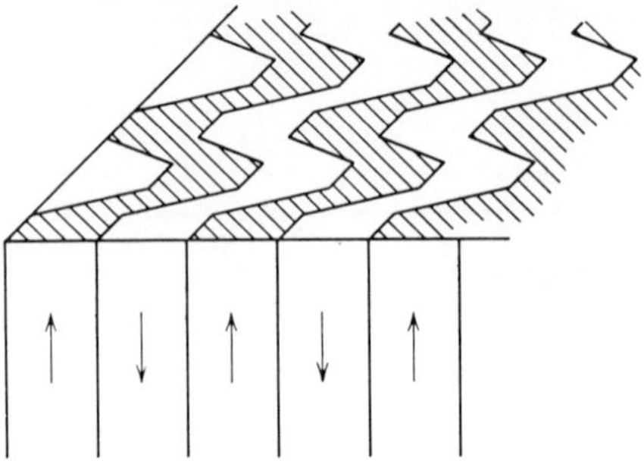

Fig. 10.7. Surface distribution of magnetic free poles of the wavy pattern.

coplanar strips. Such a wavy domain pattern is actually observed for some thin magnetic films (cf. Chapter 18).

When one sort of magnetic free pole is appearing in a limited region, the magnetostatic energy of the system can be lowered by diluting the same amount of free poles over a larger volume. Figure 10.8*a* shows a square hole in a ferromagnetic crystal with magnetic poles on its surfaces. In such a crystal the blade-shaped domains are induced on both sides of the hole as shown in Fig. 10.8*b*, so as to decrease the magnetostatic energy. The total amount of one sort of magnetic pole is exactly the same in (*a*) and (*b*) of Fig. 10.8, because, if we apply Gauss' theorem,

$$\iint I_n \, dS = -\iiint \rho_m \, dv, \tag{10.39}$$

to the closed surfaces shown by broken lines in the figure, we find that the total amount of free poles is given by $I_n S$ for both cases, where I_n is the normal component of magnetization and S is the cross-sectional area of the hole. The free pole of intensity $I_n S$ is diluted over larger surfaces

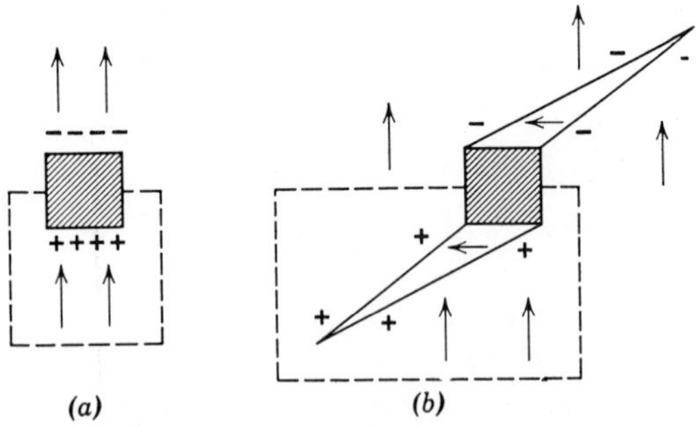

Fig. 10.8. Free pole distribution around the square hole in a ferromagnetic crystal: (*a*) without any small domains; (*b*) with blade domains.

in Fig. 10.8*b*, so that the magnetostatic energy is expected to be lower than in Fig. 10.8*a*.

Let us calculate the magnetostatic energy quantitatively for both cases. The free pole distributions of both of them can be approximated to those of the sphere and the elongated rotational ellipsoid as shown in Fig. 10.9. Using formula (10.23), we obtain for the magnetostatic energy of the sphere

$$U = \frac{I_s^2}{6\mu_0}\frac{4\pi}{3}r^3 = 5.55 \times 10^5 I_s^2 r^3 \quad (= 8.80 I_s^2 r^3 \text{ in cgs}), \qquad (10.40)$$

where r is the radius of the sphere. On the other hand, we find for the magnetostatic energy of the ellipsoid

$$U = \frac{I_s^2}{2\mu_0 k^2}(\log_e 2k - 1)\frac{4\pi}{3}\frac{r^2 l}{2}, \qquad (10.41)$$

where l is the length of the ellipsoid (or twice of the length of the blade domain), r is the radius of the cross section, and k is the dimensional ratio or the ratio of length to diameter. If we tentatively assume that $k = 10$,

$$\frac{1}{k^2}(\log_e 2k - 1) = 0.020,$$

and

$$\frac{r^2 l}{2} = 10r^3.$$

Then (10.41) becomes

$$U = 3.33 \times 10^5 I_s^2 r^3 \quad (= 5.27 I_s^2 r^3 \text{ in cgs}). \qquad (10.42)$$

Comparing (10.42) and (10.40), we find that the magnetostatic energy is obviously lowered by the dilution of the free poles. The same sort of discussion has already been presented in Chapter 9 regarding the dependence of the zigzag angle of 90° walls on the intensity of free poles induced on the wall or on the intensity of the internal stress. The dilution of free

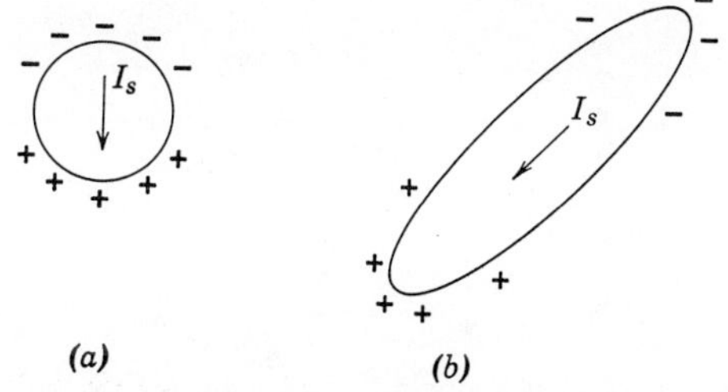

Fig. 10.9. Free pole distributions approximating those of Fig. 10.8.

poles is effected by decreasing the zigzag angle, and a decrease in the magnetostatic energy of the system results[6] (cf. Section 9.3).

10.3 μ* Correction

In the preceding discussion we imagined that the domain magnetizations are fixed in the directions of easy magnetization. Actually, however, if the internal field which is produced by free poles exerts a torque on the domain magnetizations, they will, more or less, rotate out of the directions of easy magnetization. Then the magnetostatic energy of the system is expected to be greatly changed by the redistribution of the free poles. This correction was first considered independently by Lifshitz, Néel, and Shockley.[7]

Let us suppose that the domain structure is as shown in Fig. 10.10. The magnetization makes the small incidence angle θ with the crystal surface. The surface density of magnetic poles is given by

$$\omega = \pm I_s \sin \theta. \tag{10.43}$$

If the directions of domain magnetizations are frozen in along the direction of easy magnetization, the magnetostatic energy per unit area of the crystal surface becomes, from (10.36),

$$\epsilon_m = 5.40 \times 10^4 I_s^2 d \sin^2 \theta \quad (= 0.852 I_s^2 d \sin^2 \theta \text{ in cgs}). \tag{10.44}$$

In fact, however, the direction of domain magnetization near the crystal surface is changed by the action of the strong demagnetizing field produced by the surface free poles. The magnetic free poles, which were previously confined in the crystal surface (Fig. 10.11*a*), are released and spread into the interior of the crystal as shown in Fig. 10.11*b*. The magnetostatic energy is then expected to be greatly reduced, because such dilution of free poles is a natural consequence of the repulsive interaction between the

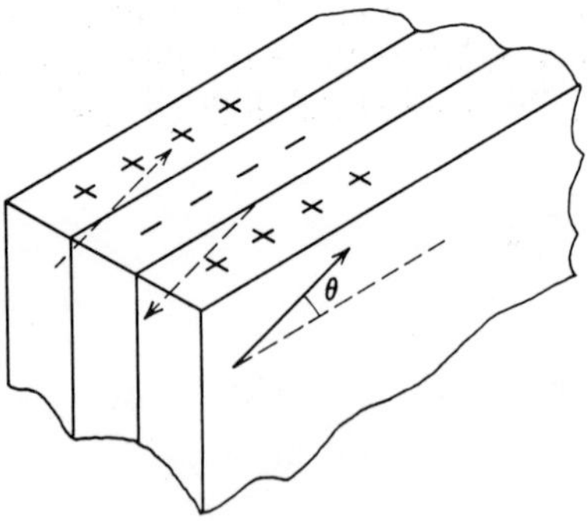

Fig. 10.10. Domain structure of a single crystal plate whose wide surface makes the angle θ with (001).

free poles. This decrease in magnetostatic energy is counterbalanced by an increase in the anisotropy energy caused by a rotation of domain magnetizations out of easy directions.

One of the methods of solving this problem is to determine the equilibrium directions of the spins from the minimum energy condition and then to calculate the magnetostatic energy for the volume distribution of magnetic free poles. The total energy of the system is given by the sum of the magnetostatic energy thus obtained and the anisotropy energy due to the rotation of spins.

A more elegant method is to replace the ferromagnetic crystal tentatively by a homogeneous magnetic material with the reversible permeability μ^* by leaving the original surface distribution of the magnetic free poles unchanged (Fig. 10.12). The asterisk with μ signifies that the permeability assumed here is different from the actual permeability μ, which is affected by wall displacement.

Here the boundary condition (10.30) should be changed to

$$\mu_0\left(\frac{\partial\phi}{\partial z}\right)_{z=+0} - \mu^*\left(\frac{\partial\phi}{\partial z}\right)_{z=-0} = -\omega. \tag{10.45}$$

This equation means that the divergence of magnetic flux, which does not include the permanent magnetization, is equal to the magnetic free pole density. Let us suppose that the potential for $\mu^* = \mu_0$ is

$$\phi = f(x, z). \tag{10.46}$$

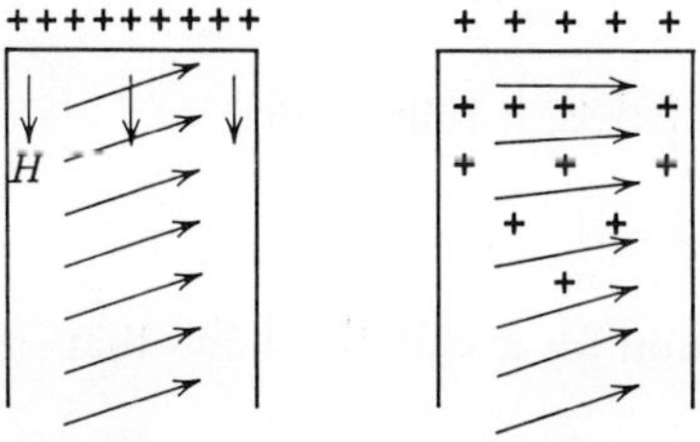

Fig. 10.11. Free pole distribution of one domain in Fig. 10.10: (*a*) when the domain magnetization is fixed in the direction of easy magnetization; (*b*) when the domain magnetization rotates under the influence of the demagnetizing field.

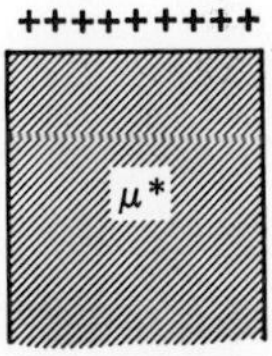

Fig. 10.12. A soft magnetic material with a magnetic pole on its surface. This configuration is equivalent to that in Fig. 10.11*b* regarding the magnetostatic energy.

When $\mu^* = \mu^*$, the potential can be expressed in the form

$$\begin{aligned} \phi &= Af(x, \alpha z) \qquad z > 0, \\ \phi &= Af(x, \beta z) \qquad z < 0, \end{aligned} \tag{10.47}$$

because the potential might no longer be symmetrical with respect to the plane $z = 0$. Putting (10.47) into (10.45), we have

$$\mu_0 A\alpha f_z(x, +0) - \mu^* A\beta f_z(x, -0) = -\omega, \tag{10.48}$$

where f_z means $\partial f(x, z)/\partial z$. Since (10.46) must satisfy (10.30) and (10.31),

$$f_z(x, +0) - f_z(z, -0) = -\frac{\omega}{\mu_0} \tag{10.49}$$

and

$$f_z(x, +0) = -f_z(x, -0). \tag{10.50}$$

Putting these two equations into (10.48), we have

$$A(\mu_0\alpha + \mu^*\beta) = 2\mu_0. \tag{10.51}$$

On the other hand, ϕ must satisfy the Laplace equation in both cases, so that for (10.47)

$$\Delta\phi = \frac{\partial^2\phi}{\partial x^2} + \frac{\partial^2\phi}{\partial z^2} = Af_{xx} + A\alpha^2 f_{zz} = 0, \tag{10.52}$$

and for (10.46)

$$\Delta\phi = f_{xx} + f_{zz} = 0. \tag{10.53}$$

When (10.52) and (10.53) are compared, it follows that

$$\alpha = 1. \tag{10.54}$$

Similarly, from the Laplace equation for $z < 0$, it follows that

$$\beta = 1. \tag{10.55}$$

Thus from (10.51) we have

$$A = \frac{2}{1 + \bar{\mu}^*}. \tag{10.56}$$

Referring to (10.35), we see that the potential at $z = 0$ thus becomes

$$\phi_{(z=0)} = \frac{4I_s d \sin\theta}{(1 + \bar{\mu}^*)\pi^2\mu_0} \sum_{n=1}^{\infty} \frac{1}{n^2} \sin n\left(\frac{\pi}{d}\right)x. \tag{10.57}$$

The magnetostatic energy is also A times as large as (10.36); thus

$$\epsilon_m = \frac{2}{1+\bar{\mu}^*}(5.40 \times 10^4) I_s^2 d \sin^2\theta$$

$$\left(= \frac{2}{1+\bar{\mu}^*} 0.852 I_s^2 d \sin^2\theta \text{ in cgs}\right). \tag{10.58}$$

The value of μ^* depends on the ease of rotation of domain magnetization from the direction of easy magnetization. For the structure shown in Fig. 10.10, we put $\theta_0 \approx \pi/2$ in formula (13.13), which expresses the permeability due to rotation magnetization; thus

$$\bar{\mu}^* = 1 + \frac{I_s^2}{2\mu_0 K_1}. \tag{10.59}$$

When the direction of easy magnetization makes a 45° angle with the crystal surface or when the magnetic poles are appearing on a 90° wall, $\theta_0 \approx \pi/4$ in (13.13); hence

$$\bar{\mu}^* = 1 + \frac{I_s^2}{4\mu_0 K_1}. \tag{10.60}$$

For the magnetic poles on a wall, both sides of the wall are filled with the magnetic material, so that, instead of (10.56), we must use

$$A = \frac{1}{\mu^*}. \tag{10.61}$$

For iron, $K_1 = 4.2 \times 10^4$ J/m³, $I_s = 2.15$ Wb/m², it turns out that $\bar{\mu}^* = 45$; hence the coefficient (10.56) gives a value as small as $A = 0.0435$. In this connection we cannot neglect this correction, even in a rough estimation of the magnetostatic energy.

For domain structures as shown by Figs. 10.4, 10.5, 10.6, and 10.7, where the domain magnetizations point normal to the crystal surface, the μ^* correction is not necessary because $\bar{\mu}^* = 1$.

Problems

10.1. An infinitely long magnetic cylinder with radius R is magnetized in the radial directions to its saturation magnetization I_s. Calculate the magnetostatic energy per unit length along its long axis.

10.2. Calculate the magnetostatic energy per unit area of the plane on which magnetic free poles of surface density $\omega = I_s \sin(\pi/d)x$ are distributed. Ignore the μ^* correction.

10.3. Consider $\pm x$ domains and $\pm y$ domains with thickness d and separated by a 90° wall plane which makes an angle ϕ with the x axis. How does the magnetostatic energy per unit area of the 90° wall change with a change of ϕ?

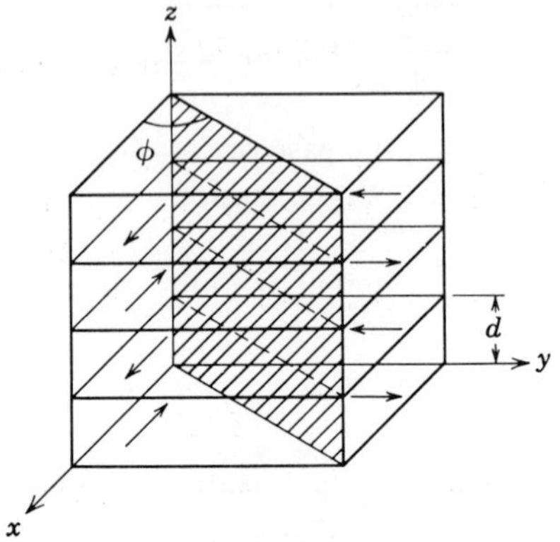

Problem 10.3.

Let the intensity of domain magnetization be I_s and the anisotropy constant be K_1, and assume that the magnetic body is infinitely large.

References

10.1. L. Landau and E. Lifshitz: *Physik. Z. Sowjetunion* **8,** 153 (1953); E. Lifshitz: *J. Phys. USSR* **8,** 53 (1944).
10.2. L. Néel: Cahiers Phys **25,** 1, 22 (1944).
10.3. H. J. Williams, R. M. Bozorth, and W. Shockley: *Phys. Rev.* **75,** 155 (1949).
10.4. C. Kittel: G.6.
10.5. J. B. Goodenough: *Phys. Rev.* **102,** 356 (1956).
10.6. S. Chikazumi and K. Suzuki: *J. Phys. Soc. Japan* **10,** 523 (1955).
10.7. W. Shockley: *Phys. Rev.* **73,** 1246 (1948).

11

Domain Structure

11.1 Domain Pattern and Its Interpretation

The most direct verification of the domain structure of ferromagnetic materials is the actual experimental observation of magnetic domain patterns. The first experiments were made by Bitter[1] and independently by Hamos and Thiessen.[2] In their method a drop of colloidal suspension, which contains fine ferromagnetic colloidal particles, is applied to the ferromagnetic crystal surface, and the pattern of magnetic structures revealed by the assembly of the colloidal particles is observed. This method is called the powder pattern method or the Bitter figure technique. Many investigators tried to observe the domain pattern by means of this technique, but no one obtained a comprehensive pattern until Williams, Bozorth, and Shockley[3] succeeded in observing beautiful images of true magnetic domains. By then, Elmore[4] had developed a number of techniques, such as electrolytic polishing and preparation of excellent magnetic colloids. At that time proper consideration was not given to the crystal orientation of the specimen, because the domain structure was thought to be an inherent property of the ferromagnetic material.

The colloidal suspension used for powder pattern method is prepared as follows (see Fig. 11.1). First dissolve 2 g of hydrated ferrous chloride, $FeCl_2 \cdot 2H_2O$, and 5.4 g of hydrated ferric chloride, $FeCl_3 \cdot 6H_2O$, in 300 cc of water and maintain the solution at $30 \sim 40°C$; then add a solution of NaOH (5 g in water at 50°C) at a constant rate with constant stirring with a mechanical stirrer. Black magnetite precipitate is deposited as a result

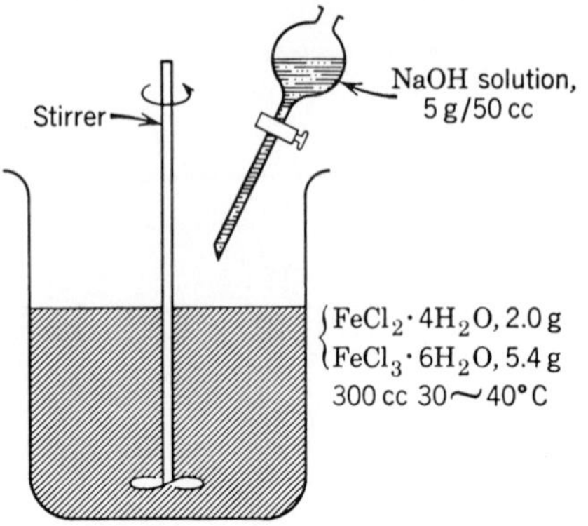

Fig. 11.1. Preparation of Fe_3O_4 precipitate.

of the reactions

$$FeCl_2 + 2FeCl_3 + 8NaOH = Fe(OH)_2 + 2Fe(OH)_3 + 8NaCl$$

$$Fe(OH)_2 + 2Fe(OH)_3 = Fe_3O_4 + 4H_2O$$

It must be noted that elevation of the temperature of the solution will result in a coarsening of the precipitated particles. After the solution has been poured out by tilting the container, repeat the procedure by adding hot water to the deposition. Then filter the deposition through a funnel-shaped filter paper and wash the deposition a few times. A new filter paper should be used for each washing. The purpose of these procedures is to remove the Na^+ and Cl^- ions completely from the deposition. Trace of Cl^- ion can be detected by using silver nitrate solution. The deposition of magnetite thus prepared can be stored in a damp state in a glass container.

Next prepare a 0.3% soap solution, using a soap of good quality which contains no sodium chloride. Add 1 cc of magnetite deposition to about 30 cc of soap solution and mix them by boiling a little or by stirring by means of ultrasonic waves. A good colloidal suspension looks like brownish red ink. Baker, Winslow, and Bittrich[5] found that coconut oil amine, the largest component of which is dodecylamine, is good as a dispersant. It is important that the pH of the solution be maintained at 7 by adding a proper amount of $1N$ HCl, and also that the stirring of the mixture be vigorous. Craik and Griffith[6] in England used $0.5 \sim 1\%$ Celacol solution with 10% glycerin as a dispersant. This colloidal suspension was found to leave magnetic particles in a dispersed state on the Celacol membrane, which covers the crystal surface after the suspension is dried up. Removing this membrane from the crystal surface, they observed details of domain structure by means of the electron microscope.

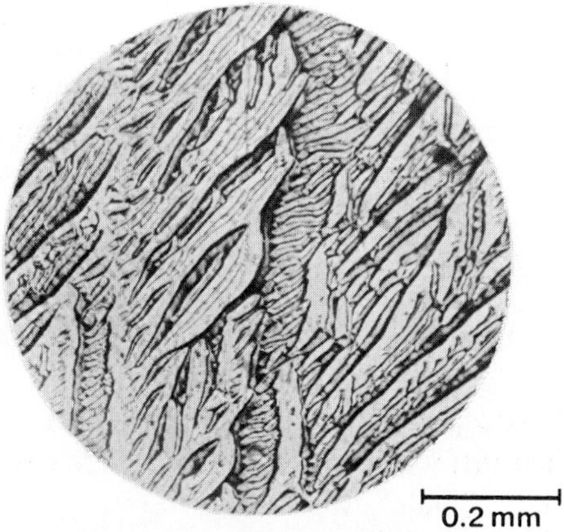

Fig. 11.2. Complicated domain pattern observed on the crystal surface which makes a fairly large angle with the direction of easy magnetization (4% Si-Fe).

Domain patterns corresponding to the interior domain structure are usually observed only on a surface whose orientation is parallel to the direction of domain magnetization. If the crystal surface tilts from the proper orientation, the surface free poles, which are induced as shown in Fig. 10.10, result in an appearance of the complicated surface domains. Figure 11.2 shows an example of such a surface domain. Even if the crystal surface has a proper orientation, it must be carefully polished by means of electrolytic polishing.

Electrolytic polishing can be achieved by using a mixture of phosphoric acid (85%) and solid chromic acid in the ratio 9 to 1. Fairly heavy currents of the order of 10–20 A/cm^2 are used to polish the specimen which is held by tweezers connected to the anode. The area of the cathode should be fairly large compared to that of the specimen. A diagram of an electric circuit is shown in Fig. 11.3. Because the polishing agent easily absorbs the moisture from the air, it is desirable to store it in a closed container. Electropolishing is also effective in preventing the polygonization of the single crystal specimen when it is polished before it is annealed. Before observation of the domain pattern the polishing agent should be completely removed from the specimen by washing it with water.

The domain pattern can be observed by using a reflection microscope of the magnification 70–150 times. A small electromagnet like that shown in Fig. 11.4 is convenient for applying a horizontal or a vertical magnetic field to the specimen. First the specimen is placed on the magnet, which is

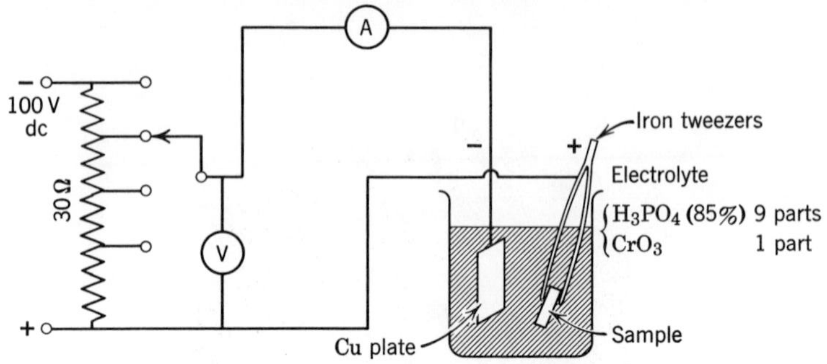

Fig. 11.3. Method of electrolytic polishing.

set under the microscope; then a drop of colloidal suspension is applied to the electropolished surface, and a thin microscope cover glass is placed on this to spread the colloidal suspension uniformly on the crystal surface (Fig. 11.5). Colloidal particles are attracted to the magnetic free poles which are appearing along magnetic domain walls; thus the domain walls are becoming visible as black lines. Figure 11.6 shows the domain pattern observed on a (001) surface of 4% Si-Fe crystal. The directions of domain magnetizations are indicated by the arrows in the figure. These are determined from the striations which elongate perpendicular to the domain magnetization. The reason is that the magnetic free poles are always distributed along the surfaces of irregularities which are perpendicular to the domain magnetization (Fig. 11.7). Actual irregularities which cause striations are thought to be non-flatness of the electropolished surface and inhomogeneity in alloy composition. Williams, Bozorth, and Shockley[3]

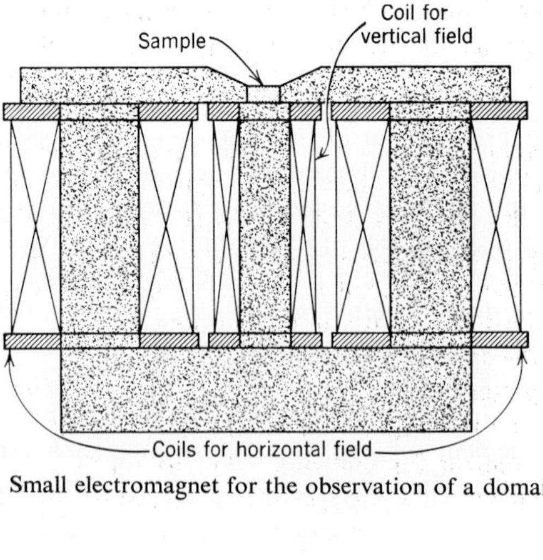

Fig. 11.4. Small electromagnet for the observation of a domain pattern.

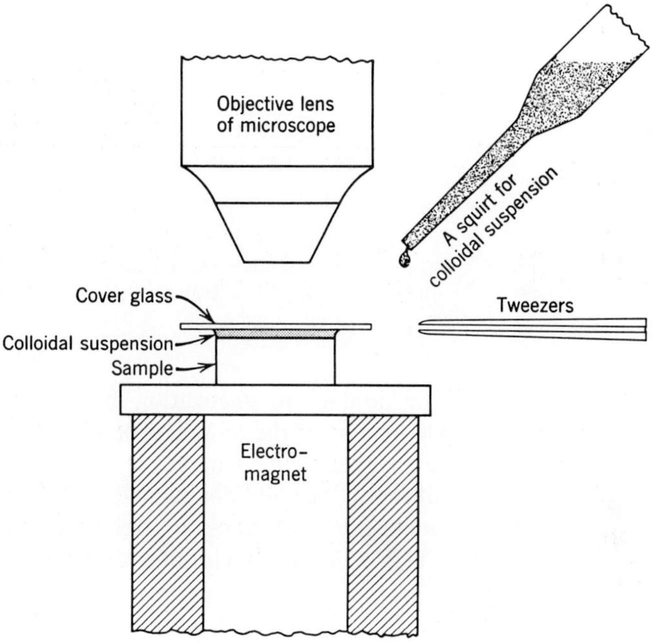

Fig. 11.5. Powder pattern method.

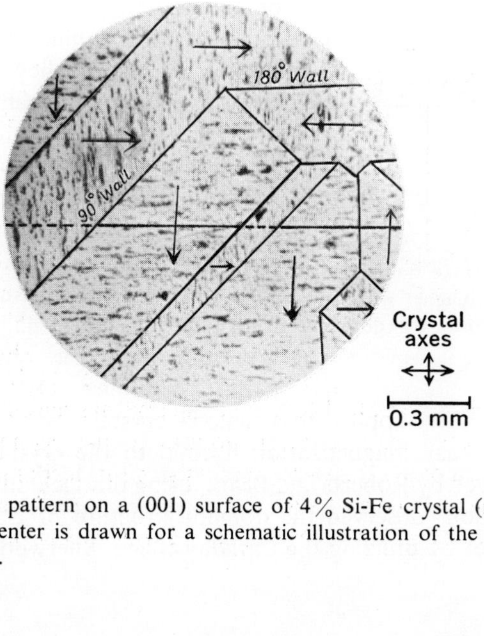

Fig. 11.6. Domain pattern on a (001) surface of 4% Si-Fe crystal (retouched). The black line at the center is drawn for a schematic illustration of the appearance of a mechanical scratch.

Fig. 11.7. Collection of colloidal particles around a spherical void.

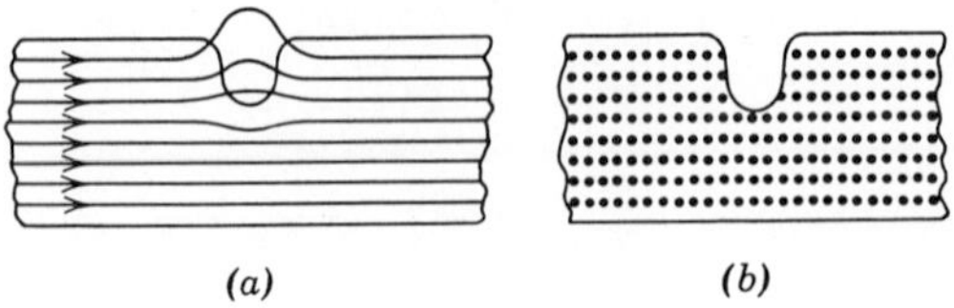

Fig. 11.8. Magnetic flux line around a mechanical scratch: (*a*) a scratch perpendicular to the domain magnetization; (*b*) a scratch parallel to the domain magnetization.

determined the direction of domain magnetization by scratching the crystal surface with a fine glass fiber. If the scratch is perpendicular to the domain magnetization, the magnetic flux emerges from the groove as shown in Fig. 11.8*a*, thus collecting the colloidal particles; if the scratch is parallel to the domain magnetization, it induces no free poles, thus being left as an invisible line (Fig. 11.8*b*). This behavior of a scratch is schematically shown in Fig. 11.6.

The sense of domain magnetization can be determined from the sense of displacement of domain walls on the application of the magnetic field the sense of which is already known. Williams et al.[3] placed a fine permanent magnet needle on the domain surface and observed an induced spike domain as shown in Fig. 11.9, from which they determined the sense of domain magnetization.

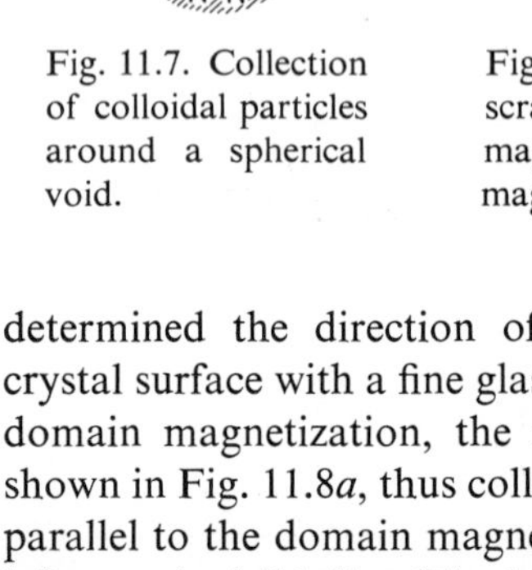

Fig. 11.9. Small spike domain induced by a thin magnet wire, indicating the sense of domain magnetization.

The magnetic Kerr effect, which is the rotation of the polarization of polarized light on reflection at the surface of a magnetized substance, and the Faraday effect, which is the rotation of the polarization during the transmission of polarized light through a magnetic substance, are both used for observing domain patterns. Figure 11.10 illustrates the magnetic Kerr effect applied to a uniaxial crystal such as MnBi which has a direction of easy magnetization normal to the crystal surface. The patterns observed by Roberts and Bean,[7] using this technique, are shown in Fig. 18.3. Dillon[8] observed the domain structure of transparent ferromagnetic garnet by utilizing the Faraday effect. Sherwood et al.[9] applied

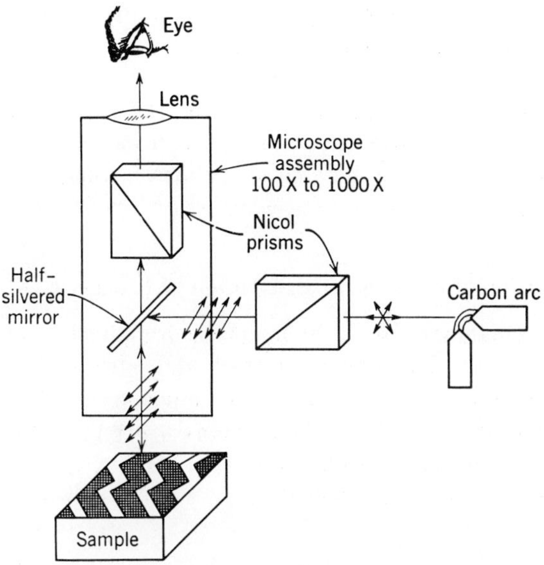

(a)

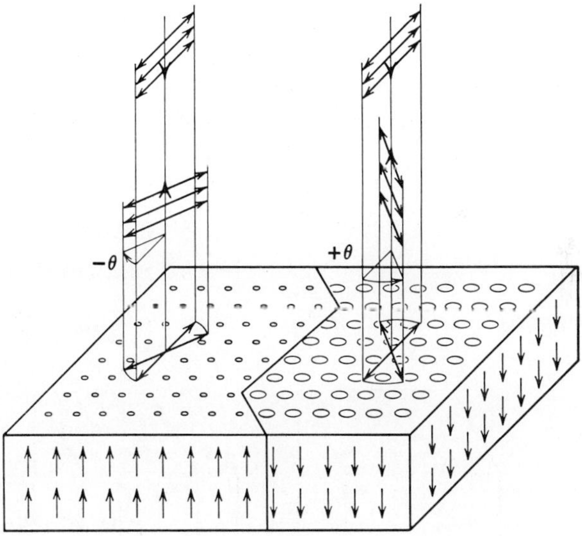

(b)

Fig. 11.10. Method of observation of domain pattern by utilizing a magnetic Kerr effect: (*a*) assembly of apparatus; (*b*) rotation of polarization of reflecting light (Roberts and Bean[7]).

the same technique to $MgFe_2O_4$, etc., by using infrared light in combination with an infrared conversion tube.

Fuller and Hale[10] succeeded in observing the magnetic structure and motion of domain walls for thin films by using electron microscopy (cf. Fig. 18.12). These direct methods, optical or electron-optical, are convenient for the investigation of the dynamic behavior of domain walls or for the observation of domain structure at elevated or low temperatures.

11.2 Calculation of a Uniform Distribution of Magnetic Domains

First we calculate the size of the magnetic domains of an infinitely large ferromagnetic plate. The direction of easy magnetization is assumed to be perpendicular to the wide surface. If the plate consists of a single domain (Fig. 11.11*a*), magnetic poles of the surface density $\pm I_s$ will appear on the top and bottom surface, respectively, so that the demagnetizing field in the plate becomes $-I_s/\mu_0$. The magnetostatic energy is then given by

$$\epsilon_m = \frac{I_s^2}{2\mu_0} l, \tag{11.1}$$

per unit area of the surface, where l is the thickness of the plate. If, however, the plate is divided into many thin domains of width d (Fig. 11.11*b*), the magnetostatic energy of the free pole distribution on one side of the plate is given by (10.36), provided that the domain width d is so small compared to the plate thickness l that the field produced by the free poles of one side of the place does not reach the other side. Then the magnetostatic energy is simply twice that of (10.36), or

$$\epsilon_m = 1.08 \times 10^5 I_s^2 d. \tag{11.2}$$

Thus the magnetostatic energy is decreased as the width of the domain, d,

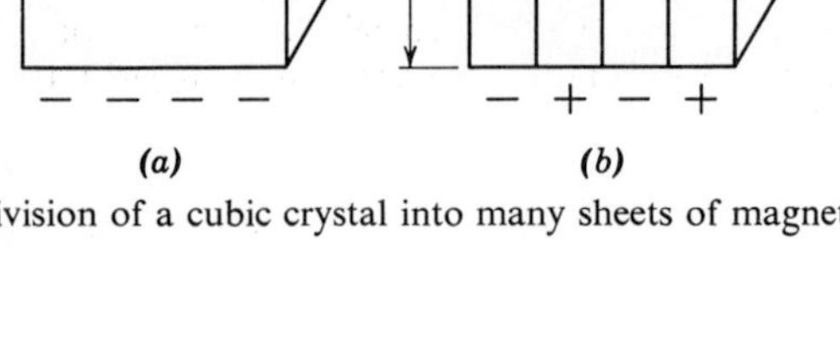

Fig. 11.11. Division of a cubic crystal into many sheets of magnetic domain.

decreases. On the other hand, the number of domain walls, and accordingly the total energy of the domain wall, increases with a decrease of d. The total area of the domain walls included in the unit area of the crystal plate is given by l/d; hence the energy of the wall becomes

$$\epsilon_w = \frac{\gamma l}{d}, \tag{11.3}$$

where γ is the surface energy of the domain wall. The equilibrium value of d is determined so as to minimize the total energy,

$$\epsilon = \epsilon_m + \epsilon_w = 1.08 \times 10^5 I_s^2 d + \frac{\gamma l}{d}. \tag{11.4}$$

That is,

$$\frac{\partial \epsilon}{\partial d} = 1.08 \times 10^5 I_s^2 - \frac{\gamma l}{d^2} = 0, \tag{11.5}$$

from which we have

$$d = 3.04 \times 10^{-3} \frac{\sqrt{\gamma l}}{I_s}. \tag{11.6}$$

For iron, $\gamma_{100} = 1.6 \times 10^{-3}$, $I_s = 2.15$, so that, if we assume $l = 1$ cm $=$ 0.01 m, (11.6) gives

$$d = 3.04 \times 10^{-3} \frac{\sqrt{1.6 \times 10^{-5}}}{2.15} = 5.6 \times 10^{-6} \text{ m}. \tag{11.7}$$

The width of the domain in this plate is about $\frac{1}{100}$ mm. The total energy then becomes

$$\epsilon = 6.56 \times 10^2 I_s \sqrt{\gamma l} \approx 5.63 \text{ J/m}^2. \tag{11.8}$$

On comparing it with the energy of a single domain (11.1) the numerical value of which is given by

$$\epsilon = \frac{I_s^2}{2\mu_0} l = \frac{2.15^2 \times 10^{-2}}{2 \times 4\pi \times 10^{-7}} = 1.8 \times 10^4 \text{ J/m}^2, \tag{11.9}$$

we find that the energy of the domain structure is about $\frac{1}{1000}$ times smaller than that of a single domain. Thus the energy of the crystal is greatly decreased by the generation of finely divided domains.

For the checkerboard type of domain structure the magnetostatic energy is given by, from (10.37),

$$\epsilon_m = 6.72 \times 10^4 I_s^2 d, \tag{11.10}$$

provided that $d \ll l$. On the other hand, the total area of the domain wall is twice as large as that of the previous one, so that

$$\epsilon_w = \frac{2\gamma l}{d}. \tag{11.11}$$

From the minimum condition of the total energy, that is,

$$\frac{\partial \epsilon}{\partial d} = 6.72 \times 10^4 I_s^{\,2} - \frac{2\gamma l}{d^2} = 0, \tag{11.12}$$

we have for the width of the domain

$$d = 5.47 \times 10^{-3} \frac{\sqrt{\gamma l}}{I_s}, \tag{11.13}$$

for which the total energy becomes

$$\epsilon = 7.37 \times 10^2 I_s \sqrt{\gamma l}. \tag{11.14}$$

Comparing this total energy with that in (11.8), we find that the energy is always lower for a plate-like domain structure than for a checkerboard type of structure, irrespective of the values of I_s, γ, and l. It is concluded, therefore, that such a checkerboard type of domain cannot be realized in actual substances. Figure 11.12*a* shows the domain pattern observed on the plane normal to the *c* axis of a cobalt single crystal. This pattern is rather similar to the checkerboard pattern or the circular pattern (cf. Fig. 10.6), contrary to the expectation from the calculation above. The reasons are that the actual domain structure consists of wedge-type reverse domains as shown by Fig. 11.13, and that the surface area of the wedge-type domain wall is not so large as it was assumed to be in the calculation above. Such a structure is effective in decreasing the magnetostatic energy without adding much wall energy. The actual domain pattern observed on a crystal surface which includes the easy direction (*c* axis) in its plane reveals such a domain structure. The crystal is divided by many

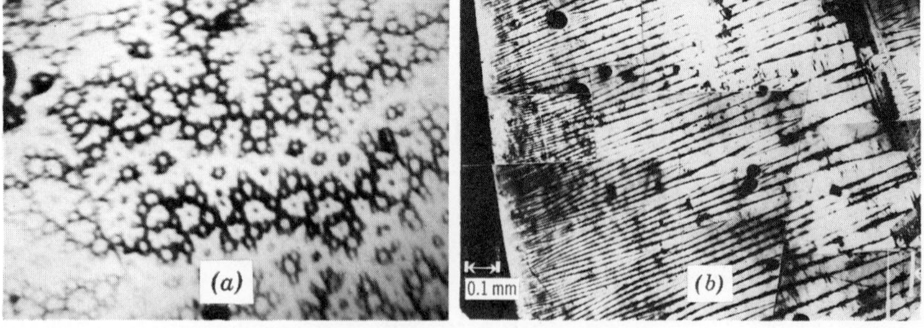

Fig. 11.12. Domain pattern observed (*a*) on a *c* plane and (*b*) on the surface parallel to the *c* axis of hexagonal cobalt (Y. Takata[11]).

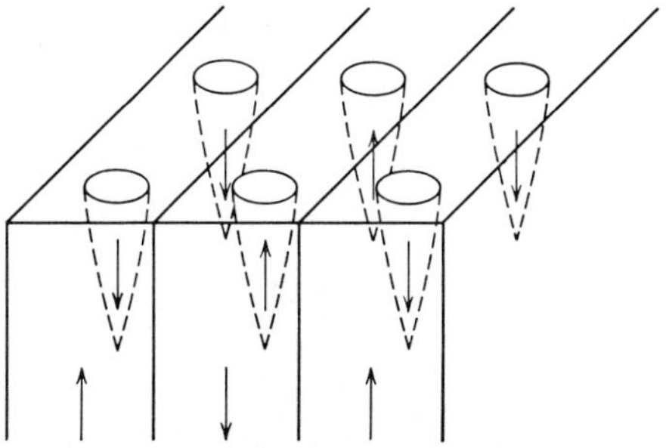

Fig. 11.13. Schematic illustration of the domain structure of cobalt.

fine domains in the vicinity of the crystal surface, while the width of the domains becomes larger as we go further from the crystal surface (Fig. 11.12*b*). Observations of domain patterns and detailed calculations of magnetostatic energy were made by Takata[11] on cobalt single crystals.

In cubic crystals which have three or more directions of easy magnetization, the appearance of the free poles at the crystal surface can be avoided by the appearance of the closure domains[3] which transport the magnetic flux from one of the underlying domains to the other (Fig. 11.14). In this case there are no magnetic free poles; hence the system does not store the magnetostatic energy. Instead, if the crystal has a finite magnetostriction, the closure domains, which tend to deform in the direction of domain magnetization, are squeezed into a space which is deformed to conform with the magnetostriction of the underlying domain (Fig. 11.15). Thus the closure domain stores magnetoelastic energy. The domain width in this case is determined by the counterbalance between the magnetoelastic energy and the wall energy. Since free elongation of the closure domain is given by $e_{xx} = 3\lambda_{100}/2$, the work necessary to compress it is $\frac{1}{2}c_{11}e_{xx}^2$, where

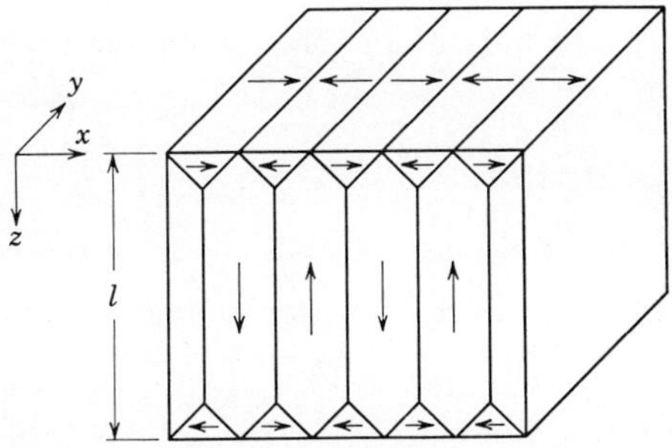

Fig. 11.14. Domain structure with closure domains.

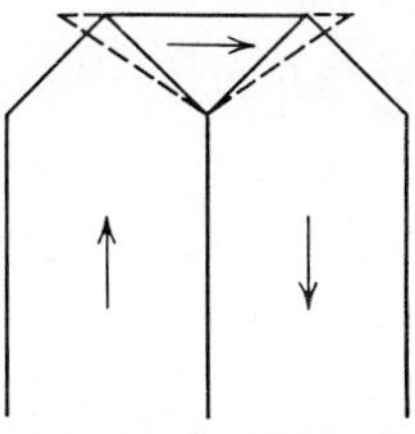

Fig. 11.15. Magnetostrictive deformation of a closure domain.

c_{11} is the elastic modulus. The magnetoelastic energy stored in the unit area of the crystal surface is then

$$\epsilon_{el} = \frac{c_{11}}{2}\left(\tfrac{9}{4}\lambda_{100}^2\right)\frac{d}{2} = \tfrac{9}{16}\lambda_{100}^2 c_{11} d. \tag{11.15}$$

The total energy,

$$\epsilon = \epsilon_{el} + \epsilon_w = \tfrac{9}{16}\lambda_{100}^2 c_{11} d + \frac{\gamma l}{d}, \tag{11.16}$$

is minimized by the condition

$$\frac{\partial \epsilon}{\partial d} = \tfrac{9}{16}\lambda_{100}^2 c_{11} - \frac{\gamma l}{d^2} = 0, \tag{11.17}$$

from which we have

$$d = \frac{4}{3}\sqrt{\frac{\gamma l}{\lambda_{100}^2 c_{11}}}. \tag{11.18}$$

Letting $l = 1$ cm, $\gamma = 1.6 \times 10^{-3}$, $\lambda_{100} = 2.07 \times 10^{-5}$, and $c_{11} = 2.41 \times 10^{11}$ for iron, we have

$$d = \frac{4}{3}\sqrt{\frac{1.6 \times 1 \times 10^{-5}}{2.07^2 \times 2.41 \times 10}} = 5.3 \times 10^{-4}\ \text{m}, \tag{11.19}$$

which is much larger than that given by (11.7). The total energy,

$$\epsilon = \tfrac{3}{2}\sqrt{\lambda_{100}^2 c_{11}\gamma l} = 6.1 \times 10^{-2}\ \text{J/m}^2, \tag{11.20}$$

is also much less than that in (11.8). Figure 11.16 shows the actual closure domains observed on the (001) surface of 4% Si-Fe crystal.

Even for a uniaxial crystal such as cobalt, there is a possibility of inducing closure domains. Since the closure domains have their magnetization pointing to the direction of hard magnetization, they store the anisotropy energy,

$$K_u = K_{u1} + K_{u2}, \tag{11.21}$$

per unit volume, where K_{u1} and K_{u2} are the anisotropy constants (cf. 7.1). The anisotropy energy stored per unit area of the crystal surface is then

$$\epsilon_a = \tfrac{1}{2}K_u d. \tag{11.22}$$

For cobalt, $K_u = 5.1 \times 10^5$ (cf. 7.2); hence

$$\epsilon_a = 2.6 \times 10^5 d, \tag{11.23}$$

while the magnetostatic energy in the absence of closure domain is calculated from (11.2) by letting $I_s = 1.99$; thus

$$\epsilon_m = (1.08 \times 10^5) \times 1.79^2 d \approx 3.46 \times 10^5 d. \tag{11.24}$$

On comparing (11.23) and (11.24), we find that the configuration accompanying closure domains is more stable than that without closure domains.

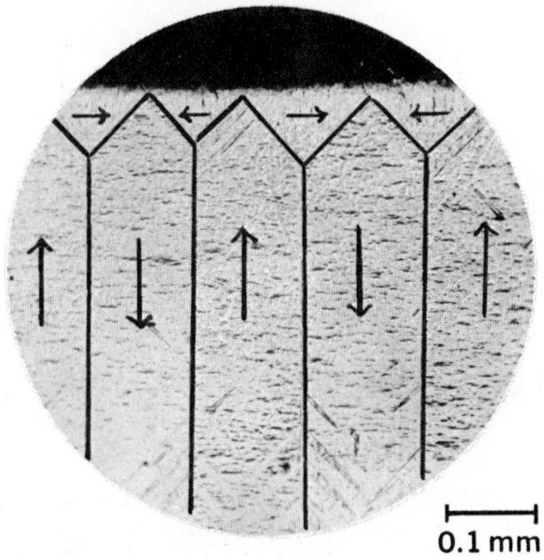

Fig. 11.16. Closure domains observed on the (001) surface of silicon-iron (Chikazumi and Suzuki) (retouched).

When the direction of domain magnetization makes a small incident angle with the crystal surface, the shape of the closure domains is fairly complicated. For instance, when the normal to the crystal surface tilts slightly from [001] toward [100], the shape of the closure domain is like trees,[3] as shown in Fig. 11.17. The structure is illustrated in Fig. 11.18. The

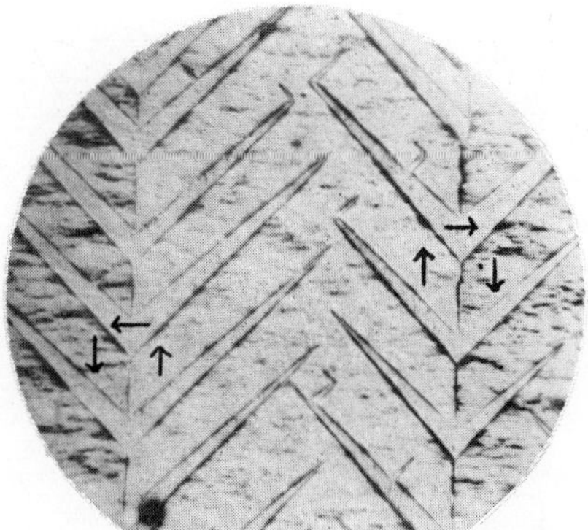

Fig. 11.17. Tree patterns observed on the crystal surface which is slightly inclined from the (001) surface (4% Si-Fe) (Chikazumi and Suzuki).

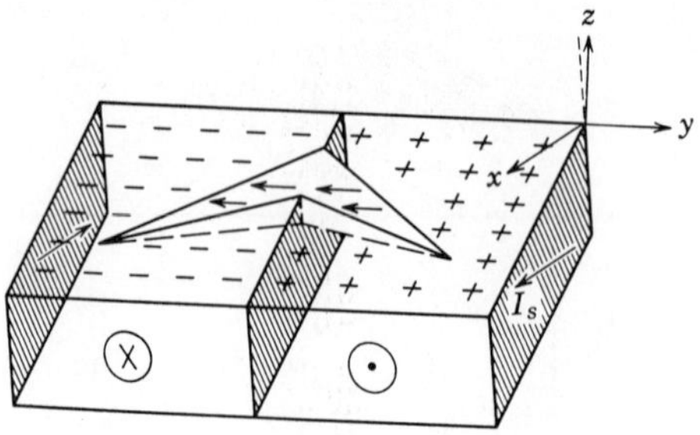

Fig. 11.18. Illustration of the tree domain.

boundary between the tree and underlying domains is possibly composed of two plane fragments the intersection of which is parallel to [110] or [1$\bar{1}$0], because in such a case the normal component of magnetization is continuous across the boundary.

The domain structure of a slab of iron single crystal which has a large surface parallel to (001) and has its long edge parallel to [110] was theoretically treated by Néel in 1944.[12] In the absence of a magnetic field this slab is composed mainly of x and y domains, whose magnetic flux is transported consistently at the crystal edge by the presence of the Q domain magnetized in [001] or [00$\bar{1}$] (Fig. 11.19*a*). If the crystal is magnetized along the long edge, the Q domains become unstable and are replaced by P domains which have their magnetization parallel to the [110] direction (Fig. 11.19*b*).

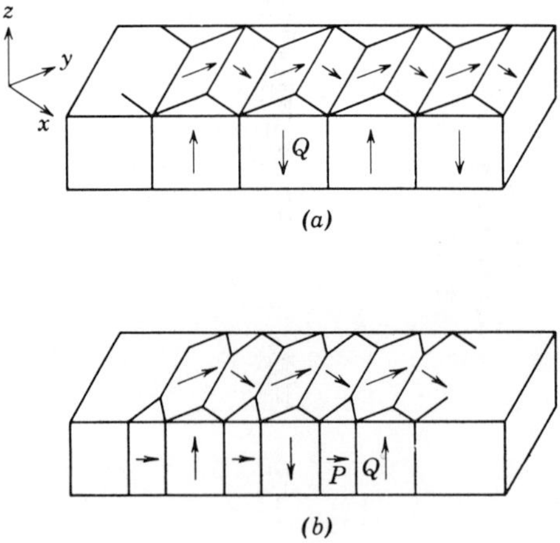

Fig. 11.19. Domain structure of an iron slab which has its long edge parallel to [110] and its wide surface parallel to (001): (*a*) without magnetic field, (*b*) a magnetic field is applied parallel to [110] (Néel[12]).

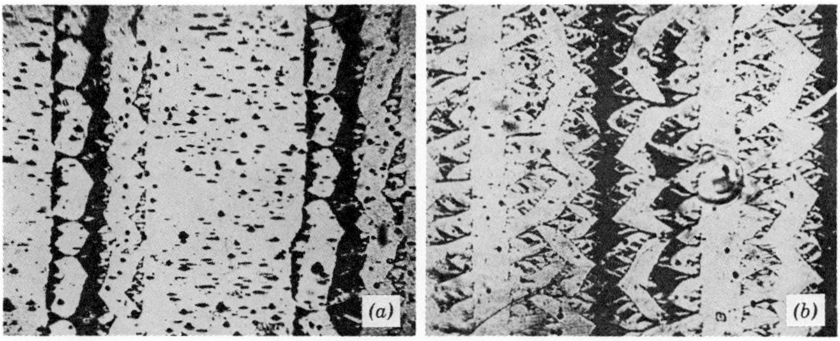

Fig. 11.20. Actual shapes of P and Q domains: (*a*) $H = 1350$ A/m ($= 17$ Oe); (*b*) $H = 2150$ A/m ($= 27$ Oe) (Bates and Mee[14]).

The actual magnetization curve and accompanying magnetostriction of the iron single crystal with this orientation was precisely measured by Lee,[13] who confirmed a beautiful agreement between his calculation and experiment. Bates and Mee[14] observed the actual shape of P and Q domains as shown in Fig. 11.20. These complicated patterns are thought to be interpreted in terms of the dependence of the wall energy on the crystallographic orientation of the walls.

11.3 Domain Structure of Non-uniform Substances

In the preceding section we treated regular domain structures in a uniform magnetic material. Such a regular domain structure can be actually attained for carefully prepared single crystals. Generally, however, the domain structures are affected by some irregularities, such as voids, inclusions, internal stresses, and crystal boundaries. In magnetically hard materials such irregularities are the main factors which govern the size and distribution of ferromagnetic domains.

First we discuss the domain structure of polycrystals. As already shown in Fig. 6.16, domain magnetizations are more or less continuous across the crystal boundary in order to decrease the magnetostatic energy. Such a continuity is, however, not perfect,[15] unless the crystal boundary takes some particular orientation with respect to the directions of easy magnetization on both sides of the crystal boundary. Let us calculate the magnetostatic energy of the average magnetic free poles which are appearing on the crystal boundaries. First we treat a uniaxial crystal with positive K_u, which has only one easy axis. Suppose that the domain magnetizations on both sides of the crystal boundary make angles θ_1 and θ_2 with respect to the normal to the crystal boundary as shown in Fig. 11.21. The surface

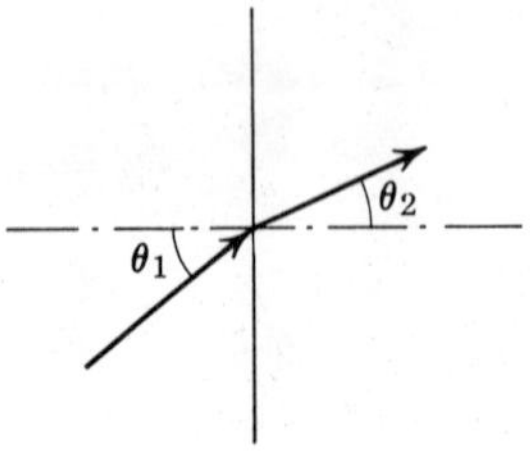

Fig. 11.21. Directions of magnetization on both sides of a crystal boundary.

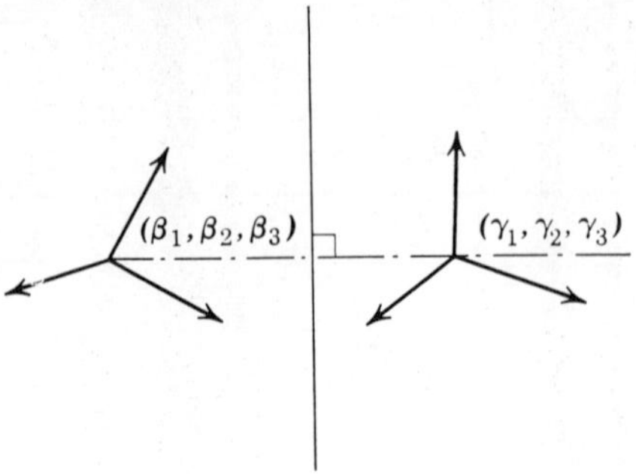

Fig. 11.22. Crystal orientations on both sides of a crystal boundary.

density of the magnetic free poles is given by

$$\omega = I_s(\cos\theta_1 - \cos\theta_2) \tag{11.25}$$

As seen in previous calculations, the magnetostatic energy is always proportional to ω^2; hence we need only to calculate the average value of ω^2 over all possible combinations of θ_1 and θ_2. If θ_1 is in the range $-\pi/2 < \theta_1 < \pi/2$, θ_2 must be also in the range $-\pi/2 < \theta_2 < \pi/2$, because otherwise both magnetizations would point face to face and an unnecessarily large free magnetic pole would result. Thus

$$\begin{aligned}\overline{\omega^2} &= I_s^2 \int_0^{\pi/2}\int_0^{\pi/2} (\cos\theta_1 - \cos\theta_2)^2 \sin\theta_1 \sin\theta_2 \, d\theta_1 \, d\theta_2 \\ &= I_s^2 \Big[\int_0^{\pi/2} \cos^2\theta_1 \sin\theta_1 \, d\theta_1 + \int_0^{\pi/2} \cos^2\theta_2 \sin\theta_2 \, d\theta_2 \\ &\quad - 2\int_0^{\pi/2}\int_0^{\pi/2} \cos\theta_1 \cos\theta_2 \sin\theta_1 \sin\theta_2 \, d\theta_1 \, d\theta_2 \Big] \\ &= I_s^2\{\tfrac{1}{3} + \tfrac{1}{3} - \tfrac{1}{2}\} = \frac{I_s^2}{6}. \end{aligned} \tag{11.26}$$

If individual crystallites are separated from each other, it follows that

$$\overline{\omega^2} = I_s^2 \int_0^{\pi/2} \cos^2\theta_1 \sin\theta_1 \, d\theta_1 = \frac{I_s^2}{3}. \tag{11.27}$$

Thus the magnetostatic energy of a poly-uniaxial crystal is about one-half that of the separated crystallites. It is expected, therefore, that the width of polycrystal domains is about $\sqrt{2}$ times larger than that of the separated crystallites, and that the total domain energy of the former is $1/\sqrt{2}$ times that of the latter.

For a cubic crystal which has three or more directions of easy magnetization the probability of having a good continuity of magnetization across the crystal boundary is much greater. Let the direction cosines of the normal to the crystal boundary be $\beta_1, \beta_2, \beta_3$ with respect to the crystal axes of the left-hand crystallite, and $\gamma_1, \gamma_2, \gamma_3$ with respect to those of the right-hand crystallite (Fig. 11.22). For crystals with positive K_1 such as iron, directions of easy magnetization are $\langle 100 \rangle$; hence the magnetization on both sides of the crystal boundary has the possibility of taking the x, y, or z direction (i or $j = 1, 2, 3$), respectively. When the magnetization takes the ith easy axis in the left-hand crystallite and the jth easy axis in the right-hand crystallite, the surface density of magnetic free pole is given by

$$\omega = I_s(\beta_i - \gamma_j). \tag{11.28}$$

Now let us discuss the relation of magnetization directions between both crystallites which minimize the surface density of magnetic free poles given by (11.28). Figure 11.23 shows the three crystal axes (x, y, z) of the right-hand crystallite, the normal $\boldsymbol{n}$ to the crystal boundary, and the direction of magnetization I_1 in the left-hand crystallite. If $\gamma_3 > \gamma_2 > \gamma_1$ and $\beta_i > \gamma_3$, the direction of magnetization in the right-hand crystallite should be parallel to the z axis because otherwise the free pole produced by I_1 would be less compensated by I_2 (magnetization in the right-hand crystallite). The free pole density in this case is given by

$$\omega = I_s(\beta_i - \gamma_3). \tag{11.29}$$

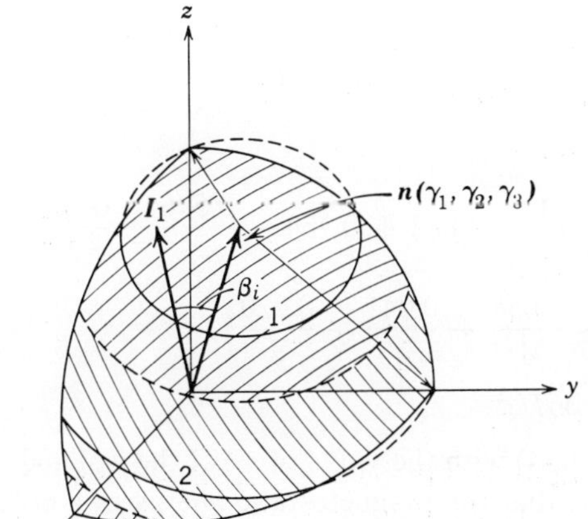

Fig. 11.23. Domain magnetization I_2 in the second crystallite takes z, y, and x directions to minimize the surface density of free poles appearing on the crystal boundary, if the domain magnetization I_1 in the first crystallite is in the region indicated by 1, 2, and 3, respectively.

As seen in this equation, the sign of ω is changed when β_i becomes less than γ_3. If β_i is further decreased and it happens that $\beta_i < (\gamma_3 + \gamma_2)/2$, the y direction becomes more favorable for I_2 than the z direction. Then

$$\omega = I_s(\beta_i - \gamma_2). \tag{11.30}$$

If β_i is further decreased and it happens that $\beta_i < (\gamma_2 + \gamma_1)/2$, the x direction becomes the most favorable one for I_2; thus

$$\omega = I_s(\beta_i - \gamma_1). \tag{11.31}$$

On averaging ω^2 with respect to β_i, we have

$$\begin{aligned}\int_0^1 \omega^2\, d\beta_i = I_s^2\Bigg[\int_{(\gamma_3+\gamma_2)/2}^{1} (\beta_i - \gamma_3)^2\, d\beta_i + \int_{(\gamma_2+\gamma_1)/2}^{(\gamma_3+\gamma_2)/2} (\beta_i - \gamma_2)^2\, d\beta_i \\ + \int_0^{(\gamma_2+\gamma_1)/2} (\beta_i - \gamma_1)^2\, d\beta_i\Bigg] \\ = I_s^2(\tfrac{1}{3} - \gamma_3 + \gamma_3^2 - \tfrac{1}{4}\gamma_3^3 - \tfrac{1}{4}\gamma_3^2\gamma_2 + \tfrac{1}{4}\gamma_3\gamma_2^2 \\ - \tfrac{1}{4}\gamma_2^2\gamma_1 + \tfrac{1}{4}\gamma_2\gamma_1^2 + \tfrac{1}{4}\gamma_1^3).\end{aligned} \tag{11.32}$$

On averaging each term in (11.32) over possible orientations of the right-hand crystallites in the range $\gamma_3 > \gamma_2 > \gamma_1$, which is shown in Fig. 11.24, we have

$$\begin{aligned}&\overline{\gamma_3} = 0.832, \quad \overline{\gamma_3^2} = 0.764, \quad \overline{\gamma_3^3} = 0.596, \\ &\overline{\gamma_3^2\gamma_2} = 0.405, \quad \overline{\gamma_3\gamma_2^2} = 0.202, \quad \overline{\gamma_2^2\gamma_1} = 0.054, \\ &\overline{\gamma_2\gamma_1^2} = 0.041, \quad \overline{\gamma_1^3} = 0.020.\end{aligned} \tag{11.33}$$

Thus (11.32) becomes

$$\begin{aligned}\omega^2 &= I_s^2\left(\frac{1}{3} - 0.832 + 0.764 - \frac{0.596}{4} - \frac{0.405}{4} + \frac{0.202}{4} \right. \\ &\left. \quad - \frac{0.054}{4} + \frac{0.041}{4} + \frac{0.020}{4}\right) \\ &= 0.067 I_s^2.\end{aligned} \tag{11.34}$$

On comparing (11.34) with the value of $\overline{\omega^2}$ of the isolated crystallites given by (11.27), we see that the magnetostatic energy of a poly-cubic crystal is about 0.20 that of the separated crystallites, so that its domain width will be $1/\sqrt{0.20} = 2.2$ times larger, and the total energy will be $\sqrt{0.20} = 0.45$ that of the separated crystallites.

Thus the magnetostatic energy of the magnetic poles appearing on the crystal boundary is small but still finite, so that the domain structure will be essentially determined by the size of the crystallites unless the size of the outside shape of the specimen is comparable to the size of the crystallites.

Next let us discuss the domain structure of a strongly stressed crystal. When a tension σ exists in a crystal, there will be induced the magnetic anisotropy given by (8.86) which rotates the local magnetization toward the direction of tension provided $\lambda > 0$. Figure 11.25 shows the domain pattern observed on the (001) crystal surface of Fe_3Al crystal to which a mechanical indentation was applied. Since this alloy has a large magnetostriction ($\lambda = 3.7 \times 10^{-5}$) and a small magnetocrystalline anisotropy, the direction of domain magnetization is mainly determined by the distribution of residual stresses in the crystal. If we assume that $\sigma = 100\ \text{kg/mm}^2 \fallingdotseq 10^9\ \text{N/m}^2$, the anisotropy constant becomes $\frac{3}{2}\lambda\sigma = 1.5 \times 3.7 \times 10^{-5} \times 10^9 \approx 5.6 \times 10^4\ \text{J/m}^3$, which is almost of the same order of magnitude as the magnetocrystalline anisotropy constant of iron. It is supposed that the domain structure of such a strongly stressed crystal may be similar to that of a poly-uniaxial crystal with respect to the point that the easy axis changes its orientation from place to place. Actually, however, there are many vague points about the distribution of the internal stress, and, in

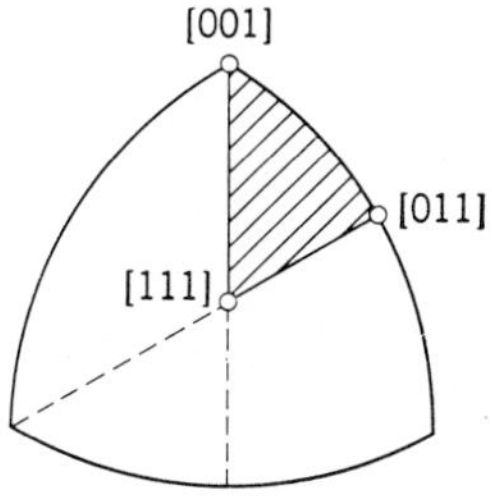

Fig. 11.24. Illustration indicating the range of integration.

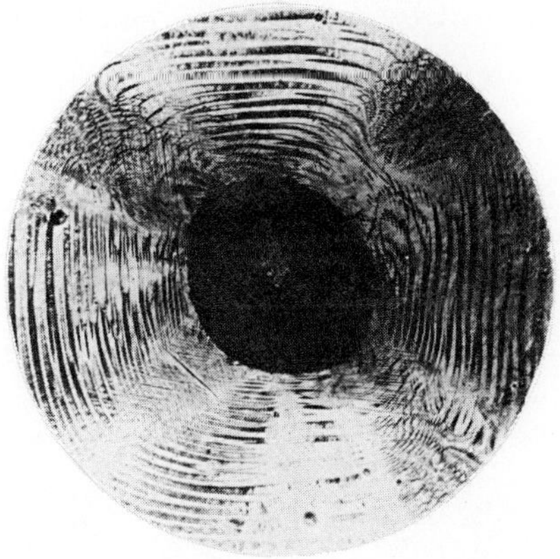

Fig. 11.25. Magnetic domains induced by strong stresses made by an indentation, for Fe_3Al (001) (Chikazumi, Suzuki and Shimizu).

addition, the magnetostriction is generally anisotropic, so the anisotropy induced by internal stresses would be fairly complicated. Domain structures of extremely soft magnetic materials such as Permalloy and Supermalloy which have a small magnetocrystalline anisotropy supposedly also belong to this category.[16]

Next we discuss the domain structure of crystals which include voids, inclusions, and precipitates. As shown in Fig. 6.13, a domain wall tends to pass the vicinity of voids or non-magnetic inclusions, because in such crystals the magnetostatic energy can be decreased by the rearrangement of the free poles. The amount of decrease in a magnetostatic energy for the type shown in Fig. 6.11*c* is about one-half that given by (10.40); that is

$$\Delta U \approx 2.8 \times 10^5 I_s^2 r^3, \tag{11.35}$$

where r is the radius of the spherical hole. If the blade-like domain is connected to the wall as shown in Fig. 6.13, the magnetostatic energy associated with the blade domain given by (10.42) or, half of it, that is,

$$\Delta U \approx 1.6 \sim 3.3 \times 10^5 I_s^2 r^3, \tag{11.36}$$

will be decreased per one hole or inclusion. As an average, we can set

$$\Delta U = 3 \times 10^5 I_s^2 r^3. \tag{11.37}$$

Now let us suppose that N spherical holes are distributed in a unit volume of the crystal. It is assumed that the crystal is composed of many $\pm y$ domains separated by 180° walls which are almost parallel to the yz plane (Fig. 11.26). Total wall energy is given by γ/d, where d is the average distance between the neighboring domain walls. On the other hand, the individual wall tends to pass through the vicinity of the holes. The number of holes which are passed by one domain wall is roughly estimated as

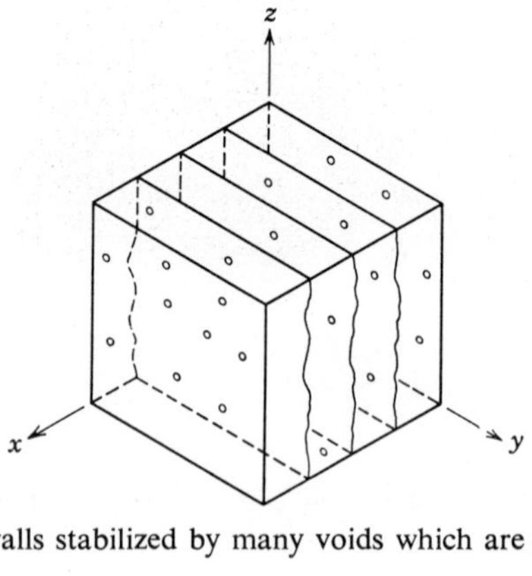

Fig. 11.26. Domain walls stabilized by many voids which are distributed in a crystal.

$N^{2/3}$, so that $N^{2/3}/d$ holes in a unit volume decrease the magnetostatic energy. The amount of decrease in total magnetostatic energy is given by $N^{2/3}\,\Delta U/d$. If, therefore,

$$N^{2/3}\,\Delta U \geq \gamma, \tag{11.38}$$

the wall energy is to be completely compensated by a decrease in a magnetostatic energy associated with the holes. For example, when holes with $r = 0.01$ mm are distributed in an iron crystal, on setting $I_s = 2.15$ in (11.37) we have

$$\Delta U = 3 \times 10^5 (2.15)^2 (10^{-5})^3 = 1.4 \times 10^{-9}\ \text{J}. \tag{11.39}$$

Since $\gamma = 1.6 \times 10^{-3}$ for iron, (11.38) becomes

$$N \geq \left(\frac{\gamma}{\Delta U}\right)^{3/2} = \left(\frac{1.6 \times 10^{-3}}{1.4 \times 10^{-9}}\right)^{3/2} = 10^9\ \text{m}^{-3}. \tag{11.40}$$

If the separation between the neighboring holes is less than $N^{-1/3} \approx 10^{-3}$ m or 1 mm, such a group of holes can generate domain walls in a crystal.

When the density of holes, inclusions, or precipitations in a substance is extremely large, the substance can be regarded as an aggregate of fine particles of the original substance separated by voids or foreign materials. If a spherical fine ferromagnetic particle is completely separated from its surroundings, the magnetostatic energy is roughly given by

$$U_m = \frac{I_s^2}{6\mu_0}\frac{4}{3}\pi r^3 \frac{d}{2r} = \frac{\pi I_s^2 r^2}{9\mu_0} d, \tag{11.41}$$

and the wall energy is given by

$$U_w = \gamma(\pi r^2)\frac{2r}{d} = \frac{2\pi\gamma r^3}{d}, \tag{11.42}$$

where d is the width of the domains. Total energy,

$$U = U_m + U_w, \tag{11.43}$$

is minimized by the condition

$$\frac{\partial U}{\partial d} = \frac{\pi I_s^2 r^2}{9\mu_0} - \frac{2\pi\gamma r^3}{d^2} = 0, \tag{11.44}$$

from which we have

$$d = \sqrt{\frac{18\gamma\mu_0 r}{I_s^2}}. \tag{11.45}$$

It is seen in this formula that the width of the domain is decreased in proportion to $\sqrt{r}$ with a decrease of the radius r of the sphere (Fig. 11.27). Below a certain critical radius r_c, the domain width d is thus expected to become less than the diameter of the sphere; in other words, the sphere is expected to be composed of a single domain. Since $d = 2r_c$ at the critical radius, (11.45) becomes

$$2r_c = \sqrt{\frac{18\gamma\mu_0 r_c}{I_s^2}}$$

or

$$r_c = \frac{9\gamma\mu_0}{2I_s^2}. \tag{11.46}$$

For iron, $I_s = 2.15$, $\gamma = 1.6 \times 10^{-3}$, so that the critical radius becomes

$$r_c = 2 \times 10^{-9}\ \text{m} = 0.002\ \mu. \tag{11.47}$$

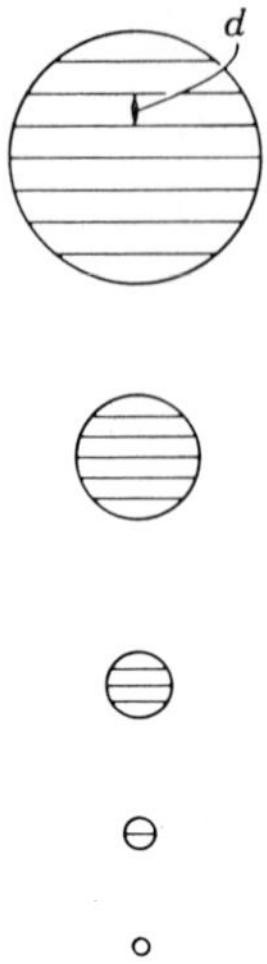

Fig. 11.27. Relation between the size of ferromagnetic spheres and that of their magnetic domains.

Such a single domain structure was first foreseen by Frenkel and Dorfman,[17] and later precise calculations were performed by Kittel,[18] Néel,[19] and Stoner and Wohlfarth.[20,21] Kittel also examined the possibility of having a single domain structure for thin wires and thin films as well as for fine particles.

If fine particles aggregate into a large block in which they are separated by non-magnetic binder, voids, inclusions, or precipitates, the magnetostatic energy per unit volume of the substance will be decreased because of the magnetostatic interaction between the particles. For instance, Ba ferrite is normally composed of fine crystal grains each of which has a uniaxial crystal anisotropy. As calculated in this section, the magnetostatic energy is about one-half that of the separated crystallites; hence the critical radius r_c is also expected to be twice as large as that of the isolated particles.

Problems

11.1. How does the width of a ferromagnetic domain in a single crystal plate with a uniaxial anisotropy constant K_u change with a change in K_u? Assume (i) that the easy axis is perpendicular to the plate surface, and then (ii) that the easy axis makes a small angle θ with the crystal surface, and also $I_s^2/\mu_0 \gg K_u$.

11.2. Calculate the width of a domain in a cobalt single crystal plate of thickness $d = 1$ cm in which the easy axis makes the small angle $\theta = 0.1$ rad with the crystal surface. Assume that $K_u = 4.1 \times 10^5\ \text{J/m}^3$, $I_s = 1.8\ \text{Wb/m}^2$, and $\gamma = 1.5 \times 10^{-2}\ \text{J/m}^2$.

11.3. Estimate the critical size of a crystallite of iron polycrystal in which the individual crystallite is composed of a single domain. Assume that the shape of the crystallite is a sphere.

References

11.1. F. Bitter: *Phys. Rev.* **38,** 1903 (1931); **41,** 507 (1932).
11.2. L. V. Hamos and P. A. Thiessen: *Z. Physik.* **71,** 442 (1932).
11.3. H. J. Williams, R. M. Bozorth, and W. Shockley: *Phys. Rev.* **75,** 155 (1949).
11.4. W. C. Elmore: *Phys. Rev.* **51,** 982 (1937); **53,** 757 (1938); **54,** 309 (1938); **62,** 486 (1942).
11.5. W. O. Baker, F. H. Winslow and G. Bittrich: G.10, p. 533.
11.6. D. J. Craik and P. M. Griffith: *British J. Appl. Phys.* **9,** 279 (1958).
11.7. B. W. Roberts and C. P. Bean: *Phys. Rev.* **96,** 1494 (1954).
11.8. J. F. Dillon, Jr.: *J. Appl. Phys.* **29,** 1286 (1958).
11.9. R. C. Sherwood, J. R. Remeika, and H. J. Williams: *J. Appl. Phys.* **30,** 217 (1959).
11.10. H. W. Fuller and M. E. Hale: *J. Appl. Phys.* **31,** 238 (1960); **31,** 308S (1959).
11.11. Y. Takata: *J. Phys. Soc. Japan* **18,** 87 (1963).
11.12. L. Néel: *J. Phys. Radium* **5,** 241, 265 (1944).
11.13. E. W. Lee: *Proc. Phys. Soc. (London)* **A66,** 623 (1953); *ibid.*, **A68,** 65 (1955).
11.14. L. F. Bates and C. D. Mee: *Proc. Phys. Soc. (London)* **A65,** 129 (1952).
11.15. W. S. Paxton and T. G. Nilan: *J. Appl. Phys.* **26,** 994 (1955).
11.16. S. Chikazumi: *Phys. Rev.* **85,** 918 (1952).
11.17. J. Frankel and J. Dorfman: *Nature* **126,** 274 (1930).
11.18. C. Kittel: *Phys. Rev.* **70,** 965 (1946).
11.19. L. Néel: *Compt. Rend.* **224,** 1488 (1947).
11.20. E. C. Stoner and E. P. Wohlfarth: *Nature* **160,** 650 (1947).
11.21. E. C. Stoner and E. P. Wohlfarth: *Phil. Trans. Roy. Soc. London* **A240,** 599 (1948).

part 4

THE MAGNETIZATION PROCESS

12

The Magnetization Curve and Domain Structure

12.1 Mechanism of Technical Magnetization

When a ferromagnetic substance is put into an increasing magnetic field, its magnetization is increased and finally reaches the saturation magnetization. Such a process is called technical magnetization, because it is essentially achieved by a change in the direction of domain magnetization and can be distinguished from a change in the intensity of spontaneous magnetization. A narrow range of magnetization starting from the demagnetized state ($I = 0$ at $H = 0$), in which the magnetization changes reversibly, is called the initial permeability range (Fig. 12.1). In this range, domain magnetizations in every domain rotate reversibly from the stable directions, and at the same time domain walls are displaced

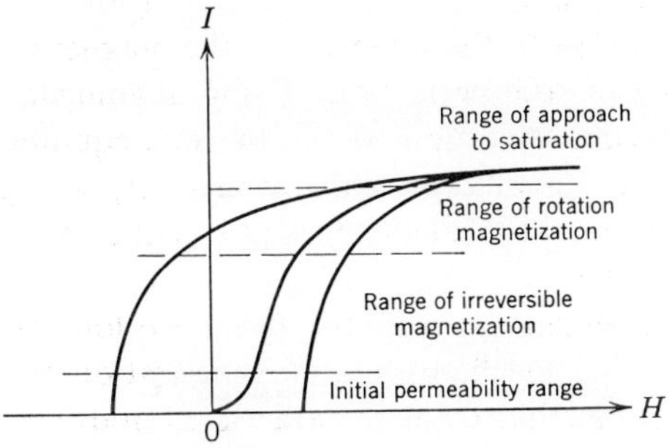

Fig. 12.1. Magnetization curve and the classification of magnetization mechanisms.

reversibly from their stable positions. We shall discuss these processes in detail in Chapter 13. As we shall see there, the rotation magnetization against magnetocrystalline anisotropy gives $\bar{\mu}_a = 29$ for iron, while soft iron actually exhibits $\bar{\mu}_a = 100 \sim 200$. This fact tells us that the reversible magnetization in this range is accomplished mainly by reversible displacements of domain walls. Since, however, ease of displacement of domain walls is essentially determined by the homogeneity of materials, the contribution of the displacement of domain walls to an initial permeability is entirely dependent on the sort of material studied.

If the magnetic field is increased beyond the initial permeability range, the intensity of magnetization increases more drastically. Since this process is irreversible, this range is called the irreversible magnetization range. This range is attained mainly by irreversible displacement of domain walls from one stable position to another. Occurrence of irreversible rotation of domains is also expected in fine particles or in an extremely heterogeneous material which contains a lot of inclusions and precipitates. Barkhausen noise (cf. Chapter 6) can be heard in this range also because many small discontinuous changes in magnetization are induced by irreversible displacement of domain walls and by irreversible rotation of local domain magnetization. The magnetothermal effect, which is the generation of heat accompanying technical magnetization, is also observable in this range, because part of the work done by the magnetic field is dissipated as heat during the process of discontinuous magnetization. If the field is increased further, the magnetization curve becomes less steep and its processes become reversible once more. In this range the displacements of domain walls have already been completed and the magnetization takes place by rotation magnetization. In this connection, this range is called the rotation magnetization range. Beyond this range the magnetization gradually approaches the saturation magnetization the range of which is called the range of approach to saturation magnetization. We discuss this subject in Section 13.3.

After the magnetization reaches the saturation value, it increases gradually in proportion to the intensity of the magnetic field. This effect results from increasing the perfection of spin alignment, which otherwise is disturbed by thermal agitation to the thermal equilibrium state. Normally this effect is very small even under moderately strong magnetic fields, except at temperatures just below the Curie point. We also discuss this subject in Section 13.3.

Mechanisms of magnetization along a hysteresis loop are similar to those discussed above. The mechanism of magnetization from saturation to remanence is rotation, that from remanence to midway of the ascendant hysteresis curve is irreversible magnetization, and that to the opposite saturation is again rotation magnetization. The total heat generated during one cycle of hysteresis, which is measured by the area surrounded by a hysteresis loop, is equal to the total energy dissipated by discontinuous magnetization processes.

The preceding description of magnetization processes pertains to relatively soft magnetic materials such as soft iron in which the displacement of domain walls takes place easily, but rotation magnetization needs a fairly large magnetic field. For fairly hard magnetic materials such a clear-cut discrimination of the mechanism of magnetization is not always possible, because wall displacement and rotation magnetization take place almost simultaneously.

The mechanism of technical magnetization also depends on the particular type of domain structure. For instance, a Permalloy wire which is cooled from a high temperature while a magnetic field is applied parallel to its long axis contains two sorts of domains whose magnetizations are parallel or antiparallel to its length (cf. Section 17.1). When a magnetic field is applied parallel to the long axis of the wire, its magnetization curve rises directly from the demagnetized state to saturation. This means that the whole process is attained exclusively by the displacement of 180° walls. Isoperm, that is, the cold-rolled 50% Fe-Ni alloy, has its easy axis perpendicular to the roll direction. When the field is applied parallel to the roll direction, the magnetization changes reversibly along a linear magnetization curve which has a constant inclination from the demagnetized state to saturation. The mechanism of magnetization in this alloy consists entirely of the rotation process. In substances with very low anisotropy or, for that matter, in any substance heated to just below the Curie point, incoherent rotations of magnetization, which should be classified between wall displacement and coherent rotation magnetization, are thought to be predominant. This phenomenon may correspond to the fact that ferromagnetic substances generally show a sharp maximum of initial permeability just below the Curie point; this is normally called the Hopkinson effect.

The mechanism of magnetization also depends on the frequency with which the magnetic field is changed. For instance, displacement of domain walls is normally the easiest mechanism under quasistatic change of magnetic field, but its movement is more easily damped than rotation of magnetization for a high frequency magnetic field. Even the rotation of magnetization becomes unable to follow the change of the magnetic field for higher frequencies, because it causes natural resonance with external frequency. We discuss this topic in Chapter 16.

We must emphasize the following two points concerning the magnetization curve: First, the point on the magnetization curve represents the intensity of magnetization which the substance has immediately after the magnetic field is applied, but actually the intensity of magnetization changes gradually with a lapse of time. When a beginner measures the magnetization curve point by point, the measured points often scatter about the normal magnetization curve. One of the reasons is thought to be that the intensity of magnetization changes during the measurement. This phenomenon is called the magnetic aftereffect or the magnetic viscosity, which we discuss in Chapter 15.

The second point concerns the meaning of the shearing correction mentioned in Section 2.2. As discussed in Section 6.2, the magnetic domain structure of an inhomogeneous substance is determined by its inherent structure. For such a substance the "true" magnetization curve can be reproduced by shearing correction from the measurement for a finite-shaped specimen. A homogeneous magnetic specimen, however, has a domain structure which depends on its size and shape, so that its magnetic properties are also expected to depend on its size and shape. In such a specimen a "true" magnetization curve obtained after shearing correction represents not the magnetic property of this substance, but the magnetic property of this particular specimen. For instance, the magnetic property of magnetic sheets, whose crystal grains are larger in diameter than the thickness of the sheet, commonly depends on its thickness.

12.2 Distribution of Domain Magnetizations at Remanence

In order to understand the real character of a magnetization curve it is necessary to know the distribution of directions of domain magnetizations at each point on the curve. This is particularly important in order to know the value of residual magnetization. It should also be recognized that the distribution of domain magnetizations is quite different between the demagnetized state and the point at the coercive force, though both states have, equally $I = 0$. Utilizing this fact, an ingenious method of remagnetizing a permanent magnet has been proposed by a Philips investigator. We discuss these matters in this section.

First let us consider a substance which has an isotropic distribution of easy axes of uniaxial magnetic anisotropy. Examples of such substances are polycrystalline cobalt, a highly stressed material, and those which are sensitive to a magnetic annealing but were actually annealed without a magnetic field. In the demagnetized state these substances have their domain magnetizations isotropically distributed as shown by the polar distribution diagram at the point O in Fig. 12.2. When a magnetic field is applied in the plus direction, the domain magnetizations which point in the minus direction are first reversed by wall displacement, thus resulting in the

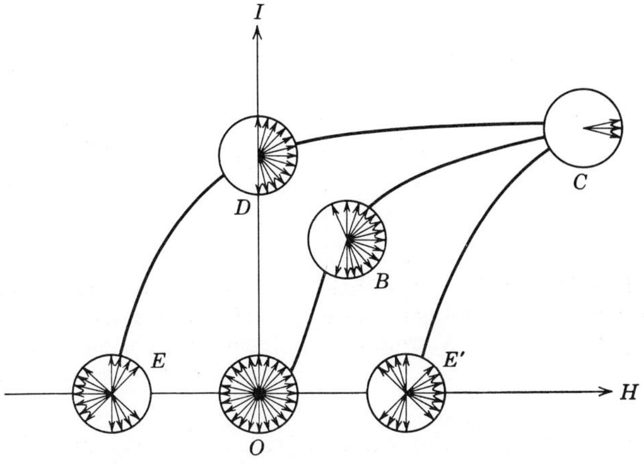

Fig. 12.2. Magnetization curve and the angular distribution of domain magnetization.

distribution shown at point B. At sufficiently strong magnetic field all magnetizations are lined up nearly parallel to the plus direction shown at point C. When the intensity of magnetic field is decreased from point C, every domain magnetization turns back to the nearest easy direction or the plus sense of that easy axis, and thus covers a half-sphere in the plus side at the remanence point D. The intensity of magnetization in this state is easily calculated as

$$I_r = \int_0^{\pi/2} I_s \cos\theta \sin\theta \, d\theta = \frac{I_s}{2}. \tag{12.1}$$

Actually most highly stressed materials do show a residual magnetization about one-half of the saturation magnetization. If the intensity of magnetic field is increased further in the minus direction, the domain magnetizations which point in the plus direction are reversed first, resulting in the distribution shown at point E. It should be noted that the distributions at E and O both correspond to $I = 0$ but still have quite different domain distributions.

In a cubic crystal with positive K_1 the easy directions are $\langle 100 \rangle$, so that domain magnetizations at remanence stay in the one of the three $\langle 100 \rangle$ directions which is closest to the plus direction as shown in Fig. 12.3. The maximum deviation from the plus direction is realized when the $\langle 111 \rangle$ axis of a crystallite is parallel to the plus direction; hence its angle is given by $\cos^{-1}(1/\sqrt{3}) = 55°$. The distribution of domain magnetization in this case, therefore, is confined to the solid angle the vertical half-angle of which is 55° (Fig. 12.4). This idea was suggested by Becker and Döring[1] in their book.

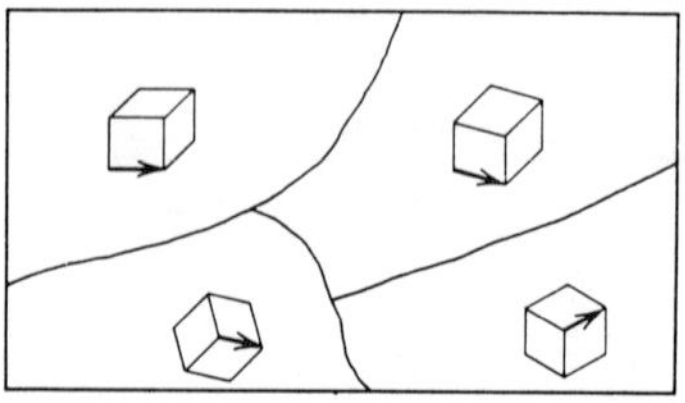

Fig. 12.3. Distribution of domain magnetizations of a cubic polycrystal with positive K_1 at its remanence state.

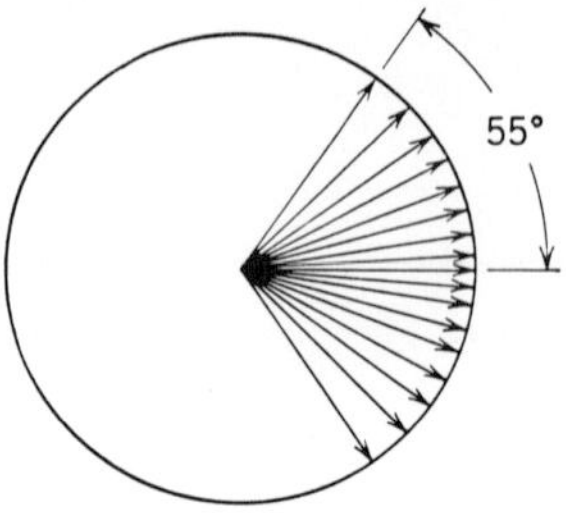

Fig. 12.4. Angular distribution of domain magnetizations of a cubic polycrystal with positive K_1 at its remanence point.

Let us now calculate the value of remanent magnetization for this distribution. As long as the direction of the field is in the hatched region in Fig. 12.5 with respect to the crystal axes of a crystallite, the domain magnetization is expected to turn back to the z direction. The contribution of this domain magnetization to the residual magnetization is $I_s \cos \theta$, where θ is the angle between the direction of H and the z axis. On averaging this value over possible orientation of H in the indicated region, we have

$$
\begin{aligned}
I_r &= I_s \overline{\cos \theta} \\
&= \frac{6I_s}{\pi}\left(\int_0^{\pi/4}\int_{-\pi/4}^{\pi/4} \cos\theta \sin\theta \, d\phi \, d\theta + \int_{\pi/4}^{\cos^{-1}\sqrt{1/3}}\int_{-\phi_0}^{\phi_0} \cos\theta \sin\theta \, d\phi \, d\theta\right) \\
&= 0.832 I_s, \qquad (12.2)
\end{aligned}
$$

where ϕ is the aximuthal angle measured from the $(0\bar{1}1)$ plane and ϕ_0 is

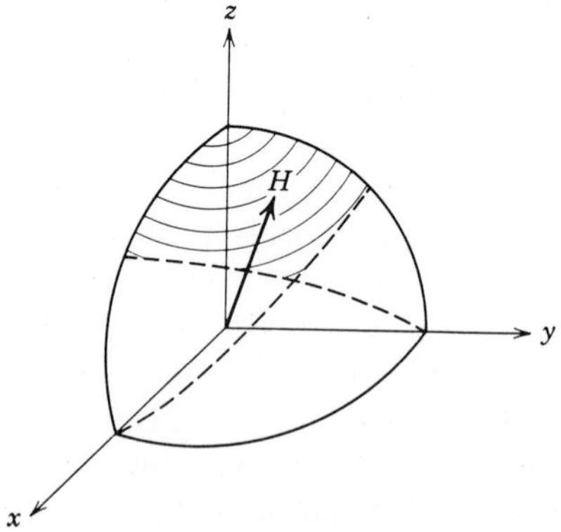

Fig. 12.5. The range of integration of the integral in equation (12.2).

given by

$$\phi_0 = \frac{\pi}{4} - \cos^{-1} \cot \theta. \tag{12.3}$$

When $K_1 < 0$, or the easy directions are $\langle 111 \rangle$, the domain magnetization turns back to the [111] direction as long as H is in the quadrant surrounded by the $+x$, $+y$, and $+z$ directions (Fig. 12.6). Let θ be the angle between H and the [111] direction; then the average residual magnetization is

$$\begin{aligned} I_r &= I_s \overline{\cos \theta} \\ &= \frac{6I_s}{\pi}\left(\int_0^{\cos^{-1}\sqrt{2/3}} \int_{-\pi/3}^{\pi/3} \cos\theta \sin\theta \, d\phi \, d\theta \right. \\ &\quad \left. + \int_{\cos^{-1}\sqrt{2/3}}^{\cos^{-1}\sqrt{1/3}} \int_{-\phi_0}^{\phi_0} \cos\theta \sin\theta \, d\phi \, d\theta \right) \\ &= 0.866 I_s, \end{aligned} \tag{12.4}$$

where ϕ is the azimuthal angle measured from the $(0\bar{1}1)$ plane about the [111] axis and ϕ_0 is given by

$$\phi_0 = \frac{\pi}{3} - \cos^{-1}\left(\frac{1}{\sqrt{2}} \cot \theta\right). \tag{12.5}$$

Thus the residual magnetization for negative K_1 is larger than that for positive K_1 because there are four easy directions when K_1 is negative, so that a domain magnetization has a greater chance to take a direction close to H than when K_1 is positive, in spite of the fact that the maximum deviation of magnetization is also 55° for $K_1 < 0$. In conclusion, the remanence of a cubic substance is 80 to 90%, larger than that of the stressed material.

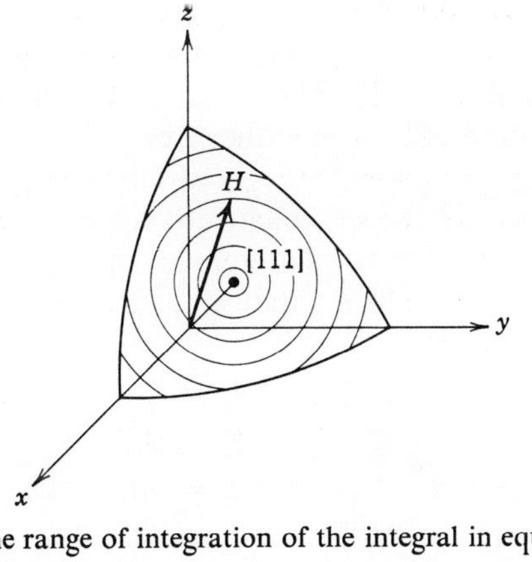

Fig. 12.6. The range of integration of the integral in equation (12.4).

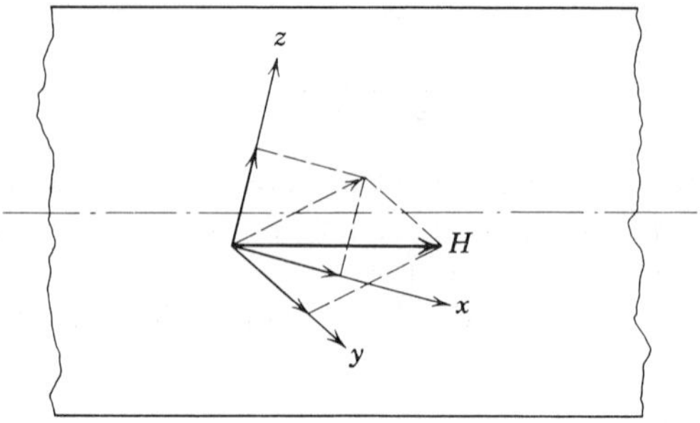

Fig. 12.7. Distribution of domain magnetization at the remanence point showing Kaya's law,

It can be said, therefore, that the shape of magnetization curves of cubic substances is generally rather rectangular.

In the treatment above, we ignored the influence of the demagnetizing field produced by free poles appearing on crystal boundaries. As stated in Section 11.3, the magnetostatic energy of a polycrystalline material is as large as about 20 $\sim$ 50% of that of an isolated crystallite; hence it is natural that individual crystallites tend to take a domain structure to reduce the magnetostatic energy. Then the residual magnetization of inhomogeneous materials tends to become less than the calculated values. Bozorth[2] observed that the residual magnetization of Permalloy becomes less than 7% of the saturation magnetization. Ideally this value should be zero, but actually, because of the difficulty of nucleation of domain walls and also because of the hindrance presented to the displacement of walls, the remanence will not be reduced to an extremely small value.

Residual magnetization is also influenced by the shape of the specimen. Kaya[3] measured residual magnetization on cylindrical iron specimens and found that the domain distribution in this case is such that the resultant magnetization averaged over x, y, and z domains points parallel to the long axis of the specimen (Fig. 12.7). The reason is that the normal component of magnetization would otherwise induce free poles on the side surface of the specimen, giving rise to a large magnetostatic energy. Let the direction cosines of the long axis of the specimen with respect to the x, y, and z axes be (l, m, n); and the total magnetic moment of x domains included in a unit volume of the specimen be I_x; that of the y domain, I_y, and that of the z domain, I_z. It follows that

$$\begin{aligned} I_x + I_y + I_z &= I_s, \\ I_x &= lI_r, \\ I_y &= mI_r, \\ I_z &= nI_r, \end{aligned} \tag{12.6}$$

from which we have

$$I_r = \frac{I_s}{l + m + n}. \tag{12.7}$$

Figure 12.8 shows a comparison of theory and the experiment made by Kaya. The agreement between the two is very good. Formula (12.7) is called the *lmn* law, or Kaya's law.

When the material exhibits an extreme magnetic anisotropy, remanence also takes a fairly extreme value. For instance, Isoperm has its easy axis perpendicular to the roll direction, so that the remanence is almost zero if the specimen was magnetized parallel to the roll direction (cf. Section 17.2). Another example is a grain-oriented magnetic sheet which has one of its easy axes parallel to the roll direction. In such a case the residual magnetization is almost 100%. Grain-oriented silicon sheet and Deltmax belong to this category.

The remanence of the 21% Fe-Ni alloy which is annealed at 490°C in the absence of a magnetic field is about 30%. Bozorth[4] explained this fact as follows: During annealing this alloy develops a uniaxial anisotropy which has its easy axis parallel to the x, y, or z axis provided $K_1 > 0$, because the domain magnetization there points in such a direction. In a place where the easy axis of the induced anisotropy is parallel to x, the domain magnetization rotates first from the field direction to the cubic easy direction, say the y axis, which is closest to the field direction, and then it splits into $+x$ and $-x$ domains by the displacement of 90° walls. This process contributes nothing to the residual magnetization. Only when the

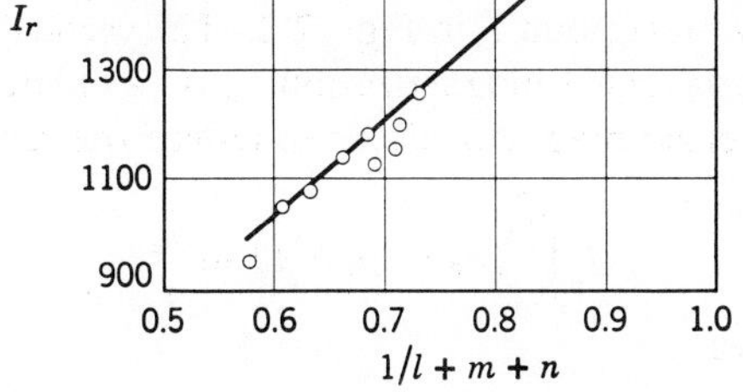

Fig. 12.8. Intensity of residual magnetization as a function of crystal orientation of cylindrical iron single crystals (after S. Kaya[3]).

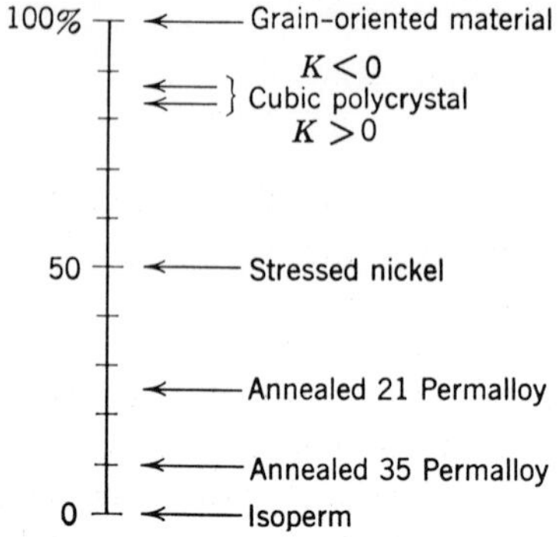

Fig. 12.9. Values of I_r/I_s for various kinds of magnetic materials.

x axis is closest to the field does it result in a residual magnetization. Since the probability for this is 1/3, the remanence in this case should be given by

$$\frac{I_r}{I_s} = \frac{83.2}{3}\,\% = 27.7\,\%. \tag{12.8}$$

Stressed iron also belongs to this category. Calculated or expected values of I_r/I_s for various materials are summarized in Fig. 12.9.

12.3 Distribution of Domain Magnetization at the Coercive Force Point

We discuss in this section the distribution of domain magnetizations which a ferromagnetic substance takes at the coercive force point. For a magnetic substance with distributed uniaxial anisotropies in which the magnetization reversal takes place through the displacement of 180° walls, the directions of domain magnetization cover a positive half-sphere at the remanence point, as already discussed in the previous section (cf. Fig. 12.2). When the field is reversed to the minus direction, the magnetization which pointed to the plus direction is first reversed, resulting in the distribution as shown by the diagram at point E in Fig. 12.2. The vertical half-angle of the distribution of the reversed magnetization θ_0 is calculated from the condition that the reversed magnetization is one-half the remanence, or

$$I_s \int_0^{\theta_0} \cos\theta \sin\theta \, d\theta = \frac{I_s}{4}. \tag{12.9}$$

On solving we have

$$\cos\theta_0 = \frac{1}{\sqrt{2}}$$

or

$$\theta_0 = 45°. \tag{12.10}$$

The last domain magnetization which is reversed at the coercive force point thus makes an angle of 45° from the direction of the field. The intensity of the field necessary to displace a 180° wall under such a condition gives the value of coercive force (cf. Section 14.2).

Let us consider here how the distribution of magnetization changes if the field is again increased from the point of coercive force to the positive direction (Fig. 12.10). This reversal of magnetization also starts with the magnetization which points in the negative direction and finishes at the positive field whose absolute intensity is equal to H_c. If we try to attain the same domain distribution from the demagnetized state, we need a fairly strong field to reverse the directions of domain magnetizations which are almost perpendicular to the field direction.

A permanent magnet is usually magnetized after it is assembled in the system of apparatus. If, however, the magnet is once magnetized and then brought to the coercive force point by applying an appropriate reversed magnetic field, it can be easily remagnetized under the application of a relatively weak magnetic field. This ingenious method of charging magnets was proposed by G. H. Weber in N. V. Philips Gloeilampen-fabrieken.[5] It should be noted that this phenomenon is independent of the mechanism of magnetization, for the domain magnetizations which have been reversed to the coercive force point are generally easily reversed again.

The description of magnetization reversal from the remanence to the coercive force point for a magnetic substance with a cubic anisotropy is

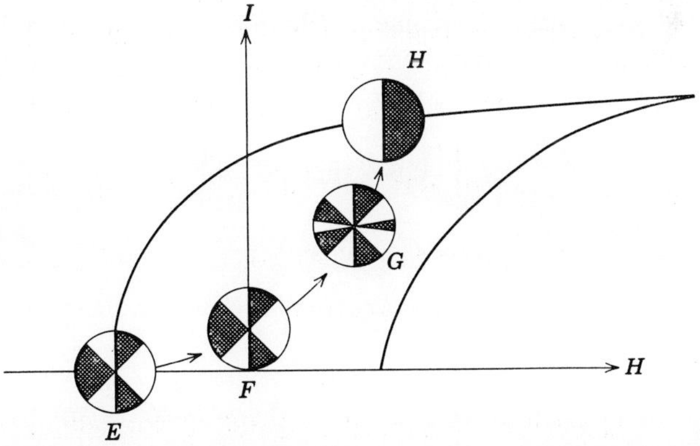

Fig. 12.10. Variation of angular distribution of domain magnetizations in the course of magnetization from the coercive force point toward the positive direction (distributed uniaxial anisotropy and wall displacement are assumed).

similar to the description above. For $K_1 > 0$, half of the residual magnetization, $0.832I_s$, is expected to be reversed until the coercive force point is attained, or

$$3I_s \int_0^{\theta_0} \cos\theta \sin\theta \, d\theta = 0.416I_s, \tag{12.11}$$

where θ_0 is the vertical half-angle of the angular distribution of reversed magnetizations. Solving this equation, we have

$$\cos\theta_0 = 0.851$$

or

$$\theta_0 = 31.7°. \tag{12.12}$$

The reason why no complicated integral involving the azimuthal angle ϕ appears in (12.11) is that θ_0 is less than 45°.

Similarly, for $K_1 < 0$, the remanence is $0.866I_s$, half of which is expected to be reversed when the coercive force point is attained. Thus

$$4I_s \int_0^{\theta_0} \cos\theta \sin\theta \, d\theta = 0.433I_s \tag{12.13}$$

or

$$\cos\theta_0 = 0.885$$

or

$$\theta_0 = 27.6°. \tag{12.14}$$

Since the obtained θ_0 is less than 35°, we need not worry about the integration with respect to the azimuthal angle ϕ again in (12.13).

In the discussion above we assumed that the mechanism of magnetization is the displacement of 180° walls. In aggregates of single domain particles, however, the magnetization takes place exclusively by rotation magnetization. As will be discussed in Section 14.1, a fine crystal grain with a uniaxial anisotropy or an elongated fine particle reverses its magnetization most easily when the magnetic field makes an angle of $\pi/4$ from the easy axis. On assuming that the reversal of magnetization is completed in the range $\theta_1 = \pi/4 - \epsilon$ to $\theta_2 = \pi/4 + \epsilon$ at the coercive force point, we have

$$I_s \int_{\theta_1}^{\theta_2} \cos\theta \sin\theta \, d\theta = \frac{I_s}{4} \tag{12.15}$$

or

$$\cos^2\theta_1 - \cos^2\theta_2 = \tfrac{1}{2}$$

or

$$\sin 2\epsilon = \tfrac{1}{2},$$

which gives

$$\epsilon = 15°. \tag{12.16}$$

The range of magnetization reversal is therefore $\theta_1 = 30°$ to $\theta_2 = 60°$. Figure 12.11 summarizes the distributions of domain magnetization at remanence and coercive force points for various kinds of magnetic materials.

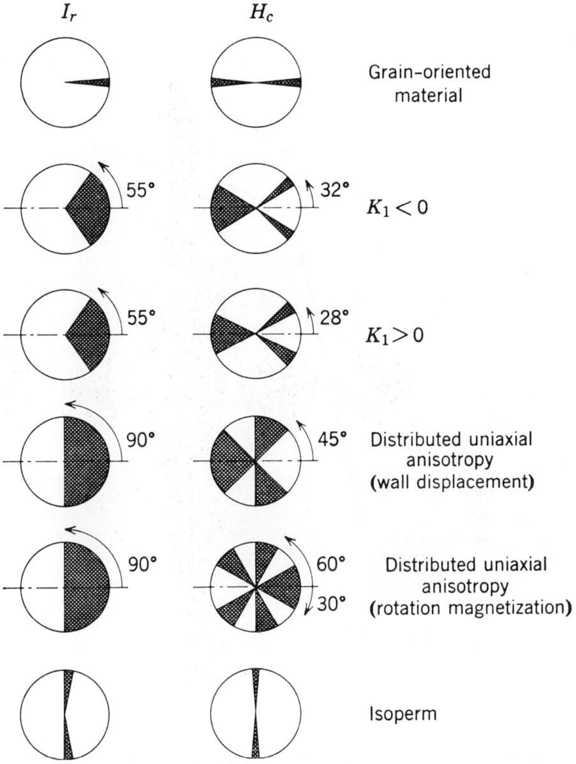

Fig. 12.11. Angular distribution of domain magnetizations at remanence and the coercive force point for various magnetic materials.

Finally let us discuss the method of demagnetization and its meaning. There are two methods of demagnetization: One is thermal demagnetization, and the other is ac demagnetization. The procedure of the former method consists in heating the specimen to its Curie point and cooling it in the absence of a magnetic field. In the latter method the specimen is magnetized by applying an alternating magnetic field and then decreasing its amplitude gradually toward zero. The mechanism of ac demagnetization for distributed uniaxial anisotropy and 180° wall displacement is shown in Fig. 12.12. When the amplitude of the magnetic field decreases, the domain magnetizations which make large angles with the fields are settled first. The final state is an isotropic distribution of domain magnetizations as shown at the origin of Fig. 12.12. It is understood from the figure that the uniformity of the angular distribution of domain magnetizations can be improved by decreasing the rate of decrease in the amplitude of the ac magnetic field.

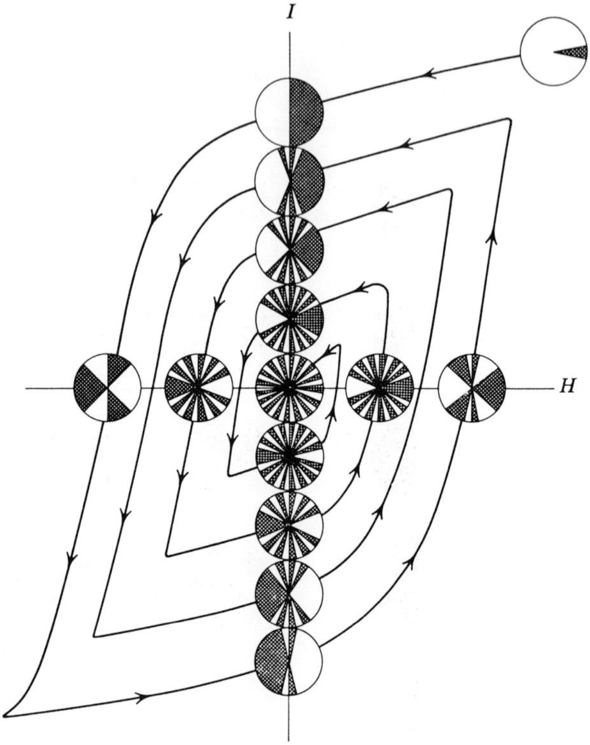

Fig. 12.12. Variation of angular distribution of domain magnetizations during the process of ac demagnetization (distributed uniaxial anistropy and wall displacement are assumed).

It should be noted that an ac demagnetization does not always result in an isotropic distribution. For instance, magnetic substances with cubic anisotropy are expected to have a domain distribution at the demagnetized state which is confined within 55° about the axis of the ac magnetic field. Thus, in order to obtain an isotropic domain distribution, we must use either thermal demagnetization or ac demagnetization with a rotating magnetic field instead of an alternating magnetic field.

Problems

12.1. Calculate the intensity of the residual magnetization relative to the saturation magnetization for an aggregate of elongated fine particles which lie in a plane with random angular distribution.

12.2 Determine the angular distribution of domain magnetizations at the coercive force point for the system described in Problem 12.1. Assume (i) occurrence of the displacement of 180° walls and (ii) occurrence of rotation magnetization.

12.3. Suppose that a magnetic substance in which the directions of easy axes of uniaxial anisotropy are distributed at random is magnetized to some extent from the demagnetized state, and then its magnetization is reduced to $I_s/4$ by the removal of the magnetic field. Determine the angular distribution of domain magnetizations at the final state. Assume (i) the occurrence of the displacement of 180° walls and (ii) the occurrence of rotation magnetization.

References

12.1. R. Becker and W. Döring: G.3, p. 288.
12.2. R. M. Bozorth: G.10, p. 502.
12.3. S. Kaya: *Z. Physik.* **84,** 705 (1933).
12.4. R. M. Bozorth: *Z. Physik.* **124,** 519 (1948).
12.5. G. H. Weber (N.V. Philips Fab.): Japan Pat. Sho 32–2125 (1957).

13

The Reversible Magnetization Process

13.1 Reversible Rotation Magnetization

We discuss in this section mechanisms of reversible rotation of domain magnetization. First we consider uniaxial anisotropy. If the field H is applied in a direction which makes an angle θ_0 with the easy axis of the uniaxial anisotropy, the energy of the system is

$$E = -K_u \cos^2(\theta - \theta_0) - I_s H \cos\theta, \tag{13.1}$$

per unit volume, where θ is the angle between the domain magnetization and the applied field (Fig. 13.1). The constant of uniaxial anisotropy, K_u, is equal to K_{u1} in (7.1) for uniaxial crystal anisotropy, and is given by $\frac{3}{2}\lambda\sigma$ if the anisotropy is caused by a stress σ. Uniaxial anisotropy is also induced by magnetic annealing, cold work, or precipitation, as will be discussed in Chapter 17. Formulas for these anisotropies and examples of each are given in Table 13.1. An elongated single domain particle also shows uniaxial anisotropy, because it stores magnetostatic energy $I_s^2/4\mu_0$ when the magnetization is perpendicular to the long axis. The magnetostatic energy of an aggregate of fine particles is thus

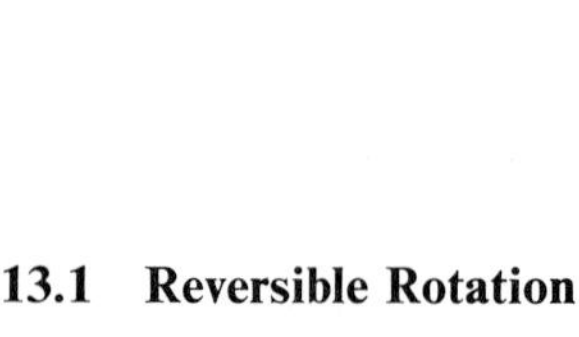

Fig. 13.1. Rotation of domain magnetization from an easy axis.

$$E_a = -\frac{I_s^2 \beta}{4\mu_0} \cos^2\theta, \tag{13.2}$$

Table 13.1. Various Uniaxial Anisotropies

Origin	Expression	Substance	K_u, J/m³ or ×10 ergs/cc	Remarks
Crystal anisotropy	K_{u1}	Cobalt	4.1×10^5	At room temperature
Magnetostrictive	$\frac{3}{2}\lambda\sigma$	Nickel	4.7×10^4	$\lambda = 31 \times 10^{-6}$, $\sigma = 100\,\text{kg/mm}^2$
Magnetic annealing	K_{uf}	24-Permalloy	2×10^2	Cooling rate, 1°/m
Roll magnetic anisotropy	K_{ur}	24-Permalloy	2×10^4	(110) roll, reduction 70%
Single domain particles	$I_s^2\beta/4\mu_0$	Iron	4.6×10^5	Elongated, $\beta = 0.5$

where β is the fractional volume occupied by fine particles. On assuming tentatively $\beta = 0.5$, we can calculate the anisotropy constant of iron fine particles as

$$K_u \approx \frac{(2.16)^2 \times 0.5}{4 \times (4\pi \times 10^{-7})} = 4.6 \times 10^5\,\text{J/m}^3, \tag{13.3}$$

which is fairly large compared to the other kinds of anisotropy listed in Table 13.1.

Now the stable direction of domain magnetization can be determined by minimizing the energy (13.1), or by

$$\frac{\partial E}{\partial \theta} = K_u \sin 2(\theta - \theta_0) + I_s H \sin \theta = 0. \tag{13.4}$$

On putting $\cos\theta = x$, (13.4) becomes

$$4x^4 + 4p\cos 2\theta_0 x^3 - (4 - p^2)x^2 - 4p\cos 2\theta_0 x + \sin^2 2\theta_0 - p^2 = 0, \tag{13.5}$$

where $p = I_s H/K_u$. After determining x from this equation, we can calculate the component of domain magnetization parallel to the applied field as

$$I = I_s \cos\theta = I_s x, \tag{13.6}$$

as a function of p or H. The result is shown graphically in Fig. 13.2. As seen in this graph, the magnetization curve for $\theta_0 = 90°$ is exactly linear. This fact can be easily verified by taking $\theta_0 = \pi/2$ in (13.4); we obtain

$$\cos\theta = \frac{I_s H}{2K_u}. \tag{13.7}$$

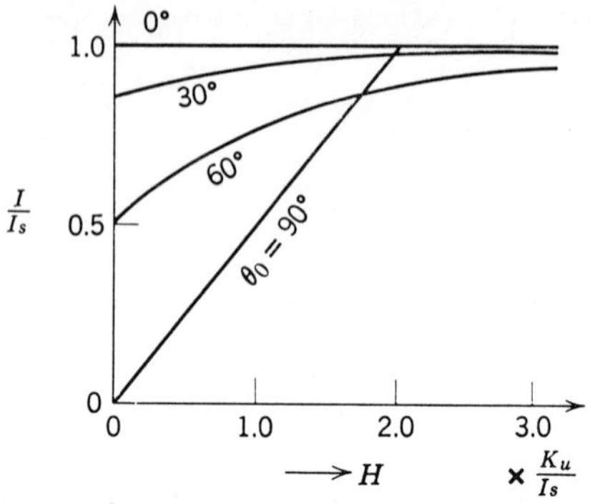

Fig. 13.2. Magnetization curves due to rotation magnetization against the uniaxial anisotropy.

On putting it into (13.6), we have

$$I = \frac{I_s^2}{2K_u} H, \tag{13.8}$$

which signifies a linear magnetization curve. Such a linear magnetization curve is actually observed for a stretched nickel wire, for a Permalloy bar which is cooled in a magnetic field directed perpendicular to the long axis, for Isoperm (which is a cold-rolled 50% Fe-Ni alloy), and for a cobalt single crystal which is magnetized perpendicular to the *c* axis. If the field is decreased from the saturated state, the magnetization is decreased almost reversibly, resulting in a very small residual magnetization. As θ_0 is decreased from 90°, the residual magnetization is increased, whereas the reversible permeability is decreased as seen in Fig. 13.2. It is interesting that magnetization is saturated at a finite field when $\theta_0 = 90°$, whereas complete saturation is rather difficult to attain when θ_0 is smaller.

For a very weak magnetic field ($H \ll K_u/I_s$), θ is nearly equal to θ_0; hence if we set

$$\Delta\theta = \theta_0 - \theta, \tag{13.9}$$

the anisotropy energy is expressed as $K_u\, \Delta\theta^2$. Then (13.4) becomes

$$2K_u\, \Delta\theta = I_s H \sin\theta_0$$

or

$$\Delta\theta = \frac{I_s H}{2K_u} \sin\theta_0. \tag{13.10}$$

The initial susceptibility χ_a is calculated from (13.6) as

$$\chi_a = \left(\frac{\partial I}{\partial H}\right)_{H=0} = I_s \sin\theta_0 \frac{\partial\theta}{\partial H}. \tag{13.11}$$

When we calculate $\partial\theta/\partial H$ by (13.10), (13.11) becomes

$$\chi_a = \frac{I_s^2 \sin^2 \theta_0}{2K_u}. \tag{13.12}$$

For cubic anisotropy with positive K_1, the anisotropy energy can be expressed as $K_1\theta^2$ for a small rotation of domain magnetization. The initial susceptibility is, therefore, expressed as

$$\chi_a = \frac{I_s^2 \sin^2 \theta_0}{2K_1}. \tag{13.13}$$

For $K_1 < 0$, the anisotropy energy is expressed as $-(2K_1/3)\,\Delta\theta^2$ as given by (7.38), so that the initial susceptibility becomes

$$\chi_a = -\frac{3I_s^2 \sin^2 \theta_0}{4K_1}. \tag{13.14}$$

When the angular distribution of domain magnetizations is isotropic, $\overline{\sin^2 \theta_0} = \frac{2}{3}$; hence (13.13) becomes

$$\bar{\mu} \approx \bar{\chi}_a = \frac{I_s^2}{3K_1\mu_0}. \tag{13.15}$$

For iron, $I_s = 2.16$ Wb/m^2 and $K_1 = 4.2 \times 10^4$ J/m^2; therefore (13.15) gives

$$\bar{\mu}_a = \frac{2.16^2}{3 \times 4.2 \times 4\pi \times 10^{-3}} \fallingdotseq 29. \tag{13.16}$$

Since, however, common soft iron exhibits a relative initial permeability of the order of $100 \sim 200$, it is concluded that rotation magnetization contributes only a small part of the actual initial permeability.

The susceptibility due to the displacement of domain walls is very sensitive to the irregularity of the substance, whereas that of the rotation magnetization depends only on the magnitude of crystal anisotropy, which is fairly insensitive to the presence of small amounts of impurities or weak internal stresses. Thus it can be said that the calculated value of initial permeability based on the rotation mechanism gives the lower limit of the actual initial permeability.

The initial permeability of Fe-Ni alloy is shown in Fig. 13.3 as a function of alloy composition for various kinds of heat treatment. After being quenched from 600°C subsequent to slow cooling from above 1000°C to 600°C (double treatment), the alloy exhibits very high permeability at about 21% Fe-Ni. The crystal anisotropy constant in this alloy system goes through zero at 24% Fe-Ni, as shown in Fig. 7.9. The magnetostriction constant also goes through zero at 19% Fe-Ni, as shown in

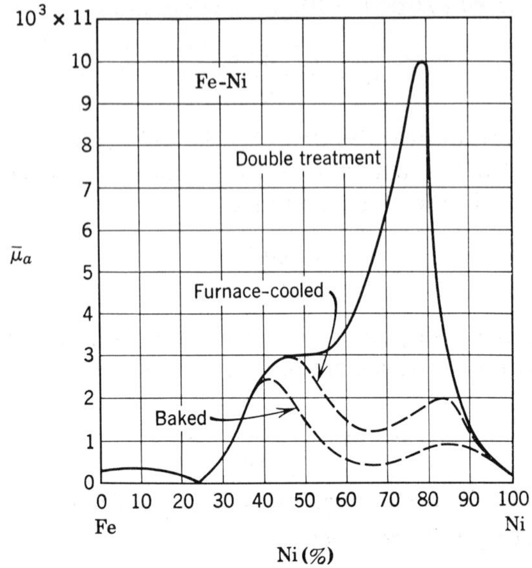

Fig. 13.3. Initial permeability of an iron-nickel alloy system (after Bozorth[1]).

Fig. 8.11. Bozorth[2] interpreted the high permeability realized at 21% Fe-Ni to be due to the combined effect of a low crystal anisotropy and a low magnetostriction around this composition. The purpose of quenching from 600°C is to prevent the development of induced magnetic anisotropy due to a local directional ordering (cf. Section 17.1).

The actual mechanism of magnetization in this alloy, however, has not been established. If we assume a coherent rotation magnetization, the highest initial permeability observed in this alloy requires the anisotropy constant to be

$$K = \frac{I_s^2}{3\mu_0\bar{\mu}_a} = \frac{1.1^2}{3 \times 4\pi \times 10^{-7} \times 10{,}000} \fallingdotseq 32. \qquad (13.17)$$

Since it is very hard to realize such a low effective anisotropy uniformly throughout the material, a coherent rotation magnetization is hardly conceivable. Displacement of broad walls or incoherent rotation magnetization[3] is considered to be the most likely mechanism of magnetization in this alloy.

13.2 Reversible Wall Displacement

When a wall is displaced in a completely uniform material, no change results in the surface energy of the wall. Upon removal of the magnetic field, the wall does not go back to its original position. In order to have a reversible displacement of the wall, therefore, it is necessary to have some irregularities which cause variations of energy for the displacement of the wall.

Let us first assume that a plane domain wall is displaced in some non-uniform material. The energy of the wall per unit area, ϵ_w, is assumed to be changing with the displacement of the wall, s, as shown in Fig. 13.4. In the absence of a magnetic field, the wall stays at some minimum point where $\partial\epsilon_w/\partial s = 0$. The energy can be expressed in the first approximation as

$$\epsilon_w = \tfrac{1}{2}\alpha s^2 \tag{13.18}$$

in the vicinity of the stable point. If a magnetic field H is applied in a direction which makes an angle θ with I_s (Fig. 13.5), the energy supplied by the magnetic field is

$$\epsilon_H = -2I_sH(\cos\theta)s \tag{13.19}$$

for a 180° wall, because the displacement s of the wall results in a change in the magnetization of $2I_s s$ per unit area of the wall. From the condition of minimizing the total energy,

$$\epsilon = \epsilon_w + \epsilon_H = \tfrac{1}{2}\alpha s^2 - 2I_sH(\cos\theta)s \tag{13.20}$$

or

$$\frac{\partial\epsilon}{\partial s} = \alpha s - 2I_sH\cos\theta = 0, \tag{13.21}$$

we have

$$s = \frac{2I_s\cos\theta}{\alpha}H. \tag{13.22}$$

As a result of the displacement s of the wall, the magnetization in the

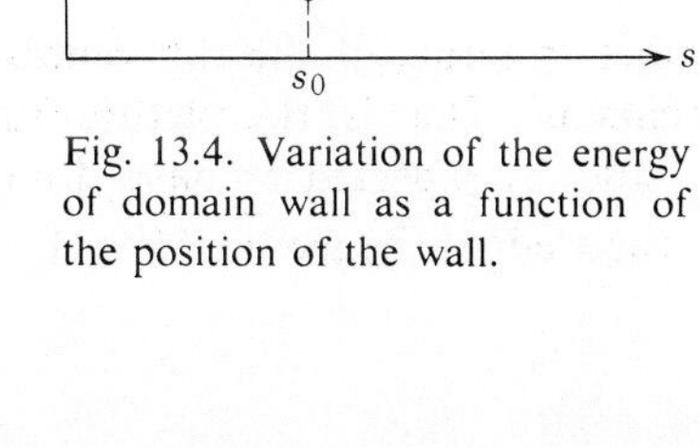

Fig. 13.4. Variation of the energy of domain wall as a function of the position of the wall.

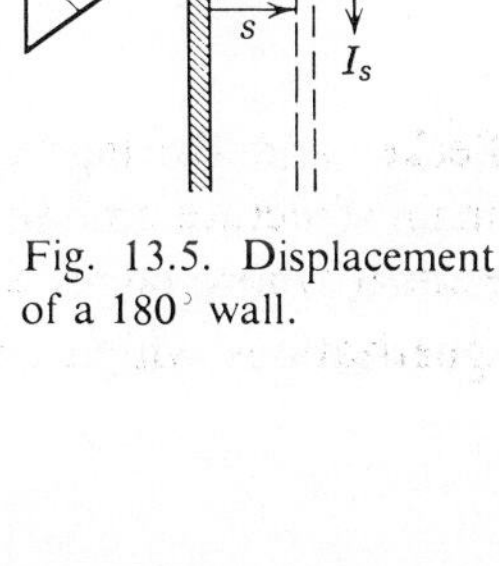

Fig. 13.5. Displacement of a 180° wall.

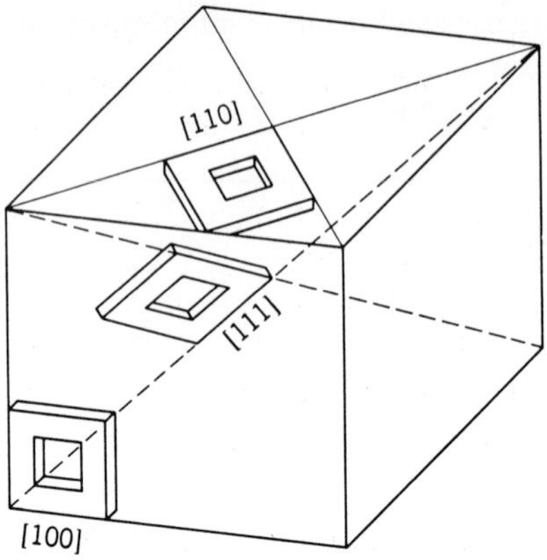

Fig. 13.6. Method of cutting three kinds of picture-frame specimens out of a single crystal.

direction of the magnetic field increases by the amount $2I_s(\cos\theta)s$, and the total magnetization is given by

$$I = \frac{4I_s^2 \cos^2\theta}{\alpha} SH, \tag{13.23}$$

where S is the total area of the 180° wall included in a unit volume. The initial susceptibility is, therefore,

$$\chi_a = \frac{4I_s^2 \cos^2\theta}{\alpha} S. \tag{13.24}$$

For a cubic crystal like iron with positive K_1, the average value $\overline{\cos^2\theta}$ for the three directions of domain magnetization is calculated as

$$\overline{\cos^2\theta} = \tfrac{1}{3}(\alpha_1^2 + \alpha_2^2 + \alpha_3^2) = \tfrac{1}{3}. \tag{13.25}$$

Hence χ_a is expected to be isotropic.

For the purpose of examining this point, H. J. Williams[4] cut three picture-frame single crystals of silicon-iron whose edges were parallel to ⟨100⟩, ⟨110⟩, or ⟨111⟩ (Fig. 13.6), put the winding coils on as is usual with ring cores, and measured their initial permeability. He found that the measured permeabilities are not the same but are in the ratio

$$\bar{\mu}_{100}:\bar{\mu}_{110}:\bar{\mu}_{111} = 1:\tfrac{1}{2}:\tfrac{1}{3}. \tag{13.26}$$

Becker and Döring[5] explained this fact in terms of the differences in domain structure among the three specimens. That is, the picture-frame specimen whose edges are parallel to ⟨100⟩ is expected to have domain magnetizations which run parallel to each edge; hence $\overline{\cos^2\theta} = 1$. On the other hand, domains in a ⟨110⟩ picture-frame specimen are expected to take the two easy axes which are closest to the direction of each edge; hence $\cos^2\theta = \frac{1}{2}[(1/\sqrt{2})^2 + (1/\sqrt{2})^2] = \frac{1}{2}$. Similarly the ⟨111⟩ picture-frame specimen may contain three kinds of domains, so $\overline{\cos^2\theta} = \frac{1}{3}[(1/\sqrt{3})^2 + (1/\sqrt{3})^2 + (1/\sqrt{3})^2] = \frac{1}{3}$. Thus we can explain the ratio in (13.26) in terms of that of $\overline{\cos^2\theta}$. We assumed implicitly that the total area of 180° wall, S, is just the same for three crystals, but this point is very doubtful.

For negative K_1 or crystals with easy axes parallel to ⟨111⟩, the average value becomes

$$\begin{aligned}\overline{\cos^2\theta} = \frac{1}{4}\bigg[&\left(\frac{1}{\sqrt{3}}\alpha_1 + \frac{1}{\sqrt{3}}\alpha_2 + \frac{1}{\sqrt{3}}\alpha_3\right)^2 + \left(\frac{1}{\sqrt{3}}\alpha_1 + \frac{1}{\sqrt{3}}\alpha_2 - \frac{1}{\sqrt{3}}\alpha_3\right)^2 \\ &+ \left(\frac{1}{\sqrt{3}}\alpha_1 - \frac{1}{\sqrt{3}}\alpha_2 + \frac{1}{\sqrt{3}}\alpha_3\right)^2 \\ &+ \left(-\frac{1}{\sqrt{3}}\alpha_1 + \frac{1}{\sqrt{3}}\alpha_2 + \frac{1}{\sqrt{3}}\alpha_3\right)^2\bigg] \\ = \tfrac{1}{3}.&\end{aligned} \tag{13.27}$$

Hence the susceptibility given by (13.24) again becomes isotropic. When the easy axes are distributed isotropically,

$$\overline{\cos^2\theta} = \int_0^{\pi/2} \cos^2\theta \sin\theta \, d\theta = \tfrac{1}{3}. \tag{13.28}$$

In any case, therefore, (13.24) gives

$$\chi_{a_{180^\circ}} = \frac{4I_s^{\,2}}{3\alpha} S. \tag{13.29}$$

A conceivable origin of the fluctuation in the energy of a domain wall as shown in Fig. 13.4 is the internal stress, as first proposed by Kondorsky[6] and developed by Kersten.[7] Although our discussion below follows Kersten's, it is not faithful to his original formulation. First we assume a sinusoidal variation of internal stress, or

$$\sigma = \sigma_0 \cos 2\pi \frac{s}{l}, \tag{13.30}$$

where l is the wavelength of spatial variation of the internal stress. The anisotropy constant is, therefore,

$$K = K_1 - \tfrac{3}{2}\lambda\sigma_0 \cos 2\pi \frac{s}{l}. \tag{13.31}$$

Here we assumed for simplicity that the functional form of the anisotropy energy is common to K_1 and $\lambda\sigma_0$ terms, though it is not exactly true. If the wavelength l is sufficiently large compared to the wall thickness δ, we can assume that the anisotropy constant does not change inside the wall; then the surface energy of the wall is given by

$$\gamma = 2\sqrt{A(K_1 - \tfrac{3}{2}\lambda\sigma_0 \cos 2\pi(s/l))}. \tag{13.32}$$

On expanding this in a power series around $s = 0$ by assuming that $K_1 \gg \lambda\sigma$, we have

$$\begin{aligned}\gamma &= 2\sqrt{AK_1}\left(1 - \frac{3\lambda\sigma_0}{4K_1}\cos 2\pi\frac{s}{l} + \cdots\right)\\ &= 2\sqrt{AK_1}\left(1 - \frac{3\lambda\sigma_0}{4K_1}\left(1 - \frac{2\pi^2 s^2}{l^2}\right) + \cdots\right).\end{aligned} \tag{13.33}$$

The second derivative of the wall energy, α, thus becomes

$$\alpha = \frac{\partial^2\gamma}{\partial s^2} = 6\pi^2\sqrt{\frac{A}{K_1}}\frac{\lambda\sigma_0}{l^2}. \tag{13.34}$$

The wall thickness δ is approximately given by

$$\delta = 3\sqrt{\frac{A}{K_1}}, \tag{13.35}$$

as seen in Fig. 9.11; hence (13.34) becomes

$$\alpha = 2\pi^2\frac{\lambda\sigma_0}{l^2}\delta. \tag{13.36}$$

The total area of the wall, S, is

$$S = \frac{3}{l}, \tag{13.37}$$

provided that the internal stress fluctuates in the y and z directions as well as in the x direction and also that walls exist at every stable position. Putting (13.36) and (13.37) into (13.29), we have

$$\chi_{a_{180°}} = \frac{2I_s^2}{\pi^2\lambda\sigma_0}\frac{l}{\delta}. \tag{13.38}$$

According to this formula, the susceptibility decreases with a decrease of the wavelength l. When, however, l becomes less than δ, (13.38) is no longer valid. If $l \ll \delta$, or if σ varies in the wall, the spin arrangement cannot follow such an abrupt change of the local anisotropy, for it would otherwise store too much exchange energy. Then the spatial change of wall energy is averaged, so the susceptibility is again increased. When $l \approx \delta$, therefore, the initial susceptibility χ_a takes the minimum value which is

$$\chi_a = \frac{2I_s^2}{\pi^2 \lambda \sigma}. \tag{13.39}$$

It is interesting that this formula gives almost the same value as the susceptibility due to rotation magnetization out of the easy axis of the anisotropy caused by stress, which is calculated by putting $K_u = \frac{3}{2}\lambda\sigma$ in (13.12), or

$$\chi_{a\mathrm{rot}} = \frac{2I_s^2}{9\lambda\sigma}. \tag{13.40}$$

Next we discuss the reversible displacement of a 90° wall. Suppose that a 90° wall is running parallel to [110], separating x and y domains (cf. Fig. 13.7), and the magnetic field H is applied in a direction which makes the angle θ with [110]. The work done by the magnetic field is expressed as

$$\epsilon_H = -\sqrt{2}\, I_s H(\cos\theta)s \tag{13.41}$$

as a function of the displacement s of the wall. Since the magnetization $\sqrt{2}\, I_s(\cos\theta)s$ is induced by this displacement along the direction of magnetic field H, the susceptibility for this case is given, by replacing the factor 4 in (13.24) by 2, by

$$\chi_a = \frac{2I_s^2 \cos^2\theta}{\alpha} S. \tag{13.42}$$

Averaging $\cos^2\theta$ over the four orientations $\langle 110 \rangle$ of 90° walls, we have again $\overline{\cos^2\theta} = \frac{1}{3}$. The susceptibility for an equal distribution of 90° walls

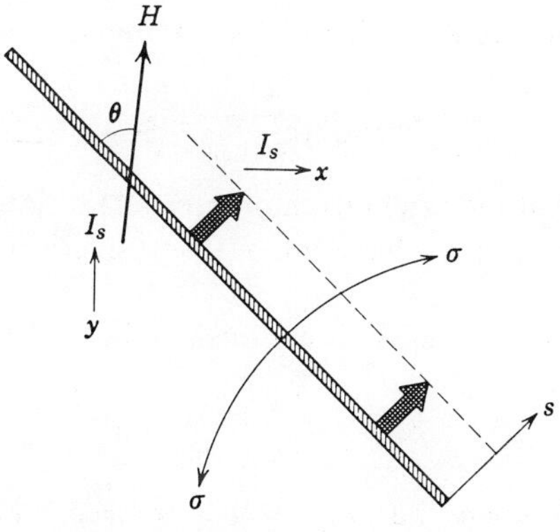

Fig. 13.7. Displacement of a 90° wall.

over the four possible orientations is thus given by

$$\chi_{a_{90^\circ}} = \frac{2I_s^2}{3\alpha} S. \tag{13.43}$$

An internal stress affects the position of a 90° wall through a change in the magnetoelastic energy of the domains on both sides of the wall, as well as through a change in the surface energy of the 90° wall itself. Suppose that the internal tension σ changes its orientation gradually from place to place as shown in Fig. 13.7. Let the direction cosines of the axis of the tension be $(\gamma_1, \gamma_2, \gamma_3)$. The magnetoelastic energies of the x and y domains are given by (8.80) and (8.81), from which we know that the x domain is stable in the place where $\gamma_1 > \gamma_2$, while the y domain is stable in the place where $\gamma_1 < \gamma_2$, provided $\lambda_{100} > 0$. When the wall is displaced from s to $s + ds$, the x domain of the volume ds is changed to y domain per unit area of the wall. Thus the magnetoelastic energy is changed by the amount

$$d\epsilon_\sigma = -\tfrac{3}{2}\lambda_{100}\sigma(\gamma_2^2 - \gamma_1^2)\, ds. \tag{13.44}$$

We assume here that the spatial change in the orientation of the internal tension is given by the sinusoidal form

$$\gamma_1^2 - \gamma_2^2 = \sin 2\pi \frac{s}{l}. \tag{13.45}$$

When the displacement is small, this gives

$$\frac{\partial}{\partial s}(\gamma_2^2 - \gamma_1^2) = -\frac{2\pi}{l}. \tag{13.46}$$

From (13.44) and (13.46), we have

$$\alpha = \frac{\partial^2 \epsilon_\sigma}{\partial s^2} = \frac{3\pi\lambda_{100}\sigma}{l}. \tag{13.47}$$

Using this expression, we have, from (13.43), the susceptibility

$$\chi_{a_{90^\circ}} = \frac{4I_s^2}{3\pi\lambda_{100}\sigma}, \tag{13.48}$$

where we assumed that the orientation of internal tension changes along all $\langle 110 \rangle$ directions and that we have 90° walls at every minimum point, so that $S = 6/l$.

For negative K_1 or $\langle 111 \rangle$ easy directions, we have similarly

$$\chi_{a_{90^\circ}} = \frac{2I_s^2}{3\pi\lambda_{111}\sigma}. \tag{13.49}$$

It is commonly known that $\lambda_{111} = 0$ is a necessary condition for a rectangular hysteresis curve for a ferrite core.[8,9] This fact may possibly be explained by considering the ease of displacement of a 90° wall, and (13.49) helps us to understand this readily. It is again interesting that formulas (13.48) and (13.49) give almost the same value as does (13.39) or (13.40).

If we assume the presence of an internal stress $\sigma \approx 50$ kg/mm^2 = 5×10^8 N/m^2, we know from (13.48) that the initial susceptibility of iron due to a 90° wall displacement (if we use $\lambda_{100} = 2.07 \times 10^{-5}$ and $I_s = 2.16$) is

$$\bar{\mu}_a \approx \bar{\chi}_a = \frac{4 \times 2.16^2}{3 \times 3.14 \times 2.07 \times 10^{-5} \times 5 \times 10^8 \times 4 \times 3.14 \times 10^{-7}}$$
$$\approx 153. \tag{13.50}$$

If the wavelength of the spatial variation of the internal stress, l, is almost the same as the wall thickness, the displacement of 180° walls contributes almost the same value to the initial permeability. Since the contribution from rotation magnetization is very small as calculated in (13.16), the resultant permeability becomes about 300. The actual value of the initial permeability of mild steel or iron is $\bar{\mu}_a \approx 100 \sim 200$, which is in fair agreement with the calculated value.

Another mechanism which possibly causes reversible displacement of domain walls is the hindrance of wall motion due to non-magnetic inclusions, precipitates, and voids, as first considered by Kersten.[10] He postulated that a domain wall tends to be trapped by inclusions or voids, because the wall can save its area at such an inclusion or void, thus reducing the total wall energy. The displacement of the wall results in an increase in the wall area, causing the restoring force on the wall. On the basis of this idea, Kersten calculated the reversible susceptibility and the coercive force by assuming a proper arrangement of inclusions or voids, and a rigid planc wall.

Néel[11] criticized these assumptions and stressed that the free poles resulting from magnetization variations produced by inclusions and internal strains play an important role in producing a restoring force on a domain wall. Later, however, Kersten[12] modified the assumptions in his previous paper and calculated an initial permeability by assuming the flexibility of domain walls. We shall follow these calculations below.

Let us suppose that a 180° wall is fixed at both ends as shown in Fig. 13.8. When the field H is applied in a direction which makes the angle θ with the direction of domain magnetization, the 180° wall tends to move so as to increase the volume of the domain with positive magnetization. If a part of the wall with area S is displaced by the distance s, the work done by the field is given by

$$U = -2I_sH(\cos\theta)Ss. \tag{13.51}$$

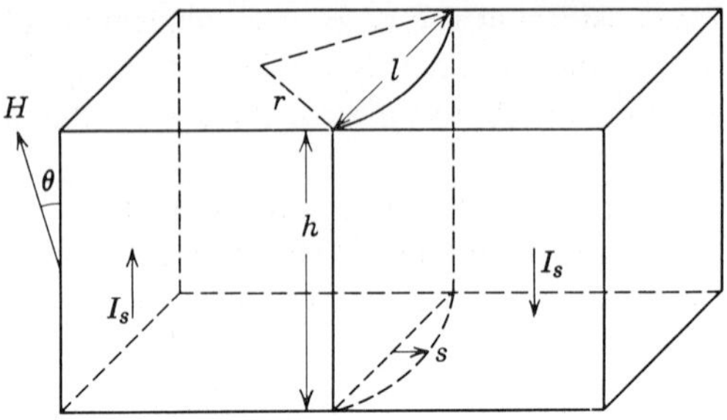

Fig. 13.8. A swelled 180° wall under the action of a magnetic field (after Kersten[12]).

Then the force exerted on a unit area of the wall is

$$p = -\frac{1}{S}\frac{\partial U}{\partial s} = 2I_s H \cos\theta. \tag{13.52}$$

Thus the action of a magnetic field is equivalent to an exertion of pressure on the magnetic wall. As a result of such an action of the magnetic field, the wall is bent into a cylindrical form as shown in Fig. 13.8; its radius of curvature, r, is related to the field H by

$$\frac{\gamma}{r} = 2I_s H \cos\theta, \tag{13.53}$$

where γ is the surface energy of the wall. The change in the volume of the domain with the positive magnetization is

$$\delta V = \tfrac{2}{3}lhs, \tag{13.54}$$

where h is the height of the wall, and l the distance between the two constrained portions of the wall (Fig. 13.8). The increase in the magnetization is thus given by

$$I = \tfrac{4}{3}I_s(\cos\theta)Ss, \tag{13.55}$$

where S is the total area of the domain wall included in a unit volume. Using the geometrical relation,

$$s = \frac{l^2}{8r}, \tag{13.56}$$

and (13.53), we find that (13.55) becomes

$$I = \frac{Sl^2 I_s^{\,2}}{3\gamma}(\cos^2\theta)H. \tag{13.57}$$

Since $\overline{\cos^2\theta} = \frac{1}{3}$ in most cases, the susceptibility becomes

$$\chi_{a_{180°}} = \frac{Sl^2 I_s^{\,2}}{9\gamma}. \tag{13.58}$$

For 90° walls, the pressure exerted on the wall is

$$p = \sqrt{2}\, I_s H \cos\theta, \tag{13.59}$$

where θ is the angle between H and [110]. The magnetization change due to the displacement of the 90° wall is given by $\sqrt{2}\, I_s(\cos\theta)s$, which can be transformed, as in (13.57), to

$$I_{90^\circ} = \frac{S_{90^\circ} l^2 I_s^{\,2}}{6\gamma_{90^\circ}} (\cos^2\theta) H. \tag{13.60}$$

Thus we have

$$\chi_{a_{90^\circ}} = \frac{S_{90^\circ} l^2 I_s^{\,2}}{18\gamma_{90^\circ}}. \tag{13.61}$$

If we assume that inclusions or voids which constrain domain walls are distributed in the form of a simple cubic lattice with lattice constant l, the total surface of the wall is given by $S = 2/l$, so that (13.58) becomes

$$\chi_{a_{180^\circ}} = \frac{2 I_s^{\,2} l}{9\gamma}. \tag{13.62}$$

For iron, $I_s = 2.15$ Wb/m², $\gamma = 1.6 \times 10^{-3}$ J/m², and, if we assume that $l = 10^{-4}$ m, we have from (13.62)

$$\bar{\mu}_a \fallingdotseq \bar{\chi}_a = \frac{2 \times (2.15)^2 \times 10^{-4}}{9 \times 1.6 \times 10^{-3} \times 4\pi \times 10^{-7}} = 51{,}000. \tag{13.63}$$

A wall displacement of this kind thus gives a fairly large permeability compared to the rotation magnetization. It has been reported[13] that well-annealed pure iron exhibits a permeability of about $\bar{\mu}_a = 25{,}000$ which would be explained more or less by this sort of mechanism of magnetization.

Finally let us discuss the temperature dependence of initial permeability. Generally speaking, initial permeability increases with an increase of temperature, has a sharp maximum at just below the Curie point, and then drops off to a very small value (Hopkinson effect). Kersten[12] also treated this problem. The temperature dependence of the surface energy of the domain wall, γ, is caused by a change in the anisotropy constant K and a change in the coefficient of exchange interaction A. Since A is proportional to S^2, as seen in (9.18) to (9.20), it follows that

$$A \propto I_s^{\,2}, \tag{13.64}$$

at a finite temperature. On the other hand, γ is proportional to $\sqrt{AK}$, so that

$$\gamma \propto I_s \sqrt{K}. \tag{13.65}$$

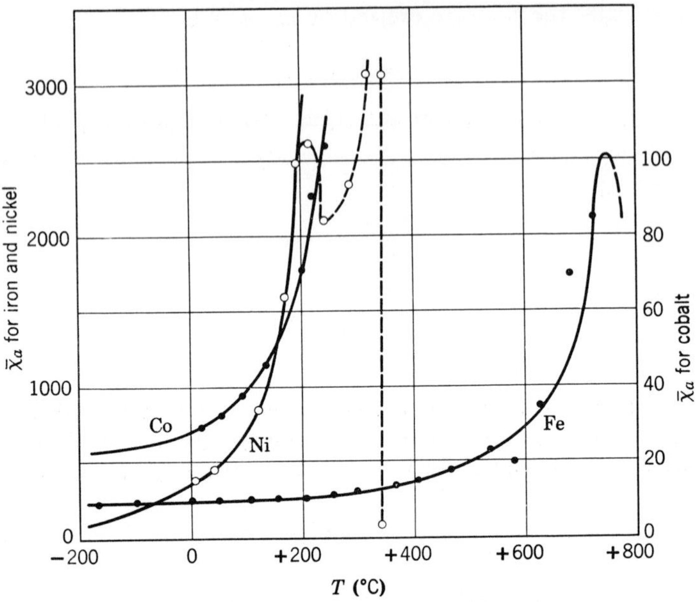

Fig. 13.9. Temperature dependence of initial permeability for iron, nickel, and cobalt, and comparison with calculated (solid) curves; (cf. 13.66, in which $K = K_{u1} + K_{u2}$ for cobalt) (after Kersten[12]).

It is expected, therefore, that the susceptibility given by (13.58) or (13.61) is related to I_s and K as

$$\chi_a \propto \frac{I_s}{\sqrt{K}}. \tag{13.66}$$

Using the actual temperature dependence of I_s and K in (13.66), Kersten explained the observed temperature dependence of the initial susceptibilities of iron, cobalt, and nickel as shown in Fig. 13.9, in which we can see excellent agreement between theory and experiment.

13.3 Law of Approach to Saturation

Under a moderately strong magnetic field, ferromagnetic substances are generally magnetized to their saturated state, where all wall displacements have been finished and the magnetization is pointed almost parallel to the magnetic field.

In this section we examine the manner of rotation of magnetization under such circumstances. Let the angle between the magnetization and the magnetic field be θ. Then the component of magnetization in the direction of the field is given by

$$I = I_s \cos\theta = I_s\left(1 - \frac{\theta^2}{2} + \cdots\right). \tag{13.67}$$

The torque exerted by a magnetic field is counterbalanced by the torque caused by the magnetic anisotropy (Fig. 13.10); thus

$$I_s H \sin\theta = -\frac{\partial E_a}{\partial \theta}, \tag{13.68}$$

where E_a is the anisotropy energy. Since θ is very small, it can be determined from (13.68) as

$$\theta = \frac{C}{I_s}\frac{1}{H}, \tag{13.69}$$

where

$$C = -\left(\frac{\partial E_a}{\partial \theta}\right)_{\theta=0}. \tag{13.70}$$

On substituting (13.69) in (13.67), we have

$$I = I_s\left(1 - \frac{b}{H^2} - \cdots\right), \tag{13.71}$$

where

$$b = \frac{1}{2}\frac{C^2}{I_s^2}. \tag{13.72}$$

Now we solve for the actual form of C for cubic anisotropy. Since the magnetization rotates along the maximum gradient of the anisotropy energy in the vicinity of H,

$$C^2 = |\text{grad } E_a|^2 = \left(\frac{\partial E_a}{\partial \theta}\right)^2 + \frac{1}{\sin^2\theta}\left(\frac{\partial E_a}{\partial \phi}\right)^2, \tag{13.73}$$

where (θ, ϕ) are the polar coordinates of the magnetization. Since E_a is normally expressed as a function of the direction cosines $(\alpha_1, \alpha_2, \alpha_3)$ of

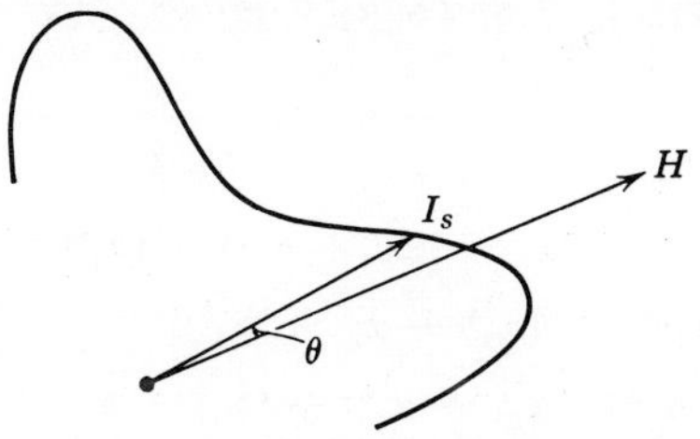

Fig. 13.10. Rotation of magnetization against the anisotropy energy.

the magnetization, which are related to (θ, ϕ) by $\alpha_1 = \sin\theta\cos\phi$, $\alpha_2 = \sin\theta\sin\phi$, $\alpha_3 = \cos\theta$,

$$\begin{aligned}\frac{\partial E_a}{\partial \theta} &= \left(\frac{\partial E_a}{\partial \alpha_1}\right)\frac{\partial \alpha_1}{\partial \theta} + \left(\frac{\partial E_a}{\partial \alpha_2}\right)\frac{\partial \alpha_2}{\partial \theta} + \left(\frac{\partial E_a}{\partial \alpha_3}\right)\frac{\partial \alpha_3}{\partial \theta} \\ &= \left(\frac{\partial E_a}{\partial \alpha_1}\right)\cos\theta\cos\phi + \left(\frac{\partial E_a}{\partial \alpha_2}\right)\cos\theta\sin\phi - \left(\frac{\partial E_a}{\partial \alpha_3}\right)\sin\theta, \qquad (13.74)\end{aligned}$$

and

$$\begin{aligned}\frac{1}{\sin\theta}\frac{\partial E_a}{\partial \phi} &= \left(\frac{\partial E}{\partial \alpha_1}\right)\frac{\partial \alpha_1}{\sin\theta\,\partial\phi} + \left(\frac{\partial E_a}{\partial \alpha_2}\right)\frac{\partial \alpha_2}{\sin\theta\,\partial\phi} + \left(\frac{\partial E_a}{\partial \alpha_3}\right)\frac{\partial \alpha_3}{\sin\theta\,\partial\phi} \\ &= -\left(\frac{\partial E_a}{\partial \alpha_1}\right)\sin\phi + \left(\frac{\partial E_a}{\partial \alpha_2}\right)\cos\phi.\end{aligned}$$

Then (13.73) becomes

$$\begin{aligned}C^2 &= \left(\frac{\partial E_a}{\partial \alpha_1}\right)^2(\cos^2\theta\cos^2\phi + \sin^2\phi) + \left(\frac{\partial E_a}{\partial \alpha_2}\right)^2(\cos^2\theta\sin^2\phi + \cos^2\phi) \\ &\quad + \left(\frac{\partial E_a}{\partial \alpha_3}\right)^2\sin^2\theta + 2\left(\frac{\partial E_a}{\partial \alpha_1}\right)\left(\frac{\partial E_a}{\partial \alpha_2}\right)(\cos^2\theta\sin\phi\cos\phi - \sin\phi\cos\phi) \\ &\quad - 2\left(\frac{\partial E_a}{\partial \alpha_1}\right)\left(\frac{\partial E_a}{\partial \alpha_3}\right)\sin\theta\cos\theta\cos\phi - 2\left(\frac{\partial E_a}{\partial \alpha_2}\right)\left(\frac{\partial E_a}{\partial \alpha_3}\right)\sin\theta\cos\theta\sin\phi \\ &= \left(\frac{\partial E_a}{\partial \alpha_1}\right)^2(\alpha_3^2 + \alpha_2^2) + \left(\frac{\partial E_a}{\partial \alpha_2}\right)^2(\alpha_3^2 + \alpha_1^3) + \left(\frac{\partial E_a}{\partial \alpha_3}\right)^2(\alpha_1^2 + \alpha_2^2) \\ &\quad - 2\left(\frac{\partial E_a}{\partial \alpha_1}\right)\left(\frac{\partial E_a}{\partial \alpha_2}\right)\alpha_1\alpha_2 - 2\left(\frac{\partial E_a}{\partial \alpha_1}\right)\left(\frac{\partial E_a}{\partial \alpha_3}\right)\alpha_1\alpha_3 - 2\left(\frac{\partial E_a}{\partial \alpha_2}\right)\left(\frac{\partial E_a}{\partial \alpha_3}\right)\alpha_2\alpha_3 \\ &= \left(\frac{\partial E_a}{\partial \alpha_1}\right)^2 + \left(\frac{\partial E_a}{\partial \alpha_2}\right)^2 + \left(\frac{\partial E_a}{\partial \alpha_3}\right)^2 - \left[\left(\frac{\partial E_a}{\partial \alpha_1}\right)\alpha_1 + \left(\frac{\partial E_a}{\partial \alpha_2}\right)\alpha_2 + \left(\frac{\partial E_a}{\partial \alpha_3}\right)\alpha_3\right]^2. \end{aligned} \qquad (13.75)$$

If we adopt the first term of (7.6) for E_a,

$$\frac{\partial E_a}{\partial \alpha_1} = 2K_1\alpha_1(1 - \alpha_1^2), \quad \frac{\partial E_a}{\partial \alpha_2} = 2K_1\alpha_2(1 - \alpha_2^2), \quad \frac{\partial E_a}{\partial \alpha_3} = 2K_1\alpha_3(1 - \alpha_3^2). \qquad (13.76)$$

Then (13.75) becomes

$$\begin{aligned}C^2 &= 4K_1^2[1 - 2(\alpha_1^4 + \alpha_2^4 + \alpha_3^4) + (\alpha_1^6 + \alpha_2^6 + \alpha_3^6)] \\ &\quad - 4K_1^2[1 - (\alpha_1^4 + \alpha_2^4 + \alpha_3^4)]^2 \\ &= 4K_1^2[(\alpha_1^6 + \alpha_2^6 + \alpha_3^6) - (\alpha_1^8 + \alpha_2^8 + \alpha_3^8) \\ &\quad - 2(\alpha_1^4\alpha_2^4 + \alpha_2^4\alpha_3^4 + \alpha_3^4\alpha_1^4)]. \end{aligned} \qquad (13.77)$$

For a polycrystal, averaging over the all possible orientations of crystallites, we have $\overline{\alpha_i^6} = \frac{1}{7}$, $\overline{\alpha_i^8} = \frac{1}{9}$, and $\overline{\alpha_i^4\alpha_j^4} = \frac{1}{105}$; hence (13.77) becomes

$$\overline{C^2} = 4K_1^2\left\{\frac{3}{7} - \frac{3}{9} - \frac{6}{105}\right\} = \frac{16}{105} K_1^2. \tag{13.78}$$

On putting this in (13.72), we obtain

$$b = \frac{8}{105}\frac{K_1^2}{I_s^2} = 0.0762\frac{K_1^2}{I_s^2}. \tag{13.79}$$

This result coincides with that obtained by Becker and Döring[14] by an entirely different method.

The law of approach to saturation, obtained experimentally, is

$$I = I_s\left(1 - \frac{a}{H} - \frac{b}{H^2} - \cdots\right) + \chi_0 H. \tag{13.80}$$

The last term, $\chi_0 H$, is caused by an increase of the spontaneous magnetization itself, while the origin of the second term, a/H, has been discussed by many investigators, as we shall see later. Czerlinsky[15] measured dI/dH in the high field region, instead of measuring I itself, because in the latter method a small error in the measurement of I may have a fatal influence on the determination of a or b in (13.80). From (13.80), we have

$$\frac{dI}{dH} = I_s\left(\frac{a}{H^2} + \frac{2b}{H^3} + \cdots\right) + \chi_0. \tag{13.81}$$

Czerlinsky measured dI/dH for iron and nickel as a function of H and plotted it as a function of $1/H^3$, as shown in Figs. 13.11 and 13.12. If $a = 0$, these curves should be linear. From the gradient of these curves, Czerlinsky determined the value of b for two specimens of iron and two specimens of nickel, from which we have

$$\begin{aligned} |K_1| &= 4.14, \quad 3.98 \times 10^4\ \mathrm{J/m^3} \quad \text{for iron,} \\ |K_1| &= 5.0, \quad 4.66 \times 10^3\ \mathrm{J/m^3} \quad \text{for nickel.} \end{aligned} \tag{13.82}$$

These values agree fairly well with those given by (7.7) and (7.8).

It is seen in Figs. 13.11 and 13.12 that the curves deviate from linear relations in the vicinity of the origin or in the high field region. This fact indicates the presence of the term a/H. The origin of this term was attributed to the stress field about dislocations by W. F. Brown,[16] and to non-magnetic inclusions or voids by Néel.[17] Because the calculations of both investigators are very complicated, we simply discuss here the basic origin of the a/H term. As pointed out by Néel, if the expression including the a/H term is valid until the magnetic field becomes infinitely strong, the

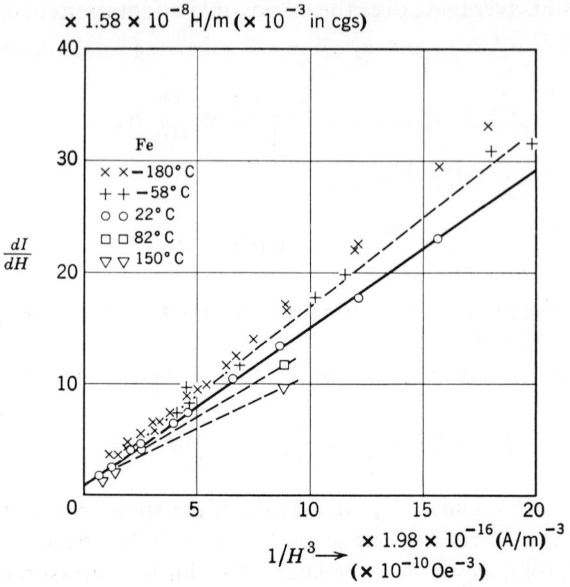

Fig. 13.11. Law of approach to saturation for iron (after Czerlinsky[15]).

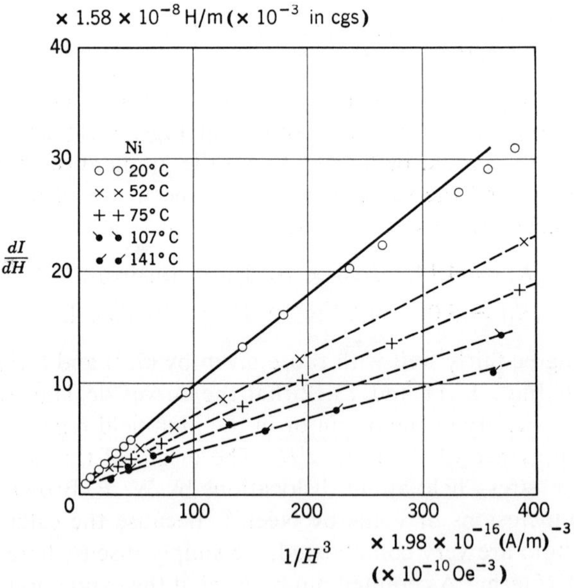

Fig. 13.12. Law of approach to saturation for nickel (after Czerlinsky[15]).

work necessary to magnetize the specimen to complete saturation diverges as shown by

$$W = \int_{I}^{I_s} H\,dI = \int_{H}^{\infty} H\left(\frac{dI}{dH}\right) dH = \int_{H}^{\infty} I_s\left(\frac{a}{H} + \frac{2b}{H^2} + \cdots\right) dH$$

$$= I_s\left(\left|a \log H\right|_{H}^{\infty} - \left|\frac{2b}{H}\right|_{H}^{\infty} + \cdots\right) = \infty. \tag{13.83}$$

Thus we must conclude that the term a/H is valid only within some finite range of strength of the field. Since the constant restoring force leads to the b/H^2 terms, we must assume the presence of the restoring force which increases with the approach of magnetization to saturation. For instance, small spike domains, which remain around voids inclusions at high field, will diminish in volume with an increase of the field strength. If local internal stresses caused by dislocations or lattice vacancies fix the magnetization firmly, the magnetization surrounding these points will form transition layers which are similar to the ordinary domain wall. The thickness of such a transition layer will be decreased with an increase of field strength. In both cases the change in magnetization is first proportional to $1/H^{1/2}$ and finally to $1/H^2$. It is naturally expected, therefore, that the change becomes proportional to $1/H$ in the intermediate field range. In any event the term a/H is expected to be a good measure of the inhomogeneity of magnetic substances.

Finally we discuss the last term, $\chi_0 H$, in (13.80). The origin of the term is an increase in spontaneous magnetization by the external magnetic field. This corresponds to the displacement of line 2 downward in Fig. 4.4, which results in an upward displacement of the intersection point between curves 1 and 2 in that figure. The change in the spontaneous magnetization is thus calculated to be

$$\frac{dI}{dH} = NM\frac{\partial L(\alpha)}{\partial \alpha}\frac{d\alpha}{dH} = NML'(\alpha)\left(\frac{M}{kT} + \frac{Mw}{kT}\frac{dI}{dH}\right), \tag{13.84}$$

from which we can find

$$\chi_0 = \frac{dI}{dH} = \frac{NM^2L'(\alpha)}{k(T - 2\Theta L'(\alpha))}. \tag{13.85}$$

The actual value of χ_0 can be obtained from the extrapolated point on the curves to the ordinate in Figs. 13.11 and 13.12. The experimental values thus obtained are normally about ten times larger than the value calculated by (13.85). These values were found to be dependent on the amount of impurities included in the materials. Becker and Döring[18] considered the reason to be that spins in the impurities or a part of spins of the matrix separated by impurities will be thermally agitated, giving rise to the larger value of χ_0.

Problems

13.1. Calculate the initial permeability of the aggregate of elongated single domain fine particles lying randomly on a plane with the packing fraction β. Assume that the field is applied (i) parallel and (ii) perpendicular to the plane.

13.2. A single crystal of cobalt shows the initial susceptibility $\bar{\chi}_a = 20$ parallel to the c axis, and $\bar{\chi}_a = 5$ perpendicular to the c axis. Determine the value of $\bar{\chi}_a$ for the field directions which make angles of 30°, 45°, and 60° with the c axis. What value of $\bar{\chi}_a$ is expected for a polycrystal which is made from the same material?

13.3. Formulate the law of approach to saturation for the polycrystal which is composed of randomly oriented uniaxial cristallites with the uniaxial anisotropy constant K_{u1}, by assuming the rotation magnetization.

References

13.1. R. M. Bozorth: G.10, p. 114.
13.2. R. M. Bozorth: *Rev. Mod. Phys.* **25,** 42 (1953).
13.3. S. Chikazumi: *Phys. Rev.* **85,** 918 (1952).
13.4. H. J. Williams: *Phys. Rev.* **52,** 747, 1004 (1937).
13.5. R. Becker and W. Döring: G.3, p. 153.
13.6. E. Kondorsky: *Physik. Z. Sowjetunion* **11,** 597 (1937).
13.7. M. Kersten: *Physik. Z.* **39,** 860 (1938).
13.8. H. P. J. Wijn, E. W. Gorter, C. J. Esveldt, and P. Goldmans: *Philips Tech. Rev.* **16,** 49 (1954).
13.9. C. Guillaud: G.26, p. 165.
13.10. M. Kersten: *Grundlagen einer Theorie der ferromagnetischen Hysterese und der Koerzitivkraft.* (S. Hirzel, Leipzig; reprinted J. W. Edwards, *Ann. Arbor,* 1943).
13.11. L. Néel: *Ann. Univ. Grenoble* **22,** 299 (1946).
13.12. M. Kersten: *Z. Angew. Phys.* **7,** 313 (1956); **8,** 382, 496 (1956).
13.13. R. M. Bozorth: G.10, p. 60.
13.14. R. Becker and W. Döring: G.3, p. 171.
13.15. E. Czerlinsky: *Ann. Physik.* **V13,** 80 (1932).
13.16. W. F. Brown: *Phys. Rev.* **60,** 139 (1941).
13.17. L. Néel: *J. Phys. Radium* **9,** 184 (1948).
13.18. R. Becker and W. Döring: G.3, p. 176.

14

The Irreversible Magnetization Process

14.1 Irreversible Rotation Magnetization

Irreversible magnetization is one of the striking features of ferromagnetism. Let us begin with a discussion of irreversible rotation magnetization. Generally speaking, a rotation of magnetization against the magnetic anisotropy requires a fairly strong magnetic field, so that domain magnetizations are usually reversed by the displacement of domain walls before irreversible rotation magnetization occurs. In single domain particles, however, the irreversible rotation magnetization is the only conceivable mechanism of flux reversal, for these particles have no domain walls.

Let us consider a uniaxial anisotropy whose easy axis makes the angle θ_0 with the x axis (Fig. 14.1). If the field is applied in the negative sense of the x direction, it causes rotation of the domain magnetization I_s which makes

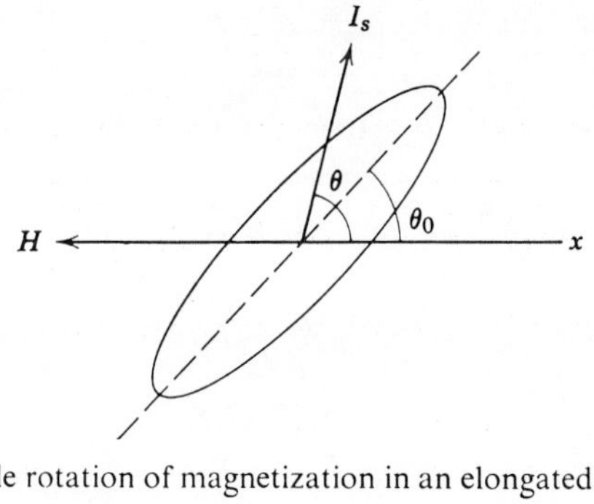

Fig. 14.1. Irreversible rotation of magnetization in an elongated single domain particle.

an angle θ with the x axis. The energy of this system is

$$E = -K_u \cos^2(\theta - \theta_0) + I_s H \cos\theta. \tag{14.1}$$

The equilibrium direction of I_s is obtained by the extremum energy condition,

$$\frac{\partial E}{\partial \theta} = K_u \sin 2(\theta - \theta_0) - I_s H \sin\theta = 0. \tag{14.2}$$

If this equilibrium is stable, it must be that

$$\frac{\partial^2 E}{\partial \theta^2} > 0.$$

If it is unstable, it must be that

$$\frac{\partial^2 E}{\partial \theta^2} < 0.$$

With an increase of the field intensity H, I_s rotates gradually and then suddenly rotates toward the direction of the field when the equilibrium state becomes unstable. At this instant, it must be that

$$\frac{\partial^2 E}{\partial \theta^2} = 0,$$

or, from (14.2),

$$\frac{\partial^2 E}{\partial \theta^2} = 2K_u \cos 2(\theta - \theta_0) - I_s H_0 \cos\theta = 0, \tag{14.3}$$

where H_0 is the critical field. On solving (14.2) and (14.3), we have

$$\begin{aligned} \sin 2(\theta - \theta_0) &= p \sin\theta, \\ \cos 2(\theta - \theta_0) &= \frac{p}{2} \cos\theta, \end{aligned} \tag{14.4}$$

where

$$p = \frac{I_s H_0}{K_u}. \tag{14.5}$$

Eliminating sin or cos $2(\theta - \theta_0)$ from the preceding equations, we obtain

$$\sin\theta = \sqrt{\frac{4 - p^2}{3p^2}}, \quad \cos\theta = \pm 2\sqrt{\frac{p^2 - 1}{3p^2}}. \tag{14.6}$$

On putting these values into (14.4), we can solve for $\sin 2\theta_0$:

$$\sin 2\theta_0 = \frac{1}{p^2}\left(\frac{4 - p^2}{3}\right)^{3/2}. \tag{14.7}$$

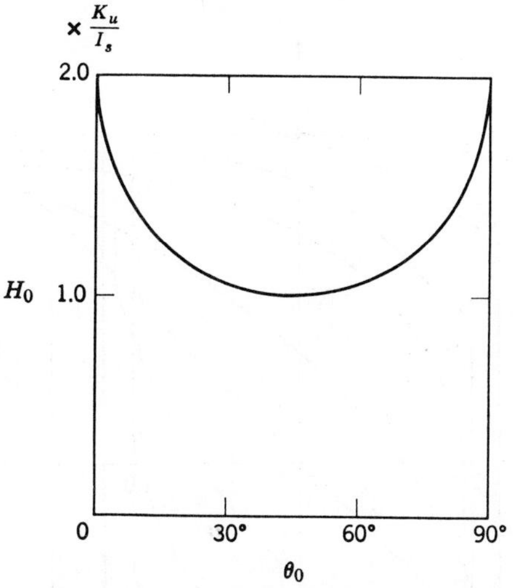

Fig. 14.2. Dependence of the critical field H_0 on the orientation of the applied field (cf. 14.7).

Figure 14.2 shows p solved as a function of θ_0. It is seen in this graph that the critical field is minimum when $\theta_0 = 45°$. When the field is 45° from the easy axis, the reversal of magnetization takes place most easily. The critical field is, then,

$$H_0 = \frac{K_u}{I_s}. \tag{14.8}$$

The critical field becomes larger and larger as θ_0 deviates from 45° and finally becomes

$$H_0 = \frac{2K_u}{I_s} \tag{14.9}$$

at $\theta = 0°$ and 90°.

Such a rotation magnetization results in the magnetization curves shown in Fig. 14.3 for various values of θ_0. The curved portions correspond to reversible rotation, and the vertical lines correspond to irreversible rotation. For $\theta_0 > 45°$ the range of reversible rotation covers a fairly large portion of the magnetization curves. If a magnetic field is applied in the negative direction and then removed, before the occurrence of irreversible rotation, the magnetization comes back to the original residual magnetization. In the case of magnetic recording tapes which are composed of elongated fine magnetic particles, it is desirable that the magnetizing signal reverses the magnetization of particles as effectively as possible. The presence of the particles which make large angles with the direction of

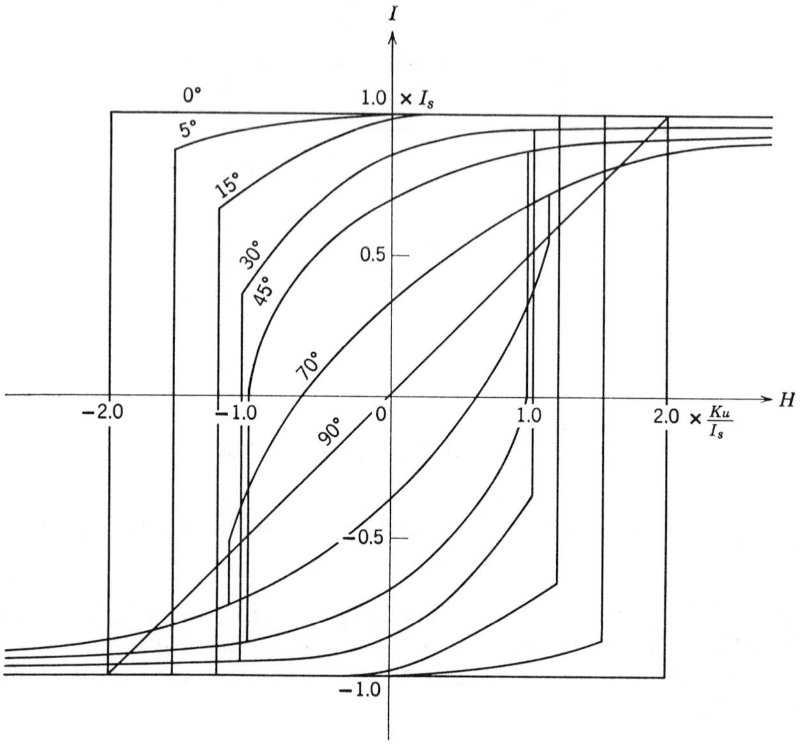

Fig. 14.3. Magnetization curves of substances with uniaxial anisotropy; irreversible rotation magnetization is assumed (numerical values signify the values of θ_0).

the field is not effective in producing a large residual magnetization. The long axes of particles in a magnetic tape, therefore, are preferably aligned parallel to the direction of the signal field.

Next we discuss irreversible rotation against cubic crystal anisotropy. The functional form of the cubic anisotropy is fairly complex, and it is not so easy to treat the general case. Here we treat irreversible rotation through 180° from an easy axis by a field applied antiparallel to the original direction of magnetization. Since the anisotropy energy can be expressed as in (7.36), provided $K_1 > 0$, the energy of the system is expressed, similarly to (14.1), as

$$E = K_1\theta^2 + I_sH\cos\theta \tag{14.10}$$

At the critical field H_0, it must be that

$$\frac{\partial E}{\partial \theta} = 2K_1\theta - I_sH_0\sin\theta = 0 \tag{14.11}$$

and

$$\frac{\partial^2 E}{\partial \theta^2} = 2K_1 - I_s H_0 \cos\theta = 0, \tag{14.12}$$

from which we have

$$H_0 = \frac{2K_1}{I_s} \tag{14.13}$$

for $K_1 > 0$.

If $K_1 < 0$, the anisotropy energy is given by (7.38) in the vicinity of the easy axis; hence the energy is

$$E = -\tfrac{2}{3}K_1\theta^2 + I_s H \cos\theta. \tag{14.14}$$

From the conditions

$$\frac{\partial E}{\partial \theta} = -\tfrac{4}{3}K_1\theta - I_s H_0 \sin\theta = 0 \tag{14.15}$$

and

$$\frac{\partial^2 E}{\partial \theta^2} = -\tfrac{4}{3}K_1 - I_s H_0 \cos\theta = 0, \tag{14.16}$$

we have the critical field

$$H_0 = -\frac{4}{3}\frac{K_1}{I_s} \tag{14.17}$$

for $K_1 < 0$.

Thus the order of magnitude of the critical field for irreversible rotation is given by K/I_s, where the saturation magnetization is about $I_s = 1 \sim 2$ Wb/m^2 and the anisotropy constant is $K = 10^4 \sim 10^5$ J/m^3 for usual ferromagnetic substances. It follows, therefore, that $H_0 = 10^4 \sim 10^5$ A/m ($= 10^2 \sim 10^3$ Oe), which is a fairly large value for the coercive force of permanent magnetic materials. This fact is utilized in many recently developed permanent magnets, such as O.P. magnet, Ba-ferrite, and metal powder magnet, which are composed of single domain fine particles. We discuss this subject again in Section 22.3.

14.2 Irreversible Displacement of the Domain Wall

In the average ferromagnetic substance which has domain structure, irreversible magnetization takes place exclusively through the irreversible displacement of domain walls.

First we consider a plane domain wall which travels in the medium where the wall energy fluctuates as shown in Fig. 14.4. The effect of an externally applied magnetic field H is equivalent to the exertion of a pressure p ($= 2I_s H \cos\theta$), as already discussed (cf. 13.52). The wall is displaced from

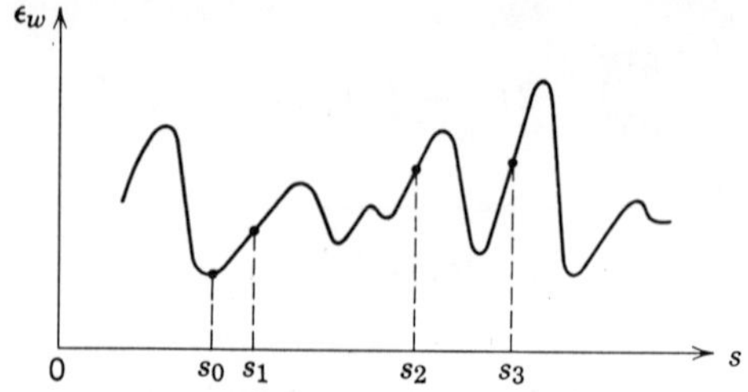

Fig. 14.4. Variation of the wall energy as a function of position of a domain wall.

the minimum position s_0 to s_1, where the restoring force of the wall is just counterbalanced by the pressure, or

$$\frac{\partial \epsilon_w}{\partial s} = p = 2I_s H \cos \theta. \tag{14.18}$$

If we assume that the gradient of the energy ϵ_w has a maximum at s_1, further increase of the field intensity will result in an irreversible displacement of the wall from s_1 to, say s_2 where the gradient is larger and thus can counterbalance the stronger pressure (cf. Fig. 14.4). If the field is reduced from this state, the wall will come back, not to s_0, but to the minimum point which is closest to s_2. If the largest maximum of the gradient exists at s_3, the wall will finally reach s_3 and then will discontinuously move to the final goal upon an additional increase of H. Thus the critical field H_0 for one domain wall is

$$H_0 = \frac{1}{2I_s \cos \theta}\left(\frac{\partial \epsilon_w}{\partial s}\right)_{\max}. \tag{14.19}$$

If the fluctuation of the energy ϵ_w is due to the spatial variation of the internal stress as shown by (13.30), the surface energy of the wall is given by (13.32); then

$$\left(\frac{\partial \gamma}{\partial s}\right)_{\max} = \frac{6\pi\lambda\sigma_0}{l}\left(\sqrt{\frac{A}{K_1 - \frac{3}{2}\lambda\sigma_0 \cos 2\pi(s/l)}} \sin 2\pi \frac{s}{l}\right)_{\max}$$

$$= \frac{6\pi\lambda\sigma_0}{l}\sqrt{\frac{A}{K_1}} = \frac{2\pi\lambda\sigma_0}{l}\,\delta, \tag{14.20}$$

where δ is the thickness of the wall (cf. 13.35). Since the wall does not change its area ($\epsilon_w = \gamma$), (14.19) becomes

$$H_0 = \frac{\pi\lambda\sigma_0}{I_s \cos \theta}\,\frac{\delta}{l}. \tag{14.21}$$

Thus the critical field will increase with a decrease of the wavelength of the stress variation l. If, however, l becomes less than δ, the effect of the stress will be smoothed over by the action of the exchange interaction, as discussed before, so that the critical field will again be decreased. Thus the maximum value of H_0 will be realized when $l \approx \delta$, or

$$(H_0)_{\max} = \frac{\pi \lambda \sigma_0}{I_s \cos \theta}. \tag{14.22}$$

If we tentatively assume that $\lambda \doteqdot 10^{-5}$, $I_s = 1$ Wb/m², $\cos \theta \sim 1$, and $\sigma_0 = 100$ kg/mm² $\doteqdot 10^9$ N/m² for average magnetic materials, we have

$$(H_0)_{\max} = \frac{\pi \times 10^{-5} \times 10^9}{1 \times 1} = 3 \times 10^4 \text{ A/m} \quad (= 400 \text{ Oe}), \tag{14.23}$$

which is nearly equal to the maximum value of the coercive force actually attained by metal permanent magnets.

Experimental investigations of the dependence of coercive force on the magnitude of internal stress were made by Kersten.[1] He obtained the results shown in Fig. 14.5 for nickel bars which are stressed in various degrees. Curve *a* represents a nickel wire which was hard-drawn in various degrees, and curve *b* represents a hard-worked nickel specimen which was annealed to release the internal stress. The values of internal stress were determined from the initial permeability. The linear relation between H_c and σ_0 as shown in (14.21) is well reproduced in both curves. The difference in the proportionality factor is considered to be due to the difference in the wavelength of the internal stress l in the two substances.

Now we suppose that the wall is constrained at two points and that it expands under the action of the pressure caused by a magnetic field

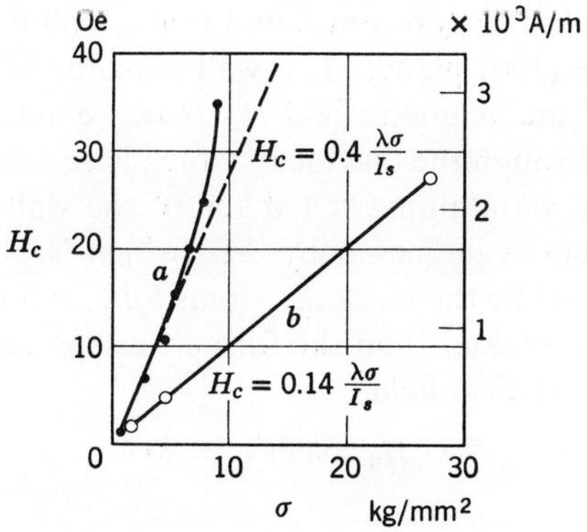

Fig. 14.5. Dependence of the coercive force on the magnitude of internal stress for a hard-worked nickel (after Kersten[1]).

(Fig. 14.6). The radius of curvature of the wall is given by (13.53), in which we see that the radius r should decrease with an increase of H. Actually, r is reduced as the wall expands as shown by curves a, b, and c in Fig. 14.6, but, if the wall expands beyond curve c, where r is equal to one-half of the separation of two constraining points l, the radius of curvature, r, is again increased with further expansions of the wall. The wall will, therefore, expand discontinuously after it passes through curve c.

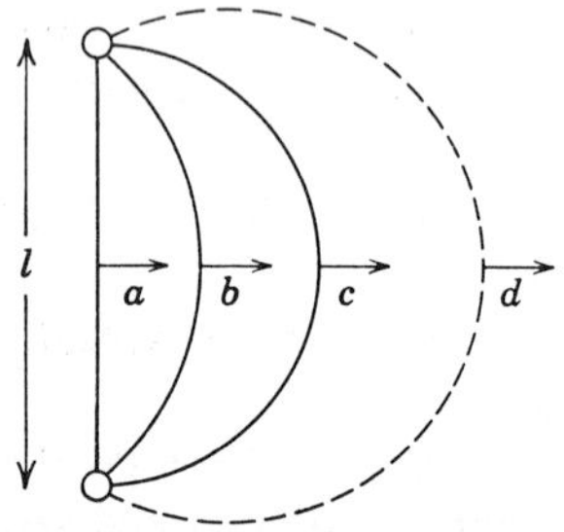

Fig. 14.6. Reversible and irreversible expansion of a wall.

The critical field H_0 is obtained by setting $r = l/2$ in (13.53); then we have

$$H_{0_{180°}} = \frac{\gamma}{I_s l \cos\theta}. \tag{14.24}$$

For a 90° wall, it follows in a similar way that

$$H_{0_{90°}} = \frac{\sqrt{2}\,\gamma}{I_s l \cos\theta}. \tag{14.25}$$

With this model, Kersten[2] explained the dependence of H_c on the composition of iron-nickel alloy and also the temperature dependence of H_c for iron.

The origins of constraint of domain walls assumed in the description above are voids, non-magnetic inclusions, and spike domains shown in Fig. 6.11. Kersten assumed in his paper that the origin may be dislocations. At any rate, if the constraining force of the wall is sufficiently strong, the coercive force is simply determined by the surface tension of the wall, independently of the manner of constraint. If, however, the constraining force is not so strong, the wall will leave the constraining points before expanding to the final shape (curve c in Fig. 14.6). Let us consider that the constraining points are spatially distributed in the manner of a simple cubic lattice with a lattice constant l, and that a wall is constrained by the points lying on the (100) plane. The wall expands in a cylindrical form under the action of the magnetic field H. Now we tentatively assume that the wall proceeds through the distance r which is equal to the radius of the constraining points maintaining the shape of the wall as it is. Then the energy of the system is increased by ΔU, which is given by (11.37) per constraining point, while the work, $2I_sH(\cos\theta)l^2r$, is done by the magnetic field. If the latter is greater than the former, a separation of the wall will be realized. At the critical field,

$$2I_sH_0\cos\theta l^2 r = \Delta U$$

or

$$H_0 = \frac{\Delta U}{2I_s\cos\theta l^2 r}. \tag{14.26}$$

Using expression (11.37) for ΔU, we have

$$H_0 = 1.5 \times 10^5 \frac{I_s r^2}{\cos \theta l^2}. \tag{14.27}$$

For iron, $I_s = 2.15$, and, assuming that $r = 10^{-5}$ (= 0.01 mm), $l = 10^{-4}$ (= 0.1 mm), and $\cos \theta \sim 1$, we have

$$H_0 = (1.5 \times 10^5) \times (2.15) \times 10^{-2} \approx 3 \times 10^3 \text{ A/m } (\fallingdotseq 36 \text{ Oe}). \tag{14.28}$$

On the other hand, a discontinuous expansion of the wall occurs at the field given by (14.24), which is calculated, by putting $\gamma = 1.6 \times 10^{-3}$, to be

$$H_0 = \frac{1.6 \times 10^{-3}}{(2.15) \times 10^{-4}} = 8 \text{ A/m } (= 0.1 \text{ Oe}). \tag{14.29}$$

Comparing (14.29) and (14.28), we conclude that in this particular case the discontinuous expansion of the wall will take place before the wall leaves the constraining points. If the radius of the constraining points is about 100 times smaller than the value assumed above, the wall will easily leave such a point before the occurrence of any appreciable expansion.

As seen in the expressions of H_0 for various cases, H_0 due to wall displacement is inversely proportional to $\cos \theta$. That is, the coercive force is minimum when the field is applied parallel to the easy axis. On denoting this value by $H_{0\parallel}$, we generally express the angular dependence of the coercive force as

$$H_0 = \frac{H_{0\parallel}}{\cos \theta_0}, \tag{14.30}$$

which is shown graphically in Fig. 14.7. Comparing this graph with Fig. 14.2, we notice that this mechanism of magnetization reversal is quite different from irreversible rotation magnetization in various respects: (1) the magnetization reversal starts from $\theta_0 = \pi/4$ for irreversible rotation, while it starts from $\theta_0 = 0$ for irreversible wall displacement, as already pointed out in Section 12.3; and (2) the whole irreversible rotation is attained in a field of finite intensity, while the irreversible wall displacement requires an extremely high field for large θ_0. The magnetization curves in Fig. 14.8 are drawn by assuming irreversible wall displacement for various orientations of uniaxial anisotropy.

Now suppose that a signal field H is applied to the aggregate of elongated fine particles and then removed. Some residual magnetization will be left as a result of irreversible magnetization. Figure 14.9 gives such a

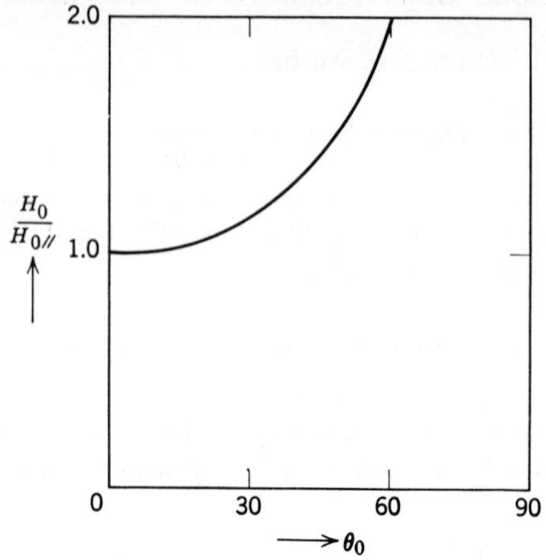

Fig. 14.7. Dependence of the critical field H_0 on the orientation of the applied field for irreversible wall displacement.

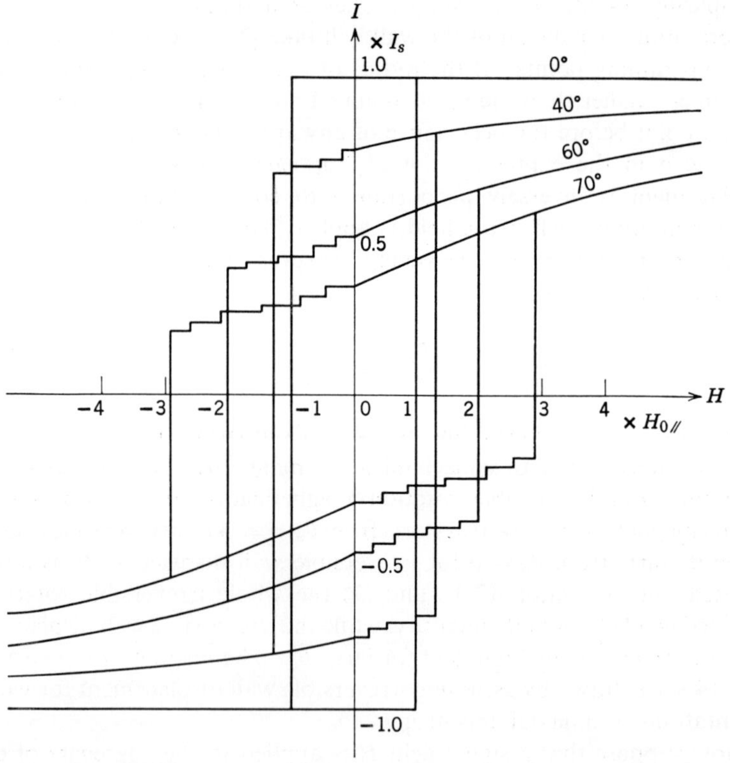

Fig. 14.8. Magnetization curves due to reversible rotation and irreversible wall displacement for a substance with uniaxial anisotropy (numerical values are those of θ_0; $H_{0\|} = 0.2K_u/I_s$ is assumed).

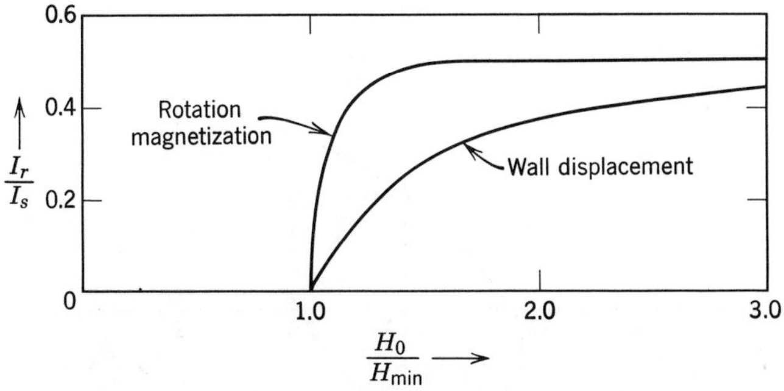

Fig. 14.9. Residual magnetization as a function of the intensity of the signal magnetic field.

residual magnetization normalized by I_s as a function of the signal field normalized by the minimum field H_{min} which is equal to K_u/I_s for rotation magnetization and to $H_{0\parallel}$ for wall displacement. As seen in the graph, rotation magnetization results in a sharp rise in the I_r–H curve. The main reason is the abundance of the particles which play a part in the initial reversal of magnetization. Such a sharp rise in residual magnetization is useful for a memory device such as a magnetic tape recording.

14.3 Shape of Magnetization Curves

The theoretical derivation of a hysteresis curve was first tried by Ewing, who used a model consisting of an assembly of magnetic compasses. Since, however, the actual mechanism of magnetization is very complicated, it is not so easy to construct a real hysteresis curve on theoretical grounds. Here we try to construct several hysteresis curves by assuming the preferential occurrence of a simple magnetization process.

First we deal with an assembly of single domain fine particles which have uniaxial magnetic anisotropy. The magnetization curves for individual particles have already been shown in Fig. 14.3. On averaging these curves for isotropic distribution of long axes of particles, we have the hysteresis curve shown in Fig. 14.10, where the solid curves represent the magnetization due to reversible rotation magnetization. It is interesting that the reversible magnetization covers such a wide range that, if we change the field toward the positive direction from the coercive force point, the magnetization is expected to go back again to the residual magnetization. If, however, the K_u/I_s is not single-valued but is scattered around the average value, we must average the curve in Fig. 14.10 by changing the scale of the abscissa. In such a case, irreversible rotations start in the range of the second quadrant.

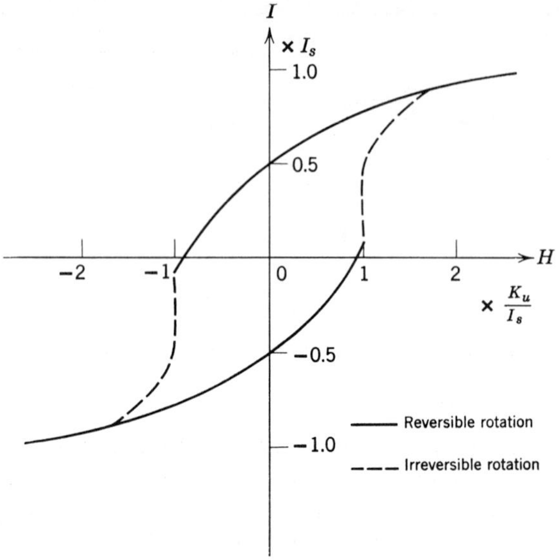

Fig. 14.10. Magnetization curve of the aggregate of single domain fine particles with uniaxial anisotropy.

The various constants for the curve in Fig. 14.10 are as follows:

$$
\begin{aligned}
&\text{Residual magnetization} && I_r = 0.5I_s, \\
&\text{Coercive force} && H_c = 1.0\,\frac{K_u}{I_s}, \\
&\text{Maximum susceptibility} && \chi_{\max} = 0.56\,\frac{I_s^{\,2}}{K_u}, \\
&\text{Hysteresis loss} && W_h = 1.98K_u.
\end{aligned}
\tag{14.31}
$$

Next we discuss irreversible displacement of domain walls for distributed uniaxial anisotropy. On averaging over the curves in Fig. 14.8 for variously oriented easy axes, we have the hysteresis curve shown in Fig. 14.11, where the solid curves again represent a reversible rotation magnetization and the broken curves are the large irreversible jumps of domain walls (large Barkhausen jumps), while the dotted curves signify small irreversible jumps) which occur in advance of the large Barkhausen jumps. It must be noted that the rotation process is characterized by K_u/I_s, while the wall displacement is characterized by $H_{0\|}$, which is independent of K_u/I_s. We tentatively assumed in Fig. 14.11 that $H_{0\|} = 0.2K_u/I_s$, in which case a large part of the magnetization in the second quadrant takes place by irreversible displacement of the wall. If we assume that $H_{0\|} > K_u/I_s$, we must expect that a considerable amount of reversible rotation is in the magnetization curve in the second quadrant. The assumption that $H_{0\parallel} \gg K_u/I_s$ is nonsense, because in such a case irreversible rotations take place before the occurrence of irreversible wall displacement. The constants of this curve are as follows:

$$\begin{aligned} &\text{Residual magnetization} && I_r = 0.5I_s, \\ &\text{Coercive force} && H_c = 1.3H_{0\parallel}, \\ &\text{Maximum susceptibility} && \chi_{\max} = 0.2\frac{I_s}{H_{0\parallel}}, \\ &\text{Hysteresis loss} && W_h = 3.21I_sH_{0\parallel}. \end{aligned} \tag{14.32}$$

Next we discuss the hysteresis curve of the substances in which a cubic crystal anisotropy predominates over the other kinds of anisotropies. Since the spin distribution is confined within the range $\theta \leq 55°$, the critical field is limited in the range

$$H_0 \leq \sqrt{3}\, H_{0\parallel}, \tag{14.33}$$

as is expected from (14.30). Therefore the magnetization reversal takes place more easily than in the case of uniaxial anisotropy, and the magnetization curve is expected to rise with considerable steepness. On assuming that

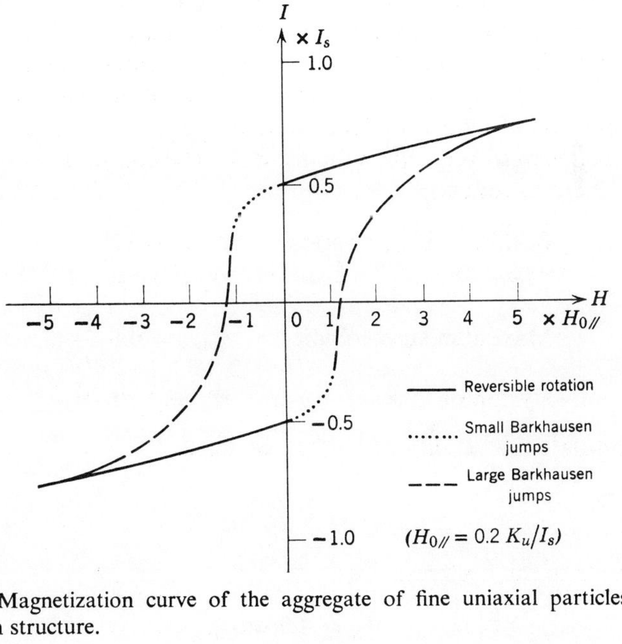

Fig. 14.11. Magnetization curve of the aggregate of fine uniaxial particles with a multidomain structure.

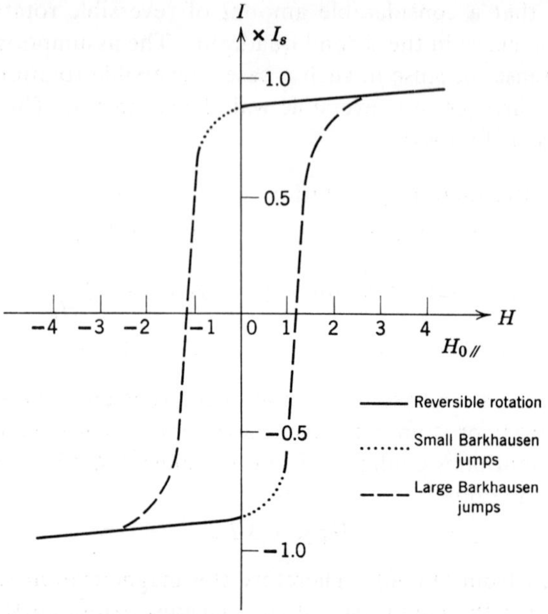

Fig. 14.12. Magnetization curve of a cubic substance ($K_1 > 0$).

all processes of magnetization reversal take place exclusively through the displacement of 180° walls, we have the average magnetization curve shown in Fig. 14.12 for $K_1 > 0$. This curve is a good representation of the hysteresis curve of average cubic magnetic metals or alloys which are annealed for the purpose of removal of the internal stress. The shape of the curve will not be changed very much for $K_1 < 0$. The constants of the curve in Fig. 14.12 are as follows:

$$\begin{aligned} &\text{Residual magnetization} \quad && I_r = 0.832 I_s, \\ &\text{Coercive force} && H_c = 1.2 H_{0\|}, \\ &\text{Maximum susceptibility} && \chi_{\max} = 0.5 \frac{I_s}{H_{0\|}}, \\ &\text{Hysteresis loss} && W_h = 4.30 I_s H_{0\|}. \end{aligned} \tag{14.34}$$

If we postulate the validity of (14.22), we have

$$H_{0\|} = \frac{\pi \lambda \sigma_0}{I_s}, \tag{14.35}$$

which leads to the hysteresis loss,

$$W_h = 4.30 \pi \lambda \sigma_0. \tag{14.36}$$

The reader will wonder, when he first looks at this result, why the hysteresis loss is expressed in terms of magnetostrictive anisotropy, in spite of our postulation of the superiority of crystal anisotropy. The reason is that the irreversible process of magnetization is not influenced by the magnitude of the crystal anisotropy; rather it is determined by the fluctuation of internal stress. In contrast, (14.31) is expressed in terms of K_u, because we considered the irreversible rotation against the uniaxial anisotropy.

We assumed above the occurrence of 180° wall displacements, but in actual materials displacements of 90° walls cannot be ignored. For instance, local stress will easily produce the new domains which make contact with neighboring domains through 90° walls as well as 180° walls. Free magnetic poles which appear at crystal boundaries will also induce new closure domains which have their magnetizations perpendicular to the originally existing domains, and thus they are surrounded by 90° walls (cf. Fig. 6.10). In these cases the displacement of 90° walls plays an important role in reversing the magnetization in the process of magnetization from remanence to the coercive force point. The coercive force here is governed by λ_{100} for $K_1 > 0$ and by λ_{111} for $K_1 < 0$. Actually it was found in connection with ferrite cores for computer memory that $\lambda_{111} = 0$ is a necessary condition for obtaining a rectangular hysteresis curve for ferrites with negative K_1.[3]

A hysteresis curve of constricted form such as that shown in Fig. 22.2 is often observed for materials which respond to magnetic annealing, when they are cooled in the absence of magnetic field. Taniguchi[4] explained this phenomenon in terms of the fixing of domain walls by local directional order. He considered that, if a domain wall exists during the process of cooling, two or more sorts of constituent elements of the alloy will form a directional order, so as to stabilize the direction of spins in the domain wall as they are (cf. Section 17.1). After cooling, the wall is stabilized at its original position, so the magnetization curve starts with low permeability and then jumps to saturation magnetization as soon as the wall gets out of the stabilized position (Fig. 14.13). In the practical use of Perminvar, the magnetization is limited in the initial portion of the magnetization curve where the permeability is practically constant. This fine quality is destroyed by the application of a strong magnetic field which may drive walls out of their stabilized positions.

A similar calculation was made by Néel[5] and De Vries[6] in terms of diffusion of interstitial carbon or nitrogen atoms in body-centered cubic iron and experimentally investigated by Bindels, Bijvoet, and Rathenau[7] and Brissonneau and Moser[8] for silicon-iron crystals. They observed a constricted hysteresis curve for these crystals.

The shape of the magnetization curve is generally leaf-like, as shown in

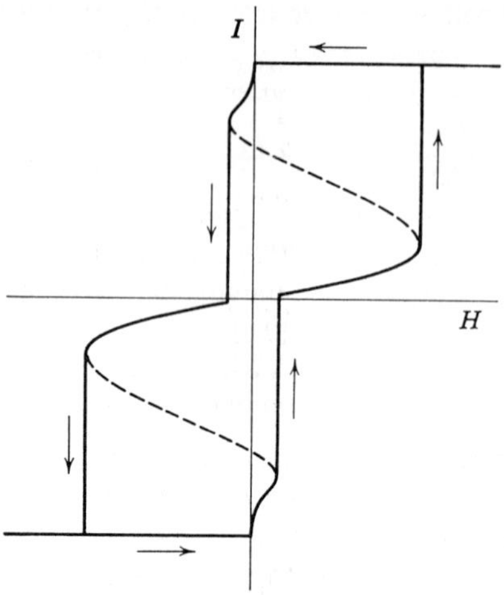

Fig. 14.13. A constricted hysteresis loop (Taniguchi[4]).

Fig. 14.14 for a small amplitude of magnetization. This phenomenon was first investigated by Lord Rayleigh[9] in 1887. He expressed the initial portion of the magnetization curve as

$$I = \chi_a H + \tfrac{1}{2}\eta H^2, \tag{14.37}$$

where χ_a is the initial susceptibility and η is the Rayleigh constant. As a first approximation, reversible magnetization can be described by the first term in (14.37), and the second term can be ascribed to an irreversible magnetization. If, therefore, the field is decreased from some positive value of H, the sign of the second term should be changed, giving rise to a magnetization curve which is concave downwards. The hysteresis loop is then expressed as

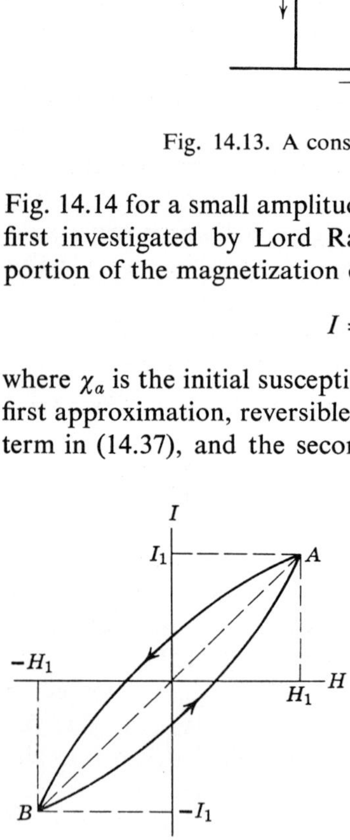

Fig. 14.14. A Rayleigh loop.

$$I + I_1 = \chi_a(H + H_1) + \tfrac{1}{2}\eta(H + H_1)^2 \tag{14.38}$$

for the ascendant branch and as

$$I - I_1 = \chi_a(H - H_1) - \tfrac{1}{2}\eta(H - H_1)^2 \tag{14.39}$$

for the descendant branch. Since $I = I_1$ at $H = H_1$, we have, from (14.38),

$$I_1 = \chi_a H_1 + \eta H_1^2. \tag{14.40}$$

On putting this relation in (14.38) and (14.39), we have

$$\begin{aligned} I &= (\chi_a + \eta H_1)H + \tfrac{1}{2}\eta(H^2 - H_1^2), \\ I &= (\chi_a + \eta H_1)H - \tfrac{1}{2}\eta(H^2 - H_1^2), \end{aligned} \tag{14.41}$$

for the ascendant and descendant branches. We call this the Rayleigh loop. The hysteresis loss of this loop is

$$W_h = \oint I \, dH = \tfrac{4}{3}\eta H_1^3. \tag{14.42}$$

It is seen in this relation that the hysteresis loss increases in proportion to the third power of the amplitude of the applied magnetic field.

If the magnetization changes along such a loop, the shape of the waveform is deformed as shown in Fig. 14.15, even if the waveform of the applied field is an exact sinusoidal one. On analyzing the waveform of I, we find that the phase of the fundamental harmonic is delayed by the angle δ and also that the third harmonic appears. If we express this phenomenon in terms of B, we have

$$B = B_\omega \sin(\omega t - \delta) + B_{3\omega} \sin 3\omega t + \cdots. \tag{14.43}$$

After simple calculation we have

$$\tan \delta = \frac{4}{3\pi} \frac{\eta H_1}{\mu_a + \eta H_1}, \tag{14.44}$$

$$k = \frac{3B_{3\omega}}{B_\omega} = \frac{4}{5\pi} \frac{\eta H_1}{\mu_a}. \tag{14.45}$$

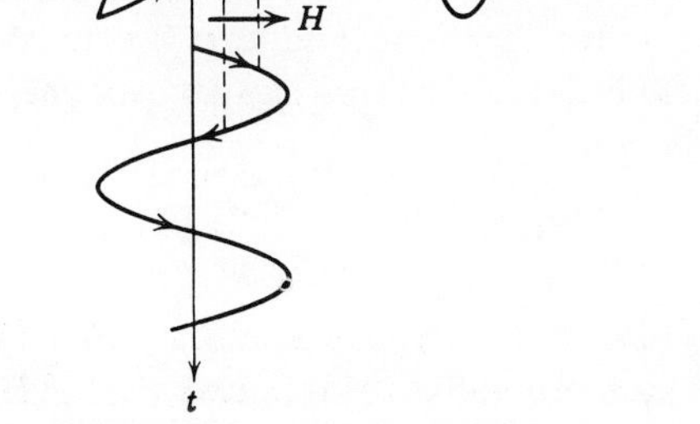

Fig. 14.15. Distortion of the wave form due to the magnetization along the Rayleigh loop.

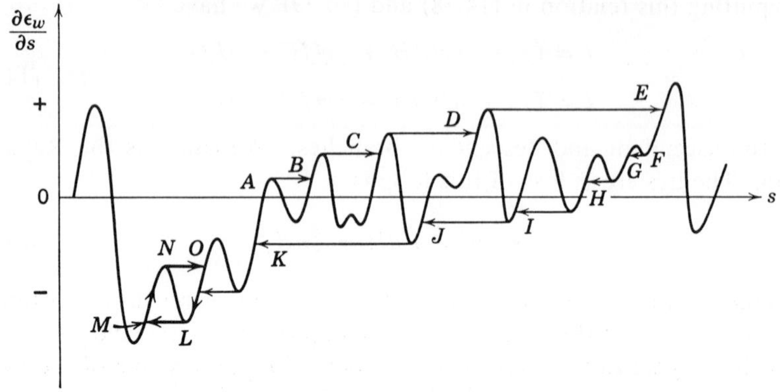

Fig. 14.16. Irreversible displacements of domain wall along the Rayleigh loop.

The former is called the loss factor, which is finite in this case because of the presence of hysteresis. The latter, k, is called the Klirr factor, and it represents the ratio of the third harmonic to the fundamental one. The presence of the third harmonics is harmful because it not only destroys the waveform but also gives rise to a loss if three-phase ac lines are delta-connected, for then the voltages induced by the third harmonics are shorted in the delta circuit.

The origin of the Rayleigh loop is the same as that of the usual hysteresis. Let us now consider the mechanism in terms of wall displacement. As discussed in Section 14.2, the wall is sustained at the place where $\partial\epsilon_w/\partial s = 2I_sH\cos\theta$ (cf. 14.18). In Fig. 14.16, $\partial\epsilon_w/\partial s$ is plotted as a function of s. At $H = 0$ the wall stays at a point, say A, where $\partial\epsilon_\omega/\partial s = 0$. With an increase of the field, this point rises along the hump and is discontinuously displaced from the maximum to the next hump, say to B. With further increase of H, the point displaces from B to $C \rightarrow D \rightarrow E$. It should be noted that each stroke of such an irreversible displacement is longer for a stronger magnetic field, for the humps which are lower than those that have already been got over by the wall are not useful in sustaining the wall.

Now we can derive Rayleigh's second term by means of this model. Let the local critical field by which the wall can get over one hump be H_0, which, from (14.19), is given by

$$H_0 = \frac{1}{2I_s\cos\theta}\left(\frac{\partial\epsilon_w}{\partial s}\right)_{\max}. \qquad (14.46)$$

Let the number of humps, which have a critical field from H_0 to $H_0 + dH_0$, be $f(H_0)\,dH_0$. It can generally be assumed that $f(H_0)$ is the Gaussian distribution function which is spread about $H_0 = 0$. Since H_0 is limited to a small range, we can postulate that $f(H_0)$ is a constant which we call f_0. If the field is increased from H to $H + dH$, the number of walls which are released from the humps should be proportional to $f_0\,dH$. On the other hand, each stroke of the displacement is proportional to the number of humps having H_0, which is less than the applied field H, or

$$\int_0^H f_0\,dH_0 = f_0 H. \tag{14.47}$$

The irreversible magnetization resulting from this process should be proportional to (the number of displaced walls) × (one stroke of the irreversible displacement) or

$$dI_{\text{irr}} = cf_0^2 H\,dH, \tag{14.48}$$

where c is the proportionality factor. The irreversible magnetization caused by the change of the field from 0 to H is thus given by

$$I_{\text{irr}} = \int_0^{I_{\text{irr}}} dI_{\text{irr}} = cf_0^2 \int_0^H H\,dH = \tfrac{1}{2}cf_0^2 H^2, \tag{14.49}$$

which is Rayleigh's second term.

If the field is decreased from point E in Fig. 14.16, the point displaces reversibly along the curve to point F and then jumps to $G \to H \to I \to J \to K$. The shape of the loop $AEFK$ closely resembles the shape of a Rayleigh loop.

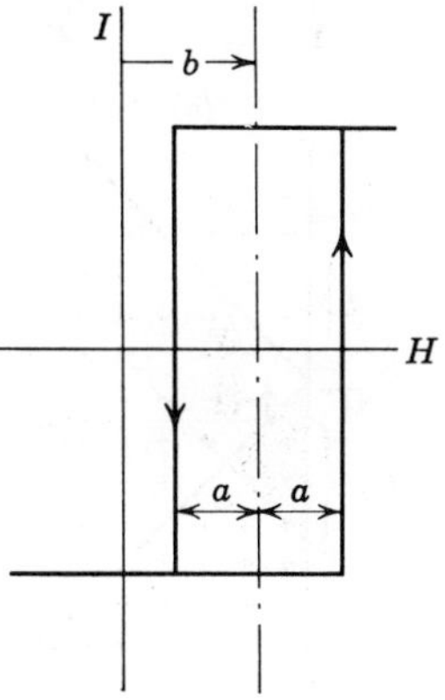

Fig. 14.17. A shifted rectangular hysteresis.

A more phenomenological explanation has been given by Weiss and Freudenreich[10] and Preisach.[11] They assumed that the material is composed of many small domains each of which has a shifted rectangular hysteresis curve with a coercive force a and a shift b (Fig. 14.17). It is assumed that the distribution of such a domain is uniform on the ab plane as shown in Fig. 14.18a, where the plus signs signify that the magnetization points in the positive direction, and the minus signs mean that the magnetization points in the negative direction. It is obvious that the magnetization is always negative in the region where $b > a$ (AOY in Fig. 14.18a) in the absence of the applied magnetic field, because the rectangular part of the hysteresis is shifted to the right side of the ordinate and $H = 0$ results in $I = -I_s$. By the same reasoning, the magnetization should be positive in the region BOY'. In the region AOB, each domain can take either positive or negative magnetization, depending on the history of magnetization. If we apply a field H_1 in the positive direction, the boundary AOB shifts upwards to $A'O'B'$ by a vertical distance H_1. Similarly, if we apply a field $-H_1$, the boundary should shift to $A''O''B''$ (Fig. 14.18c). Now, if we

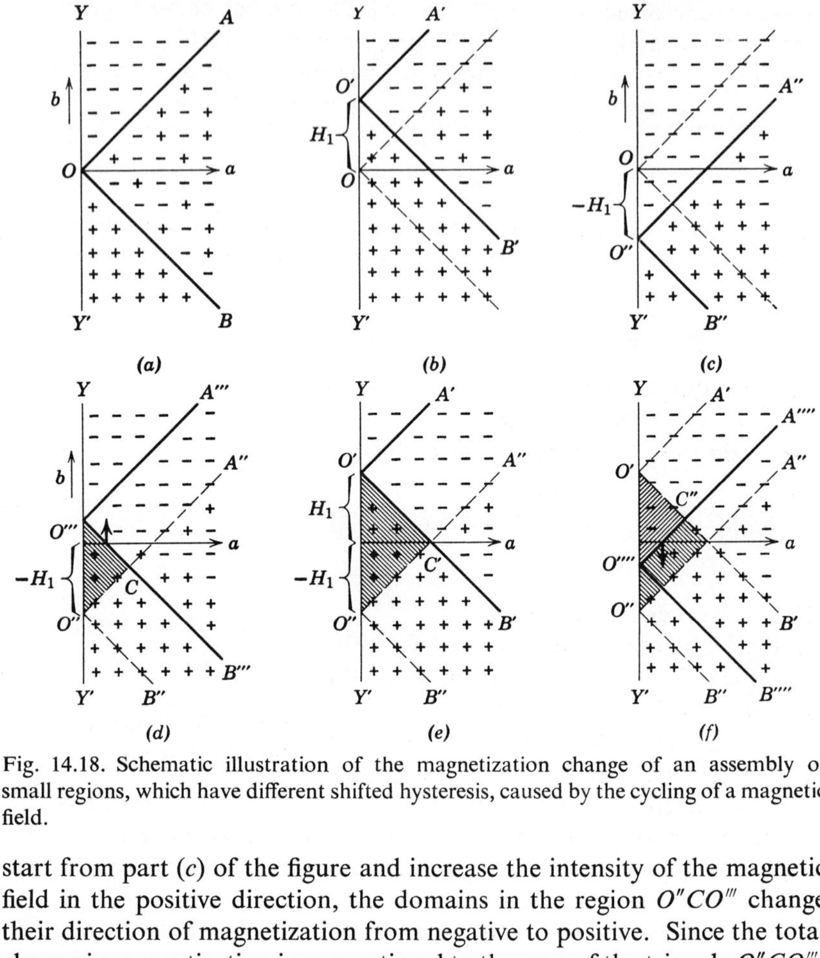

Fig. 14.18. Schematic illustration of the magnetization change of an assembly of small regions, which have different shifted hysteresis, caused by the cycling of a magnetic field.

start from part (*c*) of the figure and increase the intensity of the magnetic field in the positive direction, the domains in the region $O''CO'''$ change their direction of magnetization from negative to positive. Since the total change in magnetization is proportional to the area of the triangle $O''CO'''$, we expect the magnetization change to be proportional to the square of the increased intensity of the magnetic field. Consequently we can deduce the second term in (14.38). After reaching $H = +H_1$ (Fig. 14.18*e*), the field is again reduced; then the magnetization corresponding to the area $O'C''O''''$ is reversed, resulting in the second term in (14.39).

This model can be applied to an aggregate of small fine particles each of which exhibits a shifted hysteresis under the action of a disturbance field from surrounding particles.

Néel[12] showed that this model can also be used for the domain wall model. Actually we can see a minor loop of hysteresis, *LMNO*, in Fig. 14.16 for a single wall when the applied field is cycled under the presence of a certain biasing field. The model shown in Fig. 14.18 is convenient for discussing a phenomenology of Rayleigh loop, irrespective of the actual physical mechanism of magnetization. Néel used this model for his discussion of magnetic aftereffect[13] and some problems in rock magnetism.[14] Several constants concerning the Rayleigh loop are given in Table 14.1 for various magnetic materials.

Table 14.1. Constants Related to the Rayleigh Loop (after Bozorth[15])

Substance	$\bar{\mu}_a$	$\bar{\eta}$† per A/m ($\times 10^3/4\pi$ Oe^{-1})	W_h [calculated by (14.42) for $I = 10^{-4}$ Wb/m^2 $= 1/4\pi$ gauss] J/m^3 ($\times$ 10 ergs/cc)
Iron	200	25	2.70
Iron (unannealed)	80	0.75	1.30
Pressed iron powder	30	0.013	0.39
Cobalt	70	0.13	0.03
Nickel	220	3.1	0.25
2-81 Mo Permalloy powder	120	0.006	0.003
45 Permalloy	2,300	201	0.014
4-79 Mo Permalloy	20,000	4,300	0.0005
Supermalloy	100,000	150,000	0.0001
45-25 Perminvar	400	0.0013	0.00002

† $\bar{\eta}$ means the Rayleigh constant measured by the unit μ_0. It signifies the rate of increase of $\bar{\mu}_a$ caused by an increase in the amplitude of H.

Problems

14.1. Elongated single domain particles with a circular cross section are distributed at random on the xy plane with their long axes laid in the plane. Find the intensity and direction of residual magnetization after the magnetic field whose intensity is slightly more than $I_s/4\mu_0$ ($= \pi I_s$ in cgs) is applied first parallel to the x axis, then rotated through the y, $-x$, and $-y$ directions, and finally to the x direction, along which the field is then reduced to zero.

14.2. Many small voids are distributed in the manner of a body-centered cubic lattice with the lattice constant a in a cubic ferromagnetic substance with positive K_1. Find the critical field for the 180° wall which separates x and $\bar{x}$ domains, for the 90° wall which separates x and $\bar{y}$ domains, and for the 90° wall which separates x and $\bar{z}$ domains when the magnetic field is applied parallel to the [110] direction. Denote the surface energies of 180° and 90° walls, $\gamma_{180°}$

and $\gamma_{90°}$, and the saturation magnetization I_s. Assume that the size of the voids is large enough to sustain the walls.

14.3. Starting from the demagnetized state ($I = 0$, $H = 0$), the magnetic field is changed to $+H_1$, $-\frac{1}{2}H_1$, $+\frac{1}{4}H_1, \ldots$, $(-1)^n(1/2^n)H_1$ in the Rayleigh region. How large a magnetization is left in the limit of $n \to \infty$?

References

14.1. M. Kersten: *Probleme der technischen Magnetisierungskurve*, edited by Becker (Verlag Julius Springer, Berlin; reprinted by J. W. Edwards, Ann Arbor, 1938), pp. 42–72; G.3, p. 215.
14.2. M. Kersten: *Z. Angew. Phys.* **7,** 313 (1956); **8,** 382, 496 (1956).
14.3. H. P. J. Wijn, E. W. Gorter, C. J. Esveldt, and P. Goldmans: *Philips Tech. Rev.* **16,** 49 (1954).
14.4. S. Taniguchi: *Sci. Rept. Res. Inst. Tohoku Univ.* **A8,** 173 (1956).
14.5. L. Néel: *J. Phys. Radium* **12,** 339 (1951); **13,** 249 (1952).
14.6. G. De Vries: *Physica* **25,** 1211 (1959).
14.7. J. Bindels, J. Bijvoet and G. W. Rathenau: *Physica* **26,** 163 (1960).
14.8. P. Brissonneau and P. Moser: *J. Phys. Soc. Japan* **17,** Suppl. B-I, 331 (1962).
14.9. Lord Rayleigh: *Phil. Mag.* **23,** 225 (1887).
14.10. P. Weiss and J. de Freudenreich. *Arch. Sci. (Geneva)* **42,** 449 (1916).
14.11. F. Preisach: *Z. Physik.* **94,** 277 (1935).
14.12. L. Néel: *Cahiers Phys.* **12,** 1 (1942); **13,** 18 (1943).
14.13. L. Néel: *J. Phys. Radium* **11,** 49 (1950).
14.14. L. Néel: *Advan. Phys.* **4,** 191 (1955).
14.15. R. M. Bozorth: G.10, p. 494.

15

Magnetic Aftereffect

15.1 Phenomenology of Magnetic Aftereffect

By magnetic aftereffect is meant a delayed change in magnetization accompanying a change in the magnetic field. We do not include in this category the magnetization delay due to eddy currents, which is an electromagnetic phenomenon. We also exclude the magnetization change accompanying structural change or aging of the substance, such as precipitation or naturally occurring diffusion. Magnetic aftereffect is distinguishable from a magnetic change due to aging because the former readily permits a return to the original state by purely magnetic means, whereas the latter does not.

Let us consider that a magnetic field is suddenly changed from H_1 to H_2 at $t = 0$. The magnetization, of intensity I_1, is then immediately changed by the value I_i; this is followed by the gradual change I_n, as shown in Fig. 15.1, where I_n is a function of time, or

$$I_n = I_n(t). \tag{15.1}$$

The magnitude of I_n depends not only on the magnitude of initial change of magnetization I_i, but also on the final stage of magnetization. For instance, if the final point is in the range of rotation magnetization, the value of I_n will be fairly small; if the magnetization is in the range of irreversible magnetization, this value may be fairly large.

In a simple case, $I_n(t)$ is described by a single exponential function,

$$I_n(t) = I_{n0}(1 - e^{-t/\tau}), \tag{15.2}$$

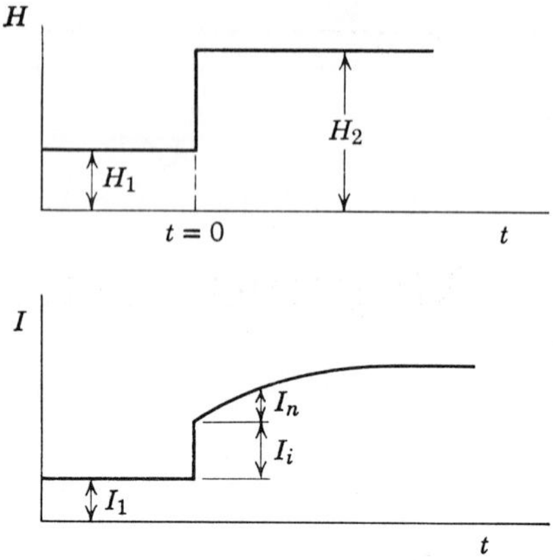

Fig. 15.1. Time change in magnetization on the occurrence of sudden change in the magnetic field.

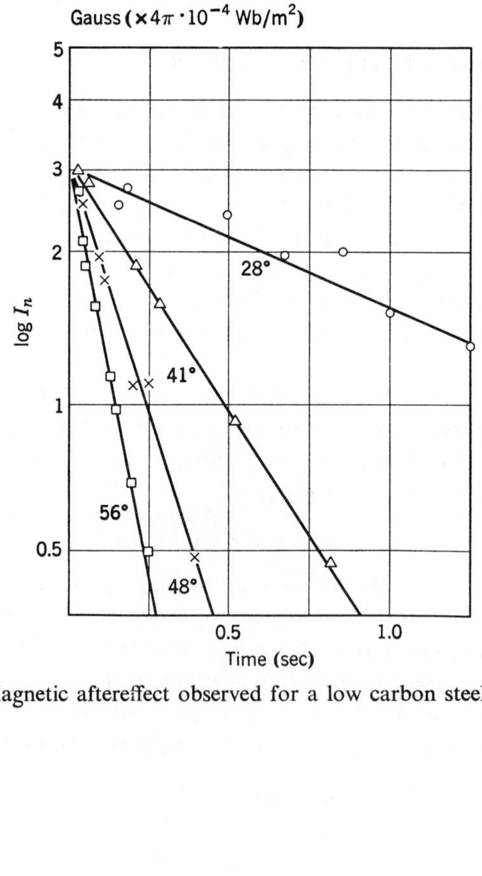

Fig. 15.2. Magnetic aftereffect observed for a low carbon steel (after Tomono[1]).

where I_{n0} is the total change in I_n from $t = 0$ to ∞. Figure 15.2 shows the experimental result for magnetic aftereffect observed by Tomono[1] for a low carbon steel; we see the manifestation of the linear relation between log I_n and t, or the validity of the single exponential relation given by (15.2). The field is changed in this experiment from $H_1 = 1.2$ A/m (= 0.015 Oe) to $H_2 = 0$, that is, in the initial permeability range. Tomono also observed that $I_{n0}/I_i = 30\%$. If we denote this ratio by ζ, we can express the magnetization change as

$$I = \chi_a H[1 + \zeta(1 - e^{-t/\tau})]. \tag{15.3}$$

Such a magnetic aftereffect can be also the origin of a loss for the ac magnetization. In order to derive this relation, we start from the differential equation

$$\frac{d}{dt}(I - \chi_a H) = -\frac{1}{\tau}[I - \chi_a H(1 + \zeta)], \tag{15.4}$$

from which we have the relation (15.3). If the ac magnetic field,

$$H = H_0 e^{i\omega t}, \tag{15.5}$$

is applied on the magnetic substance, its magnetization changes with time in the form

$$I = I_0 e^{i(\omega t - \delta)}, \tag{15.6}$$

where the loss angle δ and the amplitude of magnetization, I_0, are determined by putting (15.6) in (15.4):

$$\tan\delta = \frac{\zeta\omega\tau}{(1 + \zeta) + \omega^2\tau^2}, \tag{15.7}$$

$$I_0 = \frac{\omega\tau}{\omega\tau\cos\delta - \sin\delta}. \tag{15.8}$$

In general, we call $\tan\delta$ the loss factor.

Figure 15.3 shows the dependence of loss factor on the temperature as observed by Tomono[1] for the same material. Each curve shows a maximum at a certain temperature, which is displaced toward a higher temperature with increase of frequency, because the relaxation time τ decreases with an increase of the temperature. We see in (15.7) that $\tan\delta$ tends toward 0 with an approach of τ to 0 and also to ∞, and takes a maximum at

$$\tau = \frac{\sqrt{1 + \zeta}}{\omega}. \tag{15.9}$$

Using this relation, we can determine the value of τ from Fig. 15.3 for relatively high temperature. Figure 15.4 shows the log τ vs $1/T$ relation

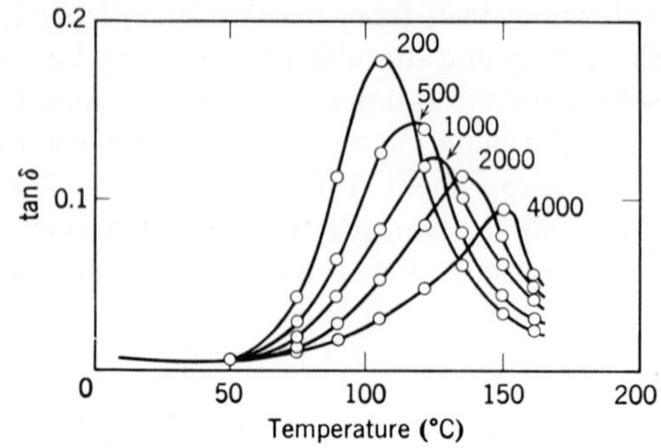

Fig. 15.3. Temperature dependence on tan δ observed for a low carbon steel. The numerical values are the frequencies of the ac field in cycles per second (after Tomono[1]).

obtained from Figs. 15.2 and 15.3. The fact that both groups of experimental points are on the same line tells us that both phenomena have the same origin.

Snoek[2] explained the nature of this phenomenon by means of the model in Fig. 15.5, which shows a ball placed on the curved concrete surface covered with a mud layer of a finite thickness. After the ball is displaced to a new position, it will sink into the layer of mud gradually, changing its equilibrium position. This corresponds to the quasistatic magnetic aftereffect. The ac magnetization corresponds to the case when the ball is oscillating back and forth around the minimum position of the concrete surface under the action of the alternating force. If the viscosity of mud is increased at low temperatures, the ball will move on the hardened surface

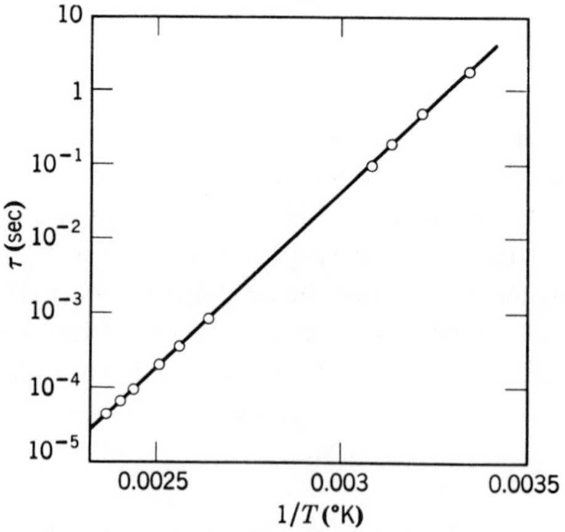

Fig. 15.4. log τ vs $1/T$ curve obtained from quasistatic and high frequency measurements (after Tomono[1]).

of mud with very low loss. On the other hand, if the viscosity is decreased at high temperatures, the ball will move through the unviscous layer of mud, sitting on the concrete surface, and again resulting in very low loss. At the intermediate temperatures, the motion of the ball is most severely damped, and a very large loss results. This is the reason why the loss factor has a maximum at an intermediate temperature.

There are various mechanisms of magnetic aftereffect, and they are discussed in the next section. The time change of the magnetization is not always described by a single exponential function for these mechanisms. It is more common that the relaxation times are distributed in some finite range. Let the volume in which the logarithm of the relaxation time τ is in the range $\log \tau$ to $\log \tau + d(\log \tau)$ be $g(\tau)\, d(\log \tau)$. Since $g(\tau)\, d(\log \tau) = [g(\tau)/\tau]\, d\tau$, the distribution function can be normalized by

$$\int_0^\infty \frac{g(\tau)}{\tau}\, d\tau = 1. \tag{15.10}$$

Then the time change of magnetization can be described by

$$I_n(t) = I_{n0}\left(1 - \int_0^\infty \frac{g(\tau)}{\tau} e^{-t/\tau}\, d\tau\right). \tag{15.11}$$

If we assume for simplicity that the distribution function $g(\tau)$ is constant from τ_1 to τ_2 and is zero outside this range, we have, from (15.10),

$$g = \frac{1}{\log \tau_2/\tau_1}, \tag{15.12}$$

for $\tau_1 < \tau < \tau_2$. If we put $t/\tau = y$, the second term of (15.11) becomes

$$\Delta I_n = I_{n0} - I_n = \frac{I_{n0}}{\log \tau_2/\tau_1} \int_{\tau_1}^{\tau_2} \frac{e^{-t/\tau}}{\tau}\, d\tau = \frac{I_{n0}}{\log \tau_2/\tau_1} \int_{t/\tau_2}^{t/\tau_1} \frac{e^{-y}}{y}\, dy. \tag{15.13}$$

If we set

$$N(\alpha) = \int_\alpha^\infty \frac{e^{-y}}{y}\, dy, \tag{15.14}$$

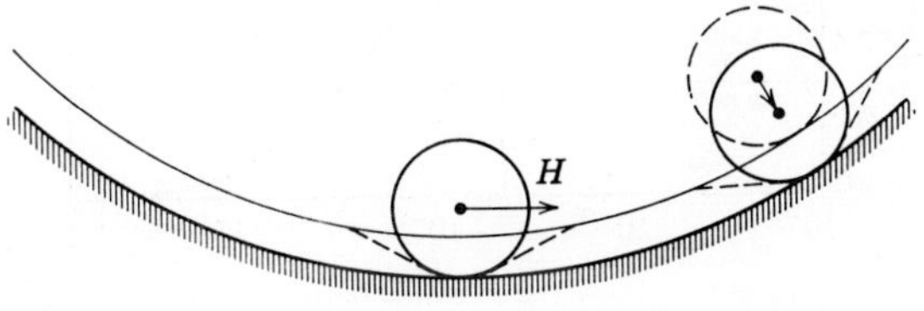

Fig. 15.5. Snoek's model of magnetic aftereffect.

(15.13) can be expressed as

$$\Delta I_n = \frac{I_{n0}}{\log \tau_2/\tau_1}\left[N\left(\frac{t}{\tau_2}\right) - N\left(\frac{t}{\tau_1}\right)\right]. \tag{15.15}$$

$N(\alpha)$ is expressed approximately by

$$N(\alpha) = -0.577 - \log \alpha + \alpha - \frac{1}{2}\frac{\alpha^2}{2!} + \frac{1}{3}\frac{\alpha^2}{3!} + \cdots \quad \text{for } \alpha \ll 1,$$

$$N(\alpha) = \frac{e^{-\alpha}}{\alpha}\left(1 - \frac{1!}{\alpha} + \frac{2!}{\alpha^2} - \cdots\right) \quad \text{for } \alpha \gg 1, \tag{15.16}$$

$$N(1) = 0.219 \quad \text{at } \alpha = 1.$$

Using these formulas, we see that ΔI_n decreases linearly with time t, or

$$\Delta I_n \cong I_{n0}\left(1 - \frac{1/\tau_1 - 1/\tau_2}{\log \tau_2/\tau_1}\, t\right) \tag{15.17}$$

for $t \ll \tau_1$, then varies in proportion to $\log t$, or

$$\Delta I_n \cong \frac{I_{n0}}{\log \tau_2/\tau_1}(\log \tau_2 - 0.577 - \log t) \tag{15.18}$$

for $\tau_1 < t < \tau_2$, and finally tends toward zero according to

$$\Delta I_n \cong I_{n0}\frac{\tau_2}{\log \tau_2/\tau_1}\frac{1}{t}\, e^{-t/\tau_2} \tag{15.19}$$

for $\tau_2 \ll t$.

Figure 15.6 shows the time variation of magnetization as a function of $\log t$ observed at 55°C for annealed carbonyl iron. The solid curve represents the function (15.15) drawn by assuming $\tau_1 = 0.0048$ sec and $\tau_2 = 0.14$ sec, and it reproduces the experiment very well. If this material is magnetized by an ac magnetic field, the loss factor may take its maximum

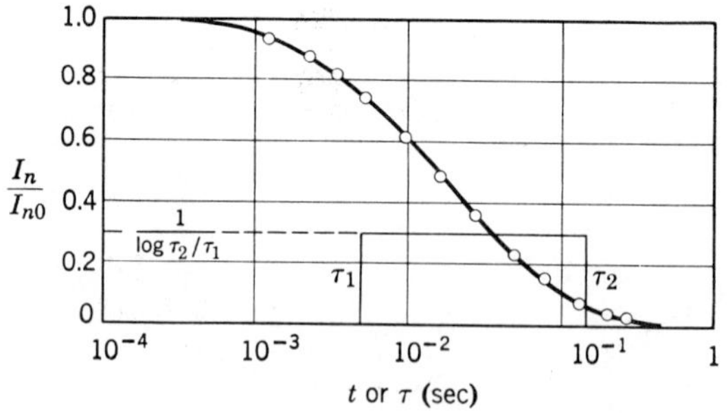

Fig. 15.6. Richter type of magnetic aftereffect (after Becker and Döring[4]).

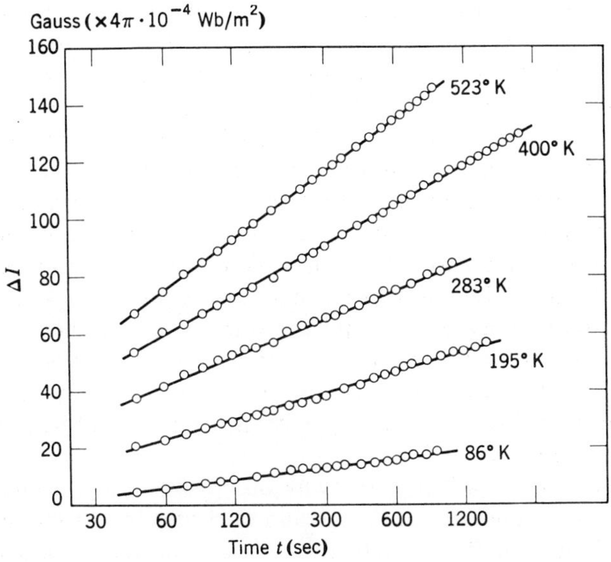

Fig. 15.7. Magnetic aftereffect for Alnico V (after Street and Wooley[5]).

at a frequency where $1/\omega$ lies between τ_1 and τ_2. Because this type of magnetic aftereffect was investigated by Richter,[3] it is referred to as the Richter type of aftereffect.

According to (15.7), tan δ should tend toward zero for a very large τ or at a very low temperature. Actually, however, as seen in Fig. 15.3, there remains a finite tan δ which is independent of frequency ω. We call such a frequency independent term the Jordan type of magnetic aftereffect. We have such a type of magnetic aftereffect if τ is distributed over a very wide range, for then the region which matches the external frequency is almost constant. If we assume that $\tau_1 \ll t \ll \tau_2$, (15.18) becomes

$$\frac{\Delta I_n}{I_n} = \text{const.} - \beta \log t. \tag{15.20}$$

This type of magnetic aftereffect often occurs in permanent magnetic materials. Figure 15.7 is the time change in magnetization observed by Street and Wooley[5] for Alnico V; it demonstrates the validity of (15.20). The actual mechanism of this phenomenon is discussed in the next section.

15.2 Mechanism of Magnetic Aftereffect

Magnetic aftereffect was first investigated by Ewing[6] in 1889. Lord Rayleigh[7] and others also made detailed studies of this effect. After a pause of about thirty years many investigations were started, stimulated by the detailed study by Richter[3] in 1937. The reader may refer to the Becker and Döring book[8] for the detailed history of investigations during this period. Although various phenomenological theories have been proposed to account for this phenomenon, the theory which is based on a physical picture was first proposed by Snoek.[2] He considered that the magnetic aftereffect observed for α-iron is due to the diffusion of carbon atoms in the body-centered cubic lattice of iron. It is known that carbon or nitrogen atoms enter into the interstitial lattice sites, as shown by the small circles in Fig. 15.8. These interstitial sites are classified as x, y, and z sites as shown in the figure. If the carbon atoms occupy, for example, x sites, they may push the nearest iron atoms away in the x direction; and this action results in an elongation of the lattice in the x direction. If the domain magnetization points in the x direction, the magnetoelastic energy will be decreased by the elongation of the lattice along the x direction, provided $\lambda > 0$, so that the carbon atoms will preferentially occupy x sites to decrease the energy. After such preferential occupation of x sites by carbon atoms is realized, an elongation of the crystal thus caused will stabilize the domain magnetization in the x direction. In order words, the depth of the potential minimum becomes larger and larger with the progress of the diffusion of carbon atoms; this process is similar to that of the model in Fig. 15.5.

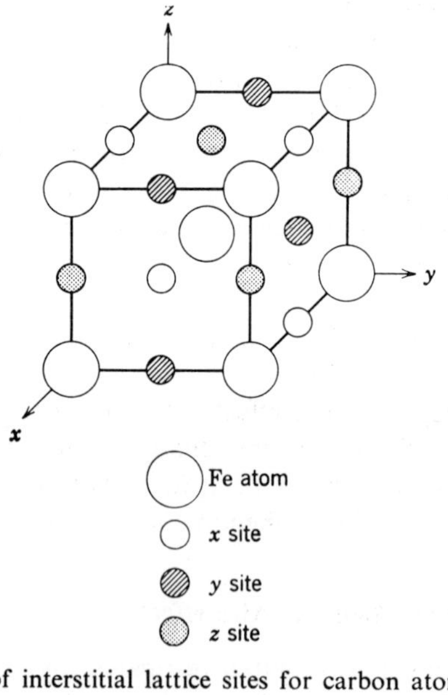

Fig. 15.8. Three kinds of interstitial lattice sites for carbon atoms in body-centered cubic iron.

Néel[9] modified Snoek's concept in certain respects and made further detailed calculations. He considered that the insertion of a carbon atom between iron atoms has a much more remarkable effect in changing the intensity of pseudo-dipolar interaction between the iron atoms than in changing the lattice parameters. De Vries, Van Geest, Gerdorf, and Rathenau[10] investigated this point experimentally by measuring induced anisotropy and time change of magnetostriction and concluded that in iron the direct pseudo-dipolar interaction is more effective than the effect of lattice elongation and also that both interactions are opposite in sign. We assume that the coefficient l of the pseudo-dipolar interaction is changed by l_c by insertion of a carbon atom. Let the total number of carbon atoms be N_c, among which N_x, N_y, and N_z atoms are assumed to occupy x, y, and z sites, respectively. The anisotropy energy related to the carbon atoms is thus given by

$$E_a = \sum_{i>j} w_{ij}$$
$$= N_x l_c(\alpha_1^2 - \tfrac{1}{3}) + N_y l_c(\alpha_2^2 - \tfrac{1}{3}) + N_x l_c(\alpha_3^2 - \tfrac{1}{3}), \qquad (15.21)$$

where $(\alpha_1, \alpha_2, \alpha_3)$ are the direction cosines of the domain magnetization. At thermal equilibrium, the carbon atoms are distributed in proportion to the Boltzmann factor, so that, for $l_c \ll kT$,

$$N_{x\infty} = \frac{N_c}{3} \exp\left[(\tfrac{1}{3} - \alpha_1^2)\frac{l_c}{kT}\right],$$
$$N_{y\infty} = \frac{N_c}{3} \exp\left[(\tfrac{1}{3} - \alpha_2^2)\frac{l_c}{kT}\right], \qquad (15.22)$$
$$N_{z\infty} = \frac{N_c}{3} \exp\left[(\tfrac{1}{3} - \alpha_3^2)\frac{l_c}{kT}\right].$$

Now we consider the process by which carbon atoms change their position from the given distribution N_x, N_y, and N_z at $t = 0$ to thermal equilibrium at $t = \infty$ given by (15.22). We postulate that a carbon atom must get over a potential barrier the height of which is given by Q, while transferring from one site to the neighboring site (Fig. 15.9). This height is also influenced by the orientation of the domain magnetization. It is given by $Q + (\frac{1}{3} - \alpha_1^2)l_c$, $Q + (\frac{1}{3} - \alpha_2^2)l_c$, and $Q + (\frac{1}{3} - \alpha_3^2)l_c$ for the three kinds of lattice sites. The number of carbon atoms which transfer from one sort of site to the other sorts should be proportional to (the number of carbon atoms present in the same sort of sites) $\times$ exp ($-$height of barrier). Since the rate of increase in N_x is given by one-half of the carbon atoms which escape

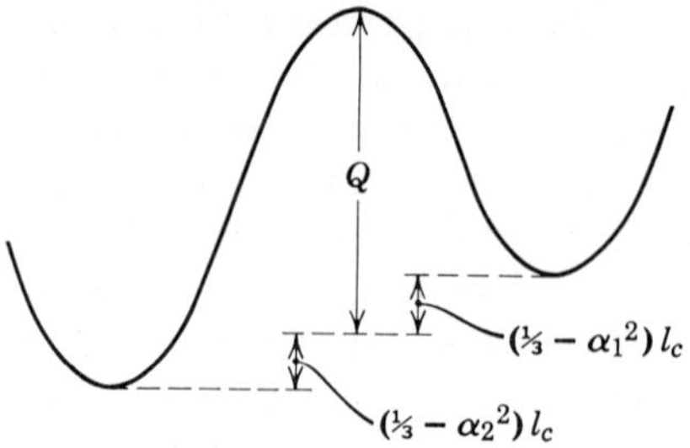

Fig. 15.9. The potential barrier between x and y sites.

from y sites and from z sites less the number of atoms which escape from x sites, we have

$$\frac{dN_x}{dt} = \tfrac{1}{2}cN_y \exp\left[-\frac{Q + (\frac{1}{3} - \alpha_2{}^2)l_c}{kT}\right] + \tfrac{1}{2}cN_z \exp\left[-\frac{Q + (\frac{1}{3} - \alpha_3{}^2)l_c}{kT}\right]$$
$$- cN_x \exp\left[-\frac{Q + (\frac{1}{3} - \alpha_1{}^2)l_c}{kT}\right]$$
$$= \frac{c}{6}N_c e^{-Q/kT}\left(\frac{N_y}{N_{y\infty}} + \frac{N_z}{N_{z\infty}} - 2\frac{N_x}{N_{x\infty}}\right). \qquad (15.23)$$

Assuming that $l_c \ll kT$, we expand the exponential functions in (15.22) in a power series of l_c/kT, and, by neglecting the higher order terms and also by using the relation $N_x + N_y + N_z = N_c$, we have

$$\frac{dN_x}{dt} = -\tfrac{3}{2}ce^{-Q/kT}(N_x - N_{x\infty}). \qquad (15.24)$$

Since N_y and N_z change with time in a similar way, the anisotropy energy given by (15.21) changes with time as

$$\frac{dE_a}{dt} = -\tfrac{3}{2}ce^{-Q/kT}(E_a - E_{a\infty}). \qquad (15.25)$$

First we discuss reversible rotation magnetization on the basis of this model. We assume that the originally existing anisotropy energy, such as a crystal anisotropy, is large compared to the anisotropy induced by carbon atoms. Then the domain magnetization rotates instantaneously to the equilibrium direction upon the application of a magnetic field, and gradually approaches the final direction. Since the angle of the final rotation is small, $E_{a\infty}$ in (15.25) can be regarded as constant. Then (15.25) is easily solved as

$$E_a = E_{a\infty} + (E_{a0} - E_{a\infty})e^{-t/\tau}, \qquad (15.26)$$

where E_{a0} is the anisotropy energy at $t = 0$ and

$$\tau = \frac{2}{3c} e^{Q/kT}. \tag{15.27}$$

Thus the time change in the anisotropy is, in this case, described by a single relaxation time. Tomono's experiment as shown in the graph Fig. 15.4, accurately represents relation (15.27). From the inclination of the log τ vs $1/T$ curve, we have the activation energy Q, which is equal to 0.99 eV in this experiment. This value agrees well with the activation energy for diffusion of carbon atoms in body-centered cubic iron.

For the displacement of domain walls the situation is not so simple. After the domain wall reaches the equilibrium position, carbon atoms in the domain wall rearrange their lattice sites so as to stabilize the local orientation of spins inside the wall. Such a rearrangement of carbon atoms changes the surface energy of the wall through a change in the anisotropy energy,[11] and the wall gradually changes its position. If the length of this gradual displacement is larger than the thickness of the wall, the orientation of spins in the wall are changed by an angle comparable to π, so that $E_{a\infty}$ in (15.25) must be regarded as a function of time. Néel[9] made a detailed calculation for such a gradual propagation of the wall. We refer to the magnetic aftereffect which is caused by diffusion of carbon or nitrogen atoms as the diffusion aftereffect.

Néel[12] proposed another type of magnetic aftereffect called the thermal fluctuation aftereffect. This is the phenomenon caused by thermal fluctuations of domain magnetization. Let us consider an elongated single domain particle which is magnetized first in the positive direction, and then is subjected to a field applied in the negative direction. If the intensity of the field is less than the critical field H_0, the magnetization will stay in the positive direction, in which state the energy of the system is given by

$$U_+ = v(-K_u + I_s H), \tag{15.28}$$

where v is the volume of the particle. If the magnetization is reversed to the negative direction, the energy becomes

$$U_- = v(-K_u - I_s H). \tag{15.29}$$

The potential barrier between the two states can be easily calculated from (14.1) and (14.2), by letting $\theta_0 = 0$, as $U_{max} = vI_s^2H^2/4K_u$ at $\cos\theta = I_sH/2K_u$.

At temperature T each spin is subject to thermal agitation the energy of which is $kT/2$ per degree of freedom of motion. The coherent rotation of all spins included in this particle is also thermally activated and is endowed with the energy $kT/2$. Since usually the height of the potential barrier is much larger than $kT/2$, such coherent rotation is not able to overcome the potential barrier. If, however, the volume of the particle is so small that the height of the potential barrier vK_u at $H = 0$ is of the same order of magnitude as $kT/2$, the domain magnetization is expected to be activated to overcome the potential barrier. This state is similar to paramagnetism and was named superparamagnetism by C. P. Bean.[13] For the critical volume v_0, $v_0K_u \sim kT/2$, so that the value of v_0 is calculated for room temperature, $T = 273°\text{K}$, and $K_u = 10^5 \text{ J/m}^3$ as

$$v_0 = \frac{kT}{2K_u} = \frac{3.77 \times 10^{-21}}{2 \times 10^5} = 1.9 \times 10^{-26}. \tag{15.30}$$

If we assume that the particle is spherical in shape, its radius is

$$r_0 = \left(\frac{3}{4\pi} v_0\right)^{1/3} = 1.7 \times 10^{-9} \text{ m} \fallingdotseq 0.002\,\mu. \tag{15.31}$$

When a magnetic field of intensity H is applied in the negative direction, the height of the potential barrier as measured from U_+ and U_- is changed to $v(2K_u - I_sH)^2/4K_u$ and $v(2K_u + I_sH)^2/4K_u$, respectively. Since the former is less than the latter, the number of the particles whose magnetization is activated from the plus direction toward the minus direction is greater than the number which are activated in the opposite direction. Thus the rate of increase in the number of particles magnetized in the plus direction is given by

$$\frac{dN_+}{dt} = -c'\left\{N_+ \exp\left[-\frac{v(2K_u - I_sH)^2}{4K_ukT}\right] - N_- \exp\left[-\frac{v(2K_u + I_sH)^2}{4K_ukT}\right]\right\}, \tag{15.32}$$

where N_- is the number of particles magnetized in the minus direction. Néel considered c' to be determined by the precessional speed of coherent rotation of the spin system caused by a thermal distortion of the crystal lattice through the change in magnetostrictive anisotropy or in demagnetizing field. If the field is increased to just below the critical field, $H_0 = 2K_u/I_s$, the first term becomes sufficiently large compared to the second term to permit us to neglect the second term, and we have

$$\frac{dN_+}{dt} = -\frac{1}{\tau} N_+, \tag{15.33}$$

where

$$\tau = \frac{1}{c'} \exp \frac{v(2K_u - I_sH)^2}{4K_ukT}. \tag{15.34}$$

Thus the thermal activation of the flux reversal can be performed even for particles having a volume larger than the critical volume v_0 as long as H is close enough to the critical field. Although the form of the time change in N_+ given by (15.33) is quite similar to that of (15.25), the substantial difference between the two is that the activation energy in (15.34) includes H. If the volumes of the particles, v, or the values of K_u are scattered around average values, τ given by (15.34) is expected to cover a very wide range. Néel estimated that a particle of volume 1×10^{-24} m^3 exhibits the relaxation time $\tau \approx 10^{-1}$ sec at room temperature, whereas a particle of volume 2×10^{-24} m^3 exhibits $\tau \approx 10^9$ sec (several tens of years) under the same conditions.† Thus the condition $\tau_1 \ll t \ll \tau_2$ is always valid for a practical duration of measurement, and the time change in magnetization in this case is expected to be proportional to $\log t$ as shown by (15.18). If we let $K_{u\,\max}$ and $v_{\max}$ denote the maximum values of K_u and v, it follows from (15.34) that

$$\tau_1 = \frac{1}{c'},$$

$$\tau_2 = \frac{1}{c'}\left[\exp \frac{v_{\max}(2K_{u\,\max} - I_s H)^2}{4K_{u\,\max}kT}\right]. \tag{15.35}$$

On putting these two expressions into (15.18), we obtain

$$\Delta I_n = \text{const.} - \frac{4K_u I_{n0} kT}{v_{\max}(2K_{u\,\max} - I_s H)^2} \log t. \tag{15.36}$$

It is known from this formula that the inclination of the ΔI_n vs $\log t$ curve increases with an increase of temperature T, as actually observed for Alnico (Fig. 15.7). A thermal fluctuation aftereffect can also be considered for the domain wall displacement. Néel[11] considered that the thermal agitation of the local spin system gives rise to a fluctuation in local magnetic fields which may cause irreversible displacement of unstable domain walls.

The substantial difference between the diffusion aftereffect and the thermal fluctuation aftereffect is that the history of the magnetization distribution in the former type can be retained by a mechanism other than magnetic, such as the distribution of carbon atoms, while it can be retained only by magnetic means in the latter type. In diffusion aftereffect, therefore, if the magnetization is changed from state A to state B and then again changed from B to C, the time change corresponding to the change from A to C is expected to occur in superposition to the time change

† It should be noted that c' in (15.34) depends also on v (cf. Néel[12]), but its dependence is small.

corresponding to the change from B to C. This is called the superposition principle in magnetic aftereffect. In contrast, for the thermal fluctuation aftereffect any memory of the former state would be destroyed by changing the magnetic state; hence the superposition principle is not valid for this aftereffect. With this in mind, Néel[15] called the former type the reversible aftereffect, and the latter the irreversible aftereffect.

15.3 Disaccommodation

It is commonly observed that the permeability of a magnetic substance changes with time after the application of a magnetic field or a mechanical stress. Figure 15.10 shows the time decrease in permeability measured by Snoek[16] for Mn-Zn ferrite. We distinguish this phenomenon from the usual magnetic aftereffect, because it refers to a time change in the ease of magnetization, not in the intensity of magnetization itself. Snoek named this phenomenon disaccommodation.

Snoek used the model shown in Fig. 15.5 to explain this phenomenon. When the ball is displaced to a new place, it can be moved fairly easily on the surface of the mud layer by an external force, whereas after some period of time the ball will sink into the mud layer and will thus lose its mobility. We can regard the ball in this model as representing a domain magnetization which rotates under the action of the magnetic anisotropy energy a part of which can be changed by the diffusion of carbon atoms. We can also regard it as representing a domain wall propagating in the magnetic medium, in which the surface energy of the wall can be changed by the diffusion of carbon atoms.

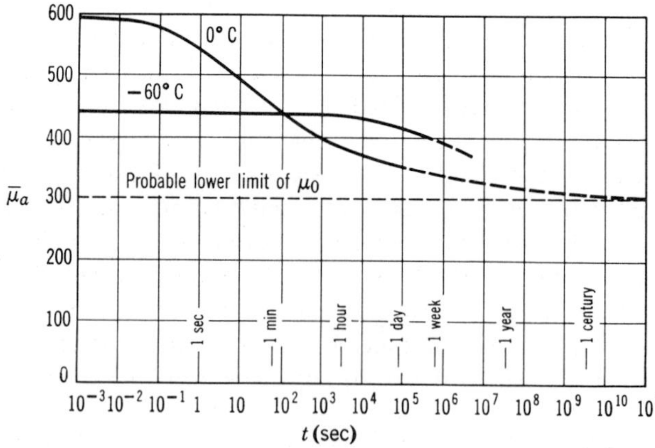

Fig. 15.10. Disaccommodation observed for Mn-Zn ferrite (after J. L. Snoek[16]).

First we discuss the mechanism of disaccommodation due to rotation magnetization. The susceptibility in this case is given by (13.13) for $K_1 > 0$. After some period of time, carbon atoms will diffuse into the favorable sites, stabilizing the domain magnetization as it is. If we assume that the domain magnetization points in the [100] direction, the final distribution of the carbon atoms is calculated from (15.22) as

$$\begin{aligned} N_{x\infty} &= \frac{N_c}{3} e^{-2l_c/3kT}, \\ N_{y\infty} &= \frac{N_c}{3} e^{l_c/3kT}, \\ N_{z\infty} &= \frac{N_c}{3} e^{l_c/3kT}. \end{aligned} \tag{15.37}$$

On putting these results into (15.21), we have the anisotropy energy,

$$E_a = \frac{N_c l_c^2}{9kT}(1 - 3\alpha_1^2) \fallingdotseq \text{const.} + \frac{N_c l_c^2}{3kT}\Delta\theta^2, \tag{15.38}$$

where $\Delta\theta$ is the angle of deviation of the domain magnetization from the [100] direction. Then the susceptibility becomes

$$\chi_a = \frac{I_s^2 \sin^2\theta_0}{2(K_1 + N_c l_c^2/3kT)}. \tag{15.39}$$

If we assume that the concentration of carbon atoms 0.01%, or $N_c/N = 10^{-4}$, where N is the total number of iron atoms included in a unit volume, which is $8.5 \times 10^{28}\ \text{m}^{-3}$, and also that $Nl_c \approx 10^7$ (cf. Section 7.2) and $T = 300°\text{K}$, we can estimate that

$$\frac{N_c l_c^2}{3kT} \approx \frac{10^{-4} \times (10^7)^2}{3 \times 8.5 \times 10^{28} \times 3.8 \times 10^{-21}} \approx 11, \tag{15.40}$$

which is smaller than the usual value of K_1 by a factor 10^{-3}. For iron, $K_1 = 4.2 \times 10^4$; hence the fractional change in the susceptibility is given by

$$\frac{\Delta\chi_a}{\chi_a} = \frac{N_c l_c^2/3kT}{K_1} = 2.6 \times 10^{-4}. \tag{15.41}$$

This value is too small to account for the observed values $20 \sim 80\%$. The functional form of the time change is the same as that of the diffusion aftereffect as given by (15.26), and its relaxation time is given by (15.27).

Let us consider next the displacement of domain walls. After a wall has remained at the same place for a long time, each spin is stabilized by a local anisotropy given by (15.38). If the wall is displaced by the distance Δs, which is smaller than the thickness of the wall δ, the spins in the wall rotate by the angle

$$\Delta\theta = \frac{\pi\,\Delta s}{\delta}, \tag{15.42}$$

if we postulate uniform rotation of the spins across the wall. The local anisotropy energy changes its value by

$$\Delta E_a = \frac{\pi^2 N_c l_c^{\,2}}{3\delta^2 kT}\,\Delta s^2, \tag{15.43}$$

so that the surface energy of the wall is changed by

$$\Delta\gamma = \Delta E_a \times \delta = \frac{\pi^2 N_c l_c^{\,2}}{3\delta kT}\,\Delta s^2. \tag{15.44}$$

Comparing this expression with (13.18), we find that the second derivative at the potential minimum is

$$\alpha = \alpha_0 + \frac{2\pi^2 N_c l_c^{\,2}}{3\delta kT}, \tag{15.45}$$

where α_0 is considered to be caused by spatial fluctuation in the surface energy γ and is given by (13.36) if it is due to the internal stress. Since the susceptibility is inversely proportional to α (cf. 13.24), the susceptibility change is

$$\frac{\Delta\chi_a}{\chi_a} = 1 - \frac{3\lambda\sigma_0 kT}{N_c l_c^{\,2}}\left(\frac{\delta}{l}\right)^2, \tag{15.46}$$

provided

$$\alpha_0 \ll \frac{2\pi^2 N_c l_c^{\,2}}{3\delta kT}.$$

If we tentatively assume that $\lambda\sigma_0 \sim 10^4$ J/m³ and $\delta/l = 0.01$ for annealed iron, we have

$$\frac{\Delta\chi_a}{\chi_a} = 1 - \frac{3 \times 10^4 \times 8.5 \times 10^{28} \times 3.8 \times 10^{-21} \times 10^{-4}}{10^{-4} \times (10^7)^2} = 90\% \tag{15.47}$$

Thus we can explain the experimental values by this model.

The disaccommodation in ferrites was first observed by Snoek[16] for Mn-Zn ferrite in 1947. He suggested that electron displacement between ferric and ferrous ions in the octahedral sites of the spinel lattice could be the cause of this phenomenon. Further investigations were made by Enz[17] for a single crystal of Mn-Fe ferrite; by Miyahara and Yamadaya[18] for Mn-Zn ceramic ferrite; by Ohta[19] for Mn-Zn, Ni-Zn, Mg-Zn and Cu-Zn ferrites. Ohta made it clear that disaccommodation is remarkable in the composition ranges with excess Fe_2O_3 for all these ferrites. Although the electrical resistivity is decreased in these ranges, indicating the occurrence of electron diffusion in the octahedral sites, it was concluded that the true origin would be not electron diffusion but the migration of lattice vacancies in the 16*d* sites. The evidence in support of this conclusion include (1) the activation energy for disaccommodation is 0.5 to 0.8 eV, while that for electron diffusion is 0.1 eV; (2) the amount of lattice vacancies increases with increase of excess Fe_2O_3; (3) disaccommodation is reduced by annealing in nitrogen atmosphere, which reduces the amount of vacancies and instead increases the amount of ferrous ions. Ohta and Yamadaya[20] also observed induced anisotropy, which has the same origin as disaccommodation, by cooling an Mn-Zn ferrite sphere down to liquid air temperature in a magnetic field; they found the magnitude of the induced K_u to be as small as 10 ergs/cc. One possible mechanism of induced anisotropy is that vacancies tend to occupy one sort of octahedral sites which have a trigonal axis mostly favored by the direction of magnetization, so as to reduce the local anisotropy energy. At low temperatures the positions of such vacancies are fixed, giving rise to induced anisotropy. At room temperature, however, they can migrate to new positions with a finite relaxation time upon a rotation of magnetization vector, giving rise to disaccommodation. Yanase[21] postulated that the local anisotropy in this case might be a magnetic dipole-dipole interaction between magnetic ions in octahedral sites.

In ferrites there are also two other relaxation phenomena.[22] One is electron diffusion, which is important at low temperatures, as observed by Wijn and van der Heide[23] for Ni-Zn ferrites. The other is ionic migration at relatively high temperatures, which causes a magnetic annealing effect (cf. Chapter 17).

For details of disaccommodation of metals and alloys the reader may refer to reviews on this subject.[24,25]

Problems

15.1. Determine the ratio of the rates of time change in magnetization at $t = 10^{-3}$ sec, $t = 1$ sec, and $t = 100$ sec for the Richter-type magnetic aftereffect, with $\tau_1 = 10^{-2}$ sec and $\tau_2 = 10$ sec.

15.2. After magnetizing a single crystal of carbon-iron parallel to [010] for a long time to its saturation, the magnetization is switched to [100]. Solve for the time change in the anisotropy energy, assuming diffusion aftereffect and also $l_c \ll kT$.

15.3. Suppose that an aggregate of aligned, long, fine particles with distributed sizes is magnetized parallel to the aligned axis and is then subjected to a field which points in the negative direction. How fast is the magnetization change if the intensity of the field is maintained at 98% of the critical field, as compared to the case when the field is maintained at 90% of the critical field? Assume the occurrence of the thermal fluctuation aftereffect.

15.4. Suppose that a single crystal of carbon-iron is magnetized parallel to [010] for a long time and then the domain magnetization is rotated to [100] at $t = 0$. Discuss the time change in the susceptibility measured by applying a weak alternating magnetic field parallel to [010] and [001]. Assume the occurrence of rotation magnetization and also that $K_1 \gg N_c l_c^2/3kT$.

References

15.1. Y. Tomono: *J. Phys. Soc. Japan* **7,** 174, 180 (1952).
15.2. J. L. Snoek: *Physica* **5,** 663 (1938); G.7, §16, p. 46.
15.3. G. Richter: *Ann. Physik.* **29,** 605 (1937).
15.4. Ref. G.3, p. 254.
15.5. R. Street and J. C. Wooley: *Proc. Phys. Soc.* (*London*) **A62,** 562 (1949).
15.6. J. A. Ewing: *Proc. Roy. Soc.* (*London*) **46,** 269 (1889).
15.7. Lord Rayleigh: *Phil. Mag.* **23,** 225 (1887).
15.8. Ref. G.3, p. 242.
15.9. L. Néel: *J. Phys. Radium* **13,** 249 (1952).
15.10. G. De Vries, D. W. Van Geest, R. Gersdorf, and G. W. Rathenan: *Physica* **25,** 1131 (1959).
15.11. G. De Vries: *Physica* **25,** 1211 (1959).
15.12. L. Néel: *Ann. géophys.* **5,** 99 (1949); *Compt. Rend.* **228,** 664 (1949); *Rev. Mod. Phys.* **25,** 293 (1953).
15.13. C. P. Bean: *J. Appl. Phys.* **26,** 1381 (1955).
15.14. L. Néel: *J. Phys. Radium* **11,** 49 (1950).
15.15. L. Néel: *J. Phys. Radium* **12,** 339 (1951).
15.16. Ref. G.7; §17, p. 54.
15.17. U. Enz: *Physica* **24,** 609 (1958).
15.18. S. Miyahara and T. Yamadaya: *J. Phys. Soc. Japan* **14,** 1635 (1959).
15.19. K. Ohta: *J. Phys. Soc. Japan* **16,** 250 (1961).
15.20. K. Ohta and T. Yamadaya: G.37, p. 291.
15.21. A. Yanase: *J. Phys. Soc. Japan* **17,** 1005 (1962).
15.22. S. Krupicka: G.37, p. 304.
15.23. H. P. J. Wijn and H. van der Heide: *Rev. Mod. Phys.* **25,** 98 (1953).
15.24. P. Brissonnean: *J. Phys. Radium* **19,** 490 (1958).
15.25. G. W. Rathenau: G.33, p. 168.

16

Dynamic Properties of Magnetization

16.1 Eddy Current Loss

The most important origin of losses accompanying the magnetization of magnetic metals is the eddy current. Let us consider first a long cylindrical ferromagnetic metal which is magnetized parallel to its long axis (Fig. 16.1*a*). Applying the integral form of the law of electromagnetic induction,

$$\oint E_s \, ds = -\iint \frac{dB_n}{dt} \, dS, \tag{16.1}$$

to the circular circuit with radius r drawn about the center axis of the cylinder, we have, for $r < r_0$ (the radius of the cylinder),

$$2\pi r E(r) = -\pi r^2 \frac{dI}{dt}$$

or

$$E(r) = -\frac{r}{2}\frac{dI}{dt}. \tag{16.2}$$

If the magnetization is increased at the constant rate dI/dt, the current density is then simply given by

$$i(r) = -\frac{r}{2\rho}\frac{dI}{dt}, \qquad (\rho = \text{resistivity}) \tag{16.3}$$

which leads to the power loss per unit volume of the material,

$$P = \frac{1}{\pi r_0^2}\int_0^{r_0} 2\pi E(r) i(r) r \, dr$$

$$= \frac{1}{2\rho r_0^2}\left(\frac{dI}{dt}\right)^2 \int_0^{r_0} r^3 \, dr = \frac{r_0^2}{8\rho}\left(\frac{dI}{dt}\right)^2. \tag{16.4}$$

Thus the eddy current loss is proportional to the square of the rate of magnetization change. This is also true for an alternating magnetization. That is, the power loss increases in proportion to the square of the frequency as long as the flux penetration is complete. It is also seen in (16.4) that the power loss is proportional to r_0^2; this means that the loss will be decreased by fine division of the material. It is natural that the loss should be inversely proportional to the resistivity. This is the reason why ferrites are more useful in the high frequency region than ferromagnetic metals and alloys.

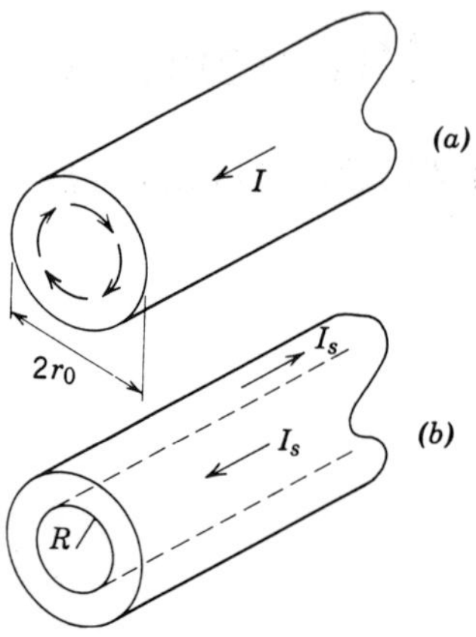

Fig. 16.1. Eddy current in a cylindrical specimen magnetized (*a*) homogeneously, and (*b*) by wall displacement.

If dI/dt is very large, the eddy current becomes strong enough to give rise to a magnetic field which is comparable to the applied field. Since the eddy current which flows along a circle of radius r produces a magnetic field only inside this circle, the integrated magnetic field produced by the total eddy current is strongest at the center and becomes weaker and weaker as it approaches the surface of the specimen. Since a magnetic field produced by the eddy current opposes the change in magnetization, the magnetization is damped away inside the material. It is calculated that the amplitude of magnetization is decreased to $1/e$ of that at the specimen surface at the skin depth, which is

$$s = \sqrt{\frac{2\rho}{\omega\mu}}, \tag{16.5}$$

where ω is the angular frequency of the alternating magnetic field. This value is independent of the size or shape of the specimen as long as the skin depth is sufficiently small compared to the diameter or thickness of the specimen. For iron ($\rho = 1 \times 10^{-7}$ Ωm) with $\bar{\mu} = 500$, magnetized by a 50-cps alternating magnetic field, the skin depth is calculated to be

$$s = \sqrt{\frac{2 \times 1 \times 10^{-7}}{2\pi \times 50 \times 500 \times 4\pi \times 10^{-7}}} = 1.0 \times 10^{-3}\ \mathrm{m} = 1\ \mathrm{mm}. \tag{16.6}$$

The iron cores of an ac machine are usually made of laminated thin sheets of magnetic metals for the purpose of attaining complete penetration of magnetic flux into the interior of the core.

Next we consider a cylindrical specimen which contains one cylindrical domain wall (Fig. 16.1*b*). The specimen is separated into two magnetic domains by this 180° wall. Inside the wall, or for $r < R$, where R is the radius of the wall, there is no flux change; hence

$$E(r) = 0. \qquad (r \leq R) \tag{16.7}$$

Outside the wall, or for $r > R$, it follows from (16.1) that

$$2\pi r E(r) = -4\pi I_s R \frac{dR}{dt}$$

or

$$E(r) = -2I_s R \frac{dR}{dt}\frac{1}{r}. \qquad (r \geq R) \tag{16.8}$$

Thus the power loss per unit volume is given by

$$P = \frac{1}{\pi r_0^{\,2}} \int_R^{r_0} \frac{E^2(r)}{\rho} 2\pi r\, dr = \frac{8I_s^{\,2}R^2}{\rho r_0^{\,2}} \left(\frac{dR}{dt}\right)^2 \log \frac{r_0}{R}. \tag{16.9}$$

If we use the rate of magnetization change,

$$\frac{dI}{dt} = \frac{4I_s}{r_0^{\,2}} R \frac{dR}{dt}, \tag{16.10}$$

for (16.9), we have

$$P = \frac{r_0^{\,2}}{2\rho}\left(\log\frac{r_0}{R}\right)\left(\frac{dI}{dt}\right)^2. \tag{16.11}$$

Thus the power loss depends on the size of the wall. On the average,

$$\overline{\log \frac{r_0}{R}} = \frac{1}{r_0}\int_0^{r_0}\left(\log\frac{r_0}{R}\right) dR = 1, \tag{16.12}$$

so that

$$\bar{P} = \frac{r_0^{\,2}}{2\rho}\left(\frac{dI}{dt}\right)^2. \tag{16.13}$$

Comparing this and (16.4), we find that the power loss is four times larger than that for homogeneous magnetization change. The reason is that the eddy currents are localized at the wall and such localization gives rise to a larger power loss, because the power loss is proportional to i^2.

This fact was first experimentally verified by Williams, Shockley, and Kittel.[1] They used a picture-frame specimen of 3% silicon-iron which contains a 180° wall which runs parallel to each arm of the specimen

(Fig. 16.2). The velocity of the domain wall was measured as a function of the applied field and found to be expressed by

$$v = c(H - H_0), \tag{16.14}$$

where c is the proportionality factor and H_0 is the critical field for displacement of the wall. The proportionality between v and H can easily be inferred from the model shown in Fig. 16.1*b*. When the cylindrical wall with radius R is expanding with velocity dR/dt, the power supplied by the magnetic field is

$$P = H\frac{dI}{dt} = \frac{4I_sHR}{r_0^2}\frac{dR}{dt}. \tag{16.15}$$

Since there is no change in the potential energy, this energy must be consumed as heat produced by the eddy current. Thus, on equating (16.15) with (16.9), we have

$$v = \frac{dR}{dt} = \frac{\rho}{2I_sR\log(r_0/R)}H, \tag{16.16}$$

which shows the proportionality between v and H. Williams et al.[1] calculated the distribution of eddy currents for the rectangular cross section in which a plane domain wall is moving as shown in Fig. 16.3*a*, and they showed that the calculated power loss is in good agreement with the observed one. They also found that the wall shrinks into the form of a cylinder, as shown in Fig. 16.3*b* when a strong magnetic field is applied,

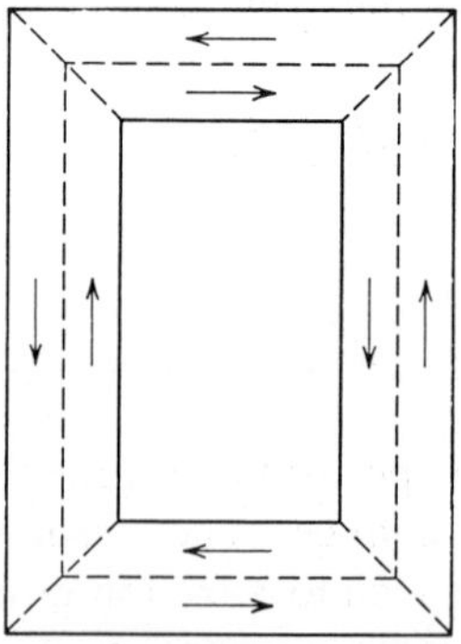

Fig. 16.2. A single crystal of Si-Fe in the picture-frame shape and its domain structure (Williams et al.[1]).

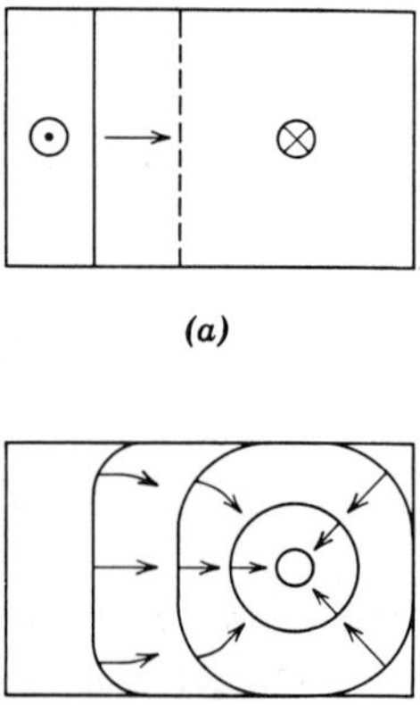

(*a*)

(*b*)

Fig. 16.3. Cross-sectional view of the wall running (*a*) at low speed and (*b*) at high speed.

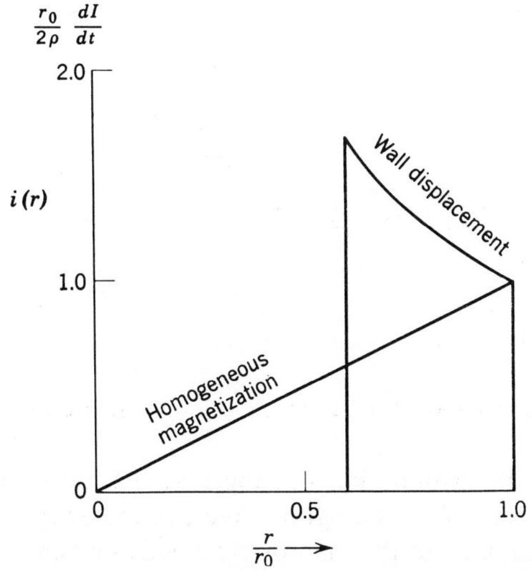

Fig. 16.4. Distribution of eddy current in a cylindrical specimen for homogeneous magnetization and for wall displacement.

because the displacement of the wall is strongly damped in the interior of the specimen. The cylindrical wall vanishes very rapidly, since the surface tension of the wall helps to diminish its area. There was very good agreement between calculation and experiment for this cylindrical wall also.

Before the success of this experiment, the calculation of eddy current losses made under the assumption of a homogeneous magnetization gave only one-half or one-third of the observed value, which was known as the eddy current anomaly. It is now known from this experiment that the anomaly is due to the ignoring of the localization of eddy currents at the domain walls.

The calculation of the distribution of eddy currents is very difficult for an actual substance which contains many domain walls, for it depends not only on the shape and the distribution of domain walls, but also on the outside shape of the specimen. Figure 16.4 shows a comparison of the eddy current distribution between the two cases in Fig. 16.1. It is seen that the eddy current changes discontinuously at the location of the domain wall. This discontinuity, Δi is easily found by letting $r = R$ in (16.8):

$$i = \frac{2I_s}{\rho} v. \tag{16.17}$$

This relation is generally valid for a 180° wall running at velocity v. Let us now assume that the distribution of eddy currents is shown by curve a in Fig. 16.5 for homogeneous magnetization. Apparently the actual eddy

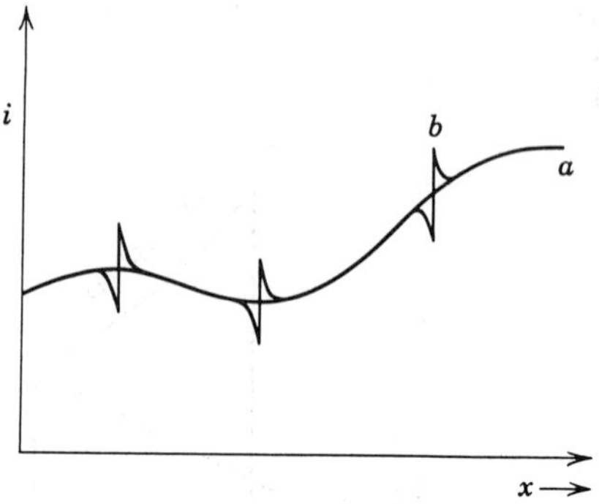

Fig. 16.5. Spatial variation of macroscopic and microscopic eddy currents.

current is expected to exhibit sharp changes at the locations of domain walls as shown by curve b in the figure. We call the eddy current due to homogeneous magnetization the macroscopic eddy current (curve a), i_{ma}, and the deviation from macroscopic eddy current due to wall displacement the microscopic eddy current (difference between curves a and b), i_{mi}. The power loss is then given by

$$P = \rho \int (i_{ma} + i_{mi})^2 \, dv$$

$$= \rho \int i_{ma}^2 \, dv + \rho \int i_{mi}^2 \, dv + 2\rho \int i_{ma} i_{mi} \, dv \tag{16.18}$$

where the integrations should be made over a unit volume of the specimen. If the spatial variation of the macroscopic eddy current is gentle compared to that of the microscopic eddy current, as it is in Fig. 16.5, the third term should vanish, since plus and minus values of i_{m1} cancel each other. For such a case we can calculate the power loss as the sum of those of the macroscopic and the microscopic eddy currents. If there are a large number of domain walls, and accordingly the velocity of an individual wall is small, the individual microscopic eddy currents become small, as seen in (16.17). For this case we can approximate the power loss by a macroscopic eddy current. This is a natural conclusion because the presence of a large number of domain walls means that the magnetization is quite homogeneous. If, however, the paths of the eddy currents are complicated because of the presence of many non-conducting inclusions, or if the separation of domain walls is comparable to the size of the specimen, not only can we ignore the third term in (16.18), but we also can hardly distinguish the two categories of macroscopic and microscopic eddy currents.

16.2 Various Types of Loss and Resonance

The desirable properties for soft magnetic materials are high permeability and low loss. In this section we summarize the various types of losses and resonances which affect the permeability of soft magnetic materials in various ranges of frequencies. If a magnetic material is magnetized by the ac magnetic field $H = H_0 e^{i\omega t}$, the magnetic flux density B is generally delayed by the phase angle δ because of the presence of loss and is thus expressed as $B = B_0 e^{i(\omega t - \delta)}$. The permeability is then

$$\mu = \frac{B}{H} = \frac{B_0 e^{i(\omega t-\delta)}}{H_0 e^{i\omega t}} = \frac{B_0}{H_0} e^{-i\delta} = \mu' - i\mu'', \tag{16.19}$$

where

$$\mu' = \frac{B_0}{H_0} \cos \delta, \qquad \mu'' = \frac{B_0}{H_0} \sin \delta. \tag{16.20}$$

μ' expresses the component of B which is in phase with H, so it corresponds to the normal permeability: if there are no losses, we should have $\mu = \mu'$. μ'' expresses the component of B which is delayed by the phase angle 90° from H. The presence of such a component requires a supply of energy to maintain the alternating magnetization, regardless of the origin of the delay. The ratio μ'' to μ' becomes, from (16.20),

$$\frac{\mu''}{\mu'} = \frac{(B_0/H_0) \sin \delta}{(B_0/H_0) \cos \delta} = \tan \delta \tag{16.21}$$

and is thus given by the loss factor. The quality of soft magnetic materials is often measured by the factor $\mu/\tan \delta$ or μQ.

The most important loss in ferromagnetic substances is the hysteresis loss. If the amplitude of magnetization is very small or if it is in the Rayleigh region, the loss factor due to the hysteresis loss depends on the amplitude of magnetic field as shown by (14.44). A loss factor of this type can be easily distinguished from the other types of loss by reducing H toward zero.† The hysteresis loss becomes less important in the high frequency range, because the wall displacement, which is the main origin of the hysteresis, is mostly damped in this range and is replaced by the rotation magnetization, as will be discussed later.

The next important loss for ferromagnetic metals and alloys is the eddy current loss. Since a power loss of this type increases in proportion to the square of the frequency, as discussed in the preceding section, it plays an important role in the high frequency range. One of the means to eliminate

† The real physical origin of the hysteresis loss is eddy current or other damping phenomena associated with irreversible wall displacement or irreversible rotation magnetization (cf. Sections 16.4 and 16.5).

the eddy current loss is to reduce the size of the materials. For instance, efforts have been made to obtain thin Permalloy sheets by means of cold-rolling to avoid eddy current. Vacuum evaporation and electroplating are also effective in preparing very thin metal films which are used as the high speed switching elements of memory devices. Dust cores or magnetic cores composed of fine metallic particles are also made for the purpose of reducing eddy current. The most effective means of avoiding eddy current, however, is to use ferromagnetic insulators instead of magnetic metals and alloys. On this account ferrites have been extensively used in the high frequency range. The average resistivity of ferrites is about 10^4 Ωm (= 10^6 Ωcm), so we need not worry about the eddy current loss. Since, however, the resistivity of magnetite (Fe_3O_4) is fairly low ($\rho = 10^{-4}$ Ωm = 10^{-2} Ωcm), the inclusion of excess Fe ions in various mixed ferrites results in a decrease in resistivity and accordingly in an increase in eddy current loss in the very high frequency range.

Magnetic aftereffect also gives rise to a magnetic loss for high frequency magnetization as discussed in Section 15.1. The relation between tan δ and the relaxation time τ is given by (15.7), which shows a maximum at a certain value of τ. Wijn and van der Heide[2] observed such a maximum of tan δ at the frequency 100 kc/sec at low temperatures for Ni-Zn and Mn-Zn ferrites. The origin of this phenomenon is considered to be the diffusion of electrons between Fe^{2+} and Fe^{3+}. It is estimated from the value of the activation energy that the maximum of tan δ will be shifted at room temperature to a frequency of several hundreds of megacycles. Since, however, tan δ generally increases very rapidly in this frequency range because of the occurrence of natural resonance, as will be discussed below, the observation of a maximum of tan δ due to electron diffusion would be fairly difficult. The magnetic aftereffect due to ion diffusion in ferrites is expected to give rise to a maximum of tan δ at room temperature only at frequencies as low as several cycles per second, so that it will give rise to a negligibly small loss in the high frequency range.

Brockman et al.[3] found that the permeability of a large magnetic core made from Mn-Zn ferrite with a cross section 1.25×2.5 cm^2 drops very rapidly at 1.5 Mc/s (Fig. 16.6). This frequency was found to be dependent on the size of the core. These investigators explained this phenomenon in terms of the build-up of an electromagnetic standing wave inside the core. The velocity of the electromagnetic wave propagating in the material with relative permeability $\bar{\mu}$ and relative dielectric constant $\bar{\epsilon}$ is reduced by the factor $1/\sqrt{\bar{\epsilon}\bar{\mu}}$ as compared with that in vacuum; hence the wave-length is

$$\lambda = \frac{c}{f\sqrt{\bar{\epsilon}\bar{\mu}}}, \tag{16.22}$$

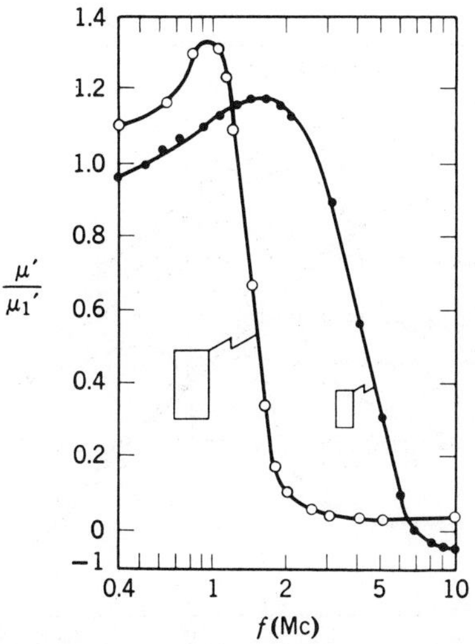

Fig. 16.6. Dimensional resonance of Mn-Zn ferrite (Brockman et al.[3]).

where c is the velocity of light and f is the frequency. For Mn-Zn ferrite, $\bar{\mu} \sim 10^3$, $\bar{\epsilon} \sim 5 \times 10^4$; if we assume $f = 1.5$ Mc, the wavelength λ is estimated to be $\lambda \simeq 2.6$ cm. If, therefore, the dimension of the core is equal to an integer multiple of the wavelength λ, the electromagnetic wave will resonate within the core, giving rise to a standing wave. It is commonly known that μ' and μ'' vary with frequency as shown in Fig. 16.15, if some kind of resonance is induced at frequency f_0 (H_z is replaced by f, and H_r by f_0). The form of the experimental curve in Fig. 16.6 is recognized to be of the resonance type. This phenomenon is called the dimensional resonance.

Generally, permeability drops off and magnetic loss increases in a very high frequency region because of the occurrence of a magnetic resonance. Figure 16.7 shows the frequency dependence of μ' and μ'' observed for Ni-Zn ferrites with various compositions. It is seen in this graph that the ferrite with high permeability tends to have its permeability decrease at a relatively low frequency. Snoek[4] explained this fact in terms of the natural resonance or the resonance of rotation magnetization under the action of the anisotropy field. The resonance frequency for such a natural resonance can be obtained by setting $H = 0$ in (7.34) as

$$\omega = \nu H_a, \tag{16.23}$$

where ν is the gyromagnetic constant

$$\nu = 1.105 \times 10^5 g \text{ m/A sec} \qquad \text{(cf. (3.11))}. \tag{16.24}$$

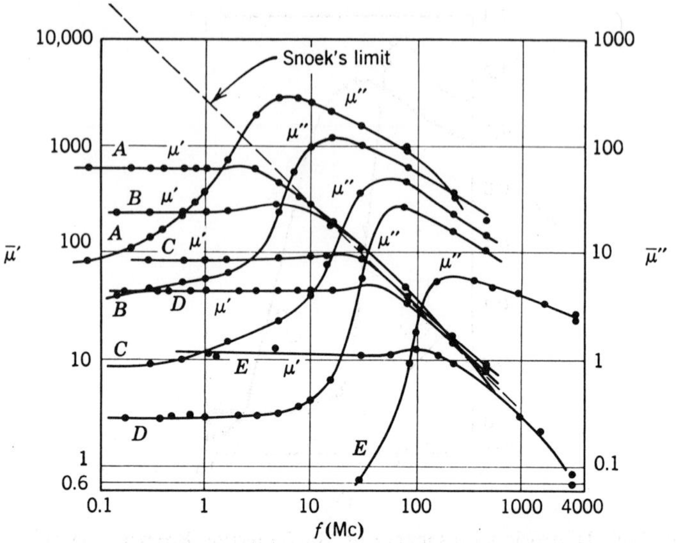

Fig. 16.7. Frequency dependence of the real and imaginary parts of the initial permeability for polycrystalline Ni-Zn ferrite (compositional ratio NiO:ZnO = 17.5:33.2 (*A*), 24.9:24.9 (*B*), 31.7:16.5 (*C*), 39.0:9.4 (*D*), 48.2:0.7 (*E*), remaining part, Fe_2O_3) (Gorter[G21]).

If we assume that $K_1 = -5 \times 10^2$ J/m³ and $I_s = 0.3$ Wb/m² for Ni-Zn ferrite, we can calculate the anisotropy field from (7.39) as

$$H_a = \frac{4 \times 5 \times 10^2}{3 \times 3 \times 10^{-1}} = 2.2 \times 10^3 \text{ A/m}. \tag{16.25}$$

Using this value and assuming that $g = 2$, we have, from (16.23),

$$\omega = (1.105 \times 10^5) \times 2 \times (2.2 \times 10^3) = 5 \times 10^8 \tag{16.26}$$

or

$$f = \frac{\omega}{2\pi} \approx 8 \times 10^7 \text{ cps} = 80 \text{ Mc/s}. \tag{16.27}$$

This value is in the range of frequency where the natural resonance occurs as seen in Fig. 16.7.

Thus the resonance frequency ω is related to the anisotropy constant K_1 by the relation which can be obtained by putting the anisotropy field for $K_1 > 0$,

$$H_a = \frac{2K_1}{I_s}, \tag{16.28}$$

into (16.23):

$$\omega = \frac{2\nu K_1}{I_s}. \tag{16.29}$$

Thus we know that the resonance frequency shifts toward the high frequency side with an increase of K_1. On the other hand, the permeability decreases with an increase of K_1, as we know from the relation (cf. 13.15)

$$\mu = \frac{I_s^2}{3K_1}. \tag{16.30}$$

On multiplying (16.29) by (16.30) we obtain

$$\omega\bar{\mu} = \frac{2}{3}\frac{\nu I_s}{\mu_0}, \tag{16.31}$$

which is independent of K_1. We have the same relation for $K_1 < 0$. Again assuming that $I_s = 0.3$ Wb/m², and by using $\omega = 2\pi f$, we have

$$f\bar{\mu} = 5.6 \times 10^9 \text{ cps} = 5600 \text{ Mc/s}. \tag{16.32}$$

The broken line in Fig. 16.7 is obtained by connecting the points where μ' drops off to one-half the maximum value. The condition given by (16.32) coincides approximately with this line. It was expected that no ferrite could attain points higher than this line, as long as the presence of cubic magnetocrystalline anisotropy is assumed.

This limitation has been overcome by an ingenious invention of researchers[5] at Philips Gloeilampenfabrieken. They made a magnetic oxide called Ferroxplana, which has the magnetoplumbite type of crystal structure. The uniaxial anisotropy constant K_u of this material is negative; hence its magnetization is stable in the plane (c plane) perpendicular to the c axis. The anisotropy in the c plane is fairly small and is represented by the anisotropy field H_{a1}, while that corresponding to the rotation out of the c plane is represented by H_{a2}. Then the resonance frequency is calculated, similarly to (3.33), as

$$\omega = \nu\sqrt{H_{a1}H_{a2}}. \tag{16.33}$$

The permeability of this material is essentially determined by the smaller anisotropy field H_{a1}, and this is calculated by using the relation (7.37) in (13.13). Thus we have

$$\omega\bar{\mu} = \frac{\nu I_s}{\mu_0}\sin^2\theta_0\sqrt{\frac{H_{a2}}{H_{a1}}}, \tag{16.34}$$

which gives a fairly large value compared to (16.31) when $H_{a2} \gg H_{a1}$. Two examples are given in Table 16.1, where we see that the resonant frequencies which are actually observed are in good agreement with the calculated ones. The frequency behavior of permeability is given in Fig. 16.8 for Ferroxplana, Co_2Z. It is seen in this graph that the resonant

Table 16.1. Various Constants and Resonant Frequency of Two Types of Ferroxplana[6]

Ferrite	$\bar{\chi}_a$ at Low Frequency	I_s Wb/m² (= $10^4/4\pi G$)	H_{a1}, A/m	H_{a2}, A/m	f_{res}, Mc/s Theory	f_{res}, Mc/s Observed
$Co_2Ba_3Fe_{24}O_{41}$	11	0.335	8.92×10^3 (= 112 Oe)	1.03×10^6 (= 13,000 Oe)	3,400	1,400
$Mg_2Ba_2Fe_{12}O_{22}$	9	0.150	4.94×10^3 (= 62 Oe)	0.797×10^6 (= 10,000 Oe)	2,200	1,000

frequency of Co_2Z is about five times larger than that of the Ni-Zn ferrite (Ferroxcube 4E). It has also been reported by some researchers[7] at Philips Gloeilampenfabrieken that the permeability can be increased by a factor of two or three by orienting the crystallites, as we expected from (16.34) which shows that $\omega\mu$ can be increased by increasing the angle θ_0.

Rado[8] pointed out the possibility of the occurrence of the resonant oscillation of domain walls and explained the maximum of μ'' observed for some ferrites. We discuss this subject in Section 16.5.

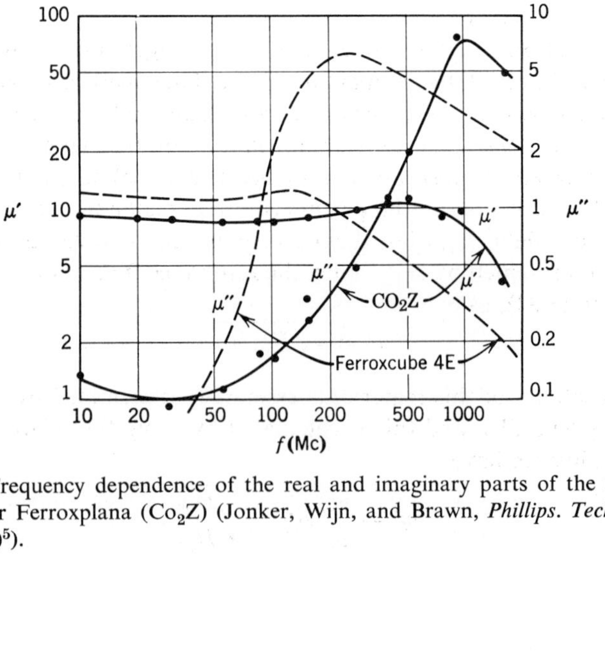

Fig. 16.8. Frequency dependence of the real and imaginary parts of the initial permeability for Ferroxplana (Co_2Z) (Jonker, Wijn, and Brawn, *Phillips. Tech. Rev.* **18,** 150 (1956–7)[5]).

16.3 Spin Dynamics

In this section we discuss the dynamic character of the spin system and, in particular, the switching mechanism as it is applied to memory devices of electronic computers. As we discussed in Section 3.2, a gyroscope performs a precession motion under the action of an external torque, but it tends to rotate as a rigid body if its free precession motion is restricted by some boundary conditions. Such a circumstance can well be expressed by the Landau-Lifshitz[9] equation of motion,

$$\frac{d\boldsymbol{I}}{dt} = -\nu[\boldsymbol{I}, \boldsymbol{H}] - \frac{4\pi\mu_0\lambda}{I^2}[\boldsymbol{I}, [\boldsymbol{I}, \boldsymbol{H}]], \tag{16.35}$$

where $\boldsymbol{I}$ is the magnetization vector and $\boldsymbol{H}$ the magnetic field vector. The first term represents the precession motion of the magnetization; the magnetization moves in the direction perpendicular to both $\boldsymbol{I}$ and $\boldsymbol{H}$, or in the direction $-[\boldsymbol{I}, \boldsymbol{H}]$. The factor ν is the gyromagnetic constant given by (16.24). The second term signifies that a damping force acting on the precession motion gives rise to a motion in the direction perpendicular to $\boldsymbol{I}$ and also to the direction of the damping force $[\boldsymbol{I}, \boldsymbol{H}]$. The coefficient λ, which has the dimension $\sec^{-1}$, expresses the degree of damping action and is thus called the relaxation frequency. The equation of motion may also be written as

$$\frac{d\boldsymbol{I}}{dt} = -\nu[\boldsymbol{I}, \boldsymbol{H}] + 4\pi\mu_0\lambda\left\{\boldsymbol{H} - \frac{(\boldsymbol{I}, \boldsymbol{H})\boldsymbol{I}}{I^2}\right\}, \tag{16.36}$$

as shown by Kittel.[10] Since the second term in the braces represents the component of $\boldsymbol{H}$ parallel to $\boldsymbol{I}$, the resultant vector composed of the two terms in the braces represents the component of $\boldsymbol{H}$ perpendicular to $\boldsymbol{I}$. This component of $\boldsymbol{H}$ is effective in exerting a couple of force on the magnetization and so is effective in giving rise to the rotation of $\boldsymbol{I}$ toward the direction of $\boldsymbol{H}$ with the cooperation of the damping action on the precession motion. It is easily verified that (16.36) is mathematically equivalent to (16.35).

In the discussion above, it was implicitly assumed that the first term, which is to be called the inertia term, is larger than the second term, the so-called damping term. If we put

$$\frac{4\pi\mu_0\lambda}{\nu I} = \alpha, \tag{16.37}$$

our assumption is equivalent to $\alpha^2 \ll 1$. Strictly speaking, the damping motion should act not only on the precession motion, but also on the motion induced by the second term in (16.35) or (16.36). In other words, the damping should act on the resultant motion of the magnetization, $d\boldsymbol{I}/dt$. Thus the magnetization performs a precession motion under the action of the external force and the damping so that the exact equation of motion should be given by[11]

$$\frac{d\boldsymbol{I}}{dt} = -\nu\left[\boldsymbol{I}, \boldsymbol{H} - \frac{\alpha}{\nu I}\frac{d\boldsymbol{I}}{dt}\right]. \tag{16.38}$$

This equation was first derived by Gilbert[11] and used by Kikuchi[12] in his discussion of the switching rate of magnetization. Equations (16.35) and (16.36) can be derived from this equation by neglecting the higher order terms in α^2.

First we consider the precession motion, neglecting the damping action. Suppose that a static magnetic field H is applied parallel to the $-z$ direction. The equation of motion is given by

$$\frac{d\boldsymbol{I}}{dt} = -\nu[\boldsymbol{I}, \boldsymbol{H}], \tag{16.39}$$

which can be written for each component of Cartesian coordinates:

$$\begin{aligned} \frac{dI_x}{dt} &= \nu I_y H, \\ \frac{dI_y}{dt} &= -\nu I_x H, \\ \frac{dI_z}{dt} &= 0. \end{aligned} \tag{16.40}$$

On solving these equations, we have

$$\begin{aligned} I_x &= I_s \sin\theta_0 e^{i\omega_0 t}, \\ I_y &= I_s \sin\theta_0 e^{i\omega_0 t + i(\pi/2)}, \\ I_z &= I_s \cos\theta_0, \end{aligned} \tag{16.41}$$

where

$$\omega_0 = \nu H. \tag{16.42}$$

This solution represents the precession motion of magnetization with constant inclination from the z axis (Fig. 16.9).

If a finite damping force acts on this precession motion, the motion will decay unless energy to keep up the precession is supplied from the outside.

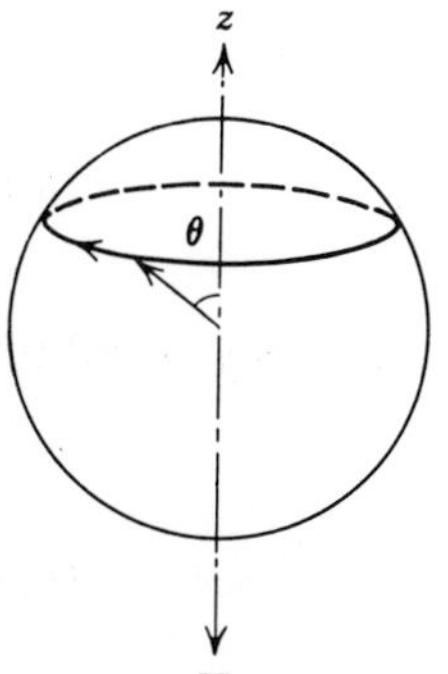

Fig. 16.9. The motion of spin in the absence of damping action.

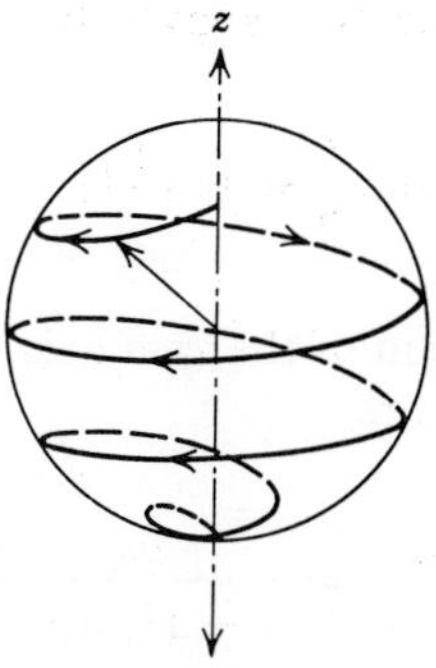

Fig. 16.10. The motion of spin for small damping.

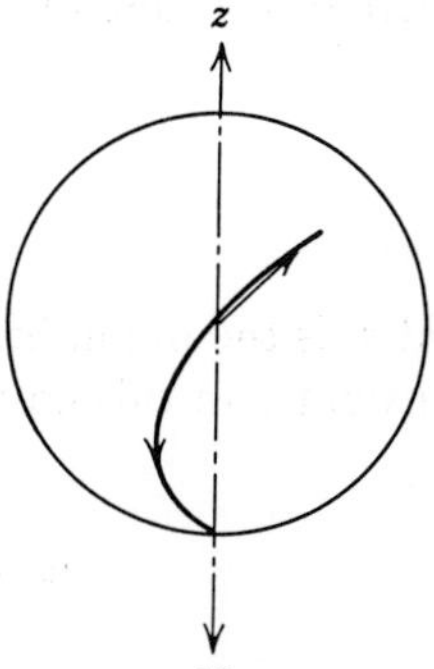

Fig. 16.11. The motion of spin for large damping.

We rewrite (16.38) for each component of the Cartesian coordinates as

$$\begin{aligned}
\frac{dI_x}{dt} &= \omega_0 I_y + \alpha \frac{I_y}{I_s}\frac{dI_z}{dt} - \alpha \frac{I_z}{I_s}\frac{dI_y}{dt}, \\
\frac{dI_y}{dt} &= -\omega_0 I_x + \alpha \frac{I_z}{I_s}\frac{dI_x}{dt} - \alpha \frac{I_x}{I_s}\frac{dI_z}{dt}, \\
\frac{dI_z}{dt} &= \qquad + \alpha \frac{I_x}{I_s}\frac{dI_y}{dt} - \alpha \frac{I_y}{I_s}\frac{dI_x}{dt}.
\end{aligned} \tag{16.43}$$

On solving these equations with respect to dI_x/dt, dI_y/dt, and dI_z/dt, we obtain

$$\begin{aligned}
\frac{dI_x}{dt} &= \frac{\omega_0}{1+\alpha^2} I_y + \frac{\omega_0\alpha}{1+\alpha^2}\frac{I_x I_z}{I_s}, \\
\frac{dI_y}{dt} &= -\frac{\omega_0}{1+\alpha^2} I_x + \frac{\omega_0\alpha}{1+\alpha^2}\frac{I_y I_z}{I_s}, \\
\frac{dI_z}{dt} &= -\frac{\omega_0\alpha}{1+\alpha^2} I_s + \frac{\omega_0\alpha}{1+\alpha^2}\frac{I_z^2}{I_s}.
\end{aligned} \tag{16.44}$$

If we start from (16.35) or (16.36), we have the equations which are obtained by replacing $1 + \alpha^2$ in the denominator of each term of (16.44) by 1. Both solutions coincide when $\alpha^2 \ll 1$. In general cases including $\alpha^2 \geqq 1$, however, we must use (16.44). Solving (16.44), we obtain

$$\begin{aligned}
I_x &= I_s \sin\theta\, e^{i\omega t}, \\
I_y &= I_s \sin\theta\, e^{i\omega t + i(\pi/2)}, \\
I_z &= I_s \cos\theta,
\end{aligned} \tag{16.45}$$

where θ is a function of time and is given by

$$\tan\frac{\theta}{2} = \tan\frac{\theta_0}{2}\, e^{t/\tau}, \tag{16.46}$$

and θ_0 is the initial inclination of the spin axis. The frequency ω and time constant τ in this equation are

$$\omega = \frac{\omega_0}{1+\alpha^2} = \frac{\omega_0}{1+(1/\omega_0\tau_0)^2} \tag{16.47}$$

$$\tau = \tau_0(1+\alpha^2) = \tau_0\left[1+\left(\frac{1}{\omega_0\tau_0}\right)^2\right], \tag{16.48}$$

where ω_0 is the resonance frequency given by (16.42) and τ_0 is

$$\tau_0 = \frac{1}{\alpha\omega_0} = \frac{I_s}{4\pi\lambda\mu_0 H}. \tag{16.49}$$

If $\alpha^2 \ll 1$, we know from (16.47), (16.48), and (16.49) that $|1/\omega| \ll |\tau|$; hence the spin performs a number of precessions before it finally points to the $-z$ direction (Fig. 16.10). If $\alpha^2 \gg 1$, it turns out that $|1/\omega| \gg |\tau|$, and the spin rotates directly toward the $-z$ direction without making many precessions (Fig. 16.11). This switching motion of spins becomes more viscous as the relaxation frequency λ becomes large. The switching time is expected to be large also if the relaxation frequency is too small, because the spin performs too many precessions. The fastest switching is, therefore, attained for a moderate value of the relaxation frequency. In Fig. 16.12 is plotted the relaxation time τ as a function of τ_0 which includes the relaxation frequency λ (cf. 16.49). The figure shows that the relaxation time τ is a minimum when

$$\tau_0 = \frac{1}{\omega_0} \tag{16.50}$$

or

$$\lambda = \frac{\nu I_s}{4\pi\mu_0}. \tag{16.51}$$

This condition is called critical damping. The minimum value of τ is given by

$$\tau_{\min} = \frac{2}{\omega_0} = \frac{2}{\nu H_z}, \tag{16.52}$$

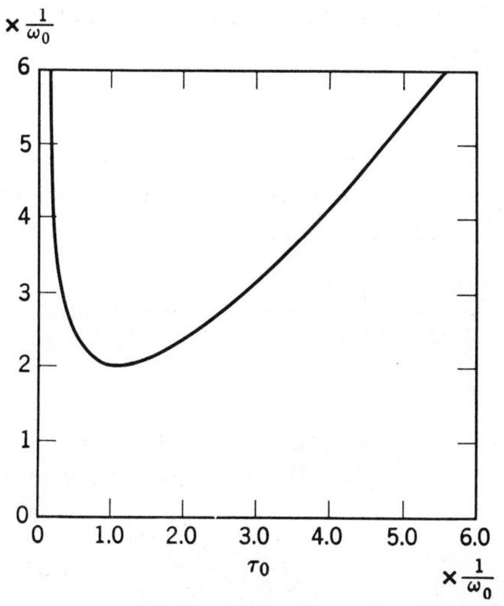

Fig. 16.12. Dependence of the relaxation time for switching of the magnetization on the relaxation frequency.

which can be estimated by assuming a strong magnetic field $H_z = 1.6$ MA/m (= 20,000 Oe) and $g = 2$, as

$$\tau_{\min} = \frac{2}{2 \times 1.105 \times 10^5 \times 1.6 \times 10^6} = 5.7 \times 10^{-12} \text{ sec.} \tag{16.53}$$

The condition of critical damping (16.51) is, for $I_s = 1$ Wb/m^2,

$$\lambda = 1.4 \times 10^{10} \text{ sec}^{-1}. \tag{16.54}$$

The relaxation frequency determined for nickel ferrite from the width of the resonance line is $10^7 \sim 10^8$ sec^{-1} (cf. 16.70), which is insufficient to realize critical damping. One means to attain critical damping is to utilize eddy current damping. Consider a thin metal wire for instance. The demagnetizing field due to the total eddy currents is calculated from (16.3) as

$$H_d = \int_0^{r_0} i(r)\, dr = -\frac{r_0^2}{4\rho}\frac{dI}{dt}. \tag{16.55}$$

On the other hand, it is easily seen in (16.38) that the damping action is equivalent to the presence of the demagnetizing field,

$$H_d = \frac{\alpha}{\nu I_s}\frac{dI}{dt} = \frac{4\pi\mu_0\lambda}{\nu^2 I_s^2}\frac{dI}{dt}. \tag{16.56}$$

On comparing (16.55) and (16.56), we have the relaxation frequency due to eddy currents given by

$$\lambda = \frac{\nu^2 I_s^2 r_0^2}{16\pi\mu_0\rho}. \tag{16.57}$$

Using the values $\nu = -2.21 \times 10^5$ m/A sec, $\mu_0 = 4\pi \times 10^{-7}$, $I_s = 1$ Wb/m², $\rho = 1 \times 10^{-7}$ Ωm, we obtain

$$\lambda = \frac{(2.21)^2 \times 10^{10} \times 1^2}{16 \times 3.14 \times 4 \times 3.14 \times 10^{-7} \times 1 \times 10^{-7}} r_0^2 = 7.73 \times 10^{21} r_0^2. \tag{16.58}$$

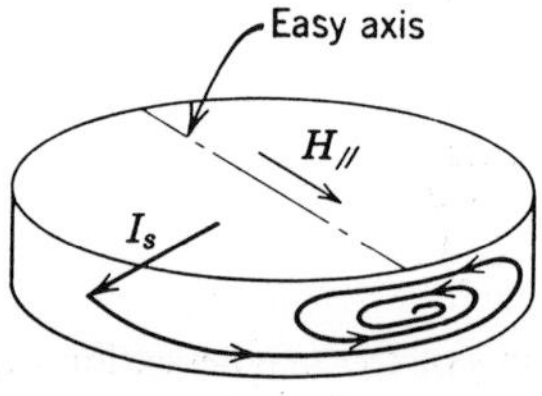

Fig. 16.13. Movement of spin for a thin film switching.

In order to attain the value of λ given by (16.54), the radius of the thin wire, r_0, must be

$$r_0 = \sqrt{\frac{1.4 \times 10^{10}}{7.73 \times 10^{21}}} = 1.35 \times 10^{-6} \text{ m}$$
$$= 1.35\ \mu. \tag{16.59}$$

Metal thin films and wires have been used for the elements of computer memory devices, because of their moderate damping action due to eddy currents.

The minimum relaxation time given by (16.52) depends on the intensity of magnetic field. The order of magnitude of H_z which is commonly used for a memory device is $H_z = 80$ A/m (= 1 Oe). It must be noted, however, that the demagnetizing field will normally help such a switching action. For a thin wire the shape anisotropy field $I_s/2\mu_0$ is added to the external field after more than half of the total flux is reversed (cf. 3.31). For a thin film the effective field is given by $(HI_s/\mu_0)^{1/2}$ (cf. 3.33). The actual mode of spin rotation in this case is schematically shown in Fig. 16.13, where the switching is practically finished within one cycle of the distorted precession motion. The apparent switching time is, therefore, expected to be much faster than the calculation above indicates. If, however, the remaining oscillatory motion has an undesirable effect for the next switching, we should adopt critical damping. Such an oscillation was actually observed by Wolf.[13] For a detailed consideration of spin rotation in a thin film the reader may refer to the calculation made by Kikuchi.[12]

16.4 Various Modes of Ferro-, Ferri-, and Antiferromagnetic Resonance

The fundamental concept of spin resonance has already been described in Chapter 3. There we assumed that all the spins forming the spontaneous magnetization maintain perfect parallelism. We call this the uniform

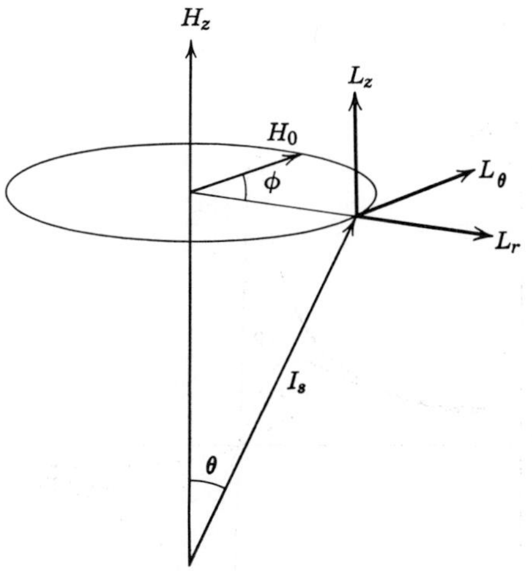

Fig. 16.14. Torque L exerted on magnetization I_s by the rotational field H_0.

mode or the Kittel mode. Starting from (16.39), we have the uniform mode of precession expressed by (16.41), and the angular precession frequency given by (16.42). If a damping force acts on the precession motion, the motion can be maintained only if energy is supplied to the system by an oscillating or a rotating magnetic field. Let H_0 be the amplitude of the rotating field, and ω its angular velocity about the z axis. In order to give rise to a finite couple of force on the magnetization vector, the rotating field vector H_0 must make a finite azimuthal angle ϕ with the magnetization vector I_s (Fig. 16.14). The components of the couple of force acting on the magnetization vector are then expressed, with respect to the cylindrical coordinates, as

$$\begin{aligned} L_r &= -I_s H_0 \cos\theta \sin\phi, \\ L_\theta &= I_s H_0 \cos\theta \cos\phi, \\ L_z &= I_s H_0 \sin\theta \sin\phi. \end{aligned} \tag{16.60}$$

If we apply the static magnetic field H_z parallel to the z axis, the Landau-Lifshitz equation (16.35) becomes, for each component,

$$\begin{aligned} \frac{dI_r}{dt} &= \nu I_s H_0 \cos\theta \sin\phi - 4\pi\lambda\mu_0 H_z \sin\theta\cos\theta, \\ \frac{dI_\theta}{dt} &= -\nu I_s H_0 \cos\theta\cos\phi + \nu I_s H_z \sin\theta, \\ \frac{dI_z}{dt} &= -\nu I_s H_0 \sin\theta\sin\phi + 4\pi\lambda\mu_0 H_z \sin^2\theta. \end{aligned} \tag{16.61}$$

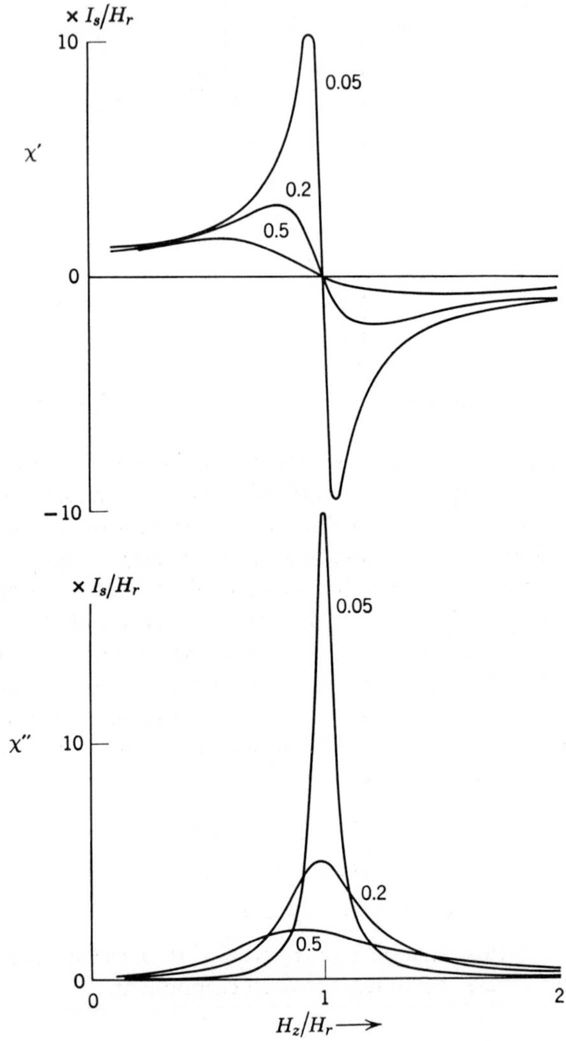

Fig. 16.15. Dependence of the real and imaginary part of the rotational susceptibility on the intensity of the dc field about the resonance field (numerical figures in the graph are the values of α).

In the equilibrium state,

$$\frac{dI_r}{dt} = 0,$$

$$\frac{dI_\theta}{dt} = I_s \omega \sin\theta, \tag{16.62}$$

$$\frac{dI_z}{dt} = 0.$$

On comparing these with equations (16.61), we have

$$-\nu I_s H_0 \sin\phi + 4\pi\lambda\mu_0 H_z \sin\theta = 0,$$

$$-H_0 \cos\theta \cos\phi + H_z \sin\theta = \frac{\omega}{\nu} \sin\theta, \tag{16.63}$$

from which we obtain

$$\sin\theta = \frac{\nu I_s}{4\pi\lambda\mu_0} \frac{H_0}{H_z} \sin\phi \tag{16.64}$$

and

$$\tan\phi = \frac{4\pi\lambda\mu_0}{\nu I_s} \frac{H_z}{H_z - H_r} \tag{16.65}$$

for $\theta \ll \pi$, where H_r is the resonance field,

$$H_r = \frac{\omega}{\nu}. \tag{16.66}$$

The real and imaginary parts of the susceptibility are expressed as

$$\chi' = \frac{I_s}{H_0} \sin\theta \cos\phi,$$

$$\chi'' = \frac{I_s}{H_0} \sin\theta \sin\phi, \tag{16.67}$$

or, from (16.64),

$$\chi' = \frac{\nu I_s^2}{4\pi\lambda\mu_0 H_z} \sin\phi \cos\phi,$$

$$\chi'' = \frac{\nu I_s^2}{4\pi\lambda\mu_0 H_z} \sin^2\phi. \tag{16.68}$$

These susceptibilities are plotted against the dc magnetic field H_z in Fig. 16.15 for various values of α (cf. 16.37). The width of the absorption curve is larger for large values of α or λ. The width of the absorption

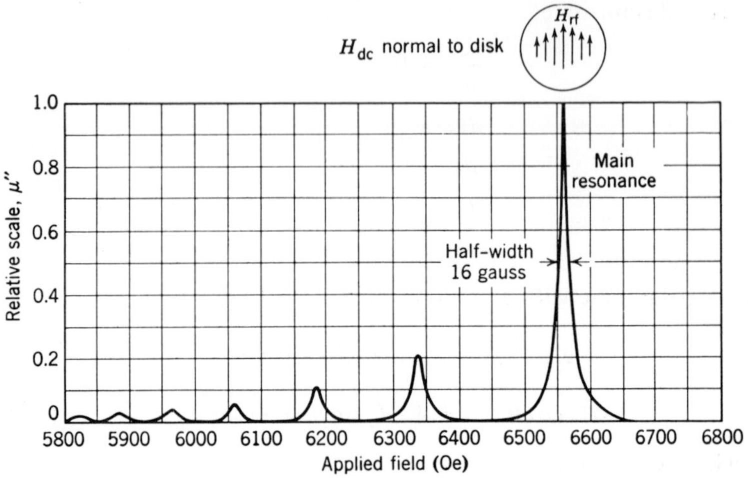

Fig. 16.16. Multiple absorption peaks in a (100) disk of Mn ferrite. The rf field variation across the disk is indicated (Walker[16]).

curve at half the height of the maximum value or a half-value width can be calculated by putting $\phi = 45°$, which makes the value of χ'' in (16.68) half of its maximum value, into (16.65) as

$$\Delta H = 2(H - H_r) = \frac{8\pi\lambda\mu_0}{\nu I_s + 4\pi\lambda\mu_0} H_r. \tag{16.69}$$

For Mn-Zn ferrite, $I_s = 0.25$ Wb/m² (≈ 200 gauss), $H_r = 2.55 \times 10^5$ A/m ($= 3200$ Oe), $\Delta H = 5.59 \times 10^3$ ($= 70$ Oe), and $g = 2$, and we obtain

$$\lambda = 3.88 \times 10^7 \text{ sec}^{-1}. \tag{16.70}$$

The actual mechanisms of the line width will be discussed after the various modes of ferromagnetic resonance have been presented.

It was observed by White, Solt, and Mercereau[14] and Dillon[15] that a number of peaks are excited, as shown in Fig. 16.16, when the specimen is placed at the point where the rf field is sufficiently inhomogeneous to vary over the specimen. This fact was attributed by Walker[16] to the excitation of the non-uniform modes of precession, shown in Fig. 16.17. In this figure the components of spins projected in the z plane at a certain instant are shown for various cross sections of the spherical specimen. All the spins precess together in the same sense. This particular configuration represents the mode denoted by (4, 3, 0), while the uniform mode, or Kittel mode, corresponds to (1, 1, 0). These modes are often referred to as Walker modes.

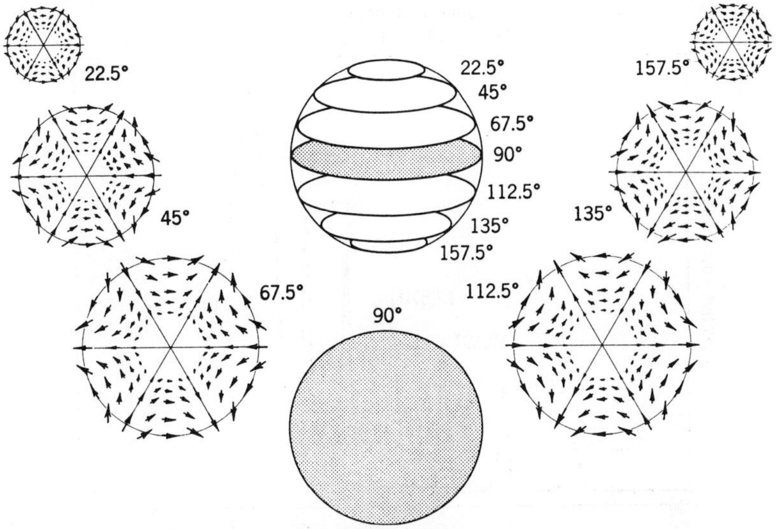

Fig. 16.17. The configuration of transverse magnetization in each cross section of a spherical specimen corresponding to the (4, 3, 0) mode (Walker[16]).

As for usual ac magnetic fields, the precession motion caused by the rf field is limited to a skin depth for conductors. Actually Kip and Arnold,[17] who first measured the magnetic anisotropy of iron single crystals by means of ferromagnetic resonance, found that the line shape is very much improved by electropolishing the surface. This means that the precession motion is mostly restricted to a very thin surface layer. In insulators, such as ferrites, the displacement current plays an important role in giving rise to an electromagnetic mode. An example was cited in Section 16.2 in connection with dimensional resonance.

Rado and Weertman[18] pointed out that the wavelength of such an electromagnetic mode is so small for a rf field that the adjacent spins make a fairly large angle and hence store a considerable amount of exchange energy. In other words, an exchange interaction between the adjacent spins supplies an additional stiffness to the spin waves.

Excitation of spin waves was first theoretically investigated by Kittel[19] and actually observed by Seavey and Tannenwald[20] for thin metallic films. In this experiment a static magnetic field was applied normal to the plane of the film, and the rf magnetic field was applied parallel to the plane. Figure 16.18 shows the absorption lines thus observed; each peak can be interpreted as corresponding to the excitation of various modes of standing waves. Here the spin is assumed to be pinned at the film surface, because of the surface anisotropy (cf. Chapter 17) or some other reasons.[21,22]

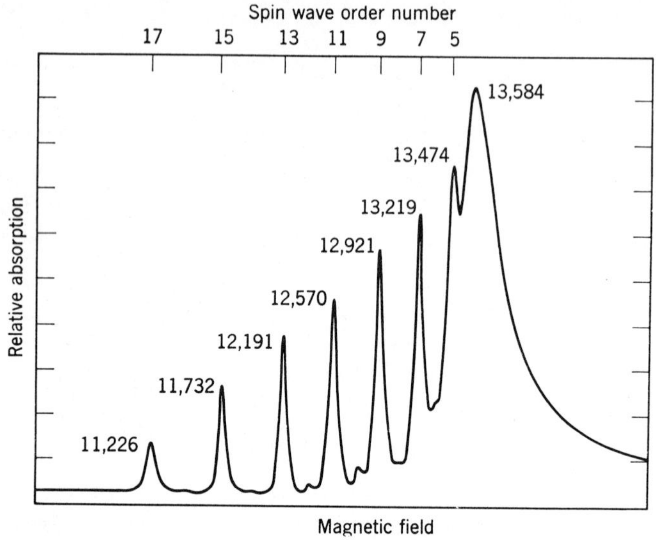

Fig. 16.18. Multiple absorption peaks due to spin wave resonance in 3900 Å film of Permalloy. Numerical values in the figure are the resonance fields in oersteds or in $(10^3/4\pi)$ A/m.

Line width mechanisms have been discussed by many investigators. The most comprehensive channel of the dissipation of the energy from the precessing spin system is that through the eddy currents induced by the precessing magnetization. In ferrites the conductivity is mainly caused by a hopping motion of electrons between Fe^{2+} and Fe^{3+} ions in the octahedral sites, as first pointed out by Verwey.[23] Besides the effect of eddy currents, the hopping motion of electrons causes another route of energy dissipation. A regular arrangement of Fe^{2+} and Fe^{3+} ions in the octahedral sites causes a large uniaxial anisotropy, as actually observed for magnetite at low temperature (Section 17.3). This anisotropy forces the magnetization to rotate parallel to the direction of easy magnetization, and, for the same reason, rotation of magnetization may cause a change in the arrangement of two kinds of ions such as to rotate the easy direction toward the direction of magnetization. Thus the precession motion of the magnetization is expected to cause a hopping of electrons, which is accompanied by a loss of energy. The energy dissipation should be a maximum when the relaxation time of electron transfer matches the precession frequency. Actually, a maximum in the line width was observed by Yager, Galt, and Merritt[24] for a single crystal of Ni-Fe ferrite magnetized parallel to the $\langle 111 \rangle$ direction during the process of heating from low temperature.

The mechanism discussed above is not valid for highly insulating materials. Another possibility of having a finite line width would be—one may naturally suppose—the effect of distributed anisotropy or inhomogeneity of the material. Since a magnetic anisotropy affects the intensity of resonance magnetic field, as discussed in Chapter 7, the resonance field is expected to be dispersed for polycrystalline materials, where the orientation of the easy magnetization differs from grain to grain. Grain boundaries, non-magnetic inclusions, voids, and other irregularities may induce demagnetizing fields the intensity and direction of which differ from place to place; hence they could also be the origin of the dispersion of the resonance magnetic field. It has, however, been pointed out by Geschwind and Clogston[25] that the broadening of line width thus induced would be narrowed by a strong magnetic dipolar interaction between the local magnetizations. This is to say that such an interaction may smooth the fluctuations in the amplitude of the precession motion over the specimen and thus make the precessions take place in unison. Actually these investigators examined this effect by using a specimen of hemispherical shape in which the local demagnetizing field differs by more than 4.54×10^4 A/m (= 570 Oe), and found that the actual line-width is only 3.99×10^3 A/m (= 50 Oe). Since the spherical specimen showed almost the same line width, the origin of the line width should be sought at another source. This dipolar narrowing cannot be expected, as they pointed out, for a very thin rod or disk, for which the line width should be comparable to the inhomogeneity of the material.

A possible and plausible mechanism of the energy dissipation from the resonant system is the scattering of spin waves, as first treated by Clogston, Suhl, Walker, and Anderson.[26] If there are some fluctuations in the local resonant field over the specimen, the uniform mode of precession, which corresponds to the wave number $k = 0$, will naturally induce spin waves with finite k. These excited spin waves will relax with the lattice at a relatively fast rate during the propagation and thus transfer energy to the lattice. The real origin of the fluctuation was first thought to be pseudo-dipolar interaction, but later on Callen and Pittelli[27] considered it to be one-ion anisotropy. Then the scattering probability of spin waves would be expected to be dependent on the crystallographic direction of the applied dc field. Actually Schnitzler, Folen, and Rado[28] observed the dependence of line width on the crystallographic direction of magnetization. Inhomogeneity of the anisotropy and the demagnetizing fields as discussed above could also be the cause of excitation of spin waves.

Next we discuss antiferromagnetic resonance, which was first treated theoretically by Nagamiya[29] and independently by Kittel.[30] Neglecting damping terms, we have the equations of motion for the two sublattices:

$$\frac{d\boldsymbol{I}_A}{dt} = -\nu[\boldsymbol{I}_A, \boldsymbol{H} + \boldsymbol{H}_{aA} + \boldsymbol{H}_{mA}],$$
$$\frac{d\boldsymbol{I}_B}{dt} = -\nu[\boldsymbol{I}_B, \boldsymbol{H} + \boldsymbol{H}_{aB} + \boldsymbol{H}_{mB}], \tag{16.71}$$

where $\boldsymbol{H}$ is the applied field, $\boldsymbol{H}_{aA}$ and $\boldsymbol{H}_{aB}$ the anisotropy fields acting on the A and B sublattices, respectively, and $\boldsymbol{H}_{mA}$ and $\boldsymbol{H}_{mB}$ the molecular fields which are

$$\boldsymbol{H}_{mA} = w_{AA}\boldsymbol{I}_A + w_{AB}\boldsymbol{I}_B$$
$$\boldsymbol{H}_{mB} = w_{BA}\boldsymbol{I}_A + w_{BB}\boldsymbol{I}_B \tag{16.72}$$

as shown in (5.1) and (5.2). For simplicity, we assume the uniaxial anisotropy which has its easy axis parallel to the z axis, so that the intensity of the anisotropy field is given by H_a and $-H_a$ for A and B sublattices respectively. We also assume that the field H_z is applied externally parallel to the z axis. Then each of the components of (16.71) becomes

$$\frac{dI_{Ax}}{dt} = -\nu I_{Ay}(H_z + H_m + H_a) - \nu I_{By}H_m,$$
$$\frac{dI_{Ay}}{dt} = \nu I_{Bx}H_m + \nu I_{Ax}(H_z + H_m + H_a),$$
$$\frac{dI_{Az}}{dt} = -\nu w_{AB}(I_{Ax}I_{By} - I_{Ay}I_{Bx}),$$
$$\frac{dI_{Bx}}{dt} = -\nu I_{By}(H_z - H_m - H_a) + \nu I_{Ay}H_m, \tag{16.73}$$
$$\frac{dI_{By}}{dt} = -\nu I_{Ax}H_m + \nu I_{Bx}(H_z - H_m - H_a),$$
$$\frac{dI_{Bz}}{dt} = -\nu w_{AB}(I_{Bx}I_{Ay} - I_{By}I_{Ax}),$$

where H_m is the absolute intensity of the molecular field acting on one sublattice from the other sublattice; that is,

$$H_m = |w_{AB}I_A| = |w_{AB}I_B|. \tag{16.74}$$

Since the two sublattice magnetizations are thus under the action of couples of forces of different intensities, $\boldsymbol{I}_A$ and $\boldsymbol{I}_B$ are expected to tilt from the z axis with different angles θ_A and θ_B (Fig. 16.19), in order that both magnetizations may precess with the same frequency. If $\theta_A > \theta_B$, the exchange force, the anisotropy field, and the applied field are all acting in the same direction on $\boldsymbol{I}_A$, whereas the anisotropy field acts on $\boldsymbol{I}_B$ oppositely to the exchange and applied fields. Then the point of the vector $\boldsymbol{I}_A$ moves faster than that of the vector $\boldsymbol{I}_B$. Since the point of $\boldsymbol{I}_A$ traces out a larger circumference than does that of $\boldsymbol{I}_B$, the angular velocities of $\boldsymbol{I}_A$ and $\boldsymbol{I}_B$ can be expected to be equal under a certain combination of θ_A and θ_B. Thus we can assume that

$$\begin{aligned} I_{Ax} &= I \sin\theta_A \cos\omega t, \\ I_{Ay} &= -I \sin\theta_A \sin\omega t, \\ I_{Az} &= I\cos\theta_A, \\ I_{Bx} &= I \sin\theta_B \cos\omega t, \\ I_{By} &= -I \sin\theta_B \sin\omega t, \\ I_{Bz} &= -I\cos\theta_B, \end{aligned} \qquad (16.75)$$

where I is the absolute intensity of the sublattice magnetization. On putting these equations into (16.73), we have the relations

$$\begin{aligned} &\omega \sin\theta_A \\ &\quad = -\nu[\sin\theta_A(H_z + H_m + H_a) + \sin\theta_B H_m], \\ &\omega \sin\theta_B \\ &\quad = -\nu[\sin\theta_B(H_z + H_m + H_a) - \sin\theta_A H_m], \end{aligned} \qquad (16.76)$$

from which we obtain

$$\omega = -\nu[H_z \pm \sqrt{H_a(H_a + 2H_m)}] \qquad (16.77)$$

and

$$\frac{\sin\theta_A}{\sin\theta_B} = 1 + \frac{H_a}{H_m} \pm \sqrt{2\frac{H_a}{H_m}}. \qquad (16.78)$$

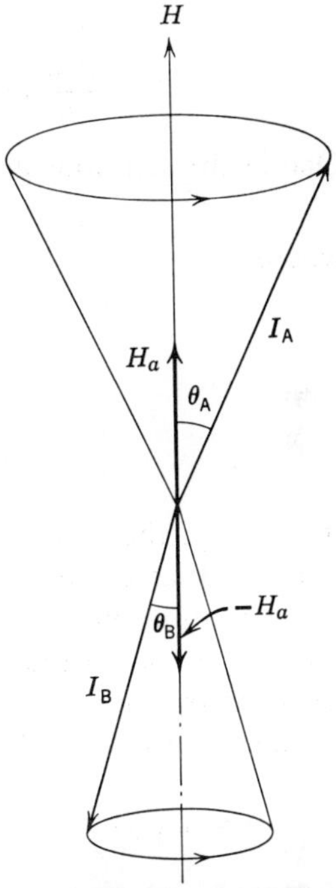

Fig. 16.19. Antiferromagnetic resonance.

It is interesting to note that resonance can occur even when $H_a = 0$, as is known from (16.77). However, it turns out that $\theta_A = \theta_B$, so that such a mode of resonance cannot be excited by the external rf field.

The theory was further extended by Yosida[31] to the more general type of magnetic anisotropy. He treated the resonance of $CuCl_2 \cdot 2H_2O$, which has a fairly small exchange interaction that allows the observation of resonance in the microwave region, and obtained beautiful agreement with experiment.

Magnetic resonance in ferrimagnetics[32,33] is essentially the same as that in ferromagnetics, as long as the molecular field is sufficiently large compared to the applied or to the anisotropy field. The resonance mode, in which the two sublattice magnetizations are flexed, has a very large resonance frequency, lying in the infrared region. Since the g factors are generally different between the magnetic ions in the A and B sublattices, the equation of motion for the two sublattice magnetizations can be written as

$$\begin{aligned} \frac{d\boldsymbol{I}_{\mathrm{A}}}{dt} &= -\nu_{\mathrm{A}}[\boldsymbol{I}_{\mathrm{A}}, (\boldsymbol{H} + \boldsymbol{H}_{a\mathrm{A}} + w_{\mathrm{AB}}\boldsymbol{I}_{\mathrm{B}})], \\ \frac{d\boldsymbol{I}_{\mathrm{B}}}{dt} &= -\nu_{\mathrm{B}}[\boldsymbol{I}_{\mathrm{B}}, (\boldsymbol{H} + \boldsymbol{H}_{a\mathrm{B}} + w_{\mathrm{AB}}\boldsymbol{I}_{\mathrm{A}})]. \end{aligned} \tag{16.79}$$

One of the solutions gives the frequency

$$\omega = \nu_{\mathrm{eff}}(H + H_{a\,\mathrm{eff}}), \tag{16.80}$$

where

$$\nu_{\mathrm{eff}} = -\frac{I_{\mathrm{A}} + I_{\mathrm{B}}}{P_{\mathrm{A}} + P_{\mathrm{B}}} \tag{16.81}$$

and

$$H_{a\,\mathrm{eff}} = \frac{(\boldsymbol{H}_{a\mathrm{A}}, \boldsymbol{I}_{\mathrm{A}}) + (\boldsymbol{H}_{a\mathrm{B}}, \boldsymbol{I}_{\mathrm{B}})}{I_{\mathrm{A}} + I_{\mathrm{B}}}. \tag{16.82}$$

If $I_{\mathrm{A}} > I_{\mathrm{B}}$, there is a normal resonance frequency, which is almost the same as that of ferromagnetic resonance. For a ferrimagnetic with a compensation point, ν_{eff} and, accordingly, g_{eff} go down to $-\infty$ and then up to $+\infty$, when the angular momenta of the two sublattices are compensated by one another.

The other solution has the frequency

$$\omega = w_{\mathrm{AB}}(\nu_{\mathrm{A}} I_{\mathrm{A}} + \nu_{\mathrm{B}} I_{\mathrm{B}}) - \frac{\nu_{\mathrm{A}} + \nu_{\mathrm{B}}}{2} H - \frac{\nu_{\mathrm{A}} H_{a\mathrm{A}} + \nu_{\mathrm{B}} H_{a\mathrm{B}}}{2}, \tag{16.83}$$

if $I_{\mathrm{A}} > I_{\mathrm{B}}$, which is normally in the submillimeter range. Near the compensation point the frequency becomes

$$\omega = \nu\{H \pm \sqrt{H_m[(\omega/\nu_{\mathrm{A}} - \omega/\nu_{\mathrm{B}}) + (H_{a\mathrm{A}} + H_{a\mathrm{B}})]}\} \tag{16.84}$$

which falls in the microwave region. This possibility was first pointed out by Wangsness[34] and first observed by McGuire.[35] Recent work on resonance absorption for a single crystal of gadolinium iron garnet done by Geschwind and Walker[36] confirmed a beautiful agreement between theory and experiment. Further details on magnetic resonances are available to the reader in a number of good reviews.[37–40]

16.5 Equation of Motion for Domain Walls

It was first pointed out by Döring[41] in 1948 that the movement of the domain wall exhibits an inertia, despite the lack of any mass displacement. In view of this property, the equation of motion for the 180° domain wall is

$$m \frac{d^2 s}{dt^2} + \beta \frac{ds}{dt} + \alpha s = 2I_s H, \tag{16.85}$$

where m is the mass of the wall per unit area, β the damping coefficient, α the restoring coefficient as expressed in (13.18). The term $2I_sH$ on the right side of the equation represents the pressure acting on the 180° wall and should be replaced by $\sqrt{2}\, I_sH$ for 90° walls.

The mass of a domain wall has its origin in the angular momenta of the spins forming the wall. As already described in Chapter 9, spins in the wall normally are confined in the xy plane as shown in Fig. 16.20, if the wall is standing still. If we apply a magnetic field in the x direction, this field pushes the wall upwards and forces the spins in the wall to rotate clockwise. This force, however, induces the precession of each spin, which

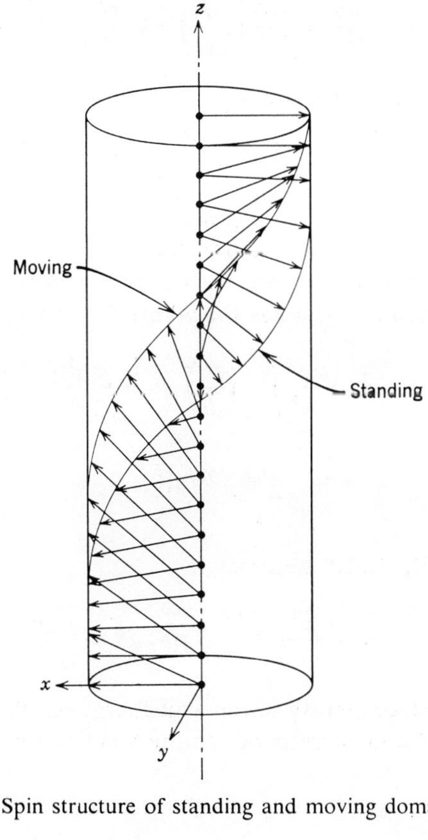

Fig. 16.20. Spin structure of standing and moving domain walls.

results in a rotation of spins out of the xy plane. This rotation of spins in the wall results in the appearance of magnetic free poles which give rise to the demagnetizing field,

$$H_z = -\frac{I_z}{\mu_0}, \tag{16.86}$$

where I_z is the z component of magnetization induced in the wall. This demagnetizing field acts on each spin so as to induce precession motion in the xy plane, which results in a displacement of the wall in the z direction. The rotational velocity of this precession motion is, from (16.42),

$$\frac{d\theta}{dt} = \nu H_z = -\frac{\nu I_z}{\mu_0}. \tag{16.87}$$

On the other hand, this angular velocity can be related to the translational velocity of the wall, v, by

$$\frac{d\theta}{dt} = \left(\frac{\partial\theta}{\partial z}\right)\left(\frac{dz}{dt}\right) = \left(\frac{\partial\theta}{\partial z}\right)v. \tag{16.88}$$

Comparing these equations, we find

$$I_z = -\frac{\mu_0}{\nu}\frac{\partial\theta}{\partial z}v. \tag{16.89}$$

That is, the wall must have a z component of magnetization in order to proceed with a finite velocity. And, if the wall has such a z component of magnetization, it should proceed with a finite velocity even without the external field. Then the wall has the additional energy

$$\gamma_v = -\frac{1}{2}\int_{-\infty}^{+\infty} I_z H_z \, dz = \frac{\mu_0 v^2}{2\nu^2}\int_{-\infty}^{\infty}\left(\frac{d\theta}{dz}\right)^2 dz = \frac{\mu_0 v^2}{2\nu^2}\int_{-\pi/2}^{\pi/2}\left(\frac{d\theta}{dz}\right) d\theta. \tag{16.90}$$

Since

$$\frac{d\theta}{dz} = \sqrt{\frac{g(\theta)}{A}}, \tag{16.91}$$

as shown in (9.27), (16.90) becomes

$$\gamma_v = \frac{\mu_0 v^2}{2\nu^2\sqrt{A}}\int_{-\pi/2}^{\pi/2}\sqrt{g(\theta)}\, d\theta = \frac{\mu_0 \gamma v^2}{4\nu^2 A}, \tag{16.92}$$

where γ is the surface density of the wall energy (cf. 9.29). This energy is proportional to v^2 and corresponds to the kinetic energy of the wall. We can express it as

$$\gamma_v = \tfrac{1}{2}mv^2, \tag{16.93}$$

where m is the virtual mass of the wall per unit area and is given, through comparison of (16.93) and (16.92), by

$$m = \frac{\mu_0 \gamma}{2\nu^2 A}. \tag{16.94}$$

For a 180° wall in iron, with $\gamma = 1.6 \times 10^{-3}$, $A = 1.49 \times 10^{-11}$, $\nu = 2.21 \times 10^5$, and $\mu_0 = 4\pi \times 10^{-7}$, we have

$$\begin{aligned} m &= \frac{(4\pi \times 10^{-7}) \times (1.6 \times 10^{-3})}{2 \times (2.21 \times 10^5)^2 \times (1.49 \times 10^{-11})} \\ &= 1.38 \times 10^{-9}\ \text{kg/m}^2\ (= 1.38 \times 10^{-10}\ \text{g/cm}^2). \end{aligned} \tag{16.95}$$

Next we consider the damping term $\beta\,(ds/dt)$ in the equation of motion (16.85). If we neglect the first and third terms, (16.85) becomes

$$\frac{ds}{dt} = \frac{2I_s}{\beta} H. \tag{16.96}$$

For the eddy current damping we can compare (16.96) and (16.16); we obtain

$$\beta = \frac{4I_s^2 R \log (r_0/R)}{\rho}. \tag{16.97}$$

For iron, with $I_s = 2.16$, $\rho = 1 \times 10^{-7}$, and $\log (r_0/R) \approx 1$,

$$\beta = \frac{4 \times 2.16^2}{1 \times 10^{-7}} R \approx 1.9 \times 10^8 R. \tag{16.98}$$

If we assume that the radius R is as small as $R = 10^{-6}\ (= 1\ \mu)$, $\beta = 1.6 \times 10^2$.

For highly insulating materials such as ferrites, the origin of the wall damping should be the same as that acting on the precession motion of the magnetization. Formally it can be expressed as shown by Kittel[10] in terms of the relaxation frequency λ which appears in the Landau-Lifshitz equation (16.35). Since the motion of the spins in the wall is mainly determined by the z component of the field H_z, because of its small relaxation, (16.35) gives

$$\frac{dI_z}{dt} = 4\pi\lambda\mu_0 H_z. \tag{16.99}$$

Thus the internal magnetic field does, in one second, the work

$$H_z \frac{dI_z}{dt} = 4\pi\lambda\mu_0 H_z^2, \tag{16.100}$$

which is dissipated as heat. The power loss per unit area of the wall is then calculated to be

$$P_w = 4\pi\lambda\mu_0 \int_{-\infty}^{\infty} H_z^{\,2}\, dz$$
$$= \frac{4\pi\,\mu_0 \lambda v^2}{\nu^2} \int_{-\infty}^{\infty} \left(\frac{d\theta}{dz}\right)^2 dz$$
$$= \frac{2\pi\,\mu_0 \lambda \gamma v^2}{\nu^2 A}. \qquad (16.101)$$

On the other hand, the external magnetic field supplies the power to the traveling wall;

$$P_w = 2I_s Hv$$
$$= \beta v^2. \qquad \text{(cf. 16.96)} \qquad (16.102)$$

On comparing (16.101) and (16.102), we have

$$\beta = \frac{2\pi\mu_0\lambda\gamma}{\nu^2 A}. \qquad (16.103)$$

For nickel ferrite with $\lambda = 3.9 \times 10^7\ \text{sec}^{-1}$, $\gamma = 4.75 \times 10^{-4}$, $A = 9.0 \times 10^{-12}$, $\nu = 2.21 \times 10^5$, and $\mu_0 = 4\pi \times 10^{-7}$,

$$\beta = \frac{2\pi(4\pi \times 10^{-7}) \times (3.9 \times 10^7) \times (4.75 \times 10^{-4})}{(2.21 \times 10^5)^2 \times (9.0 \times 10^{-12})} = 0.334. \qquad (16.104)$$

This value is fairly small compared to that obtained from eddy current damping. Galt[42] measured the velocity of the domain wall for single crystals of magnetite and nickel ferrite, which have a picture-frame shape as shown in Fig. 16.21, and confirmed the validity of (16.14). He determined that $c = 0.24\ \text{m}^2/\text{sec A}$ (= 1900 cm/sec Oe) for magnetite and $c = 2.5\ \text{m}^2/\text{sec A}$ (= 20,000 cm/sec Oe) for nickel ferrite. On comparing (16.14) and (16.96), we have

$$c = \frac{2I_s}{\beta}. \qquad (16.105)$$

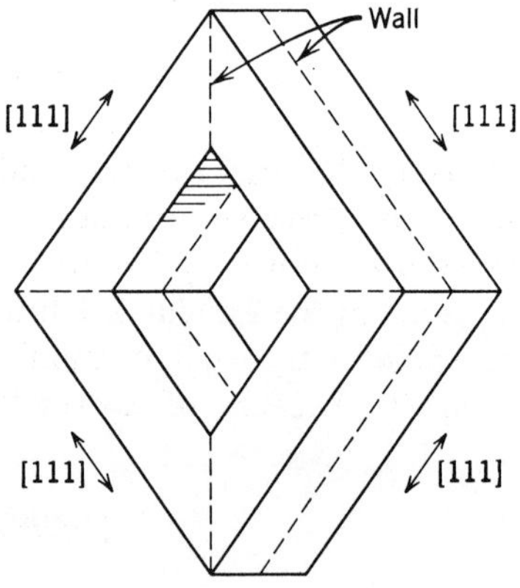

Fig. 16.21. Ferrite single crystal in a picture-frame shape (Galt[42]).

Using this relation, we can determine $\beta = 4.48$ for magnetite and $\beta = 0.26$ for nickel ferrite. The latter value is of the same order of magnitude as that given by (16.104). The value of β for magnetite is still as large as 4.06 even if the effect of eddy current is subtracted from the observed value. The reason for such a large value of β is considered to be the hopping motion of electrons between Fe^{2+} and Fe^{3+} ions in the 16d sites.

The third term in (16.85) signifies the restoring force acting on the wall. Several examples are given in Section 13.2. If we neglect the second and the last terms, (16.85) becomes

$$m\frac{d^2s}{dt^2} = -\alpha s. \tag{16.106}$$

On solving this differential equation, we have

$$s = s_0 \sin \omega_0 t, \tag{16.107}$$

where

$$\omega_0 = \sqrt{\frac{\alpha}{m}}. \tag{16.108}$$

This means that the wall will vibrate about the equilibrium position. The factor α is related to the permeability μ by (13.29) and is expressed as

$$\alpha = \frac{4I_s^{\,2}}{3\mu} S \qquad (\mu \approx \chi_a). \tag{16.109}$$

The mass m is given by (16.94), where γ is expressed, from the comparison of (9.33) and (9.34), as

$$\gamma = 4\pi\frac{A}{\delta}, \tag{16.110}$$

so that it becomes

$$m = \frac{2\pi\mu_0}{\nu^2\delta}. \tag{16.111}$$

On putting (16.109) and (16.111) into (16.108), we obtain

$$\omega_0\sqrt{\bar{\mu}} = 0.46\sqrt{S\delta}\,\frac{\nu I_s}{\mu_0}. \tag{16.112}$$

This relation is similar to the relation (16.31) derived for rotation magnetization except for the factor $\sqrt{S\delta}$. Since S is the total area and δ is the thickness, the quantity $S\delta$ signifies the fractional volume of the wall included in a unit volume of the specimen. For instance, if the spacing of the walls is 0.1 mm, it turns out that $S\delta \approx 10^{-4}$ for $\delta = 10^{-5}$ mm; hence the critical frequency is expected to be lower by the factor 10^2 than in the case of rotation magnetization. The reason is that the angular velocity of spins in the wall is much larger than that of rotation magnetization for a given amplitude of magnetization. In order that the magnetization will be changed by I_s per second, the rotation magnetization requires the angular velocity $\pi/2$ rad/sec, while the wall displacement requires $\pi/2S\delta$ rad/sec. For the example mentioned above, the velocity is about 10^4 times larger for the wall displacement than for the rotation magnetization.

By the same reasoning, the power loss is also considered to be larger for the displacement of walls than for the rotation magnetization. Since the power loss in the wall is proportional to ω^2 or $(S\delta)^{-2}$, the total loss is expected to be proportional to $(S\delta)^{-1}$. For a given amplitude of magnetic field, therefore, the displacement of the wall is more rapidly damped than is the rotation magnetization. Thus the rotation magnetization becomes more important in the high frequency range. This situation, however, depends on the value of $S\delta$. If this value could be increased to approach 1, the wall displacement would survive to the high frequency region. This means that the specimen would be filled with domain walls, so the wall displacement would become practically equivalent to the incoherent rotation magnetization. On the contrary, if $S\delta$ is very small, the difference between the two mechanisms would be remarkable. It was observed by Galt[42] that the relative permeability of a frame-shaped single crystal of magnetite drops from 5000 to 1000 in the region of 1 kc to 10 kc. One of the reasons for this is its high original value, but there is also no doubt that it is partly because $S\delta$ is as small as 10^{-6}, since this specimen contains only one domain wall.

Problems

16.1. Suppose that a 180° wall is traveling with velocity v from the top surface to the bottom surface of an infinitely wide plate of thickness d which is made from a ferromagnetic metal with resistivity ρ and saturation magnetization I_s. Calculate the eddy current loss per unit area of the plate as a function of the position of the wall.

16.2. Find the time required to rotate the magnetization by applying the magnetic field $H_z = -10^2$ A/m for a spherical specimen with saturation magnetization $I_s = 1$ Wb/m^2 and relaxation frequency $\lambda = 1 \times 10^8$ sec^{-1}, from $\theta = 60°$ to $\theta = 120°$, where θ is the angle between the z axis and the direction of magnetization. Calculate also the number of precession rotations which occur during this period of time. Assume that $g = 2$.

16.3. Calculate the line width for a uniform mode of precession observed with microwaves of wavelength 3 cm for a spherical specimen with $\lambda = 10^8$ sec^{-1}, $I_s = 1$ Wb/m^2, and $g = 2$.

16.4. How far does the 180° wall, which is initially traveling with the velocity 10 m/sec, move after the magnetic field is switched off? Assume that the relaxation frequency is 1×10^8 sec^{-1}.

References

16.1. H. J. Williams, W. Shockley, and C. Kittel: *Phys. Rev.* **80,** 1090 (1950).
16.2. H. P. J. Wijn and H. van der Heide: *Rev. Mod. Phys.* **25,** 98 (1953).
16.3. F. G. Brockman, P. H. Dowling, and W. G. Steneck: *Phys, Rev.* **77,** 85 (1950).

16.4. J. L. Snoek: *Physica* **14,** 207 (1948).
16.5. G. H. Jonker, H. P. J. Wijn, and P. B. Brawn: *Philips Tech. Rev.* **18,** 145 (1956–57).
16.6. J. Smit and H. P. J. Wijn: G.32 p. 279.
16.7. A. L. Stuijts and H. P. J. Wijn: *J. Appl. Phys.* **29,** 468 (1958).
16.8. G. T. Rado: *Rev. Mod. Phys.* **25,** 81 (1953).
16.9. L. Landau and E. Liftshitz: *Phys. Z. Sowjetunion* **8,** 153 (1935); E. Lifshitz: *J. Phys. USSR* **8,** 337 (1944).
16.10. C. Kittel: *Phys, Rev.* **80,** 918 (1950).
16.11. T. L. Gilbert and J. M. Kelley: G.20, p. 253; T. L. Gilbert: *Phys. Rev.* **100,** 1243 (1955).
16.12. R. Kikuchi: *J. Appl. Phys.* **27,** 1352 (1956).
16.13. P. Wolf: *Z. Physik.* **160,** 310 (1960); *J. Appl. Phys.* **32,** 95S (1961); W. Dietrich, W. E. Proebster, and P. Wolf: *IBM J. Res. Develop.* **4,** 189 (1960.)
16.14. R. L. White, I. H. Solt, and J. E. Mercereau: *Bull. Am. Phys. Soc. Ser.* II **1,** 12 (1956); R. L. White and I. H. Solt: *Phys. Res.* **104,** 56 (1956).
16.15. J. F. Dillon, Jr.: *Bull. Am. Phys. Soc. Ser. II* **1,** 125 (1956).
16.16. L. R. Walker: *Phys. Rev.* **105,** 390 (1957); *J. Appl. Phys.* **29,** 318 (1958).
16.17. A. F. Kip and R. D. Arnold: *Phys. Rev.* **75,** 1556 (1949).
16.18. G. T. Rado and J. R. Weertman: *Phys. Rev.* **94,** 1386 (1954); *Phys. Chem. Solids* **11,** 315 (1959).
16.19. C. Kittel: *Phys. Rev.* **110,** 1295 (1958).
16.20. M. H. Seavey, Jr., and P. E. Tannenwald: *Phys. Rev. Letters* **1,** 168 (1958); *J. Appl. Phys.* **30,** 227S (1959).
16.21. P. E. Wigen, C. F. Kooi, M. R. Shanabarger, U. K. Cumming, and M. E. Baldwin: *J. Appl. Phys.* **34,** 1137 (1963).
16.22. E. Hirota: *Natl. Tech. Rept.* (*Matsushita Elec. Ind. Co., Osaka*), **9,** 1 (1963).
16.23. E. J. W. Verwey: *Nature* **144,** 327 (1939).
16.24. W. A. Yager, J. K. Galt, and F. R. Merritt: *Phys. Rev.* **99,** 1203 (1955); J. K. Galt: *Proc. I.E.E.* (*London*) **104B,** 189 (1957).
16.25. S. Geschwind and A. M. Clogston: *Phys. Rev.* **108,** 49 (1957).
16.26. A. M. Clogston, H. Suhl, L. R. Walker, and P. W. Anderson: *Phys. Rev.* **101,** 903 (1956); *Phys. Chem. Solids* **1,** 129 (1956).
16.27. H. B. Callen and E. Pittelli: *Phys. Rev.* **119,** 1523 (1960).
16.28. A. Schnitzler, V. Folen, and G. Rado: *J. Appl. Phys.* **31,** 348S (1960).
16.29. T. Nagamiya: *Progr. Theoret. Phys.* (*Kyoto*) **6,** 342 (1951).
16.30. C. Kittel: *Phys. Rev.* **82,** 565 (1951).
16.31. K. Yosida: *Progr. Theoret. Phys.* (*Kyoto*) **7,** 25, 425 (1952).
16.32. N. Tsuya: *Progr. Theoret. Phys.* (*Kyoto*) **7,** 263 (1952).
16.33. R. K. Wangsness: *Phys. Rev.* **93,** 68 (1954).
16.34. R. K. Wangsness: *Phys. Rev.* **97,** 831 (1955).
16.35. T. R. McGuire: *Phys. Rev.* **97,** 831 (1955).
16.36. S. Geschwind and L. R. Walker: *J. Appl. Phys.* **30,** 163S (1959).
16.37. N. Bloembergen: *Proc. IRE* **44,** 1259 (1956).
16.38. G. T. Rado: *J. Appl. Phys.* **32,** 129S (1961).
16.39. J. Smit and H. P. J. Wijn: G.32, p. 87.
16.40. T. Nagamiya, K. Yosida, and R. Kubo: *Advan. Phys.* **4,** 1 (1955).
16.41. W. Döring: *Z. Naturforsch.* **3a,** 373 (1948).
16.42. J. K. Galt: *Phys. Rev.* **85,** 664 (1952); *Rev. Mod. Phys.* **25,** 93 (1953); *Bell System Tech. J.* **33,** 1023 (1954).

part 5

SPECIAL TOPICS

17

Induced Magnetic Anisotropy

17.1 Magnetic Annealing Effect

By induced magnetic anisotropy is meant the magnetic anisotropy induced by some treatment which has more or less directional properties. For instance, most ferromagnetic alloys exhibit uniaxial anisotropy when they are heat-treated in a magnetic field. We call this the magnetic annealing effect. Many ferromagnetic alloys also exhibit fairly strong anisotropies when they are cold-rolled or cold-worked. We call this the roll magnetic anisotropy. If a magnetic field is applied when a specimen is cooled through its phase transition point, a fairly large magnetic anisotropy is sometimes induced. We treat these problems in this chapter.

The magnetic annealing effect was first observed by G. A. Kelsall[1] for Fe-Ni alloys and was investigated by Dillinger and Bozorth[2] in detail. Figure 17.1 shows the effect of magnetic annealing on the shape of the magnetization curve for 21.5% Fe-Ni alloy. Curve *A* is measured by applying a magnetic field parallel to the field which was applied during cooling, and curve *C* is measured by applying the field perpendicular to the annealing field. The difference in shape between the two curves can be attributed to the induced uniaxial anisotropy the easy axis of which is parallel to the direction of the annealing field. When the field is applied parallel to the easy axis (Fig. 17.2*a*), the magnetization takes place exclusively through displacement of 180° walls, giving rise to a magnetization curve with high maximum permeability. If, however, the field is applied perpendicular to the easy axis (Fig. 17.2*c*), the magnetization takes place through a rotation of each magnetic domain, giving rise to a

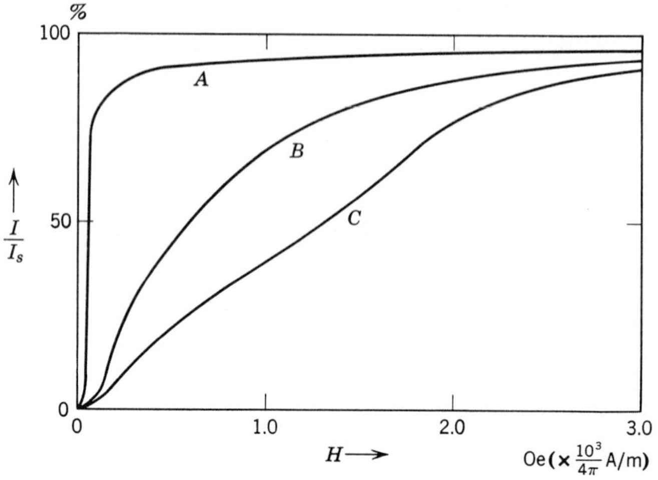

Fig. 17.1. Magnetization curves of 21.5 Permalloy which was cooled from 600°C (*A*) in a longitudinal magnetic field, (*B*) in the absence of magnetic field, and (*C*) with perpendicular (or circular) magnetic field (Chikazumi[3]).

linearly ascending magnetization curve (cf. Fig. 13.2). Curve *B* in Fig. 17.1 was obtained after the specimen was cooled in the absence of the magnetic field. In this case the uniaxial anisotropy is also induced in each domain, but its easy axis depends on the direction of domain magnetization during the heat treatment (Fig. 17.2*b*). The magnetization process is then partly the displacement of 180° walls and partly rotation magnetization. The magnetization curve is, therefore, just intermediate between the curves *A* and *C* of Fig. 17.1.

This effect is also verified by measurement of the magnetoresistance effect. Since the electrical resistance is insensitive to the sense of the domain magnetization, it will never be changed by 180° wall displacement. Actually the magnetoresistance effect is quite small for the magnetization curve *A*, while the change is large for the magnetization curve *C* in Fig. 17.1 (cf. Fig. 19.3).

The mechanism of the magnetic annealing effect was first interpreted by Bozorth.[4] When spontaneous magnetization appears below the Curie temperature, the crystal lattice tends to make a magnetostrictive deformation which, however, will be opposed by the interference of the crystal boundary or some inclusions. At high temperatures, such opposition will be relaxed through a plastic flow of crystal boundary or inclusions, which stabilizes the original orientation of domain magnetization. After cooling the specimen to room temperature, the deformation thus caused is permanently fixed, so that the rotation of domain magnetization out of the original direction will increase the elastic energy by

$$E_\lambda = -\tfrac{1}{2}(\tfrac{3}{2}\lambda)^2 c \cos^2 \phi = -\tfrac{9}{8}\lambda^2 c \cos^2 \phi, \tag{17.1}$$

where c is the elastic modulus, and ϕ the angle between domain magnetization and its original direction. This form of the energy expresses the uniaxial anisotropy, which explains the experiment qualitatively. It turns out, however, that the coefficient in (17.1) gives for 21.5% Fe-Ni alloy the numerical value

$$\begin{aligned} K_u &= \tfrac{9}{8}\lambda^2 c \\ &= \tfrac{9}{8}(2.7 \times 10^{-6})^2(2 \times 10^{11}) \\ &\approx 1.6\ \mathrm{J/m^3}, \end{aligned} \tag{17.2}$$

which is about a hundred times smaller than the observed value, $1.4 \times 10^2\ \mathrm{J/m^3}$.

Before explaining later interpretations, we shall review the history of the investigation of Permalloy. This alloy containing 21.5% Fe and 78.5% Ni, was produced in 1923 by Arnold and Elmen,[5] who found that it exhibits a high permeability only when it is quenched from 600°C. The reason for this peculiar behavior of the alloy was investigated by many researchers, who called this the Permalloy problem. In 1936 Dahl[6] found that a superlattice is formed in this alloy, and Kaya[7] investigated the magnetic properties for wide ranges of the iron-nickel system in relation to superlattice formation. The superlattice found in this alloy system is the A_3B type the atomic arrangement of which is shown in Fig. 17.3. It was found that the formation of this superlattice takes about 160 hours at 490°C. The magnetic annealing was found to be most effective at 500°C, as experimentally verified by Tomono.[8] It was also found by Chikazumi[3] that the uniaxial anisotropy induced by magnetic annealing is reduced with an approach to perfect order. Chikazumi assumed that the bond length of an AB pair is shorter than those of AA and BB pairs, on the basis of the experimental fact that the lattice parameter contracts by 5×10^{-4} with the progress of ordering or an increase in the number of AB pairs.

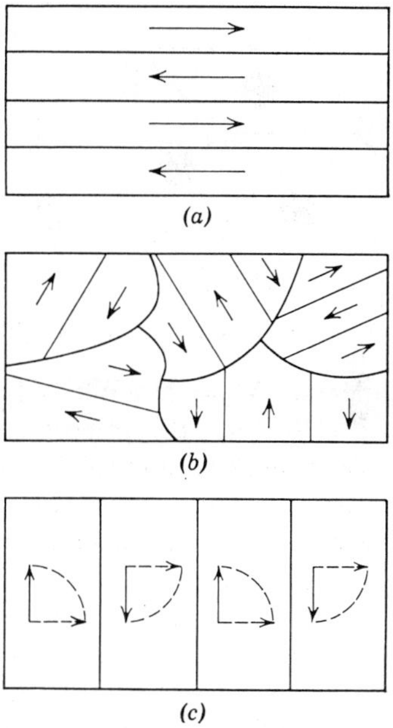

Fig. 17.2. Domain structure of a specimen magnetically annealed (*a*) in a longitudinal magnetic field, (*b*) without magnetic field, and (*c*) in a perpendicular magnetic field.

If, therefore, A and B atoms are so distributed as to align more AA and BB pairs in a certain direction, the lattice will deform spontaneously about 10^{-4}. This distribution of atom pairs, called directional order is built up during the process of magnetic annealing so as to decrease the magneto-elastic energy. Since this strain is about 10^2 times larger than the magnetostrictive deformation, the resultant anisotropy energy is also expected to be larger than that in (17.2) by the factor 10^2; thus it is in good agreement with the observed value.

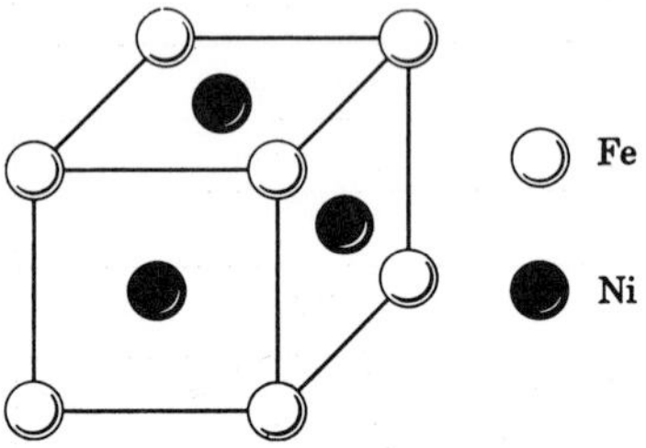

Fig. 17.3. Superstructure of Ni_3Fe.

Another theory based on the shape anisotropy of elongated order phase was proposed.[9] Since the intensity of spontaneous magnetization increases from 1.08 Wb/m^2 (= 860 gauss) to 1.13 Wb/m^2 (= 900 gauss) during the ordering process, we find, by putting the difference $\Delta I_s = 0.05$ Wb/m^2 (= 40 gauss) and $\beta = \frac{1}{2}$ into (17.24), that $K_u = 1.2 \times 10^2$ J/m^3; this value is also in good agreement with the observed value.

Both theories can explain the fact that the induced anisotropy disappears at the perfectly ordered state. For the purpose of checking these theories the compositional dependence of magnitude of the induced anisotropy was measured for iron-nickel alloys.[10] It is expected that, if the "directional order strain" theory is correct, the anisotropy should be proportional to the magnetostriction constant, and, if the shape anisotropy of the order phase is responsible, the anisotropy should be proportional to ΔI_s^2 and should take a sharp maximum at Ni_3Fe, where superlattice formation is most effective. The experimental result is, however, quite monotonic as shown by the solid curve in Fig. 17.4, which conforms to neither of the theories. This curve was finally explained by the corrected directional order theory: this idea was proposed by Néel,[11,12] Taniguchi and Yamamoto[13,14] and Chikazumi and Oomura[10] independently. The main assumption is that the pseudodipole interaction is different between AA, BB, and AB pairs; hence the directional order, whose pair distribution has lower symmetry than that of cubic, should result in uniaxial anisotropy. The detailed calculations were made by Néel and Taniguchi and Yamamoto; we shall follow Néel's calculation below.

Let the coefficients of pseudodipole interaction for AA, BB, and AB pairs be l_{AA}, l_{BB}, and l_{AB}. Then the anisotropy energy due to unbalanced distribution of three kinds of atom pairs over differently oriented pair directions is

$$E_a = \sum_i (N_{AAi} l_{AA} + N_{BBi} l_{BB} + N_{ABi} l_{AB})(\cos^2 \phi_i - \tfrac{1}{3}), \quad (17.3)$$

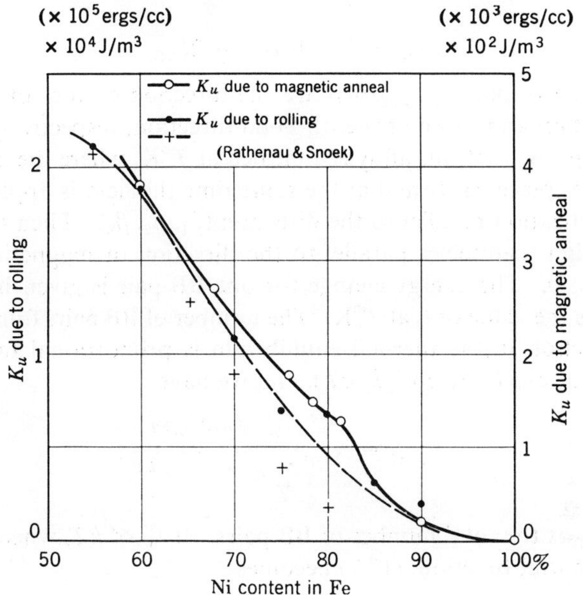

Fig. 17.4. Induced anisotropy of Fe-Ni alloys due to magnetic annealing and rolling (Chikazumi and Oomura[10]).

where i is the suffix denoting the pair directions, $N_{\mathrm{AA}i}$, $N_{\mathrm{BB}i}$, and $N_{\mathrm{AB}i}$ are the number of AA, BB, and AB pairs directed parallel to the ith direction, and ϕ_i is the angle between the domain magnetization and the ith bond direction. For the following reasons we have simple relations among $N_{\mathrm{AA}i}$, $N_{\mathrm{BB}i}$, and $N_{\mathrm{AB}i}$: If we divide each atom into z pieces, where z is the number of nearest neighbors, and attach these pieces to each bond, the AA pair has two A pieces, whereas the AB pair has one A piece. On the other hand, one A atom gives two of its pieces to the atom pairs with ith direction. Thus the total number of A pieces attached to these pairs is

$$2N_{\mathrm{AA}i} + N_{\mathrm{AB}i} = 2N_{\mathrm{A}},$$

and the number of B pieces is (17.4)

$$2N_{\mathrm{BB}i} + N_{\mathrm{AB}i} = 2N_{\mathrm{B}},$$

where N_{A} and N_{B} are the total number of A and B atoms. Using (17.4), we can express the anisotropy energy (17.3) in terms of the number of BB pairs as

$$E_a = \sum_i N_{\mathrm{BB}i} l_0 (\cos^2 \phi_i - \tfrac{1}{3}) + \text{const.}$$
$$= \sum_i N_{\mathrm{BB}i} l_0 (\alpha_1\gamma_{1i} + \alpha_2\gamma_{2i} + \alpha_3\gamma_{3i})^2 + \text{const.}, \qquad (17.5)$$

where

$$l_0 = l_{AA} + l_{BB} - 2l_{AB} \tag{17.6}$$

and $(\alpha_1, \alpha_2, \alpha_3)$ and $(\gamma_{1i}, \gamma_{2i}, \gamma_{3i})$ are the direction cosines of the domain magnetization and those of the ith bond direction, respectively.

Now suppose that this alloy is annealed at T°K, where the migration of atoms takes place easily and at the same time the field is applied to point the magnetization parallel to the direction $(\beta_1, \beta_2, \beta_3)$. Then the BB pairs tend to align themselves parallel to the direction of magnetization, provided $l_0' < 0$. The energy change for one BB pair is given by $l_0' \cos^2 \phi_i$, where l_0' is the value of l_0 at T°K. The number of BB pairs found in the ith bond direction in the thermal equilibrium is proportional to the Boltzmann factor $\exp(-l_0' \cos^2 \phi_i/kT)$, and we have

$$N_{BBi} = N_{BB} \frac{e^{-l_0' \cos^2 \phi_i/kT}}{\sum_i e^{-l_0' \cos^2 \phi_i/kT}}, \tag{17.7}$$

where N_{BB} is the total number of BB pairs. If $l_0' \ll kT$, when we expand the exponential function, (17.7) becomes

$$\begin{aligned} N_{BBi} &= \frac{2N_{BB}}{z}\left(1 - \frac{l_0' \cos^2 \phi_i'}{kT}\right) \\ &= \frac{2N_{BB}}{z}\left[1 - \frac{l_0'}{kT}(\beta_1\gamma_{1i} + \beta_2\gamma_{2i} + \beta_3\gamma_{3i})^2\right]. \end{aligned} \tag{17.8}$$

After the alloy has been quenched, these atom pairs are fixed, and they give rise to magnetic anisotropy, which is calculated, by putting (17.8) into (17.5), as

$$\begin{aligned} E_a &= \frac{2N_{BB}l_0}{z} \sum_i (\alpha_1\gamma_{1i} + \alpha_2\gamma_{2i} + \alpha_3\gamma_{3i})^2\left[1 - \frac{l_0'}{kT}(\beta_1\gamma_{1i} + \beta_2\gamma_{2i} + \beta_3\gamma_{3i})^2\right] \\ &= -\frac{2N_{BB}l_0l_0'}{zkT} \sum_i (\alpha_1\gamma_{1i} + \alpha_2\gamma_{2i} + \alpha_3\gamma_{3i})^2(\beta_1\gamma_{1i} + \beta_2\gamma_{2i} + \beta_3\gamma_{3i})^2 \\ &\quad + \text{const.} \\ &= -\frac{N_{BB}l_0l_0'}{kT}\left(k_1 \sum_i \alpha_i^2\beta_i^2 + k_2 \sum_{i>j} \alpha_i\alpha_j\beta_i\beta_j\right), \end{aligned} \tag{17.9}$$

where

$$\begin{aligned} k_1 &= \frac{2}{z}\left(\sum_i \gamma_{1i}^4 - \sum_i \gamma_{1i}^2\gamma_{2i}^2\right) \\ k_2 &= \frac{8}{z}\left(\sum_i \gamma_{1i}^2\gamma_{2i}^2\right). \end{aligned} \tag{17.10}$$

The factors k_1 and k_2 are constants specific to the crystal structures; their numerical values are shown in Table 17.1 for various crystal types. For

Table 17.1. Values of k_1 and k_2 for Various Crystal Types (cf. 17.9)

Crystal Type	k_1	k_2
Isotropic	$\frac{2}{15}$	$\frac{4}{15}$
Simple cubic	$\frac{1}{3}$	0
Body-centered cubic	0	$\frac{4}{9}$
Face-centered cubic	$\frac{1}{12}$	$\frac{4}{12}$

the isotropic type, (17.9) becomes

$$E_a = -\frac{2N_{BB}l_0l_0'}{15kT}\left(\sum_i \alpha_i^2\beta_i^2 + 2\sum_{i>j}\alpha_i\alpha_j\beta_i\beta_j\right)$$
$$= -\frac{2N_{BB}l_0l_0'}{15kT}\left(\sum_i \alpha_i\beta_i\right)^2$$
$$= -\frac{2N_{BB}l_0l_0'}{15kT}\cos^2\phi, \tag{17.11}$$

where ϕ is the angle between the domain magnetization and the direction of the annealing magnetic field. If l_0' has the same sign as l_0, the coefficient is negative, and the easy axis is built up parallel to the annealing field. For a dilute and disordered solution of B atoms, the probability of finding BB pairs is proportional to C_B^2, where C_B is the concentration of B atoms; hence

$$N_{BB} = \frac{zN}{2} C_B^2, \tag{17.12}$$

where N is the total number of atoms included in a unit volume. On using (17.12), we can express the anisotropy constant as

$$K_u = \frac{zNl_0l_0'}{15kT} C_B^2. \tag{17.13}$$

The compositional dependence of this expression explains the experimental result shown in Fig. 17.4.

Néel estimated the values of l_{AA}, l_{BB}, and l_{AB} from the compositional dependence of magnetostriction constants. Since the number of AA, BB, and AB pairs is proportional to C_A^2, C_B^2, and $2C_AC_B$ for disordered alloys, respectively, the average value of l is

$$l = C_A^2 l_{AA} + 2C_AC_B l_{AB} + C_B^2 l_{BB}. \tag{17.14}$$

Similar dependence can also be expected for $(\partial l/\partial r)$. From (8.39), we expect the magnetostriction constants λ_{100} and λ_{111} to depend on the alloy composition in a similar way to the dependence expressed in (17.14). Actually the compositional dependence of the constants λ_{100} and λ_{111} in Fe-Ni alloy can be approximated by parabolic curves, which are given by

$$\begin{aligned} \lambda_{111} \times 10^6 &= -27C_A{}^2 + 134C_AC_B + 13C_B{}^2, \\ \lambda_{100} \times 10^6 &= -55C_A{}^2 + 340C_AC_B - 245C_B{}^2. \end{aligned} \tag{17.15}$$

From a comparison of (17.14) and (17.15) we can estimate the value of l_{NiNi}, l_{FeFe}, and l_{FeNi}; and finally we have

$$Nl_0 = N(l_{NiNi} + l_{FeFe} - 2l_{FeNi}) = 3.1 \times 10^7 \text{ J/m}^3. \tag{17.16}$$

The value of Nl_0' can be estimated by assuming the $I_s{}^3$ law for the temperature dependence of l, similarly to the theoretical temperature dependence of K_u. For 21.5% Fe-Ni, $I_s = 1.13$ Wb/m² ($\fallingdotseq$ 900 gauss) at room temperature and $I_s = 0.50$ Wb/m² ($\fallingdotseq$ 400 gauss) at 530°C (700°K). Then $Nl_0' = 3.1 \times 10^7 \times (0.50/1.13)^3 = 2.68 \times 10^6$ J/m³. On putting $z = 12$, $N = 9.17 \times 10^{28}$ m^{-3}, $k = 1.38 \times 10^{-23}$ J/deg, $T = 700$°K, and the values above for l_0 and l_0' into (17.13), we obtain

$$K_u = 7.5 \times 10^4 C_B{}^2 \text{ J/m}^3. \tag{17.17}$$

The experimental curve in Fig. 17.4 is approximately expressed as

$$K_u = 2.9 \times 10^3 C_B{}^2 \text{ J/m}^3, \tag{17.18}$$

which is somewhat smaller than the theoretical expectation. One of the reasons would be the opposition of normal ordering to the formation of directional order.

Formula (17.9) also shows that the magnitude of induced anisotropy depends on the crystallographic direction of the annealing field except for the isotropic type. Figure 17.5 shows the anisotropy energy curves observed for a (110) oblate single crystal of Ni_3Fe after cooling from 600°C at a rate of 14°C/min in a magnetic field the direction of which is changed in the plane of (110).[15] These energy curves are obtained by integration of torque curves with respect to the angle and then are analyzed into the crystal and the uniaxial anisotropy energy curves by the Fourier analysis. The arrows in the figure indicate the direction of annealing field and the dots show for each the minimum points of the uniaxial anisotropy energy. It is found in these curves that magnetic annealing is most effective for $\langle 111 \rangle$ annealing, less effective for $\langle 110 \rangle$, and least effective for $\langle 100 \rangle$. This figure also shows that the minimum point coincides with the direction of the annealing field for three principal axes, whereas it tends to deviate toward $\langle 111 \rangle$ for intermediate directions of the annealing field.

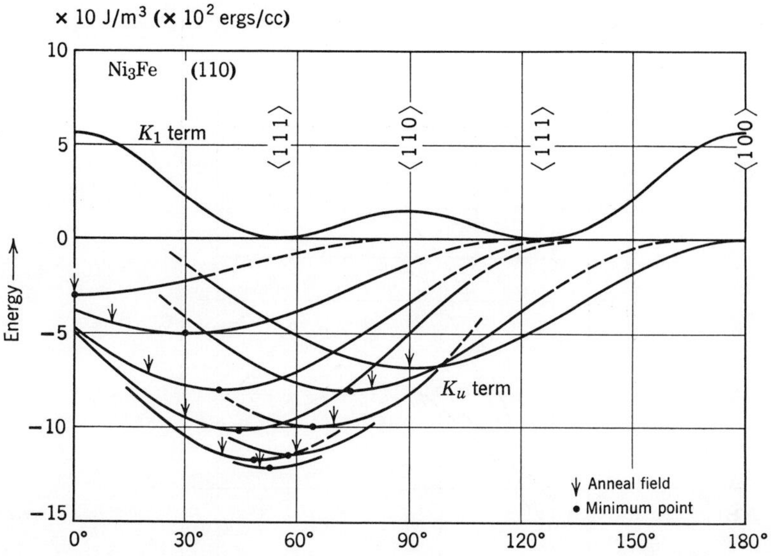

Fig. 17.5. Magnetic anisotropy energy induced by magnetic annealing of a (110) disk of Ni_3Fe (Chikazumi[15]).

For the purpose of comparison with theory, we express the anisotropy energy as

$$E_a = -K_u \cos^2 (\theta - \theta_0) \tag{17.19}$$

where θ is the angle of spontaneous magnetization as measured in $(1\bar{1}0)$ from [001] (so that we have $\alpha_1 = \alpha_2 = \sqrt{1/2} \sin \theta$ and $\alpha_3 = \cos \theta$) and θ_0 represents the minimum point. On comparing (17.19) and (17.9), we have

$$\langle 100 \rangle \qquad K_u = \frac{N_{\mathrm{BB}} l_0 l_0'}{kT} k_1, \tag{17.20}$$

$$\langle 110 \rangle \qquad K_u = \frac{N_{\mathrm{BB}} l_0 l_0'}{kT} \left(\frac{k_1}{2} + \frac{k_2}{4} \right), \tag{17.21}$$

$$\langle 111 \rangle \qquad K_u = \frac{N_{\mathrm{BB}} l_0 l_0'}{kT} \frac{k_2}{2}. \tag{17.22}$$

On putting into these formulas the theoretical values of k_1 and k_2 indicated in Table 17.1, we find that the ratio of the K_u should be 2:3:4 for the three principal axes. This explains the experimental results fairly well. Strictly speaking, however, experiment shows a larger dependence; see Fig. 17.6, in which K_u as well as K_1 is plotted as a function of direction of

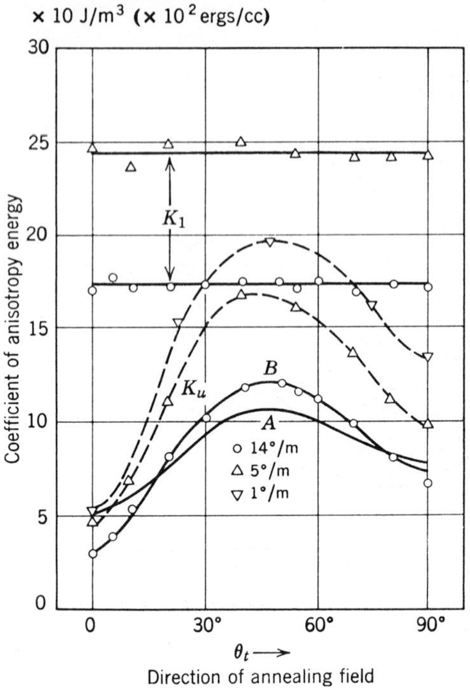

Fig. 17.6. Induced uniaxial anisotropy constant as a function of crystallographic orientation of the annealing field (Chikazumi[15]).

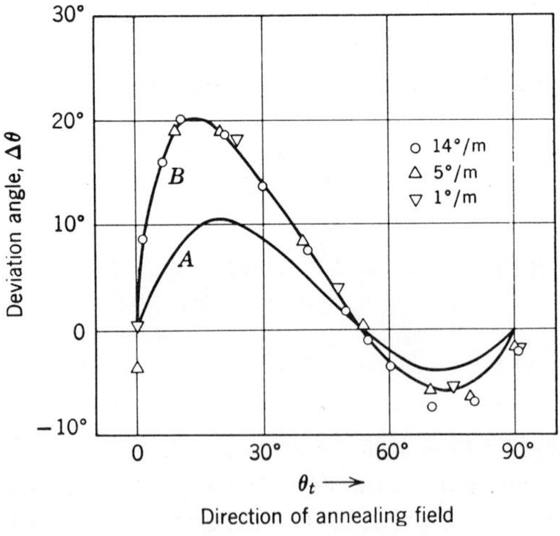

Fig. 17.7. Angle of difference, $\Delta\theta$, between the direction of annealing field and the easy axis of the induced anisotropy as a function of crystallographic orientation of the annealing field (Chikazumi[15]).

the annealing field. The angle of deviation of the minimum point from the direction of annealing field,

$$\Delta\theta = \theta_0 - \theta_t, \tag{17.23}$$

is plotted in Fig. 17.7 as a function of the direction of the annealing field. The theoretical curves A, which are derived from (17.9), explain the behavior of the experiment (curves B) qualitatively. The analysis of the experimental curves shows that the all experimental results are consistently explained by assuming $k_1:k_2 = 1:8.5$. The solid lines of curves B in Figs. 17.6 and 17.7 are drawn by postulating this ratio. The values of the ratio $k_1:k_2$ were measured by several investigators[16–20] for a couple of alloys; their results are given in Table 17.2. The values are scattered from 1.9 to

Table 17.2. Theoretical and Experimental Values of the Ratio $k_1:k_2$

Substance	$k_1:k_2$		Investigator
	Theory	Experiment	
Polycrystal	1:2	1:2	
Ni_3Fe	1:4	1:8.5	Chikazumi[15]
20% Co-Fe	1:4	1:3.0	Aoyagi et al.[16]
12.5% Co-Fe	1:4	1:2.3 ~ 2.6	Aoyagi et al.[17,18]
54% Ni-Fe	1:4	1:1.9 ~ 2.4	Aoyagi et al.[17,18]
83% Ni-Fe	1:4	1:8.3 ~ 9.6	Aoyagi et al.[17,18]
$Fe_{3.35}Al$	0:1	1:3.4	Suzuki[19]
$Fe_{4.7}Al$	0:1	1:1	Chikazumi and Wakiyama[20]

9.0 even for face-centered cubic alloys, a fact which has not yet been interpreted. It is also interesting that the value of k_1, which represents the effect for $\langle 100 \rangle$ annealing, is finite for the Fe-Al alloy, which has a body-centered cubic structure, in spite of the theoretical expectation that $k_1 = 0$. This fact can presumably be explained by assuming a second nearest neighbor interaction. Further investigations on the compositional dependence of induced anisotropy and its dependence on the annealing temperature were made by Aoyagi et al.[17] for Co-Ni and Fe-Ni and by Ferguson[21] for Fe-Co and Fe-Ni, and a detailed analysis was made by Iwata.[22]

Magnetic annealing has been found to be effective also for some ferrites, such as Co ferrite, which has been used as an O.P. magnet.[23] Actually magnetic annealing has been used for this magnet to improve the (BH) product. The induced anisotropy was measured by Iida, Sekizawa, and Aiyama[24] for the system of Fe-Co ferrites which are magnetically annealed under various partial pressures of oxygen (Fig. 17.8). They found that the material responds to magnetic annealing only when it is more or less oxidized. In other words, the magnetic annealing of ferrite is effective only in the presence of vacancies. They interpreted this phenomenon in terms of the occurrence of directional order between Co^{2+} ions and vacancies in 16*d* sites of spinels. Single crystal measurements were made for the same system by Penoyer and Bickford,[25] who found that the coefficients of the first and second terms, F and G, or k_1 and k_2 times the proportionality factor outside the parentheses in (17.9), depend on the composition in a different way, as shown in Fig. 17.9. That is, F increases in proportion to the square of the content of Co, while G increases linearly with composition. Slonczewski[26] explained this fact in terms of the one-ion anisotropy of the Co^{2+} ion. As explained in Section 7.3, the Co^{2+} ion exhibits a large uniaxial anisotropy in the 16*d* site under the action of the trigonal field. The symmetry axis of the trigonal field or the easy axis of the uniaxial anisotropy of the Co^{2+} ion is parallel to one of the ⟨111⟩ directions which is determined by the surrounding metal ions (Fig. 7.20). If the

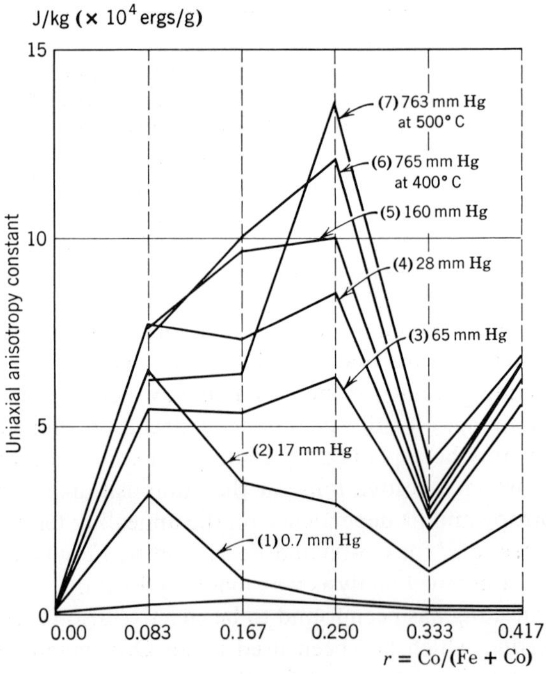

Fig. 17.8. Effect of magnetic annealing under the various partial pressures of oxygen observed for Co-Fe ferrites (Iida, Sekizawa, and Aiyama[24]).

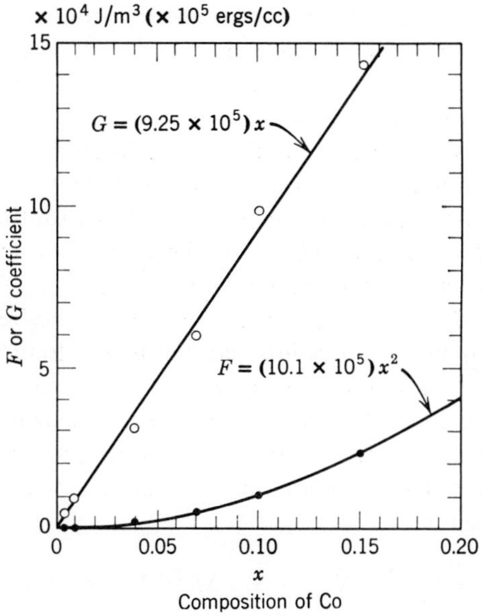

Fig. 17.9. Variation of two coefficients F and G (which are proportional to k_1 and k_2, respectively) with a change in the composition of Co in Co-Fe ferrite (Penoyer and Bickford[25]).

Co^{2+} ions are distributed equally over the four kinds of 16*d* sites, each of which has a differently oriented trigonal axis, the uniaxial anisotropies of Co^{2+} ions cancel each other. When the material is annealed in a magnetic field, Co^{2+} ions tend to occupy the site whose trigonal axis is closest to the direction of the field. Such an unbalanced distribution of Co^{2+} ions is fixed after cooling of the material down to room temperature and thus gives rise to uniaxial anisotropy. Since the symmetry axis of one-ion anisotropy is parallel to $\langle 111 \rangle$, this mechanism should contribute to an increase in k_2 or G value, as in body-centered cubic alloys. The magnitude of the induced anisotropy should be proportional to the number of available Co^{2+} ions or the concentration of Co. This model explains the linear dependence of the G value on the composition. Slonczewski interpreted the quadratic dependence of F to be due to the directional ordering of Co^{2+}-Co^{2+} pairs. The meaning of the necessity for vacancies was examined by several investigators.[27–29] According to these investigations, it is likely that the vacancy has nothing to do with the anisotropy itself, but it acts as an agent to promote the migration of ions during annealing. The effect of oxygen on magnetic annealing was also investigated for Ni ferrite.[30]

Magnetic annealing is also effective for precipitation alloys. This phenomenon has been found[31] in relation to the permanent magnets, such as MK steel, Alnico 5, and Alnico 2, the energy product (cf. Section 22.3) of which can be improved by this treatment. The mechanism of magnetic annealing in this case is considered to be due to a preferential growth of the elongated precipitation particles along the direction of the annealing field, as was proposed by Néel,[32] Shubina, and Shur,[33] and Kittel, Nesbitt, and Shockley.[34] Suppose that an elongated particle, which has a spontaneous magnetization I_s', precipitates in the matrix with the magnetization I_s (Fig. 17.10). The surface density of the free poles appearing on the surface of the particle is proportional to the difference of the magnetizations, $\Delta I_s = I_s' - I_s$. The magnetostatic energy of the free poles is small as long as the magnetization points parallel to the long axis of the precipitate, whereas it becomes fairly large if the magnetization rotates to point perpendicular to the long axis. Here the magnetization in and out of the precipitate is considered to maintain perfect parallelism because of the action of the exchange interaction. Thus the magnetostatic energy stored in a unit volume is

$$E_a = -\frac{1}{2\mu_0}(N_\perp - N_\parallel)\,\Delta I_s^2 \beta(1-\beta)\cos^2\theta, \tag{17.24}$$

where $N_\perp$ and $N_\parallel$ are the demagnetizing factors of the particle parallel and perpendicular to the long axis, β is the fractional volume of the precipitates, and θ is the angle between magnetization and the long axis of the particle. Here we assumed that the long axes of all the precipitates are parallel to each other. Actually Nesbitt and Heidenreich[35] observed by electron microscope that the precipitates of Alnico 5 are beautifully aligned when the alloy is annealed in a magnetic field. Néel[36,12] pointed out that, besides the shape anisotropy discussed above, the surface anisotropy may be also a possible origin of the induced anisotropy. That is, if the precipitate is composed of B atoms, while the matrix is composed of A atoms (Fig. 17.11), the number of AB pairs which point perpendicular to the long axis is greater than the number of those which point parallel to the long axis. If the sign of l_0 in (17.6) is negative, or if the direction of an AB pair is a hard axis, the surface anisotropy gives rise to an easy axis parallel to the long axis; whereas if $l_0 > 0$, the long axis becomes a hard direction, and then the magnetization tends to point perpendicular to the long axis.

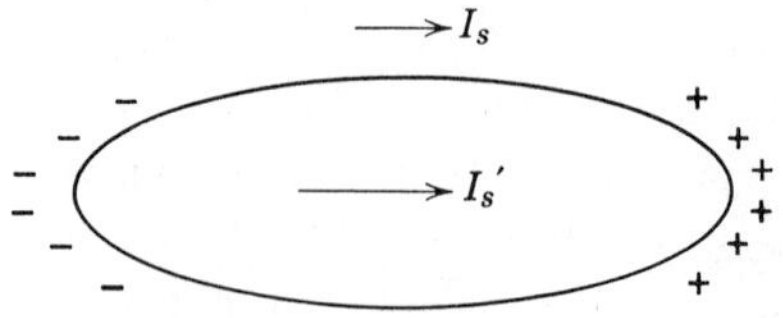

Fig. 17.10. Free pole distribution on the surface of a precipitate particle.

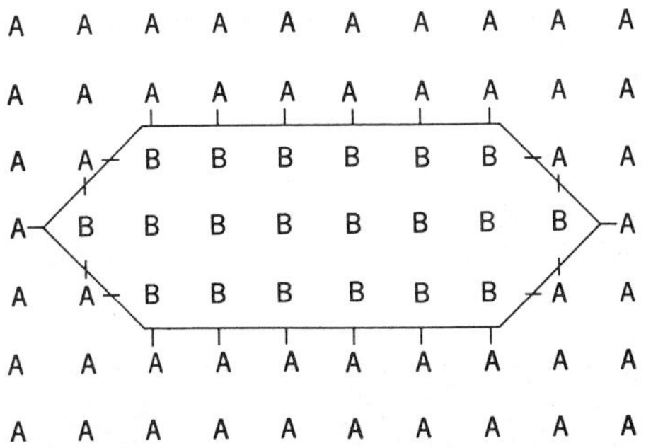

Fig. 17.11. Distribution of AB pairs on the surface of a precipitate made from B atoms surrounded by a matrix composed of A atoms.

The magnitude of this anisotropy is estimated to be $0.13 \sim 1.3 \times 10^{-3}$ J/m^2 ($= 0.13 \sim 1.3$ erg/cm^2), which could give rise to the anisotropy energy of the order of $10^3 \sim 10^4$ J/m^3 ($=10^4 \sim 10^5$ erg/cc) if a fine division of the precipitates is assumed [for example, a separation of 10^{-7} m ($= 0.1\ \mu$)]. Miyahara and Mitsui[37] and others[38] investigated the effect of magnetic annealing for the precipitation of the Co phase from 2% Co-Cu alloy. They found that magnetic annealing is most effective at the initial stage of precipitation, as it is also for Alnico 5. For further details the reader may refer to the excellent review by Graham.[39]

17.2 Roll Magnetic Anisotropy

It was discovered by Six, Snoek, and Burgers[40] in 1934 that a large magnetic anisotropy is created during the process of cold-rolling iron-nickel alloys. Utilizing this effect, they made the magnetic material called Isoperm, which has constant permeability over a wide range of the applied field. This material the composition of which is 50% Fe-Ni alloy is first strongly cold-rolled, then recrystallized into the system (001)[100] by annealing, and finally cold-rolled again to about 50% reduction. The sheet thus manufactured exhibits a large uniaxial magnetic anisotropy with its easy axis perpendicular to the roll direction, and magnetization parallel to the roll direction takes place exclusively through a rotation of domain magnetization, giving rise to a linear magnetization curve (Fig. 17.12).

Detailed investigations were made by Conradt, Dahl, and Sixtus[41] in 1940, and by Rathenau and Snoek[42] in 1941. They concluded that this anisotropy cannot be explained in terms of the magnetostrictive anisotropy given by (8.86). Néel[43,12] and Taniguchi and Yamamoto[13] extended their interpretation of magnetic annealing to this phenomenon and considered

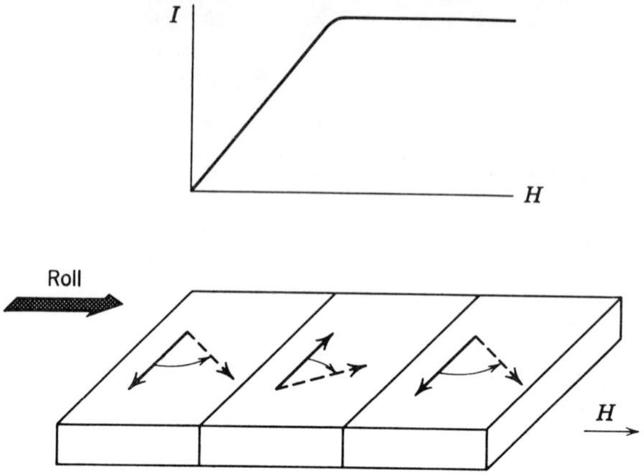

Fig. 17.12. Magnetization curve and domain structure of Isoperm.

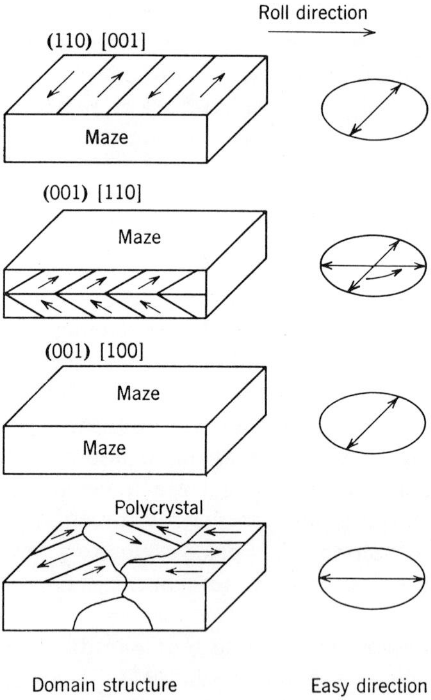

Fig. 17.13. Domain structure and easy axis of roll magnetic anisotropy as measured in a roll plane for various crystallographic orientations of rolling (Chikazumi, Suzuki, and Iwata[44]).

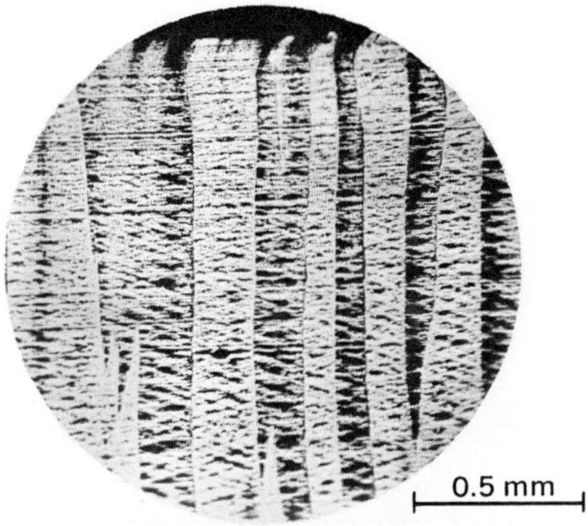

Fig. 17.14. Domain structure appearing on the roll plane after rolling by 55% reduction of an (110) [001] Ni_3Fe crystal. Roll direction is horizontal.

that A and B atoms migrate to take a stable directional order during the process of cold rolling. They considered that the plastic deformation may play an important role in bringing atoms into stable positions. There is no doubt that the directional order is the main origin of this anisotropy, because the magnitude of the induced anisotropy is also proportional to C_B^2, as shown by the broken curve in Fig. 17.4. The magnitude of the anisotropy caused by rolling is, however, about 50 times larger than that caused by magnetic annealing.

Investigations of this phenomenon utilizing single crystals were made by Chikazumi, Suzuki, and Iwata[44] for Ni_3Fe. They found that the anisotropy is strongly dependent on the crystallographic orientation of rolling in its magnitude and also in its functional form. If the rolling is done along the (110) plane and in the [001] direction, there is induced the uniaxial anisotropy which has its easy axis perpendicular to the roll direction (Fig. 17.13). This has been confirmed through the measurement of the magnetization curve and the torque curve as well as by the observation of the domain pattern, which reveals the presence of a number of magnetic domains with the magnetizations perpendicular to the roll direction (Fig. 17.14). The magnitude of the uniaxial anisotropy constant increases with the progress of rolling, as shown in Fig. 17.15. In contrast, the (001)[110] rolling first develops anisotropy with easy axis perpendicular to the roll direction, then changes its easy axis to parallel to the roll direction. The anisotropy constant is plotted against the roll reduction in Fig. 17.16, where we see

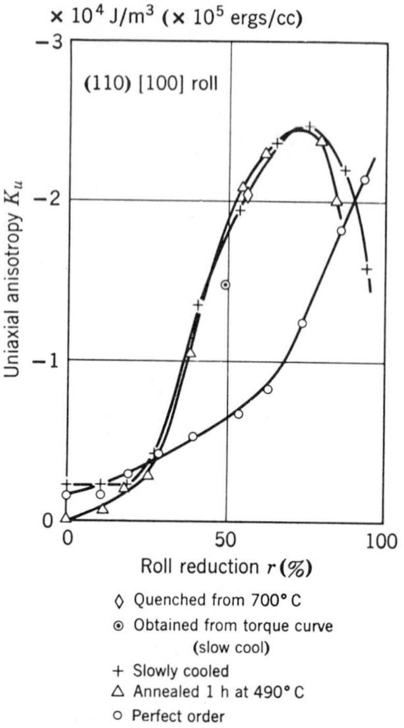

Fig. 17.15. Variation of the uniaxial anisotropy constant of roll magnetic anisotropy with the progress of (110) [100] rolling of an Ni_3Fe crystal.

that it changes its sign at about 20 ∼ 30% reduction. The domain pattern observed on the top surface is complicated, as shown in Fig. 17.17*a*. This signifies that the domain magnetization has a fairly large component normal to the crystal surface. The pattern observed on the side surface reveals that domain magnetizations make some angle with the top surface (Fig. 17.17*b*). This angle decreases with the progress of rolling (Fig. 17.17*d*), and at the same time the patterns at the top surface are elongated in the direction of rolling as shown in Fig. 17.17*c*.

In (001)[100] rolling, which is the same as that of Isoperm, the torque measurement reveals that the easy axis in a roll plane is perpendicular to the roll direction, while the actual orientation of the easy axis was found to be not in the roll plane but making some angle with the roll plane as well as the side surface, as recognized from the complicated domain patterns shown in Fig. 17.18. For polycrystalline materials the average easy axis is found, from a torque curve, to be parallel to the roll direction, as Rathenau and Snoek[42] had already pointed out. The easy axes of the induced anisotropy in individual crystallites are, however, quite different in their directions as observed in the domain pattern (Fig. 17.19).

These experimental results have been explained fairly well in terms of the directional order induced by slip deformation.[44] The slip system in face-centered cubic lattice is believed to be {111} ⟨110⟩. Actually, that the orientations of slip bands observed on the top and side surfaces are consistent with this system has been confirmed (Figs. 17.20, 17.21). Figure 17.22 shows a perfectly ordered face-centered cubic lattice of the A_3B type which is deformed by the traveling of a single dislocation. There are no BB pairs in the undeformed portion, whereas a number of BB pairs are induced along the slipped portion. These BB pairs are all parallel to the [011] direction as indicated by double lines in the figure; such an unbalanced distribution of BB pairs should give rise to uniaxial anisotropy. Since the BB pairs thus induced are again canceled by a subsequent slip along the same slip plane, to describe the number of the produced BB pairs we introduce the probability p_0 of creating an isolated dislocation, or that of non-occurrence of pair creation of dislocations. Now we consider

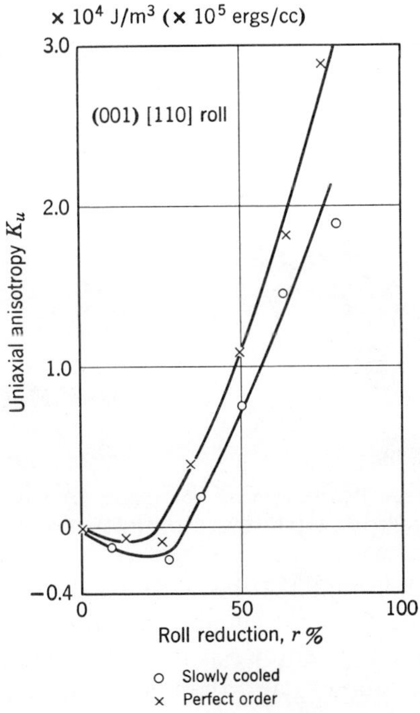

Fig. 17.16. Variation of the uniaxial anisotropy constant of roll magnetic anisotropy with the progress of (001) [110] rolling of an Ni_3Fe crystal.

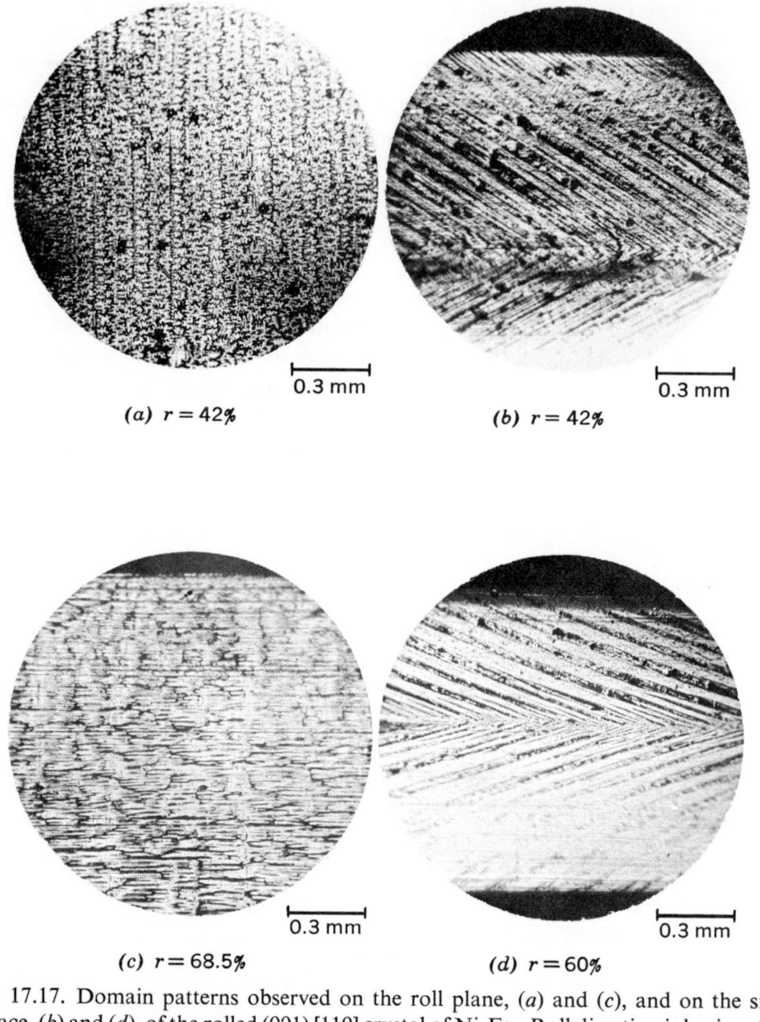

(*a*) $r = 42\%$ (*b*) $r = 42\%$

(*c*) $r = 68.5\%$ (*d*) $r = 60\%$

Fig. 17.17. Domain patterns observed on the roll plane, (*a*) and (*c*), and on the side surface, (*b*) and (*d*), of the rolled (001) [110] crystal of Ni_3Fe. Roll direction is horizontal.

that ns dislocations run through the crystal per n atom layers (s is called the slip density hereafter). Then the number of atom planes upon which the BB pairs are created is given by $p_0p'ns$, where p' is the probability of the creation of a dislocation on a new atom plane. If we denote by α those lattice sites which should be occupied by A atoms, and by β those of B atoms, the number of $\beta\beta$ pairs created in a unit area of the slipped atom

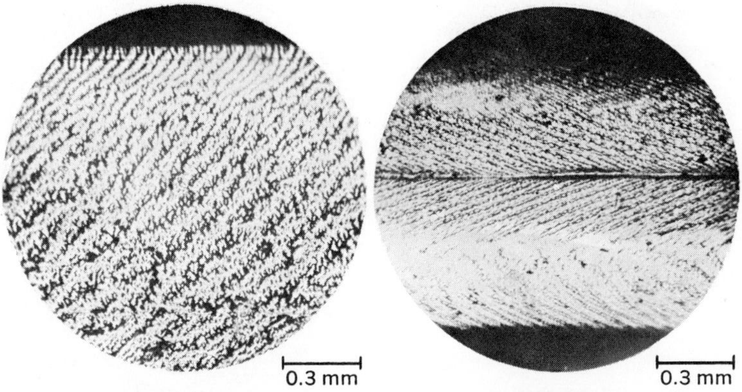

(*a*) $r = 50\%$ (*b*) $r = 50\%$

Fig. 17.18. Domain patterns observed (*a*) on the roll plane and (*b*) on the side surface of the rolled (001) [100] crystal of Ni_3Fe. Roll direction is horizontal.

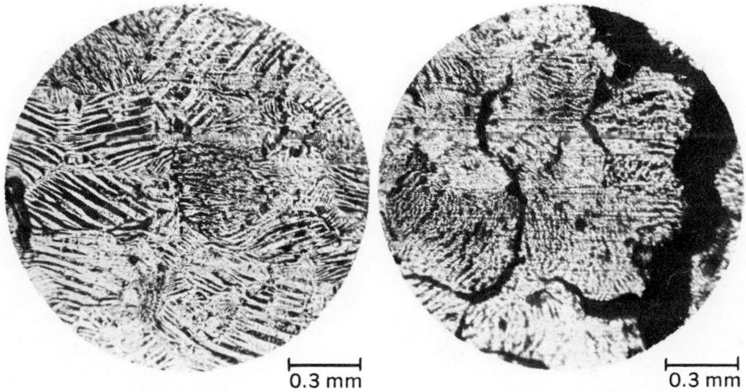

(*a*) $r = 39\%$ (*b*) $r = 30\%$

Fig. 17.19. Domain patterns of the rolled polycrystal of Ni_3Fe. Roll direction is horizontal.

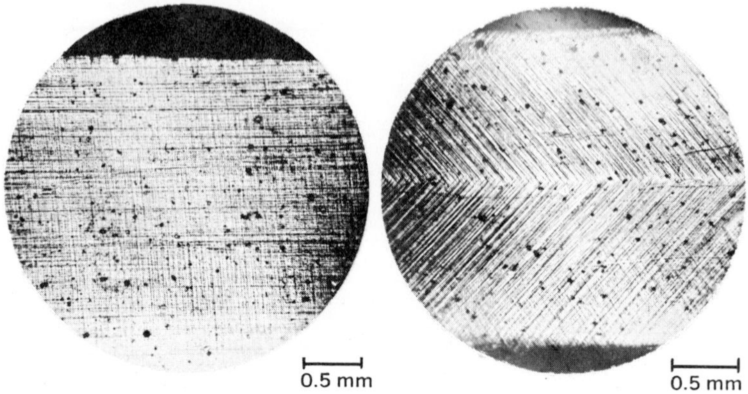

(*a*) $r = 9.1\%$ (*b*) $r = 9.1\%$

Fig. 17.20. Slip bands observed (*a*) on the rolled plane and (*b*) on the side surface of the rolled (001) [110] crystal of Ni_3Fe. Roll direction is horizontal.

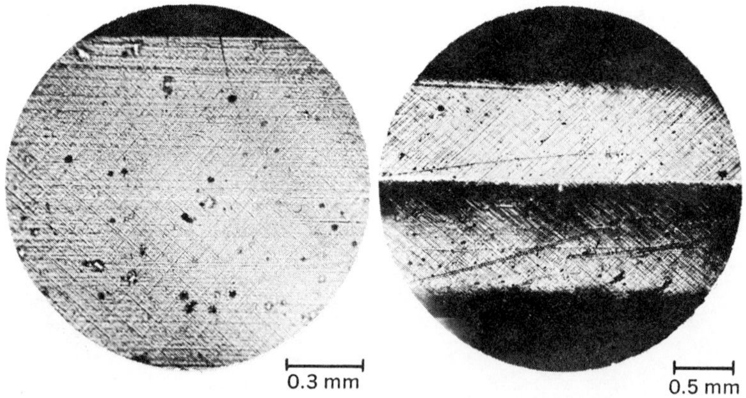

(*a*) $r = 10\%$ (*b*) $r = 10\%$

Fig. 17.21. Slip bands observed (*a*) on the rolled plane and (*b*) on the side surface of the rolled (001) [100] crystal of Ni_3Fe. Roll direction is horizontal.

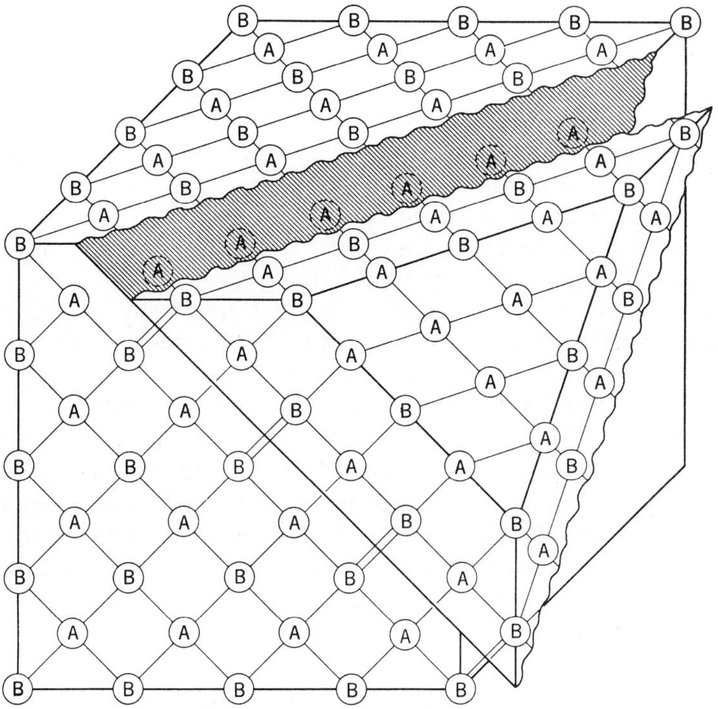

Fig. 17.22. Diagram indicating the appearance of BB pairs due to a single step slip along the (111) plane in the [01$\bar{1}$] direction in an A_3B-type superlattice.

plane can be described by $1/(\sqrt{3}\,a^2)$, where a is the lattice constant. Then the actual number of unbalanced BB pairs is given by $S^2/(\sqrt{3}\,a^2)$, where S is the long range order parameter. Considering that there are $n = \sqrt{3}/a$ atom layers in a unit length perpendicular to the slip plane, we can describe the number of BB pairs included in a unit volume as

$$N_{BBi} = \frac{S^2}{\sqrt{3}\,a^2} \frac{\sqrt{3}\,p_0 p'}{2a} |s_i| = \tfrac{1}{8} N p S^2 |s_i|, \tag{17.25}$$

where

$$p = p_0 p' \tag{17.26}$$

and the suffix i specifies the sort of slip systems. The factor 2 in the denominator of (17.25) is put in because, even if all dislocations are isolated from one another, the probability of having a disordered plane is still 1/2. The anisotropy energy thus induced is expressed by putting (17.25) into (17.5):

$$\begin{aligned} E_a &= \tfrac{1}{8} N l_0 p S^2 \Sigma\, |s_i|\, (\alpha_1 \gamma_{1i} + \alpha_2 \gamma_{2i} + \alpha_3 \gamma_{3i})^2 \\ &= \tfrac{1}{8} N l_0 p S^2 \Sigma\, |s_i|\, f_i(\alpha_1, \alpha_2, \alpha_3), \end{aligned} \tag{17.27}$$

where

$$f_i(\alpha_1, \alpha_2, \alpha_3) = \gamma_{1i}{}^2\alpha_1{}^2 + \gamma_{2i}{}^2\alpha_2{}^2 + \gamma_{3i}{}^2\alpha_3{}^2 + 2\gamma_{2i}\gamma_{3i}\alpha_2\alpha_3 + 2\gamma_{3i}\gamma_{1i}\alpha_3\alpha_1 + 2\gamma_{1i}\gamma_{2i}\alpha_1\alpha_2. \quad (17.28)$$

Since there are four slip planes with three slip directions for each, the number of slip systems is 12. The direction cosines $(\gamma_1, \gamma_2, \gamma_3)$ of the induced BB pairs and the coefficients of each term of $f_i(\alpha_1, \alpha_2, \alpha_3)$ are given for each slip system in Table 17.3. We call this type of anisotropy the Long range order-Fine slip type (L.F. type).

We can consider another type of induced anisotropy. Suppose that the upper part of the crystal shown in Fig. 17.22 continues to travel over an out-of-step boundary to bring itself onto the neighboring order domain. Then one of the three kinds of α site is replaced by a β site in the lower part of the crystal, and the direction of the BB pairs is changed from [011] to [110] or [101]. Thus, when a coarse slip takes place, the directions of BB pairs are expected to be distributed equally among the [011], [110], and [101] directions. The same thing is expected when the size of order domains is small, or, in other words, when only a short range order prevails in the crystal. The average anisotropy of these three BB pairs is given by

$$\begin{aligned} w &= \frac{l_0}{3}(\cos^2\phi_{[011]} + \cos^2\phi_{[101]} + \cos^2\phi_{[110]}) \\ &= \frac{l_0}{3}\sum_{k=1,2,3}(\gamma_{1k}\alpha_1 + \gamma_{2k}\alpha_2 + \gamma_{3k}\alpha_3)^2 \\ &= l_0(n_{2i}n_{3i}\alpha_2\alpha_3 + n_{3i}n_{1i}\alpha_3\alpha_1 + n_{1i}n_{2i}\alpha_1\alpha_2), \end{aligned} \quad (17.29)$$

where (n_{1i}, n_{2i}, n_{3i}) are the direction cosines of the normal to the slip plane of the ith system. The anisotropy given by (17.29) is a uniaxial anisotropy with the axis normal to the slipped plane. The comprehensive explanation for this is that, if the slip takes place in a crystal with a short range order, the atomic relation between both sides of the slipped plane becomes disordered, signifying an abundance of BB pairs and thus rendering the hard axis normal to the slip plane, provided that $l_0 > 0$. The number of BB pairs produced in a unit area of the slipped plane is

$$N_{\mathrm{BB}} = \tfrac{1}{16}Np'\sigma\,|s_i|, \quad (17.30)$$

where σ is the short range order parameter. The anisotropy energy is then expressed as

$$E_a = \tfrac{1}{16}Nl_0p'\sigma\sum|s_i|\,g_i(\alpha_1, \alpha_2, \alpha_3), \quad (17.31)$$

Table 17.3. Direction Cosines of BB Pairs and the Coefficients of Each Term in $f_i(\alpha_1, \alpha_2, \alpha_3)$ in the Expression of Long Range Order-Fine Slip Type of Anisotropy (cf. 17.28)

No. of Slip System	Slip Plane	Slip Direction	γ_1	γ_2	γ_3	Coefficients in $f(\alpha_1, \alpha_2, \alpha_3)$: α_1^2	α_2^2	α_3^2	$\alpha_2\alpha_3$	$\alpha_3\alpha_1$	$\alpha_1\alpha_2$
1	(111)	$[01\bar{1}]$	0	$\frac{1}{\sqrt{2}}$	$\frac{1}{\sqrt{2}}$	0	$\frac{1}{2}$	$\frac{1}{2}$	1	0	0
2		$[10\bar{1}]$	$\frac{1}{\sqrt{2}}$	0	$\frac{1}{\sqrt{2}}$	$\frac{1}{2}$	0	$\frac{1}{2}$	0	1	0
3		$[1\bar{1}0]$	$\frac{1}{\sqrt{2}}$	$\frac{1}{\sqrt{2}}$	0	$\frac{1}{2}$	$\frac{1}{2}$	0	0	0	1
4	$(11\bar{1})$	$[101]$	$\frac{1}{\sqrt{2}}$	0	$-\frac{1}{\sqrt{2}}$	$\frac{1}{2}$	0	$\frac{1}{2}$	0	-1	0
5		$[011]$	0	$\frac{1}{\sqrt{2}}$	$-\frac{1}{\sqrt{2}}$	0	$\frac{1}{2}$	$\frac{1}{2}$	-1	0	0
6		$[1\bar{1}0]$	$\frac{1}{\sqrt{2}}$	$\frac{1}{\sqrt{2}}$	0	$\frac{1}{2}$	$\frac{1}{2}$	0	0	0	1
7	$(1\bar{1}1)$	$[110]$	$\frac{1}{\sqrt{2}}$	$-\frac{1}{\sqrt{2}}$	0	$\frac{1}{2}$	$\frac{1}{2}$	0	0	0	-1
8		$[10\bar{1}]$	$\frac{1}{\sqrt{2}}$	0	$\frac{1}{\sqrt{2}}$	$\frac{1}{2}$	0	$\frac{1}{2}$	0	1	0
9		$[011]$	0	$\frac{1}{\sqrt{2}}$	$-\frac{1}{\sqrt{2}}$	0	$\frac{1}{2}$	$\frac{1}{2}$	-1	0	0
10	$(\bar{1}11)$	$[01\bar{1}]$	0	$\frac{1}{\sqrt{2}}$	$\frac{1}{\sqrt{2}}$	0	$\frac{1}{2}$	$\frac{1}{2}$	1	0	0
11		$[101]$	$\frac{1}{\sqrt{2}}$	0	$-\frac{1}{\sqrt{2}}$	$\frac{1}{2}$	0	$\frac{1}{2}$	0	-1	0
12		$[110]$	$\frac{1}{\sqrt{2}}$	$-\frac{1}{\sqrt{2}}$	0	$\frac{1}{2}$	$\frac{1}{2}$	0	0	0	-1

where

$$g_i(\alpha_1, \alpha_2, \alpha_3) = n_{2i}n_{3i}\alpha_2\alpha_3 + n_{3i}n_{1i}\alpha_3\alpha_1 + n_{1i}n_{2i}\alpha_1\alpha_2. \tag{17.32}$$

The values of (n_{1i}, n_{2i}, n_{3i}) and the coefficients of each term are listed in Table 17.4. This type of anisotropy is called the Short range order-Coarse slip type (S.C. type).

Now we can interpret the individual experimental facts in terms of these two categories of anisotropy. For (110) [001] rolling, we expect the occurrence of four kinds of slip systems, the index for each being 1, 2, 4,

Table 17.4. Direction Cosines of the Normal to the Slip Plane and Coefficient of Each Term in $g_i(\alpha_1, \alpha_2, \alpha_3)$ in the Expression for the Short Range Order-Coarse Slip Type of Anisotropy (cf. 17.32)

No. of Slip System	n_1	n_2	n_3	Coefficients in $g(\alpha_1, \alpha_2, \alpha_3)$ $\alpha_2\alpha_3$	$\alpha_3\alpha_1$	$\alpha_1\alpha_2$
1	$1/\sqrt{3}$	$1/\sqrt{3}$	$1/\sqrt{3}$	$\frac{1}{3}$	$\frac{1}{3}$	$\frac{1}{3}$
2	$1/\sqrt{3}$	$1/\sqrt{3}$	$1/\sqrt{3}$	$\frac{1}{3}$	$\frac{1}{3}$	$\frac{1}{3}$
3	$1/\sqrt{3}$	$1/\sqrt{3}$	$1/\sqrt{3}$	$\frac{1}{3}$	$\frac{1}{3}$	$\frac{1}{3}$
4	$1/\sqrt{3}$	$1/\sqrt{3}$	$-1/\sqrt{3}$	$-\frac{1}{3}$	$-\frac{1}{3}$	$\frac{1}{3}$
5	$1/\sqrt{3}$	$1/\sqrt{3}$	$-1/\sqrt{3}$	$-\frac{1}{3}$	$-\frac{1}{3}$	$\frac{1}{3}$
6	$1/\sqrt{3}$	$1/\sqrt{3}$	$-1/\sqrt{3}$	$-\frac{1}{3}$	$-\frac{1}{3}$	$\frac{1}{3}$
7	$1/\sqrt{3}$	$-1/\sqrt{3}$	$1/\sqrt{3}$	$-\frac{1}{3}$	$\frac{1}{3}$	$-\frac{1}{3}$
8	$1/\sqrt{3}$	$-1/\sqrt{3}$	$1/\sqrt{3}$	$-\frac{1}{3}$	$\frac{1}{3}$	$-\frac{1}{3}$
9	$1/\sqrt{3}$	$-1/\sqrt{3}$	$1/\sqrt{3}$	$-\frac{1}{3}$	$\frac{1}{3}$	$-\frac{1}{3}$
10	$-1/\sqrt{3}$	$1/\sqrt{3}$	$1/\sqrt{3}$	$\frac{1}{3}$	$-\frac{1}{3}$	$-\frac{1}{3}$
11	$-1/\sqrt{3}$	$1/\sqrt{3}$	$1/\sqrt{3}$	$\frac{1}{3}$	$-\frac{1}{3}$	$-\frac{1}{3}$
12	$-1/\sqrt{3}$	$1/\sqrt{3}$	$1/\sqrt{3}$	$\frac{1}{3}$	$-\frac{1}{3}$	$-\frac{1}{3}$

and 5 (Fig. 17.23). It is easily shown by a simple calculation that the slip densities of each slip system are

$$s_1 = s_4 = -\frac{r}{2}, \qquad s_2 = s_5 = \frac{r}{2}, \tag{17.33}$$

where r is the fractional reduction of the thickness. On putting (17.33) into (17.27) and (17.31), we have

$$E_a = \tfrac{1}{16} N l_0 p S^2 r \alpha_3{}^2 \tag{17.34}$$

for the L.F. type and

$$E_a = \tfrac{1}{24} N l_0 p' \sigma r \alpha_1 \alpha_2 \tag{17.35}$$

for the S.C. type. Since $l_0 > 0$ for Fe-Ni alloys, the L.F. type of anisotropy (17.34) gives the hard axis parallel to the roll direction, or the easy plane perpendicular to the roll direction. The S.C. type of anisotropy shows that easy axis is parallel to $[1\bar{1}0]$ or perpendicular to the roll

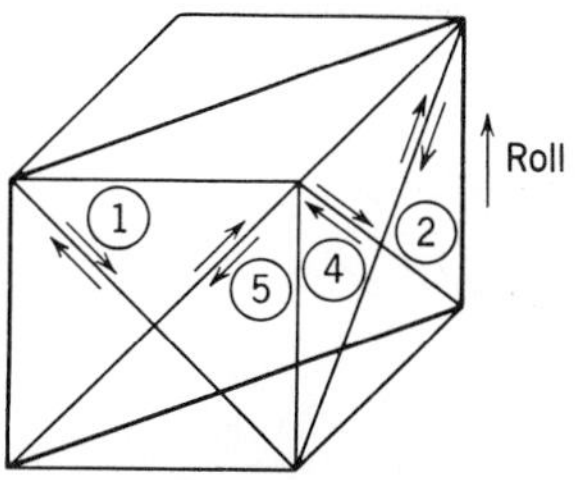

Fig. 17.23. Slip systems contributing to the slip deformation (110) [001].

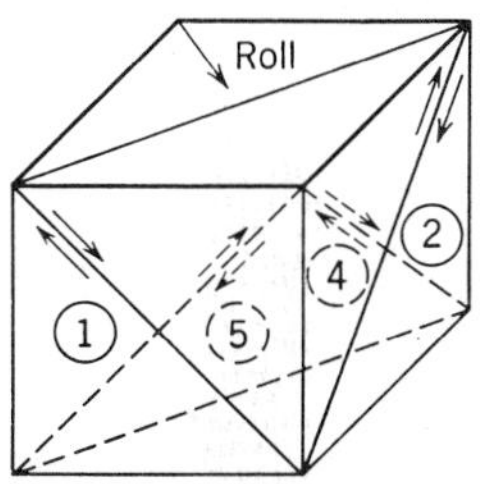

Fig. 17.24. Slip systems contributing to the slip deformation (001) [110].

direction. Thus both types support the experimental fact that the easy axis is perpendicular to the roll direction.

For (001) [110] rolling, the possible slip systems are 1, 2, 4, and 5 (Fig. 17.24), among which only 1 and 2 (or 4 and 5) actually occur in the upper half (or the lower half) of the crystal plate, as seen in the slip bands observed on the side surface of the crystal plate in Fig. 17.20*b*. The reason why such an easy glide occurs only for (001) [110] rolling is suggested to be: The external force which acts on the specimen plate during the process of rolling is, at first approximation, resolved into a tension parallel to the roll direction and a compression normal to the roll plane and is thus considered to give rise to the maximum shearing stress along the plane inclined 45° from the roll plane. In (001) [110] rolling, the actual slip plane makes an angle of 54.5° with the roll plane at the initial stage of rolling (Fig. 17.25), and the slip along this plane rotates the crystal so as to drive this angle toward 45°, making itself more favorable than the other slip plane. In contrast, the slip plane makes an angle of 35.5° with the roll plane for (110) [001] rolling (Fig. 17.25), and thus the crystal rotation resulting from the occurrence of slip along one slip plane does favor the subsequent occurrence of slip along the other slip plane, resulting in a crossed slip.

The slip densities for (001) [110] rolling are given by

$$s_1 = r, \qquad s_2 = -r \tag{17.36}$$

from which we have

$$E_a = \tfrac{1}{16} N l_0 p S^2 r \alpha_3 (2\alpha_1 + 2\alpha_2 + \alpha_3) \tag{17.37}$$

for the L.F. type and

$$E_a = \tfrac{1}{24} N l_0 p' \sigma r (\alpha_1\alpha_2 + \alpha_2\alpha_3 + \alpha_3\alpha_1) \tag{17.38}$$

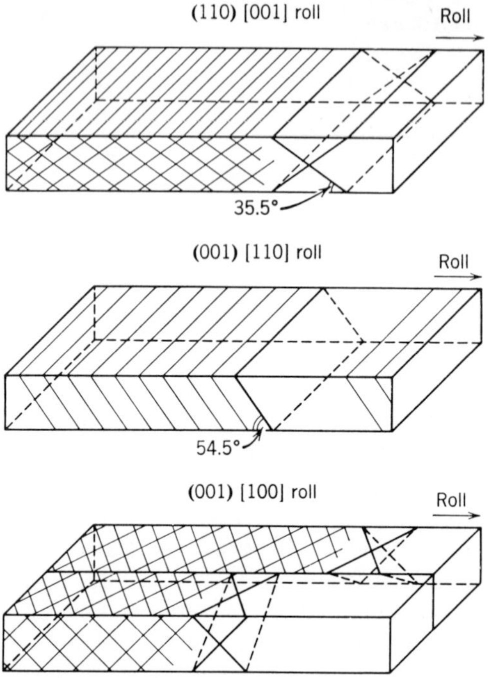

Fig. 17.25. Actual slips occurring in the rolled (110) [001], (001) [110], and (001) [100] crystals.

for the S.C. type. The L.F. type given by (17.37) does not give rise to any anisotropy energy in the roll plane for which $\alpha_3 = 0$, whereas the S.C. type does give rise to the uniaxial anisotropy with the easy axis perpendicular to the roll direction in the roll plane. The actual easy axis of the L.F. type of anisotropy is, however, parallel to $[11\bar{1}]$, and this conforms with the experimental observation of domain magnetization at the side surface (shown in Figs. 17.17*b* and 17.17*d*). With the progress of easy glide, this easy axis rotates to approach the roll plane, giving rise to uniaxial anisotropy with an easy axis parallel to the roll direction. The latter contribution to the anisotropy constant is expected to be proportional to the square of the roll reduction *r*, and we can explain the shape of the experimental curve shown in Fig. 17.16.

In (001) [100] rolling, the slip systems seem to be complicated as shown in Fig. 17.25, where we see the occurrence of a pair of slip systems which are different in different places. The anisotropy energy which stabilizes the domain magnetization perpendicular to the roll direction is expected from both types of induced anisotropy.

A more detailed discussion is to be found in the papers by Chikazumi, Suzuki and Iwata.[44] Bunge and Müller[45] proposed a similar interpretation for this phenomenon. The investigations were extended to the roll magnetic anisotropy of Fe_3Al crystal,[46] which is a body-centered cubic crystal, and also to the Ni-Co alloys.[47] In the latter, fourth order anisotropy was found to be induced.

17.3 Anisotropy Induced at the Crystal Transition Point

The third type of induced anisotropy is that induced at a point where the crystal transforms from cubic to uniaxial or other types with lower symmetry. If a magnetic field is applied in a certain direction, the crystallites which favor this direction of magnetization may be expected to grow at the expense of crystallites otherwise oriented. One of the interesting examples of this phenomenon is the low temperature transition of magnetite about which many interesting investigations have been made for a very long time.

This phenomenon was discovered by Weiss and Forrer,[48] who observed that the saturation magnetization is abruptly reduced at $-155°C$ ($= 118°K$). It was found later on that this does not signify any decrease in spontaneous magnetization but is simply due to the appearance of a very large magnetocrystalline anisotropy, as recognized from the fact that the jump in the magnetization is greatly reduced by the application of a strong magnetic field (Fig. 17.26*a*). This abrupt change in magnetocrystalline anisotropy is accompanied by an anomaly of specific heat[50] (Fig. 17.26*b*), a discontinuous increase in electrical resistivity[51] (Fig. 17.26*c*), and a fairly large deformation of the lattice[49,52] (Fig. 17.26*d*). This deformation was investigated in detail by X-ray measurement and found to correspond to the crystal transformation from cubic to orthorhombic[53] (Fig. 17.27). In this orthorhombic lattice, the equal numbers of Fe^{2+} and Fe^{3+} ions in the 16*d* sites are considered to make the ordered arrangement shown in Fig. 17.27, as first proposed by Verwey,[54] and later confirmed by Hamilton[55] by means of neutron diffraction. It was originally considered by Verwey[56] that the relatively low electrical resistivity of magnetite is due to a hopping motion of electrons among Fe^{2+} and Fe^{3+} ions in 16*d* sites. The ordered arrangement of Fe^{2+} and Fe^{3+} ions prohibits such a hopping motion of electrons and thus explains an increase in resistivity at the transition point. We call this ordered arrangement the Verwey order.

The effect of magnetic cooling through this transition point was first observed by Li.[57] If a magnetic field is applied parallel to one of the $\langle 100 \rangle$ directions during the process of cooling through the transition point, this $\langle 100 \rangle$ direction becomes the easy axis of a strong magnetic

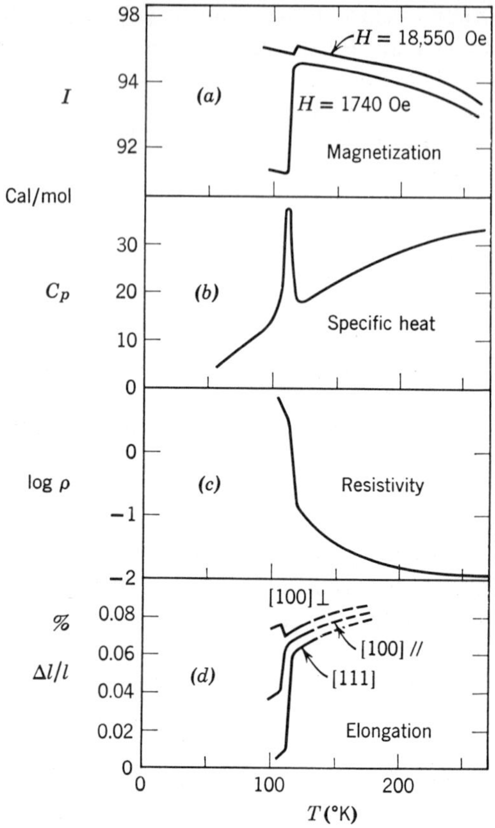

Fig. 17.26. Variation of several quantities at the low temperature transition point (Bickford[49]).

anisotropy. This anisotropy is interpreted to be caused by a build-up of the Verwey order whose *c* axis is parallel to the ⟨100⟩ direction along which the field was applied. If the direction of the field is changed to another ⟨100⟩ direction at a temperature just below the transition point, the original Verwey order is switched over to another phase of Verwey order whose *c* axis is parallel to the new ⟨100⟩ direction. This switchover effect was observed by Bickford[58] by means of ferromagnetic resonance as shown in Fig. 17.28, where we see the absorption line corresponding to the original Verwey order decreasing while that corresponding to the new Verwey order appears during the process of heating. The difference in the position of the absorption lines corresponds to the difference in the anisotropy field between both phases of the Verwey order. For a normal magnetic annealing due to the directional ordering of alloy constituents, we expect the position of the absorption line to be continuously shifted without changing the height of the line. In magnetite, however, the positions of the absorption lines are almost invariant throughout the whole process of switching over. This means that the atom arrangement is changed directly from one type of Verwey order to another one without taking on an intermediate arrangement. It is by this feature that the third type of induced anisotropy is distinguished from normal magnetic annealing, as mentioned in Section 17.1.

Magnetocrystalline anisotropy of Verwey order was measured by Williams, Bozorth, and Goertz[59] and by Calhoun.[60] The latter expressed the anisotropy energy as

$$E_a = K_a\alpha_1^2 + K_b\alpha_2^2 + K_{aa}\alpha_1^4 + K_{ab}\alpha_1^2\alpha_2^2 + K_{bb}\alpha_2^4, \quad (17.39)$$

where $(\alpha_1, \alpha_2, \alpha_3)$ are the direction cosines of magnetization with respect to the orthorhombic axes *a*, *b*, and *c*. The coefficients have been experimentally determined to be

$$K_a = 9.0 \times 10^4, \quad K_b = 4.0 \times 10^4,$$
$$K_{aa} = 10.2 \times 10^4, \quad K_{ab} = 13.0 \times 10^4, \quad (17.40)$$
$$K_{bb} = 2.2 \times 10^4,$$

where the units are joules per cubic meter.

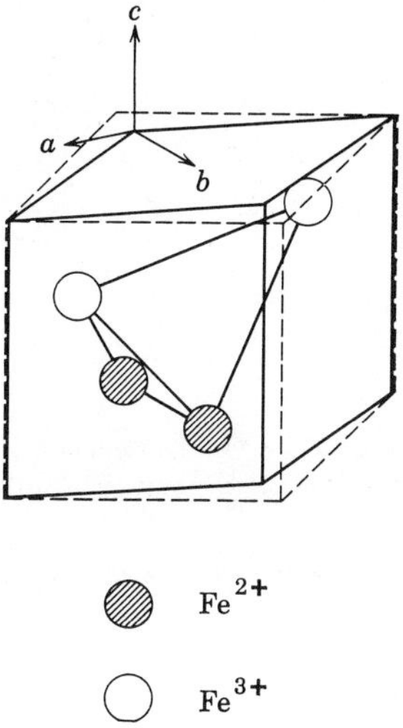

Fig. 17.27. Distribution of Fe^{2+} and Fe^{3+} ions in Verwey order in relation to the orthorhombic axes.

This anisotropy energy is plotted in Fig. 17.29 as a function of the rotational angle of the magnetization in the $(1\bar{1}0)$ plane of the cubic coordinates for six types of the Verwey order. The graph in the figure shows that, if the magnetization points parallel to the *c* axis, the energy is lowest for types 5 and 6; hence these types are expected to grow simultaneously as twins. If a preferential growth of a single type of Verwey order is desired, we simply magnetize the specimen in the direction $\theta = 40°$ measured from [100], where the energy of type 5 is the lowest of the six types of Verwey order. Actually a single phase of Verwey order was developed by Calhoun[60] and Hamilton[55] by this method. It has been reported that this procedure is not successful for a natural single crystal of magnetite. The reason for this is considered to be the presence of inclusions and lattice defects in the crystal. The low temperature transition has also been investigated for Cu ferrite[61] and Ni ferrite.[62,63]

A similar effect was also observed by Takahashi and Kono[64] for pure

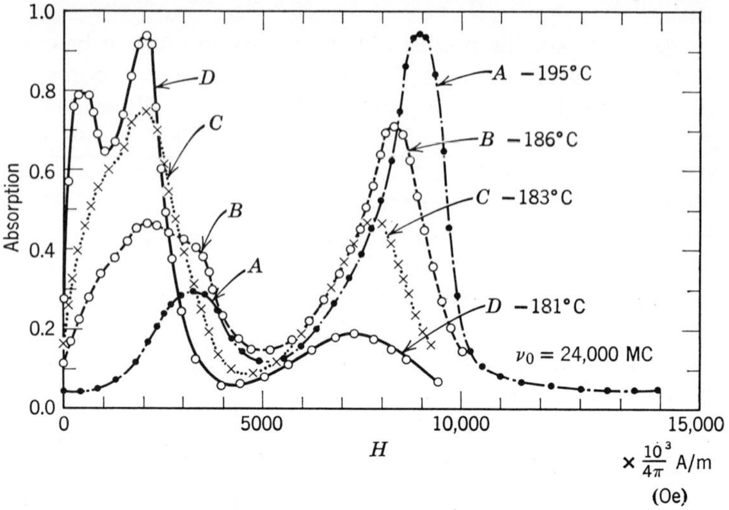

Fig. 17.28. Ferromagnetic resonance absorption lines of Fe_3O_4 observed just below the transition point (Bickford[58]).

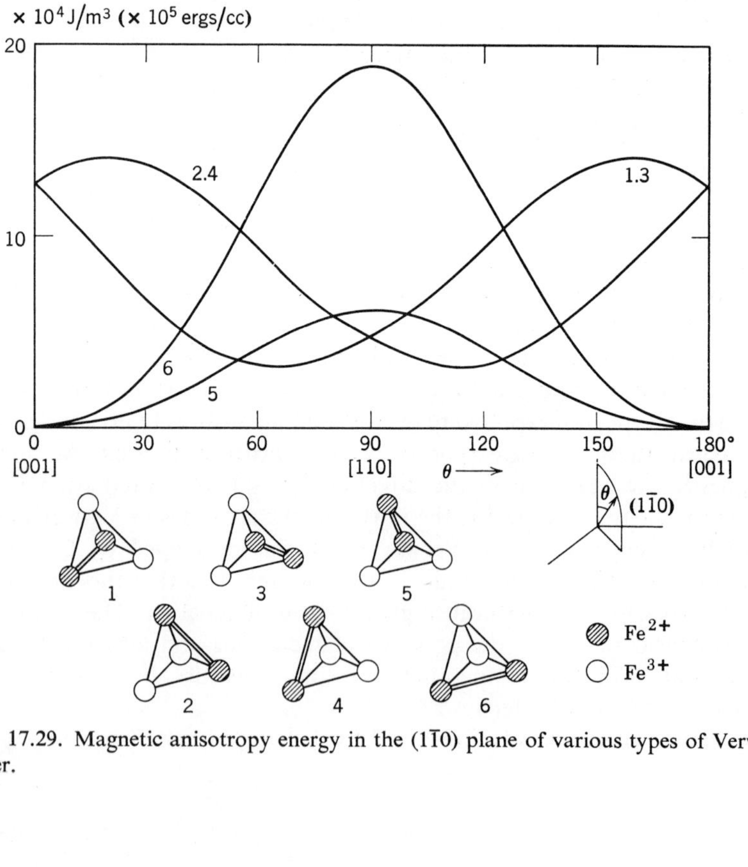

Fig. 17.29. Magnetic anisotropy energy in the (1$\bar{1}$0) plane of various types of Verwey order.

cobalt and Co-Ni alloys which transform from face-centered cubic to hexagonal at about 500°C. It is known that the hexagonal phase is developed so as to make its c axis parallel to one of the $\langle 111 \rangle$ directions of the face-centered cubic lattice. If a magnetic field is applied during the process of transition, the hexagonal phase with its c axis closest to the direction of the applied field is expected to be preferentially developed among the four possibilities, provided that the uniaxial anisotropy constant is positive. Since the anisotropy constant of cobalt changes its sign at about 260°C, the induced anisotropy develops its hard direction parallel to the annealing field at room temperature (Fig. 17.30). For a Co-Ni alloy which contains more than 5% Ni, the easy axis of the induced anisotropy becomes parallel to the direction of the annealing field, because the isotropic point is no longer above room temperature. The magnitude of the induced anisotropy is as large as 2.2×10^4 J/m³. Graham[65] obtained about ten times larger values and also measured the temperature dependence. This phenomenon is thought to be useful for the study of the mechanism of crystal transition.

Finally we briefly discuss the unidirectional anisotropy found by Meiklejohn and Bean.[66] They discovered that, if fine particles of cobalt of diameter 100 ∼ 1000 Å are oxidized slightly at their surface and cooled down to 77°K in a strong magnetic field, they exhibit a strong unidirectional magnetic anisotropy which shows only one easy direction during one complete rotation of magnetization. They interpreted this phenomenon as follows: When the particles are cooled through the Néel point of CoO, 293°K, spins in the oxide layer build up the antiferromagnetic arrangement so as to direct the end spins parallel to spins in Co particles, provided that the coupling between cobalt and the oxide layer is positive. After cooling, if the magnetic field rotates out of the original direction, the magnetization in cobalt is rotated, whereas the spins in the oxide layer maintain their original directions (Fig. 17.31), because of the nature of the antiferromagnetic arrangement and also because of the large value of the intrinsic anisotropy (5×10^5 J/m³). Thus the exchange coupling between cobalt and the oxide layer gives rise to the anisotropy energy,

$$E_a = -K_d \cos \theta, \tag{17.41}$$

where θ is the angle of magnetization measured from its stable direction. The value of K_d is as large as 1×10^5 J/m³ for the Co–CoO system.

The hysteresis loop of this material is found to shift along the abscissa as shown in Fig. 17.32. The broken line represents the loop of the same material which was cooled in the absence of a magnetic field. This figure tells us that the magnetization in the minus direction requires a much stronger magnetic field than the magnetization in the plus direction.

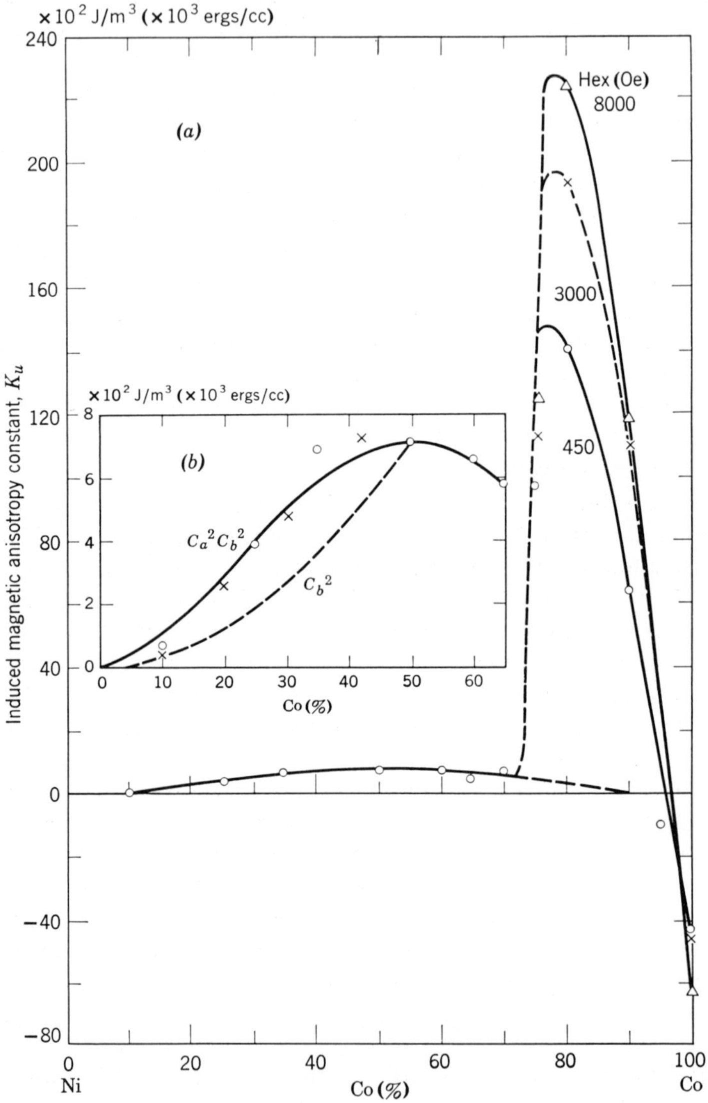

Fig. 17.30. Constant of uniaxial anisotropy as a function of the composition of Co-Ni alloy as cooled from 1000°C in a magnetic field at a rate of 3.3°/min. The numerical values are the field applied during cooling in oersteds or $(10^3/4\pi)$(A/m). (Takahashi and Kono.[64])

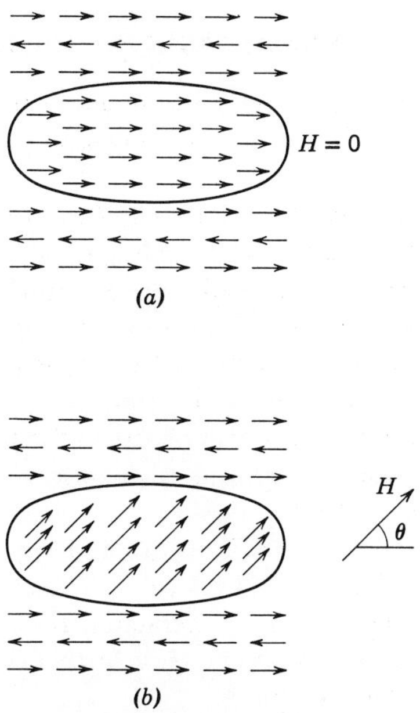

Fig. 17.31. Spin arrangement in a Co particle and in its oxide layer.

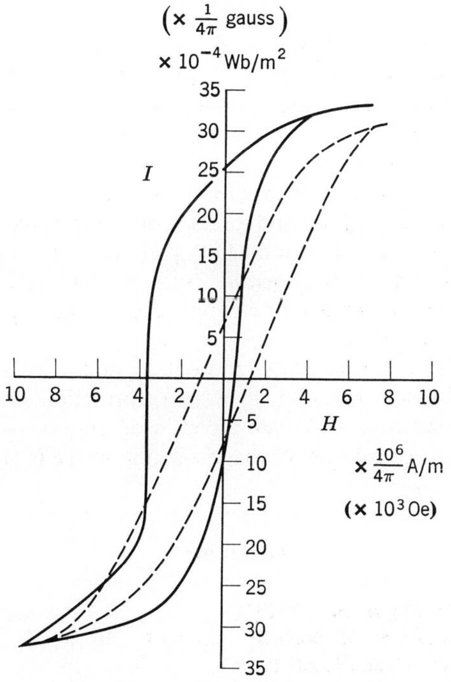

Fig. 17.32. Magnetization curve of slightly oxidized Co fine particles cooled down to 77°K with (solid line) or without (broken line) magnetic field (Meiklejohn and Bean[66]).

This is caused by the exchange interaction between the metal and the oxide layer which drives the magnetization in cobalt metal in the plus direction.

It is also reported by Meiklejohn and Bean that the rotational hysteresis (cf. Section 18.3) does not approach zero as the external field becomes infinite in contrast to the normal magnetic substance. We can understand this if we consider that part of the antiferromagnetic arrangement of spins in the oxide layer is reoriented discontinuously through the exchange interaction during the process of rotation of magnetization in cobalt. The same phenomenon was also observed for the Fe-FeO system.[67]

The exchange anisotropy was also observed for a number of alloy systems such as UMn_2,[68] Ni-Mn alloys from 20 to 30 at% Mn,[69,70] Fe-Al alloys with 30 at% Al,[71] Cu_3Mn and Ag_3Mn,[72] Co-Mn alloys from 25 to 35 at% Mn,[73] and stainless steel.[74] These phenomena are interpreted as being due to the exchange interaction between some of the magnetic atoms which constitute the alloy. For further details the reader is referred to the review by Meiklejohn.[75]

Problems

17.1. After a single crystal of a body-centered cubic alloy is cooled in a magnetic field applied parallel to [001], [110], or [111], the induced anisotropy energy is measured by rotating the magnetization in the $(1\bar{1}0)$ plane. Calculate the ratio of the uniaxial anisotropy constants for the three cases by assuming the nearest neighbor interaction.

17.2. After a binary alloy with a cubic lattice was annealed in a magnetic field applied parallel to [110], the induced anisotropy was measured by rotating the magnetization in the (001) and $(1\bar{1}0)$ planes. The anisotropy constants determined for two planes are generally different from each other. Calculate the ratio between the two values for a simple cubic, a body-centered cubic, and a face-centered cubic lattice.

17.3. Suppose that a partially ordered single crystal of a binary face-centered cubic alloy was deformed exclusively by a slip deformation along the (111) plane and in the $[1\bar{1}0]$ direction. Calculate the induced anisotropy energy expressed in terms of the angle of rotation of magnetization in the (111) plane.

References

17.1. G. A. Kelsall: *Physics* **5,** 169 (1934).
17.2. J. F. Dillinger and R. M. Bozorth: *Physics* **6,** 279 (1935); R. M. Bozorth and J. F. Dillinger: *Physics* **6,** 285 (1935).
17.3. S. Chikazumi: *J. Phys. Soc. Japan* **5,** 327, 333 (1950).
17.4. R. M. Bozorth, J. F. Dillinger, and G. A. Kelsall: *Phys. Rev.* **45,** 742 (1934).

17.5. H. D. Arnold and G. W. Elmen: *J. Franklin Inst.* **195,** 621 (1923).
17.6. O. Dahl: *Z. Metallk.* **28,** 133 (1936).
17.7. S. Kaya: *J. Fac. Sci. Hokkaido Imp. Univ.* **2,** 39 (1938).
17.8. Y. Tomono: *J. Phys. Soc. Japan* **4,** 298 (1948).
17.9. S. Kaya: *Rev. Mod. Phys.* **25,** 49 (1953).
17.10. S. Chikazumi and T. Oomura: *J. Phys. Soc. Japan* **10,** 842 (1955).
17.11. L. Néel: *Comp. Rend.* **237,** 1613 (1953).
17.12. L. Néel: *J. Phys. Radium* **15,** 225 (1954).
17.13. S. Taniguchi and M. Yamamoto: *Sci. Rept. Res. Inst. Tohoku Univ.* **A6,** 330 (1954).
17.14. S. Taniguchi: *Sci. Rept. Res. Inst. Tohoku Univ.* **A7,** 269 (1955).
17.15. S. Chikazumi: *J. Phys. Soc. Japan* **11,** 551 (1956).
17.16. K. Aoyagi, S. Taniguchi, and M. Yamamoto: *J. Phys Soc. Japan* **13,** 532. (1958).
17.17. M. Yamamoto, S. Taniguchi, and K. Aoyagi: *Sci. Rept. Res. Inst. Tohoku Univ.* **A13,** 117 (1961).
17.18. K. Aoyagi: *Sci. Rept. Res. Inst. Tohoku Univ.* **A13,** 137 (1961).
17.19. K. Suzuki: *J. Phys. Soc. Japan* **13,** 756 (1958).
17.20. S. Chikazumi and T. Wakiyama: G.37, p. 325.
17.21. E. T. Ferguson: *J. Appl. Phys.* **29,** 252 (1958); *J. Phys. Radium* **20,** 251 (1959).
17.22. T. Iwata: *Sci. Rept. Res. Inst. Tohoku Univ.* **A10,** 34 (1958).
17.23. Y. Kato and T. Takei: *J. Inst. Elec. Engrs. Japan* **53,** 408 (1933).
17.24. S. Iida, H. Sekizawa, and Y. Aiyama: *J. Phys. Soc. Japan* **10,** 907 (1955); **13,** 58 (1958).
17.25. R. F. Penoyer and L. R. Bickford, Jr.: *Phys. Rev.* **108,** 271 (1957).
17.26. J. C. Slonczewski: *Phys. Rev.* **110,** 1341 (1958).
17.27. L. R. Bickford, Jr., J. M. Brownlow, and R. F. Penoyer: *J. Appl. Phys.* **29,** 441 (1958).
17.28. T. Inoue, H. Mizuta, and S. Iida: *J. Phys. Soc. Japan* **15,** 1899 (1960).
17.29. S. Iida: *J. Appl. Phys.* **31,** 251S (1961).
17.30. Y. Aiyama, H. Sekizawa, and S. Iida: *J. Phys. Soc. Japan* **12,** 742 (1957).
17.31. D. A. Oliver and J. W. Shedden: *Nature* **142,** 209 (1947).
17.32. L. Néel: *Comp. Rend.* **225,** 109 (1947).
17.33. L. A. Shubina and J. S. Shur: *J. Tech. Phys.* (*USSR*) **19,** 88 (1949).
17.34. C. Kittel, E. A. Nesbitt, and W. Shockley: *Phys. Rev.* **77,** 739 (1950).
17.35. R. D. Heidenreich and E. A. Nesbitt: *J. Appl. Phys.* **23,** 352 (1952); E. A. Nesbitt and R. D. Heidenreich: *ibid.*, **23,** 366 (1952).
17.36. L. Néel: *Compt. Rend.* **237,** 1468 (1953).
17.37. S. Miyahara and T. Mitui: *J. Phys. Soc. Japan* **7,** 534 (1952); *J. Fac. Sci. Hokkaido Univ.* **4,** 275 (1953); T. Mitui: *J. Phys. Soc. Japan* **10,** 905 (1955), T. Mitui and S. Miyahara: *ibid.*, **10,** 1023 (1955); T. Mitui: *ibid.*, **13,** 549 (1958).
17.38. J. J. Becker: *Trans. AIME* **212,** 138 (1958); *J. Appl. Phys.* **29,** 317 (1958).
17.39. C. D. Graham: G.33, p. 288.
17.40. W. Six, J. L. Snoek, and W. G. Burgers: *Ingenieur* **49,** E195 (1934).
17.41. H. W. Conradt, O. Dahl, and K. J. Sixtus: *Z. Metallk.* **32,** 231 (1940).
17.42. G. W. Rathenau and J. L. Snoek: *Physica* **8,** 555 (1941).
17.43. L. Néel: *Compt. Rend.* **238,** 305 (1954).
17.44. S. Chikazumi and K. Suzuki: *Phys. Rev.* **98,** 1130 (1955); S. Chikazumi, K. Suzuki, and H. Iwata: *J. Phys. Soc. Japan* **12,** 1259 (1957); S. Chikazumi: *J. Appl. Phys.* **29,** 346 (1958).

17.45. H. J. Bunge and H. G. Müller: *Wiss. Z. Hochschule Verk. Dresden* **5,** 327 (1957); H. J. Bunge: *Z. Metallk.* **49,** 40 (1958).
17.46. S. Chikazumi, K. Suzuki, and H. Iwata: *J. Phys. Soc. Japan* **15,** 250 (1960); S. Chikazumi: *J. Appl. Phys.* **31,** 158S (1960).
17.47. K. Tamagawa, Y. Nakagawa, and S. Chikazumi: *J. Phys. Soc. Japan* **17,** 1256 (1962).
17.48. D. Weiss and R. Forrer: *Ann. Physik.* **12,** 279 (1929).
17.49. L. R. Bickford, Jr.: *Rev. Mod. Phys.* **25,** 75 (1953).
17.50. R. W. Miller: *J. Am. Chem. Soc.* **51,** 215 (1929).
17.51. T. Okamura: *Sci. Rept. Tohoku Imp. Univ.* **21,** 231 (1932).
17.52. C. A. Domenicali: *Phys. Rev.* **78,** 458 (1950).
17.53. S. C. Abrahams and B. A. Calhoun: *Acta Cryst.* **6,** 105 (1953).
17.54. E. J. W. Verwey, P. W. Haayman, and F. C. Romeijn: *J. Chem. Phys.* **15,** 181 (1947).
17.55. W. C. Hamilton: *Phys. Rev.* **110,** 1050 (1958).
17.56. E. J. W. Verwey: *Nature* **144,** 327 (1939).
17.57. C. H. Li: *Phys. Rev.* **40,** 1002 (1932).
17.58. L. R. Bickford, Jr.: *Phys. Rev.* **78,** 449 (1950).
17.59. H. J. Williams, R. M. Bozorth, and M. Goertz: *Phys. Rev.* **91,** 1107 (1953).
17.60. B. A. Calhoun: *Phys. Rev.* **94,** 1577 (1954).
17.61. C. Guillaud and H. Creveaux: *Compt. Rend.* **230,** 1256 (1950).
17.62. T. Okamura and J. Simoizaka: *Phys. Rev.* **83,** 664 (1951).
17.63. N. Menyuk and K. Dwight: *Phys. Rev.* **111,** 397 (1958).
17.64. M. Takahaski and T. Kono: *J. Phys. Soc. Japan* **15,** 936 (1960).
17.65. C. D. Graham, Jr.: *J. Phys. Soc. Japan* **16** (1961) 1481.
17.66. W. H. Meiklejohn and C. P. Bean: *Phys. Rev.* **102,** 1413 (1956); **105,** 904 (1957).
17.67. W. H. Meiklejohn: *J. Appl. Phys.* **29,** 454 (1958).
17.68. S. T. Lin and A. R. Kaufman: *Phys. Rev.* **108,** 1171 (1957).
17.69. J. S. Kouvel, C. D. Graham, Jr., and I. S. Jacobs: *J. Phys. Radium* **20,** 198 (1959).
17.70. J. S. Kouvel and C. D. Graham, Jr.: *J. Appl. Phys.* **30,** 312S (1959).
17.71. J. S. Kouvel: *J. Appl. Phys.* **30,** 313S (1959).
17.72. J. S. Kouvel: *J. Appl. Phys.* **31,** 142S (1960).
17.73. J. S. Kouvel: *Phys. Chem. Solids* **16,** 107 (1960).
17.74. M. H. Meiklejohn: *J. Appl. Phys.* **32,** 274 (1961).
17.75. M. H. Meiklejohn: *J. Appl. Phys.* **33S,** 1328 (1962).

18

Ferromagnetism of Thin Films, Thin Wires, and Fine Particles

18.1 Ferromagnetism of Thin Films

Interesting problems in the ferromagnetism of thin films are classified into four groups: spontaneous magnetization, domain structure, induced anisotropy, and switching mechanism.

The first problem, that of the spontaneous magnetization of thin films, began with the theory of Klein and Smith.[1] As mentioned in Chapter 4, the Bloch theory predicts that no ferromagnetism is expected to occur in one- and two-dimensional arrays of atomic magnetic moments. Klein and Smith corrected this conclusion and showed that it is not the intensity of spontaneous magnetization but the Curie point that is expected to decrease with a decrease in thickness of the film. This point appeared to be verified by the experiments of Hoffman et al.,[2] but later Neugebauer[3] showed that the temperature dependence is exactly the same as that of bulk material down to 30 Å thickness for nickel films evaporated in a vacuum 10^{-9} mm Hg. The same thing was proved by Gondo et al.[4] for nickel films evaporated in a vacuum of 10^{-5} mm Hg. It was verified theoretically by Valenta[5] that no remarkable change in spontaneous magnetization should be observed at room temperature until the thickness is decreased down to about 4 Å. Calculations were also made by Döring[6] who showed that room temperature magnetization should be 15% less for 20 Å film than for bulk nickel.

The domain structure of a thin film was first theoretically treated by Kittel[7] for a film with its easy axis perpendicular to the film. He calculated

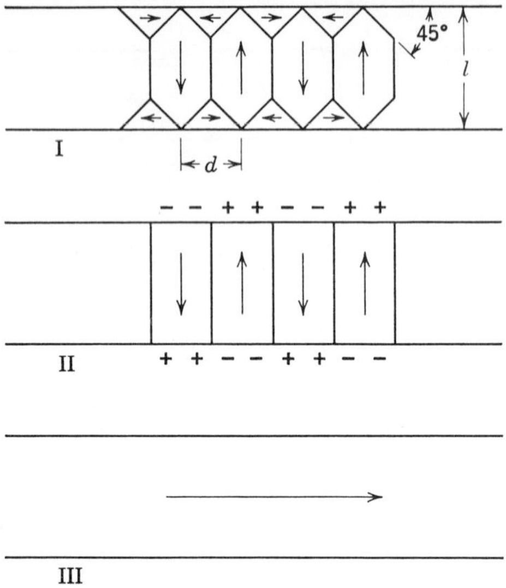

Fig. 18.1. Domain structure of magnetic thin film (the easy axis is assumed to be perpendicular to the surface of the film) (Kittel[7]).

the energy of the three types of domain structures as shown in Fig. 18.1. In film I, there are many domains with magnetization normal to the surface of the film with closure domains which transport the flux from one domain to the other. In film II, the closure domains are omitted, and, in film III, the entire film is a single domain with its magnetization parallel to the surface of the film. The energy of the system is only gradually changed with a change in thickness of film for type I, because the size of domains decreases in proportion to the root of the film thickness (cf. 11.6), whereas the energy for type III is reduced linearly with respect to the thickness, because the energy stored in this system is simply the anisotropy energy which is proportional to the total volume of the film. Figure 18.2 shows these energies calculated for cobalt as a function of film thickness. It is seen in this graph that the energy is lowest for type I if the thickness is larger than 300 Å, while type III is the most stable if the thickness is less than 300 Å. The energy of type II is always higher than those of the other two types for whole range of thicknesses. For materials with larger anisotropy constant, however, it is possible to obtain type II, because the energies of types I and III are related to the anisotropy energy and can thus be raised up relative to type II.

Figure 18.3 shows the various types of domain structures observed by Bean and Roberts[8] by means of the magnetic Kerr effect (cf. Fig. 11.10) for thin films of MnBi. The domain structure changes from the cylindrical type to the wavy type (cf. Figs. 10.6 and 10.7) as the thickness decreases from structure (*a*) to structure (*c*). It was shown by Goodenough[9] that a wavy form of the domain wall is most effective in reducing the magnetostatic energy. Figure 18.4 shows the domain structure of magnetoplumbite ($PbFe_{11}Al_1O_{19}$) observed by Williams and Sherwood,[10] who applied a magnetic field normal to the surface. As the field intensity is changed, the domains of the centipede shape are observed to change their length and move like bacteria. These workers also developed the technique of preparing an MnBi film by means of vacuum evaporation, and by using a pointed magnet wire they drew figures magnetically on the film which can be observed as black and white lines by means of the Faraday effect.

The structure of domain walls in thin films was first investigated theoretically by Néel.[11] He showed that the magnetostatic energy of free poles appearing at the place where the domain wall is intercepted by the surface of the film increases with a decrease in thickness of the film as shown by curve *a* in Fig. 18.5. He predicted that the plane of spin rotation inside the wall will be changed from parallel to the wall surface to parallel to the film surface. Then the wall will possess free poles on both sides of the wall surface, and its magnetostatic energy should be reduced with a decrease of the film thickness as shown by curve *b* in Fig. 18.5. We refer to the former type as the Bloch-type wall, and to the latter as the Néel type. We see in Fig. 18.5 that the Bloch-type wall transforms to the Néel-type wall below the critical thickness, which is estimated to be about 100 Å for iron. It is expected that Néel-type walls tend to be attracted to each other because of the free poles appearing on the side surfaces of the wall, forming a double wall. Actually such a double wall

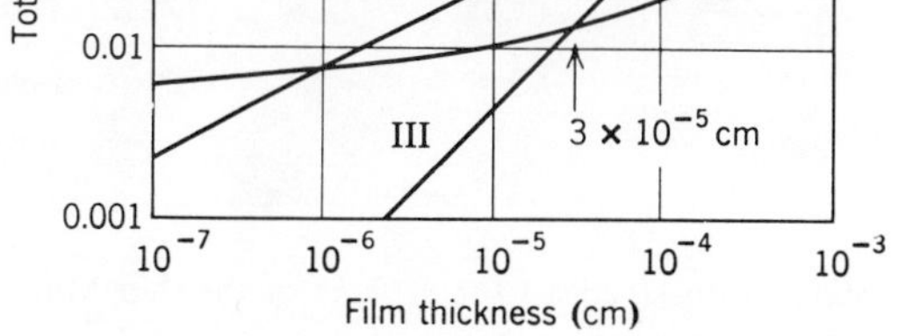

Fig. 18.2. Dependence of the total energy associated with the three kinds of domain structures (Fig. 18.1) on the thickness of the film (Kittel[7]).

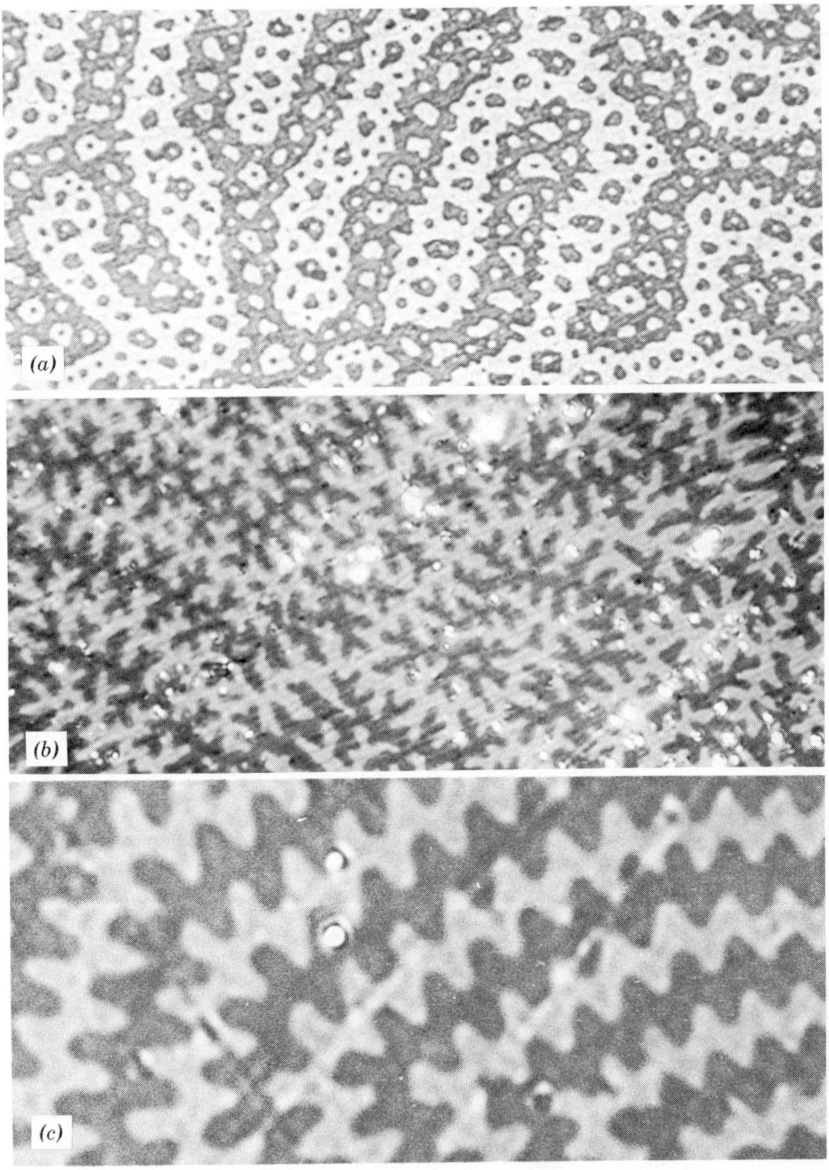

Fig. 18.3. Domain patterns observed on the *c* plane of the thin MnBi single crystal. The thickness of the crystal is greatest for (*a*), and least for (*c*) (Roberts and Bean[9]).

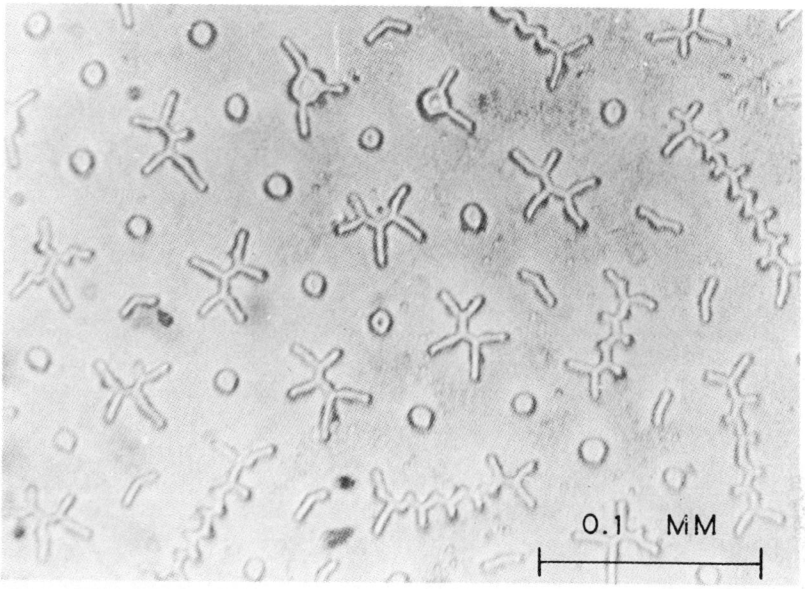

Fig. 18.4. Domain pattern observed on a thin single crystal of magnetoplumbite (Williams and Sherwood[10]).

was observed by Williams and Sherwood[12] for Mo-Permalloy film of thickness 147 Å as shown in Fig. 18.6. Since the domain magnetizations on both sides of a double wall are parallel to each other, the double wall is quite insensitive to the applied field. Hence many double walls are piled up at various places as seen in Figs. 18.6 and 18.7. The wall in an Mo-Permalloy film of 1300 Å is a Bloch-type wall (Fig. 18.8); hence the wall is fairly straight, because free poles appearing at the bent portions of the wall will effectively increase the magnetostatic energy.

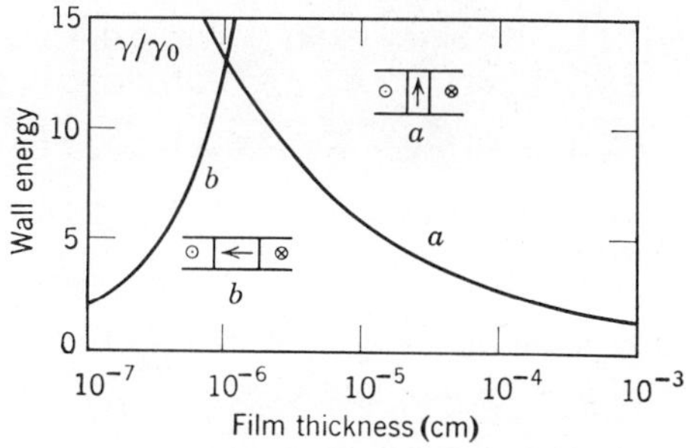

Fig. 18.5. Variation of surface density of wall energy with a change in thickness of the thin film: curve a, Bloch-type wall; curve b, Néel-type wall (Néel[11]).

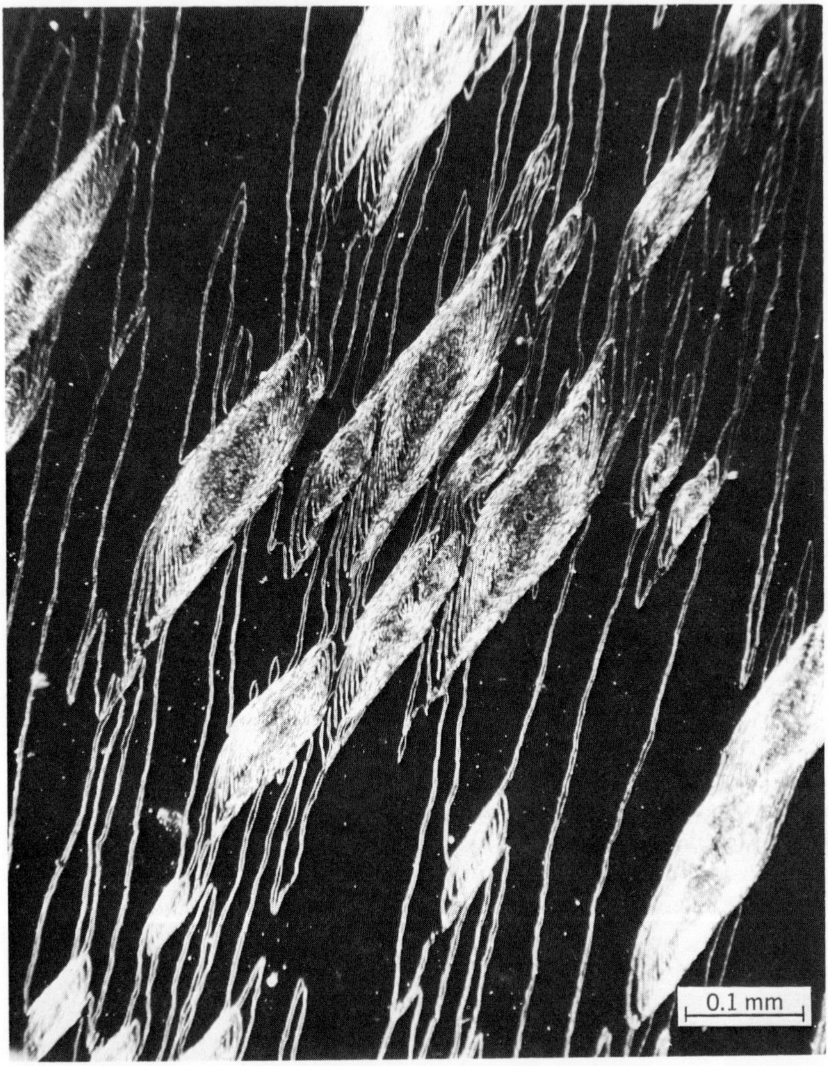

Fig. 18.6. Double walls observed for the thin film of Mo-Permalloy (the thickness is 147 Å) (Williams and Sherwood[12]).

In the intermediate range of thicknesses, the wall has many cross ties, as first observed by Huber, Smith, and Goodenough[13] (Fig. 18.9). They explained the spin structure of the wall as shown in Fig. 18.10. The type of spin rotation inside the wall is just intermediate between the Bloch and Néel types. That is, the sense of the spin rotation as well as the plane of

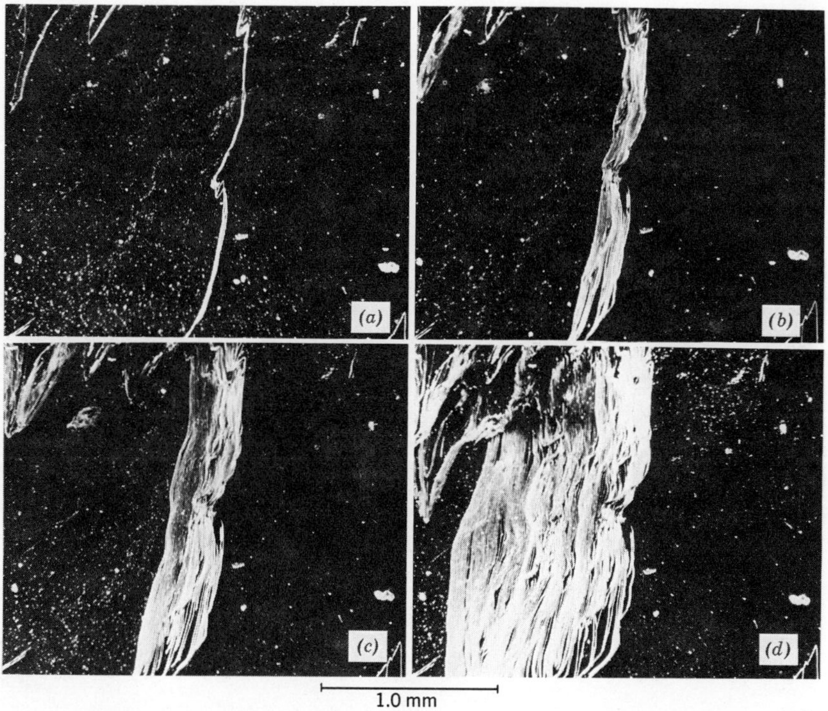

Fig. 18.7. The process of accumulation of double walls at the defect of the film (147 Å Mo-Permalloy film) (Williams and Sherwood[12]).

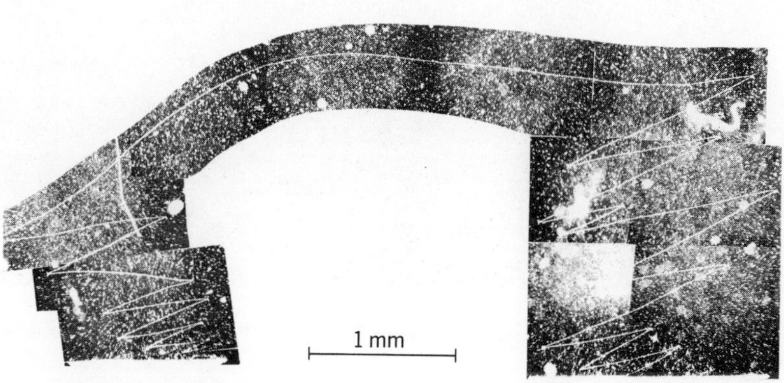

Fig. 18.8. Bloch type wall observed on 1300 Å Mo-Permalloy film (Williams and Sherwood[12]).

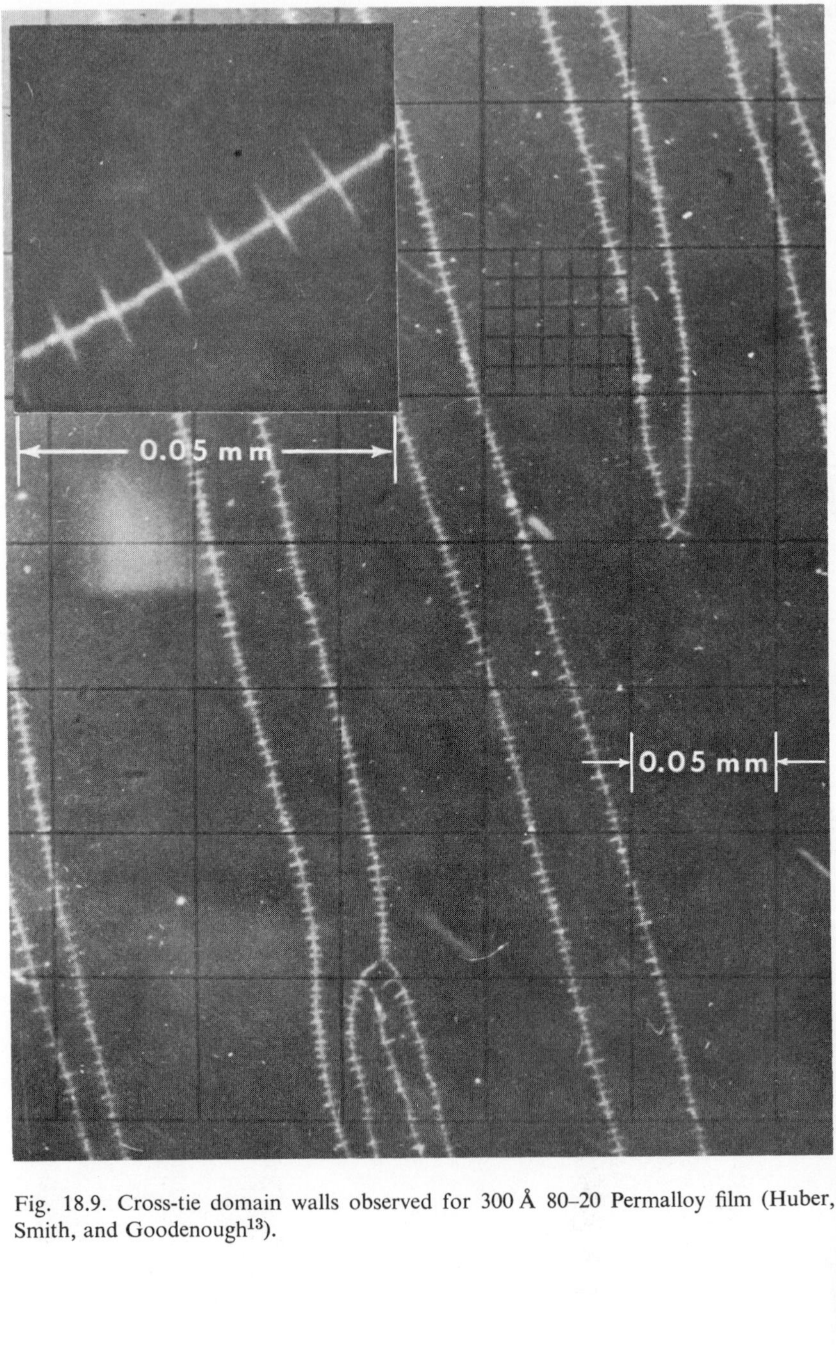

Fig. 18.9. Cross-tie domain walls observed for 300 Å 80–20 Permalloy film (Huber, Smith, and Goodenough[13]).

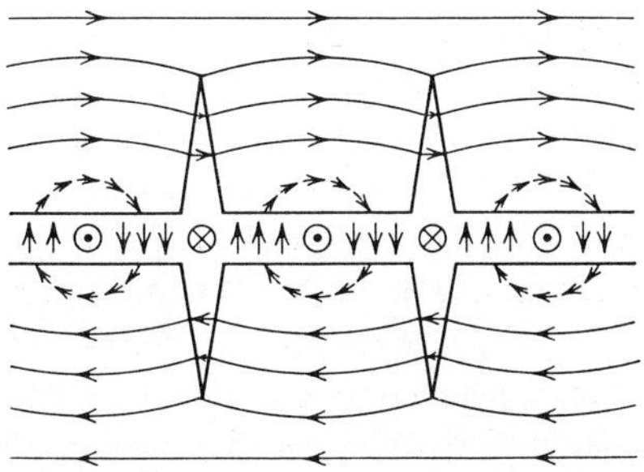

Fig. 18.10. Spin structure of a cross-tie wall.

the rotation changes from place to place. Thus the free poles appearing around the wall change sign from place to place, just as the usual magnetic domain structure does. The cross-tie structure is considered to be induced to produce the outside spin arrangement most favorable to fit the inside spin arrangement.

Recently several excellent techniques for the observation of domain structures in thin films have been developed. One makes use of the transmission electron microscope, in which the electron beam is slightly focused or defocused on passing through a part of a film where the magnetization changes its direction from plus to minus or from minus to plus (Fig. 18.11). This method was first tried by Fuller and Hale.[14] Figure 18.12 shows the pictures of Permalloy films taken by Ichinokawa.[15] Photographs (*a*) and (*b*) show domain structures which are defined by black and white walls. A weak magnetic field is applied parallel to the horizontal direction in the positive sense (*a*) and in the negative sense (*b*). We can see many striations in the domain in which the direction of magnetization is antiparallel to the direction of the applied field. These striations are thought to be caused by a deviation of local magnetization from the average direction of spontaneous magnetization.[16] Photograph (*c*) shows the cross-tie walls one of which appears as a black line with many white cross ties. This condition is explained by the model shown in Fig. 18.10. Boersch and Lambeck[17] have succeeded in observing well-defined domain structures of thin films by means of the Faraday effect as shown in Fig. 18.13, in which we can also see striations caused by local deviation of the magnetization from its average orientation.

The induced anisotropy for thin films has been investigated since old times; for the earlier work in this century the reader is referred to the paper written by Kaufman and Mayer.[18] More recently the effect of a magnetic field during evaporation was investigated by Blois;[19] this effect was also found to exist even in pure metals such as iron and nickel by Williams and Sherwood.[12] Knorr and Hoffman[20] and Smith[21] discovered that, even with no applied field, magnetic anisotropy can be induced by causing the metal vapor to strike the substrate at an oblique angle instead of normally. The induced anisotropy thus caused, called the oblique-incidence anisotropy by D. O. Smith, was measured by Pugh, Boyd, and Freedman[22] for an Fe-Ni alloy system and found to change its sign at about 10% Fe-Ni. It was observed by Smith, Cohen, and Weiss[23] that an oblique-incidence film is composed of many fine grains which are lined up perpendicular to the incidence beam (a chain structure). The chain structure is considered to be caused by the shadowing of the incident beam by already grown particles, resulting in an accumulation of

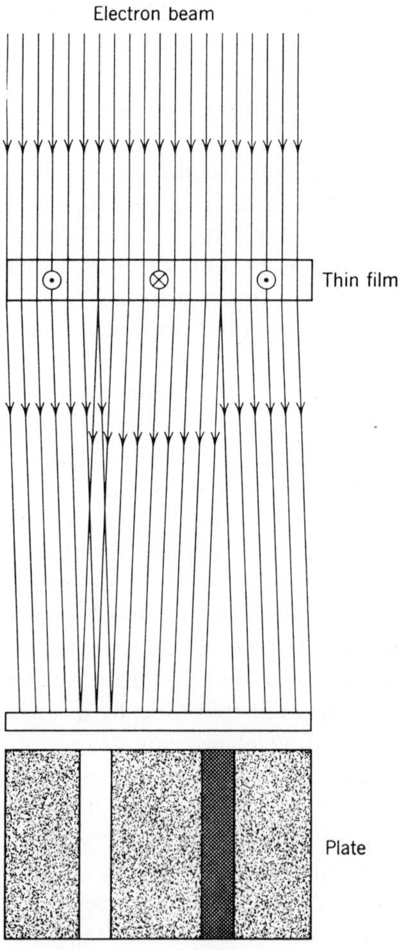

Fig. 18.11. Electron microscopic observation of domain structure in thin films.

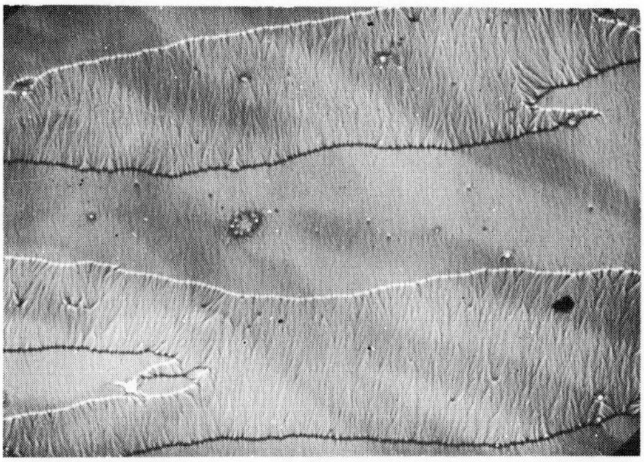

(*a*) $H \longrightarrow$

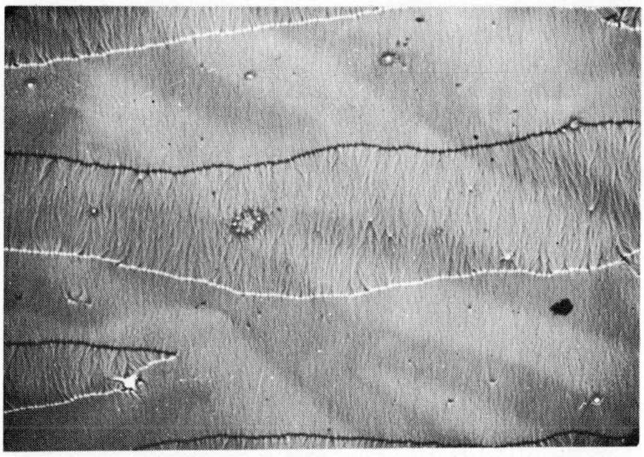

(*b*) $H \longleftarrow$

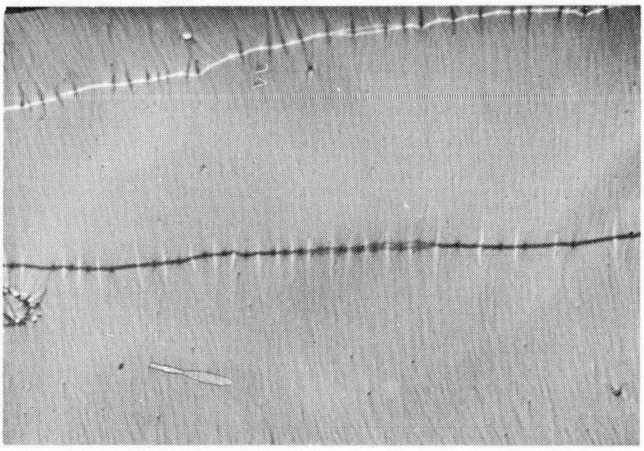

(*c*)

Fig. 18.12. Domain pictures of Permalloy film observed by means of the electron microscope: (*a*) domains observed for a film of thickness 800 Å (field is horizontal and in the positive sense); (*b*) the same domains (field is horizontal and in the negative sense); (*c*) cross-tie walls observed in the film of thickness 600 Å (after Ichinokawa[15]).

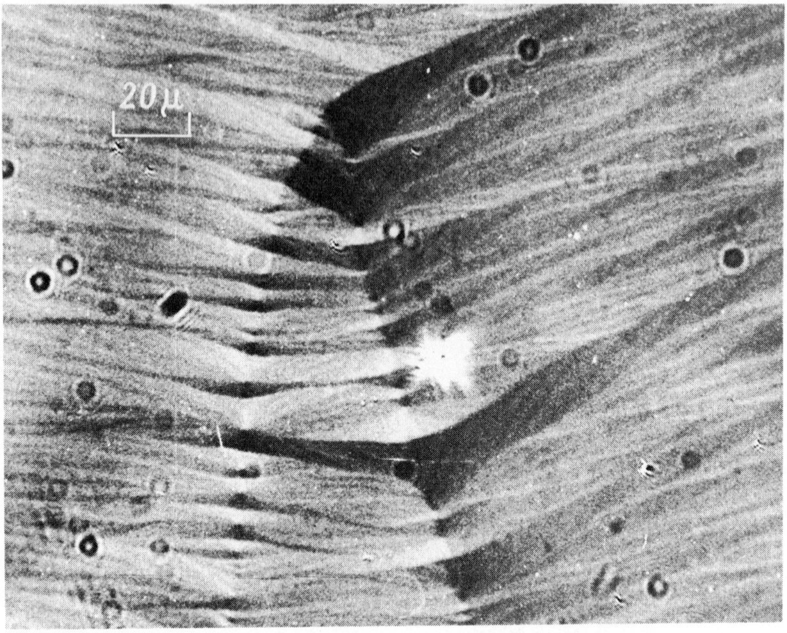

Fig. 18.13. Cross-tie walls observed by means of the Faraday effect (after Boersch and Lambeck[17]).

arriving atoms on the sides of particles rather than in front or in back of them. The magnetic anisotropy of such a chain structure was interpreted to be caused partly by the shape anisotropy and partly by an internal tension induced by a cohesion acting between the particles.

The presence of a magnetic field during evaporation was also found to be effective in producing additional anisotropy even if the direction of the incident beam is normal to the surface of the film.[24–28] Figure 18.14 shows the induced anisotropy measured by Takahashi et al.[28] as a function of composition in Fe-Ni alloys. The induced anisotropy is finite for pure nickel and is also several times larger for thin films than for bulk materials. It is observed that the anomalously large anisotropy is reduced to the normal value after the films have been annealed above 300°C in a magnetic field.[28] The easy axis of the anisotropy induced by magnetic evaporation can be easily changed by magnetizing the specimen out of the easy axis; this process can be observed even at room temperature.[26,29–32] Some workers[27] reported that this rotatable anisotropy can be removed by a slight etching of the surface of the film in acid. The anisotropy of magnetic annealed films is the same as that of bulk materials, as was verified[33] for epitaxially grown single crystal Permalloy films, except for the presence of the non-annealable anisotropy always observed before and after the annealing.

It was observed that a polycrystalline film behaves like an isotropic film except for the influence of the induced anisotropy.[34] The reason for this was postulated[34] to be that the grain size of the evaporated film is so small compared to the thickness of the wall that the spin orientation is smoothed out by exchange interaction irrespective of the fluctuation of the easy axis from grain to grain. The magnetocrystalline anisotropy observed for single crystal films,[33] however, is quite large compared to

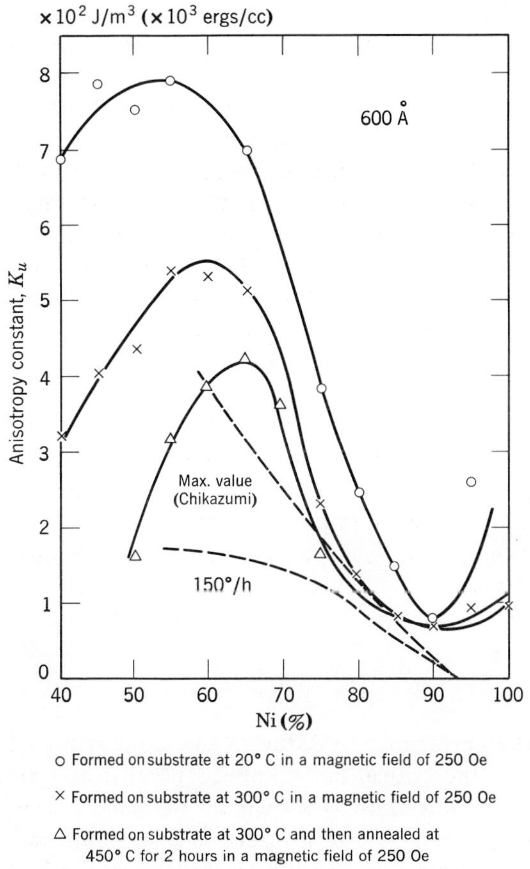

Fig. 18.14. The constant of the uniaxial anisotropy induced by magnetic evaporation in Fe-Ni films as a function of alloy composition. Circles represent films evaporated at room temperature; crosses, at 300°C; and triangles those evaporated at 300°C and then annealed 450°C for 2 hours. The broken curves are for bulk materials. (Takahashi, Watanabe, Kono, and Ogawa.[28])

that of the bulk material. Since this anomalous value is reduced to a normal value by floating off the film from the substrate, it is considered to be caused by the stress which might be produced by an epitaxial misfit or by a difference in thermal expansion between film and substrate.

The anisotropy energy for rotation of magnetization out of the plane of the film is also sometimes unexplainable in terms of the shape anisotropy of the film.[3,33,34] The pinning of spins at the film surface is also observed[35] in the experiment of spin wave resonance in thin films. The origins of this perpendicular anisotropy are considered to be the shape anisotropy, the effect of internal tension, Néel's surface anisotropy, the exchange interaction from oxide layer or from substrate, or the anisotropy caused by voids, dislocations, and other irregularities in the lattice of the film.

For the history of the technical magnetization of thin films the reader may refer to the book on thin films written by Mayer.[36] In general, the coercive force of thin films increases with a reduction of the thickness.[37] This may be caused by an increase in the surface energy of the domain wall, as suggested by Néel;[38] it may also be partly due to the effect of surface roughness which hinders the displacement of domain walls. The local anisotropy which causes a fluctuation in the orientation of magnetization may also be a possible origin giving rise to hindrances to the displacement of walls. The mechanism of magnetization reversal was investigated by Conger;[39] he placed a set of two crossed search coils on the thin film disk, connecting each of them to the x and y axes of an oscilloscope. From the shape of the locus of the oscillogram he concluded that rotation magnetization is the main mechanism of magnetization in the high frequency range. It was shown by D. O. Smith[40] and Olson and Pohm[41] that the presence of a transverse magnetic field is effective in exciting rotation magnetization. The condition for occurrence of magnetization reversal is easily calculated to be

$$H_{l0}^{2/3} + H_{t0}^{2/3} = H_a^{2/3}, \tag{18.1}$$

where H_a is the anisotropy field due to the induced uniaxial anisotropy, H_{l0} and H_{t0} are the critical fields applied parallel and perpendicular to the easy axis. This relation is shown by the solid curve in Fig. 18.15. If the transverse field is absent, in order to reverse the magnetization the longitudinal field should be larger than the anisotropy field. The critical longitudinal field is, however, reduced by the presence of the transverse magnetic field. Experimental data confirm this prediction. This fact is useful for the application of thin films in the memory devices of electronic computers (Section 23.1).

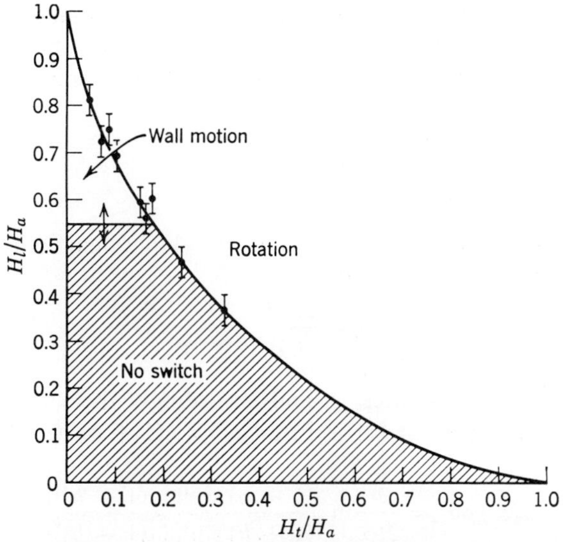

Fig. 18.15. Relationships between the longitudinal field H_l and the transverse field H_t which are necessary to cause the switching of magnetization (Olson and Pohm[41]).

18.2 Ferromagnetism of Thin Wires and Fine Particles

The factors involved in the ferromagnetism of thin wires and fine particles are also classified into three groups: the spontaneous magnetization, domain structure, and magnetization mechanism. Since these problems are interrelated, we begin with the problem of domain structures.

According to Bean,[42] there are three kinds of magnetic structures for fine particles. The first is the multidomain structure observed for the particle which is larger than the critical size. For this structure the magnetization takes place through the displacement of domain walls; hence its coercive force is generally small. The second is a single domain structure found below the critical size. Here the magnetization must be due to a rotation magnetization, and hence its coercive force is generally very large as long as the anisotropy energy, including shape anisotropy, is reasonably large. The third refers to the superparamagnetic particles (cf. Section 15.2) in which the whole assembly of spins is thermally excited to rotate in unison and to overcome the potential barrier Kv. The coercive force is again decreased by the thermal excitation the process of which, however, requires a time dependent on the size of the particle.

If the size of the particles is in the range of the third kind of magnetic structure, the intensity of spontaneous magnetization must be analyzed under the assumption of superparamagnetism. Kneller[43] solved the problem of the temperature dependence of spontaneous magnetization in this way for several examples of finely divided ferromagnetics. If the size of particles is uniform, the magnetization curve of the assembly of particles is given by

$$\frac{I}{I_s} = \coth \alpha - \frac{1}{\alpha} = L(\alpha), \tag{18.2}$$

(this equation is similar to that for the usual paramagnetism), where $L(\alpha)$ is the Langevin function and

$$\alpha = \frac{vI_sH}{kT}. \tag{18.3}$$

If the volumes of the particles, v, and accordingly the values of α are distributed, say, from $\alpha_1' = \alpha(1 - b)$ to $\alpha_2' = \alpha(1 + b)$, (18.2) becomes

$$\frac{I}{I_s} = \frac{1}{2b\alpha}\int_{\alpha(1-b)}^{\alpha(1+b)} L(\alpha')\,d\alpha' = \frac{1}{2b\alpha}\ln\left\{\frac{(1-b)\sinh[\alpha(1+b)]}{(1+b)\sinh[\alpha(1-b)]}\right\} = L^*(\alpha). \tag{18.4}$$

This function is shown in Fig. 18.16, together with the usual Langevin function. The experimental points are those observed for 29.3 at% Mn-Ni alloy annealed for 20 hours at 480°C; many finely divided ferromagnetic ordered phases are thought to be dispersed in a non-magnetic

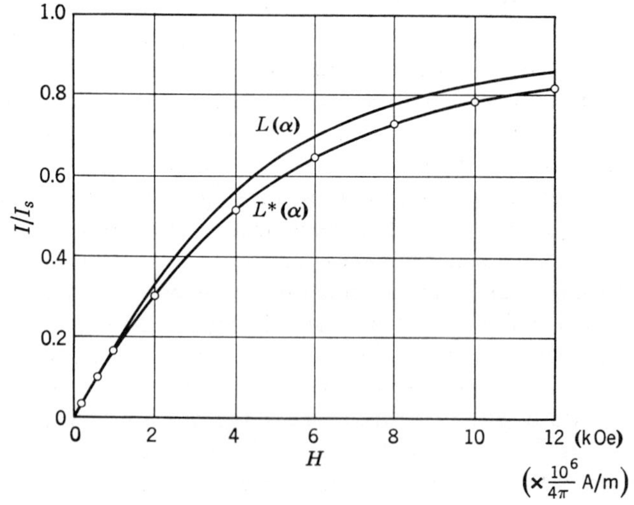

Fig. 18.16. Magnetization curve of fine particles which exhibit superparamagnetism (Kneller[43]).

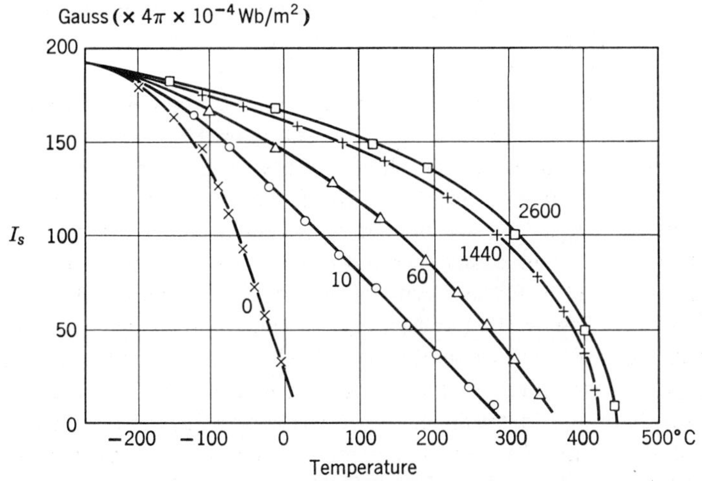

Fig. 18.17. Temperature dependence of spontaneous magnetization of the fine precipitations of Cu-Fe-Ni alloy analyzed by extrapolation of the saturation portion of magnetization curves. The numerical figures are the tempering duration in minutes at 600°C. (Kneller.[43])

disordered matrix. The solid line represents the function $L^*(\alpha)$ drawn by assuming $I_s = 0.565$ Wb/m² (= 450 gauss), $v_0 = 20.1 \times 10^{-21}$ cm³, and $b = 0.65$; it is in good agreement with experiment. From this comparison the "true" saturation magnetization can be determined.

Figure 18.17 shows an example cited by Kneller in which the saturation magnetization was determined by extrapolating the linear part near saturation toward zero magnetic field. The specimen is the 62% Cu, 6% Fe, 32% Ni alloy which is tempered for a desired time after quenching from 800°C. During the tempering the ferromagnetic phase, rich in iron and nickel, is precipitated in a non-magnetic matrix. It appears that the Curie point is fairly low at the initial stage of precipitation, where the size of the precipitated particles should be fairly small. Kneller showed, however, that this result is superficial and every curve coincides exactly with the $I_s - T$ curve for the bulk material (even for particles as small as 6.3×10^{-21} cm³, or of diameter 23 Å), if the analysis is made under the assumption of superparamagnetism.

Bean and Jacobs[44] determined the size of fine particles of iron amalgam ranging from 20 Å to 100 Å from the shape of the magnetization curves measured at 77° and 200°K. They called this method magnetic granulometry. The temperature dependence of the intensity of the intrinsic magnetization remains the same as that of bulk material down to small particles of the size 30 Å. Becker[45] also investigated the size of precipitated particles of 2% Co-Cu alloy produced during the process of aging at 650°C; he concluded that the size grows from 12 Å to 70 Å with the progress of precipitation.

Next let us briefly discuss the mechanism of magnetization in fine particles. For a single domain structure the magnetization is exclusively due to rotation magnetization which results in a high coercive force. When the particle size is decreased from the multidomain range to a single domain range, the coercive force is increased with a decrease of the particle size. The exact formulation of the variation of coercive force was first tried by Kittel,[46] who showed that H_c is expected to change inversely proportionally to the diameter of the spherical particle, and he compared his result with experiment. Néel[47] criticized this calculation, and Ohoyama[48] considers that this criticism is not entirely correct.

As a result of the fact that a high coercivity of granular materials is related to the single domain structure, the temperature dependence of the coercive force is quite different from that of other hard materials. Figure 18.18 shows the temperature dependence of the coercive force as observed by Rathenau[49] for Ba ferrite. The coercive force increases abruptly upon the appearance of ferromagnetism below the Curie point, but it decreases again with a decrease of temperature. The values of $2K/I_s$, the theoretically expected coercive force for a single domain structure, are calculated from individual values of K and I_s and are plotted in the figure. A large discrepancy at low temperature between experiment and theory signifies the appearance of a multidomain structure. This is

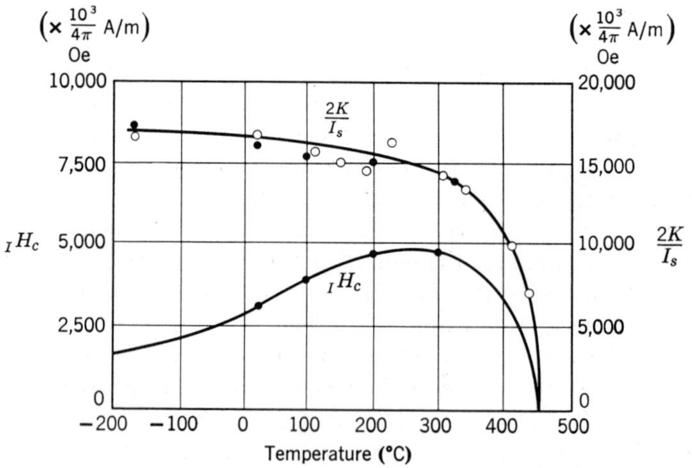

Fig. 18.18. Temperature dependence of the coercive force of Ba ferrite and its comparison to the theoretical value $2K/I_s$ (Rathenau[49]).

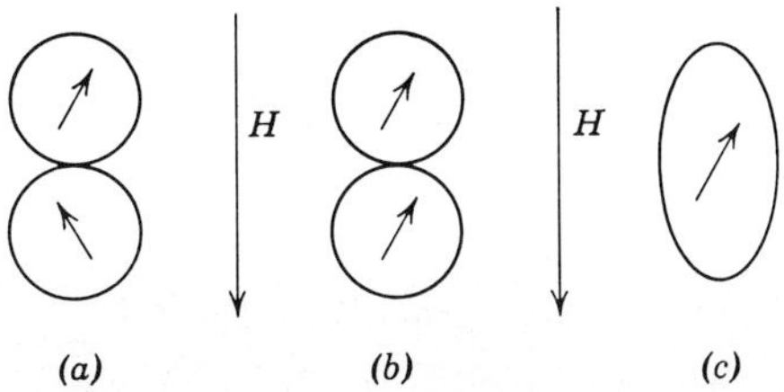

Fig. 18.19. Various modes of magnetization reversal of fine particles (Jacobs and Luborsky[50]).

quite reasonable because the critical radius r_c is expected to be proportional to $\sqrt{K}/I_s$ (cf. 11.46 and 13.65) the value of which actually decreases with a decrease of temperature.

Jacobs and Luborsky[50] investigated the magnetization mechanism of an aggregate of fine particles, and they considered the possibility of the occurrence of incoherent rotation of magnetization as illustrated in Fig. 18.19*a* and of coherent rotation also as shown in Figs. 18.19*b* and 18.19*c*. They concluded from the measurement of coercive force and also the rotational hysteresis for fine particles of Fe and Fe-Co that Fig. 18.19*a* may show the most probable mechanism of magnetization reversal. They measured the rotational hysteresis,

$$W_r = \int_0^{2\pi} L \, d\theta, \tag{18.5}$$

for various intensities of the applied magnetic field and calculated the quantity

$$R_h = \int \frac{W_r}{I_s} \, d\left(\frac{1}{H}\right). \tag{18.6}$$

This quantity is dimensionless and depends on the mechanism of magnetization and also on the distribution of fine particles. It was calculated that $R_h = 0.38 \sim 0.42$ for coherent rotations such as those in Figs. 18.19*b* and 18.19*c*, whereas $R_h = 1.0 \sim 1.5$ for the incoherent rotations shown in Fig. 18.19*a*. Actual observation shows that $R_h = 1.3 \sim 1.6$ for Fe and FeCo particles, supporting the mechanism in Fig. 18.19*a*.

Beautiful domain structures were observed by Coleman and Scott[51] and De Blois and Graham[52] for thin wires (iron whiskers which were grown by the reduction of iron bromide by hydrogen at 700° ~ 800°C). The diameter is about several tens of microns, and its cross-sectional shape is an exact square. The crystal is so perfect that its domain structure is also quite regular. Figure 18.20 shows an example of domain pattern observed on the (010) plane. Main domains are extended along the length of the whisker and are connected by closure domains at the place where the stress is introduced. Figure 18.21 shows the magnetization process

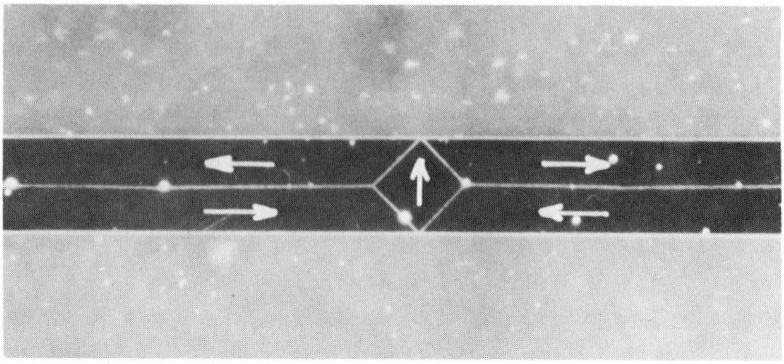

Fig. 18.20. Domain structure of iron whisker (De Blois and Graham[52]).

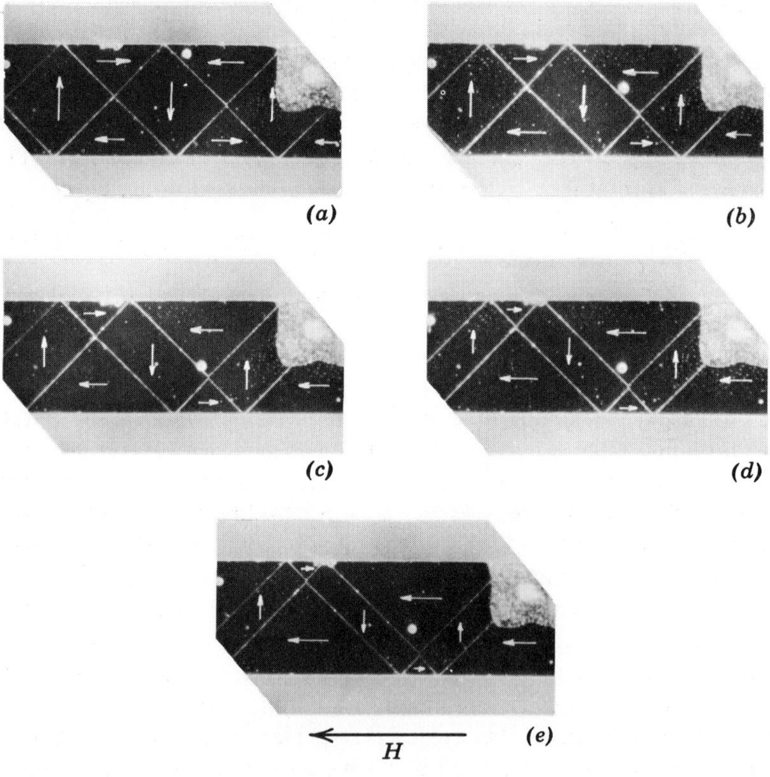

Fig. 18.21. Magnetization process of iron whisker (De Blois and Graham[52]).

of a similar domain structure. De Blois[53] wound two small secondary coils around the whisker and observed that the velocity of the displacement of the wall obeys the relation $v = c(H - H_0)$ (cf. 16.14). Because the eddy current is very weak for such a thin wire, the velocity is fairly large. The coefficient is $c = 2.5$ m²/sec A (= 20,000 cm/sec Oe) at room temperature, and $c = 0.76$ m²/sec A (= 6000 cm/sec Oe) at 77°K. These values are of the same order of magnitude as observed for Ni-Zn ferrite (Section 16.5). These researchers also found a discontinuity in the v-H curve the reason for which has not yet been clarified.

References

18.1. M. J. Klein and P. S. Smith: *Phys. Rev.* **81,** 378 (1951).
18.2. E. C. Crittenden, Jr., and R. W. Hoffman: *Rev. Mod. Phys.* **25,** 310 (1953); R. M. Hoffman and A. M. Eich: G.25, p. 78.
18.3. C. A. Neugebauer, *Structure and Properties of Thin Films* (John Wiley & Sons, Inc., New York, 1959), p. 358; *Phys. Rev.* **116,** 1441 (1959).
18.4. Y. Gondo, H. Konno, and Z. Funatogawa: *J. Phys. Soc. Japan* **16,** 2345 (1961).
18.5. L. Valenta: *Izvest. Akad. Nauk. SSR Ser. Fiz.* **21,** 879 (1957); *Czech. J. Phys.* **7,** 127, 133 (1957); also referred by C. P. Bean: *Structure and Properties of Thin Films* (John Wiley & Sons Inc., New York, 1959), p. 346.
18.6. W. Döring: *Z. Naturforsch.* **16a,** 1146 (1961).
18.7. C. Kittel: *Phys. Rev.* **70,** 965 (1946).
18.8. B. W. Roberts and C. P. Bean: *Phys. Rev.* **96,** 1494 (1954), B. W. Roberts: Memo No. MA-1, G. E. Lab. (July 1955).
18.9. J. B. Goodenough: *Phys. Rev.* **102,** 356 (1956).
18.10. H. J. Williams and R. C. Sherwood: *J. Appl. Phys.* **29,** 296 (1958).
18.11. L. Néel: *Compt. Rend.* **241,** 533 (1955).
18.12. H. J. Williams and R. C. Sherwood: *J. Appl. Phys.* **28,** 548 (1957).
18.13. E. E. Huber, D. O. Smith, and J. B. Goodenough: *J. Appl. Phys.* **29,** 294 (1958).
18.14. H. W. Fuller and M. E. Hale: *J. Appl. Phys.* **31,** 308S (1960).
18.15. T. Ichinokawa: *Mem. Sci. Eng. Waseda Univ.* **No. 25,** 80 (1961).
18.16. S. Middelhoek: Thesis (Amsterdam, 1961), p. 46.
18.17. H. Boersch and M. Lambeck: *Z. Physik.* **165,** 176 (1961).
18.18. W. Kaufman and W. Mayer: *Z. Physik.* **12,** 513 (1911).
18.19. M. S. Blois: *J. Appl. Phys.* **26,** 975 (1955).
18.20. J. G. Knorr and R. W. Hoffman: *Phys. Rev.* **113,** 1039 (1959).
18.21. D. O. Smith: *J. Appl. Phys.* **30,** 264S (1959).
18.22. E. W. Pugh, E. L. Boyd, and J. P. Freedman: *IBM J. Res. Develop.* **4,** 163 (1960).
18.23. D. O. Smith, M. S. Cohen, and G. P. Weiss: *J. Appl. Phys.* **31,** 1755 (1960).
18.24. W. Schüppel and Z. Málek: *Naturwiss.* **46,** 423 (1959).
18.25. W. Andrä, Z. Málek, and W. Schüppel: *J. Appl. Phys.* **31,** 442 (1960).
18.26. Z. Málek, W. Schüppel, O. Stemme, and W. Andrä: *Ann. Physik.* **5,** 14, 211 (1960).
18.27. Z. Málek and W. Schüppel: *Ann. Physik.* **6,** 252 (1960).

18.28. M. Takahashi, D. Watanabe, T. Kono, and S. Ogawa: *J. Phys. Soc. Japan* **15,** 1351 (1960).
18.29. E. N. Mitchell: *J. Appl. Phys.* **29,** 286 (1958).
18.30. C. D. Graham, Jr., and J. M. Lommel: *J. Appl. Phys.* **32,** 83S (1961).
18.31. R. J. Prosen, J. O. Holeman, and B. E. Gran: *J. Appl. Phys.* **32,** 91S (1961).
18.32. T. Matcovich, E. Korestoff, and A. Schmeckenbecher: *J. Appl. Phys.* **32,** 93S (1961).
18.33. S. Chikazumi: *J. Appl. Phys.* **32,** 81S (1961).
18.34. C. D. Graham, Jr.: G.33, p. 311.
18.35. R. F. Soohoo: *J. Appl. Phys.* **32,** 148S (1961).
18.36. H. Mayer: *Physik. Dünner Schichten,* I and II (Wiss. Verlag. M. B. H. Stuttgart, 1950, 1955).
18.37. R. M. Bozorth: *J. Phys. Radium* **17,** 256 (1956).
18.38. L. Néel: *J. Phys. Radium* **17,** 250 (1956).
18.39. R. L. Conger: *Phys. Rev.* **98,** 1752 (1955); G. 25, p. 610.
18.40. D. O. Smith: *J. Appl. Phys.* **29,** 264 (1958).
18.41. C. D. Olson and A. V. Pohm: *J. Appl. Phys.* **29,** 274 (1958).
18.42. C. P. Bean: *J. Appl. Phys.* **26,** 1381 (1955).
18.43. E. Kneller: *Z. Physik.* **152,** 574 (1958).
18.44. C. P. Bean and I. S. Jacobs: *J. Appl. Phys.* **27,** 1448 (1956).
18.45. J. J. Becker: *AIME Trans.* **209,** 59 (1957).
18.46. C. Kittel: *Phys. Rev.* **73,** 810 (1948).
18.47. L. Néel: *J. Phys. Radium* **17,** 250 (1956).
18.48. T. Ohoyama: *J. Phys. Soc. Japan* **12,** 827 (1957).
18.49. G. W. Rathenau: *Rev. Mod. Phys.* **25,** 297 (1953).
18.50. I. S. Jacobs and F. E. Luborsky: *J. Appl. Phys.* **28,** 467 (1957).
18.51. R. V. Coleman and G. G. Scott: *Phys. Rev.* **107,** 1276 (1957); *J. Appl. Phys.* **29,** 526 (1958).
18.52. R. W. De Blois and C. D. Graham, Jr.: *J. Appl. Phys.* **29,** 528 (1958).
18.53. R. W. De Blois: *J. Appl. Phys.* **29,** 459 (1958).

19

Various Phenomena Accompanying Magnetization

19.1 Galvanomagnetic Effect

There is a long history of investigations of the magnetoresistance effect in ferromagnetic substances; the first one was made by Thomson[1] (Lord Kelvin) in 1851. Many phenomenological investigations have been made on pure metals and several alloy systems, but the physical origin has been studied by only a few investigators.

The phenomenology of the magnetoresistance effect is quite similar to that of magnetostriction. The effect can be classified into two categories: One is the part which depends on the intensity of spontaneous magnetization, corresponding to the volume magnetostriction. The second is the resistance change caused by rotation of spontaneous magnetization, which corresponds to the usual magnetostriction.

We begin with the discussion of the temperature dependence of resistivity of ferromagnetic metals. Figure 19.1 shows a comparison of the temperature dependence of Ni and that of Pd. Pd has ten electrons outside the krypton shell, being quite similar to Ni which has ten electrons outside the argon shell. In Fig. 19.1 the scales of the coordinate are adjusted so as to match the resistivity of both metals at the Curie point of Ni. Mott[2] interpreted this phenomenon in terms of the scattering probability of the conducting electrons into 3*d* holes. If the substance is in a ferromagnetic state, half of the 3*d* shell is filled up, so that the scattering of 4*s* electrons into the plus spin state of 3*d* shell is forbidden. This scattering is, however, permitted in a non-magnetic state in which

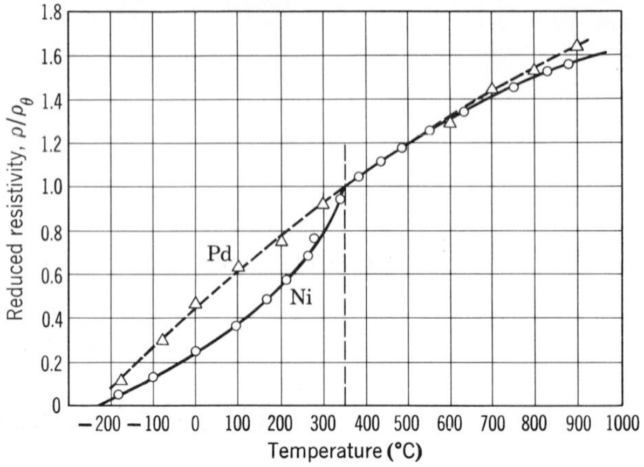

Fig. 19.1. Comparison of reduced resistivity ρ/ρ_θ between Ni and Pd (Becker and Döring[G,3]).

both the plus spin state and the minus spin state of the upper 3*d* levels are vacant. Mott explained the temperature variation of resistivity fairly well by this model.

Kasuya[3] interpreted this phenomenon from a standpoint quite different from the Mott theory. He considered that *d* electrons are localized at the lattice points and interact with conduction electrons through the exchange interaction. At 0°K the potential for conduction electrons is periodic, because the spins of 3*d* electrons of all the lattice points point in the same direction. At finite temperatures, spins of 3*d* electrons are thermally agitated and the thermal motion may break the periodicity of the potential. The 4*s* electrons are scattered by an irregularity of the periodic potential which results in additional resistivity. Kasuya postulated that the temperature dependence of the resistivity of ferromagnetic metals is composed of a monotonically increasing part due to lattice vibration and an anomalous part due to magnetic scattering, the magnitude of the latter being explained by his theory. In rare earth metals in which 4*f* electrons are forming an atomic magnetic moment, the conducting 6*s* electrons are also considered to be scattered by the irregularity in the polarization of 4*f* electrons. Actually it is shown by Colvin, Legvold, and Spedding[4] that the anomalous part of the resistivity measured for a number of rare earth metals is proportional to the square of the magnitude of the spin magnetic moment in accordance with the Kasuya theory. Since the 4*f* electrons are completely localized, the Mott theory is considered to be invalid, at least for rare earth metals.

The resistivity can be also reduced by the forced increase of spontaneous magnetization caused by an application of a strong magnetic field. Figure 19.2 shows the field dependence of the resistivity of nickel as measured parallel and perpendicular to the magnetic field. The resistivity decreases linearly with the intensity of the magnetic field for both cases, corresponding to forced magnetostriction.

This phenomenon is also interpreted by Mott's theory. Because in thermally excited states the upper part of the plus spin levels is vacant, 4*s* electrons are scattered into this vacant place. The forced increase of spontaneous magnetization, however, fills up these vacant levels with plus-spin electrons, reducing the resistivity. Kasuya's theory also explains this phenomenon: the irregularity of spin arrangement caused by thermal agitation is smoothed out to some extent by the external strong magnetic field, and thus the probability of the scattering of 4*s* electrons is reduced.

The second category of magnetoresistance effect can be described in a similar way to magnetostriction, because both quantities depend not on the sense but only on the direction of spontaneous magnetization. The change in resistivity is described as

$$\Delta\rho = \tfrac{3}{2}\Delta\rho_{100}(\alpha_1{}^2\beta_1{}^2 + \alpha_2{}^2\beta_2{}^2 + \alpha_3{}^2\beta_3{}^2 - \tfrac{1}{3}) + 3\Delta\rho_{111}(\alpha_1\alpha_2\beta_1\beta_2 + \alpha_2\alpha_3\beta_2\beta_3 + \alpha_3\alpha_1\beta_3\beta_1), \quad (19.1)$$

where $(\alpha_1, \alpha_2, \alpha_3)$ and $(\beta_1, \beta_2, \beta_3)$ are, respectively, the direction cosines of spontaneous magnetization and the direction along which the resistivity is measured. The coefficients in this formula have been determined by

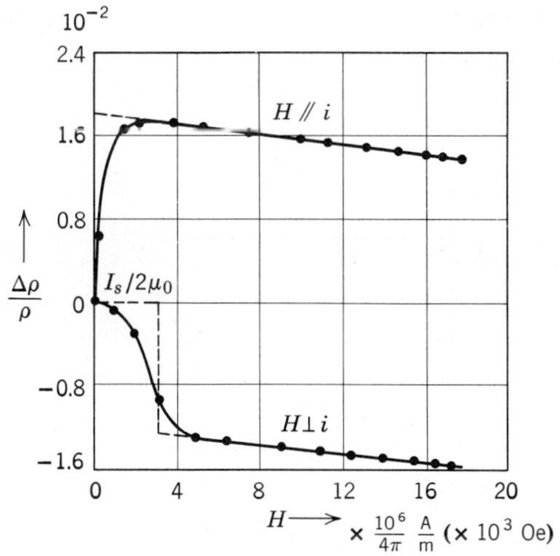

Fig. 19.2. Variation of resistivity as a function of a magnetic field (Englert[5]).

Döring[6] and Hirone and Hori[7] for iron and nickel; they used the experimental data observed by Webster,[8] Kaya,[9] Döring,[6] and Shirakawa.[10] At room temperature these values are

$$\begin{aligned} &\text{Fe:} \quad \frac{\Delta\rho_{100}}{\rho} = 0.102\%, \qquad \frac{\Delta\rho_{111}}{\rho} = 0.395\%, \\ &\text{Ni:} \quad \frac{\Delta\rho_{100}}{\rho} = 4.3\%, \qquad \frac{\Delta\rho_{111}}{\rho} = 1.9\%, \end{aligned} \tag{19.2}$$

where ρ is the resistivity at the same temperature. The magnetoresistance effect is normally measured by using a specimen in the shape of a cylindrical rod carrying an electric current along its long axis as a function of the intensity of the magnetization parallel to the current. The total change in the resistivity in this case depends on the domain distribution in the demagnetized state. Figure 19.3 shows an example measured for 21.5% Fe-Ni. Curve *a* is observed for the specimen annealed for 15 minutes at 490°C; by this treatment a random distribution of domain magnetizations is fixed by induced local uniaxial anisotropies. Curve *b* is observed after a similar heat treatment in a magnetic field. The domain magnetization is thus almost completely aligned parallel to the long axis, so that the total change is greatly reduced as compared to that described by curve *a*. In the initial range of magnetization in curve *b*, almost no change in the resistivity occurs, because the magnetization in this range takes place exclusively through the displacement of 180° domain walls. Thus the magnetoresistance effect, similarly to the magnetostriction, can be utilized to determine the distribution of domain magnetizations (cf. Section 8.1). High and low temperature measurements of the magnetoresistance effect

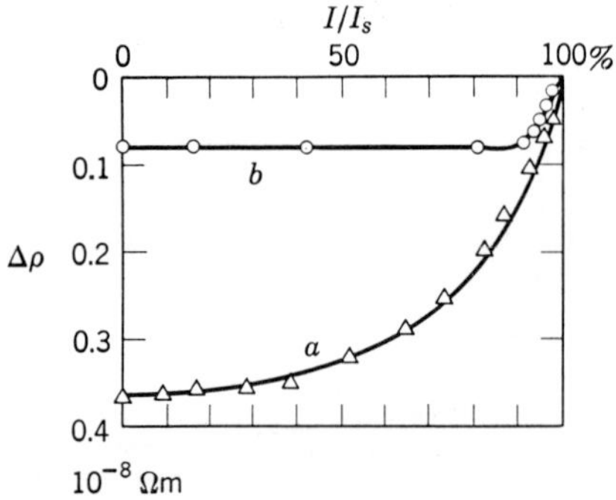

Fig. 19.3. Variation of resistivity as a function of magnetization measured for 21.5% Fe-Ni alloy which was annealed for 15 minutes at 490°C: curve *a*, annealed without a magnetic field; curve *b*, annealed with a magnetic field (Chikazumi[11]).

were made by Gondo and Funatogawa,[12] Tatsumoto and Kuwahara,[13] and others.[14] An example is shown in Fig. 19.4.

The origin of the anisotropy in the magnetoresistance effect was investigated theoretically by Kondo[15] on the basis of the localized *d* electron model. He attributed this effect to the scattering of *s* electrons by the remaining orbital magnetic moment of 3*d* electrons, which should be otherwise quenched by the crystalline field. He showed that the obtained resistance change gives a reasonable order of magnitude and that the temperature dependence is in fairly good agreement with the experiment. Figure 19.5 shows the value of $\Delta\rho$ measured by Smit[16] for a number of ferromagnetic alloys. This behavior is quite similar to the variation of the deviation of *g* values[18] from 2, which is a good measure of the remaining orbital magnetic moment. The pseudodipole interaction determined from the magnetostriction measurement, which is also a good measure of a spin-orbit interaction, also shows a similar dependence on the electron concentration.[19] These facts verify the fundamental correctness of the Kondo theory.

The magnetothermoelectric effect also was first investigated by Lord Kelvin,[20] in 1884. This effect is also classified into two categories: One is due to the change in the spontaneous magnetization, and the other is due to a technical magnetization. As for the former effect, an anomalous change in thermoelectric power was observed at Curie points during the process of heating for Ni and Ni-Cu alloy by Grew.[21] Mott[2] also explained this effect by the model described above. The part which depends on the direction of magnetization can be expressed, similarly to (19.1), by

$$\delta V = \tfrac{3}{2}\delta V_{100}(\alpha_1^2\beta_1^2 + \alpha_2^2\beta_2^2 + \alpha_3^2\beta_3^2 - \tfrac{1}{3}) + 3\delta V_{111}(\alpha_1\alpha_2\beta_1\beta_2 + \alpha_2\alpha_3\beta_2\beta_3 + \alpha_3\alpha_1\beta_3\beta_1), \quad (19.3)$$

where $(\alpha_1, \alpha_2, \alpha_3)$ and $(\beta_1, \beta_2, \beta_3)$ are, respectively, the direction cosines of spontaneous magnetization and the direction of temperature gradient along which the thermoelectromotive force is measured. Single crystal measurements were made by Funatogawa, Gondo, and Miyata[22] for iron and nickel. The coefficients of expression (19.3) determined by them are

$$\begin{aligned} &\text{Fe:} \quad \delta V_{100} = 0.70\ \mu\text{V/deg}, \quad \delta V_{111} = -0.13\ \mu\text{V/deg}, \\ &\text{Ni:} \quad \delta V_{100} = 0.57\ \mu\text{V/deg}, \quad \delta V_{111} = 0.69\ \mu\text{V/deg}, \end{aligned} \quad (19.4)$$

at room temperature. The measurements were also extended to high and low temperatures, and both coefficients were found to decrease from the room temperature value as a result of heating as well as of cooling.

The Hall effect of ferromagnetic substances was investigated first by

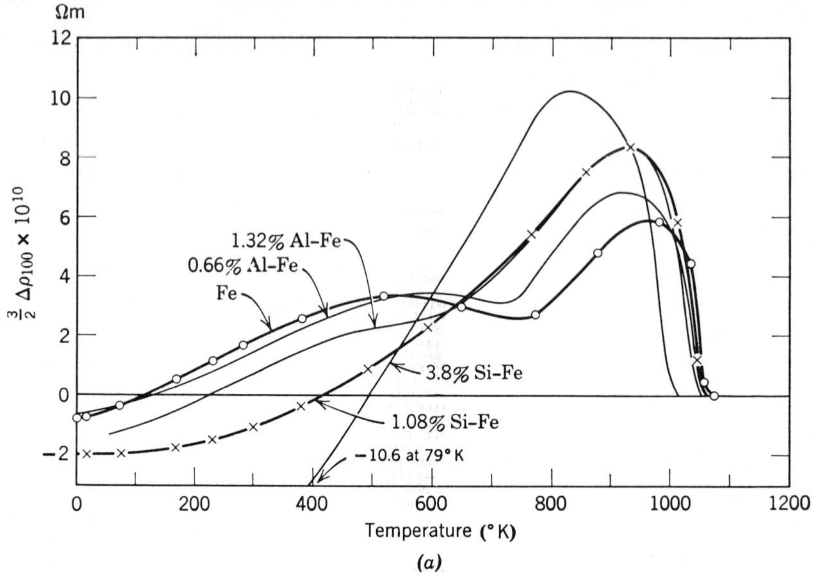

(a)

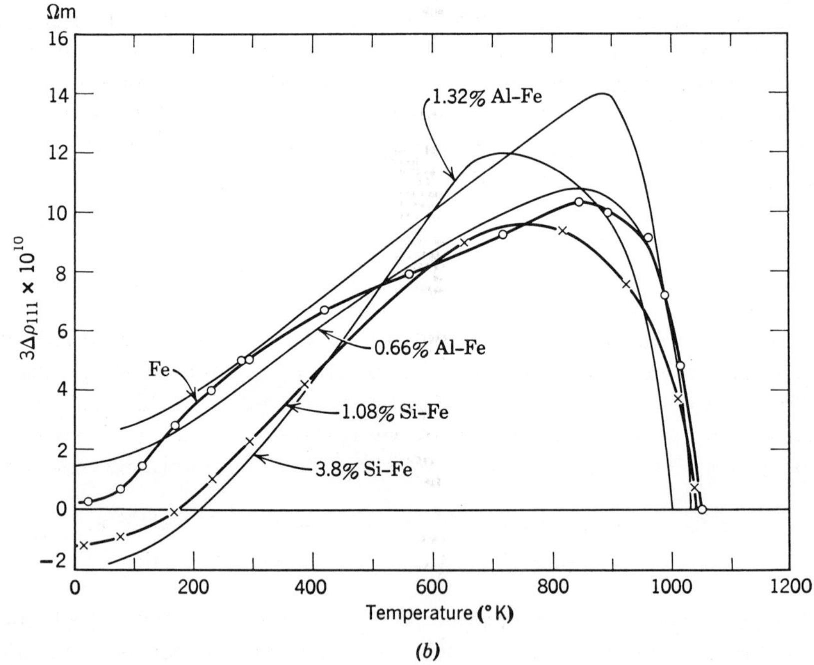

(b)

Fig. 19.4. Variation of the coefficients $\Delta\rho_{100}$ and $\Delta\rho_{111}$ of the magnetoresistance effect (cf. 19.1) as a function of temperature for Fe and its alloys (Tatsumoto, Kuwahara, and Kimura[14]).

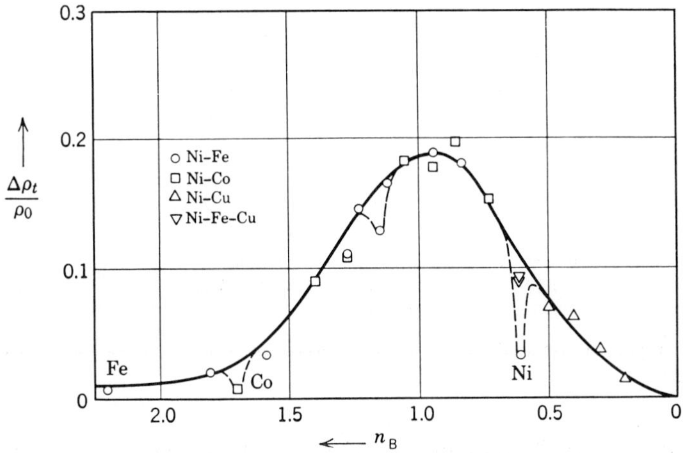

Fig. 19.5. Magnetoresistance at 20°K as a function of the mean number of Bohr magnetons per atom (Smit,[16] from Jan's paper[17]).

Kundt[23] in 1893 for Fe, Co, and Ni. He concluded that the Hall electromotive force varies almost in proportion to the intensity of magnetization. Smith and Sears,[24] however, showed later that the Hall electromotive force can be separated into two terms; one is the term proportional to the intensity of the magnetic field, and the other is the term proportional to the intensity of the magnetization. Thus the Hall electromotive force per unit electric current density can be expressed as

$$\epsilon = R_0 H + R_1 \frac{I}{\mu_0}, \tag{19.5}$$

as first proposed by Pugh.[25] The first term represents the ordinary Hall effect, and the second term the extraordinary Hall effect. From the data of Smith[26] the coefficient of the ordinary Hall effect R_0 of Ni is determined[27] as

$$R_0 = -7.6 \times 10^{-17}\ \Omega\text{m}^2/\text{A}\ (= -6.1 \times 10^{-13}\ \Omega\text{cm/Oe}) \tag{19.6}$$

at room temperature. The order of magnitude is almost the same as that of non-magnetic transition elements such as Mn or Cu. The characteristics that the temperature dependence is flat at high temperature and shows a monotonic decrease toward the lower temperatures[28] are also common to non-magnetic transition metals. The extraordinary Hall coefficient of Ni is

$$R_1 = -7.49 \times 10^{-16}\ \Omega\text{m}^2/\text{A}\ (= -7.49 \times 10^{-11}\ \Omega\text{cm/gauss})\dagger \tag{19.7}$$

† If the values in parentheses are used for (19.5), we must put $\mu_0 = 1$.

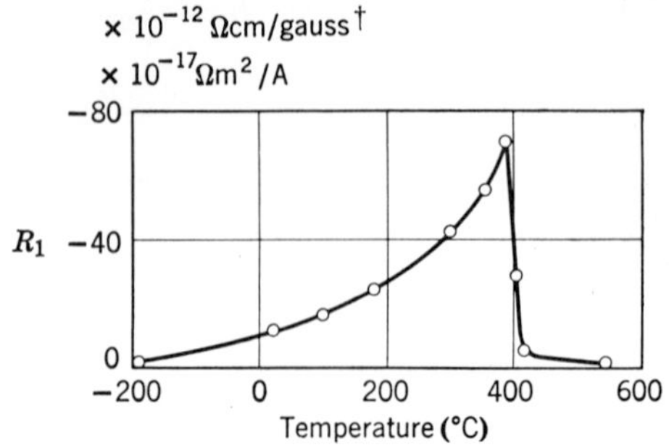

Fig. 19.6. Temperature dependence of the extraordinary Hall coefficient of Ni (Pugh and Rostoker[29]).

at room temperature. This quantity increases with increasing temperature and exhibits an anomaly at the Curie point as shown in Fig. 19.6. The measurements were extended by Pugh et al.[30,31] to include Ni, Co, Fe, Cu-Ni, and Ni-Co alloys. Single crystal measurement of the Hall effect in a ferromagnetic substance was made by Webster[32] for iron. He reported that the effect is almost isotropic.

A theoretical investigation of the extraordinary Hall effect was made by Karplus and Luttinger,[33] who classified it into two parts:

$$R_1 = R_0 + R_1'. \tag{19.8}$$

The first term is due to the internal field I/μ_0 caused by the magnetization I, and the second term is due to the direct interaction with the spontaneous polarization of $3d$ electrons. Since $R_0 \ll R_1$ in the temperature range other than very low temperatures, it can be considered that the main part of the extraordinary Hall effect is due to the latter part. Karplus and Luttinger considered that $3d$ electrons are conducting and that their orbital motion is influenced by their own spin-orbit interaction. They showed that the coefficient R_1' can be related to the resistivity ρ in this way:

$$R_1' = A\rho^2. \tag{19.9}$$

This relation holds well for various kinds of nickel alloys. In contrast to this, Kondo[15] considered that the $4s$ electrons are conducting and their orbital motions are influenced by the orbital motion of $3d$ electrons remaining unquenched. His calculation also explains the temperature dependence of the extraordinary Hall coefficient as shown in Fig. 19.6. He expected that the extraordinary Hall effect of rare earth metals could also be explained in this way.

19.2 Magnetothermal Effect

In this book, by magnetothermal effect we mean the temperature change which is caused by magnetization. Sometimes we call the temperature change caused by a forced change of spontaneous magnetization a magnetocaloric effect[34] and distinguish it from the magnetothermal effect in its narrow sense, that is, the temperature change accompanying the technical magnetization. This classification becomes, however, fairly vague near the Curie point, where both kinds of magnetization mechanism take place almost simultaneously.

First we discuss the temperature change caused by a forced magnetization. The susceptibility χ_0 for forced magnetization is given by (13.85). It obeys the Curie-Weiss law above the Curie point and decreases with a decrease of temperature below the Curie point. If the spontaneous magnetization increases by δI in the presence of the magnetic field H, the work done by the field is

$$\delta W = H\,\delta I. \tag{19.10}$$

On the other hand, the internal energy is increased by

$$\delta E = -wI\,\delta I. \tag{19.11}$$

The heat generated during this process is

$$\begin{aligned}\delta Q &= \delta W - \delta E \\ &= (H + wI)\,\delta I.\end{aligned} \tag{19.12}$$

If $T \ll \Theta$, it is considered that $H \ll wI$, so that (19.12) becomes approximately

$$\delta Q = wI\,\delta I = \tfrac{1}{2}w\,\delta(I^2). \tag{19.13}$$

If $T > \Theta$, it follows from the Curie-Weiss law and relation (4.30) that

$$H = \frac{T - \Theta}{\Theta}\,wI. \tag{19.14}$$

On putting this relation into (19.12), we have

$$\delta Q = \frac{T}{\Theta}\,wI\,\delta I = \frac{T}{\Theta}\frac{1}{2}\,w\,\delta(I^2). \tag{19.15}$$

The temperature change is given by

$$\delta T = \frac{1}{C_I}\,\delta Q, \tag{19.16}$$

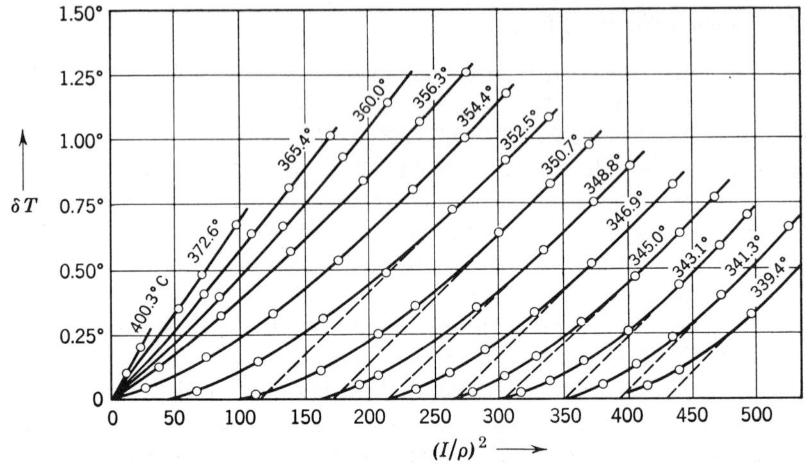

Fig. 19.7. Temperature change caused by magnetization of Ni at temperatures near the Curie point as a function of the square of magnetization (Weiss and Forrer,[35] from Becker and Döring[G.3]).

where C_I is the specific heat for constant I. It is predicted from (19.13) and (19.15) that the temperature change is proportional to the variation of I^2 below and above the Curie point. Actually this relation holds experimentally as shown by Weiss and Forrer[35] (Fig. 19.7). The inclination of the linear parts representing the proportionality of δT to δI^2 remains constant below Curie point 360°C, while it increases with an increase of temperature above the Curie point, as expected from (19.13) and (19.15).

If we express (19.13) in terms of H, it becomes

$$Q = w\chi_0 I\,\delta H = \frac{3\Theta L'(\alpha)}{T - 3\Theta L'(\alpha)}\, I\,\delta H. \qquad (19.17)$$

On the other hand, (19.15) becomes

$$\delta Q = \frac{T}{\Theta}\frac{1}{2}\, w\chi_0^{\,2}\,\delta(H^2) = \frac{T\Theta}{(T-\Theta)^2}\frac{1}{2w}\,\delta(H^2). \qquad (19.18)$$

These relations show that δQ is small at very low temperatures and at very high temperatures, and it takes a sharp maximum at the Curie point. Figure 19.8 shows the temperature change caused by an application of the magnetic field as observed by Weiss and Forrer[35] for Ni as a function of temperature. The general features are predicted by the theory. The temperature change at the Curie point is about 1°K for nickel. It is also of the same order of magnitude for iron (2°C for $H = 6 \times 10^5$ A/m = 8000 Oe).

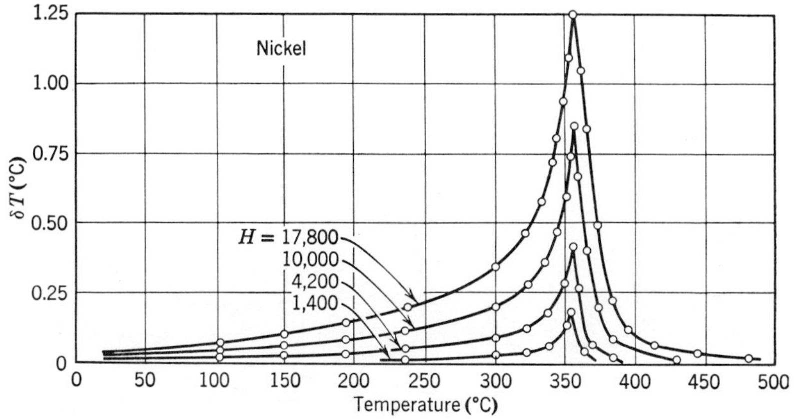

Fig. 19.8. Temperature dependence of the magnetothermal effect of Ni. Numerical values in the figure are the fields in oersteds or $(10^3/4\pi)$(A/m). (Weiss and Forrer,[35] from Bozorth[G.10].)

The temperature change accompanying the technical magnetization is very small and is about 0.001°C for one cycle of magnetization. The investigations of this phenomenon started about 1930 by Ellwood,[36] Townsend,[37] and Okamura[38] for carbon steel, nickel, and pure iron, respectively. Then Bates and his co-workers[39] made detailed investigations for iron, cobalt, and nickel ferrites[40] and permanent magnet materials.[41]

Figure 19.9 shows the temperature change as measured by Townsend[37] for one-cycle magnetization of nickel, where we see a generation of heat at the coercive force points in addition to the occurrence of reversible

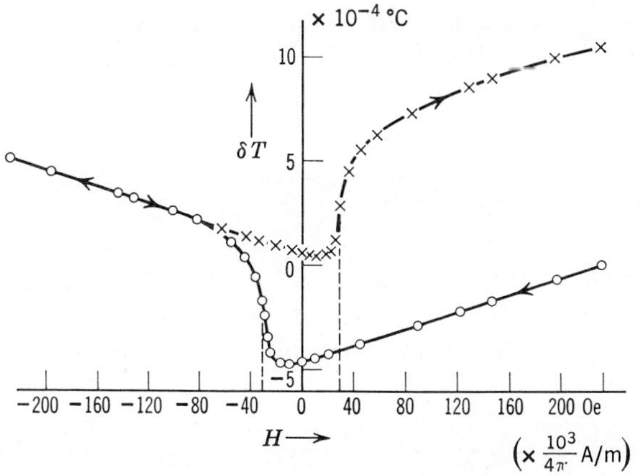

Fig. 19.9. Temperature change of Ni caused by its magnetization as a function of the applied field (Townsend[37]).

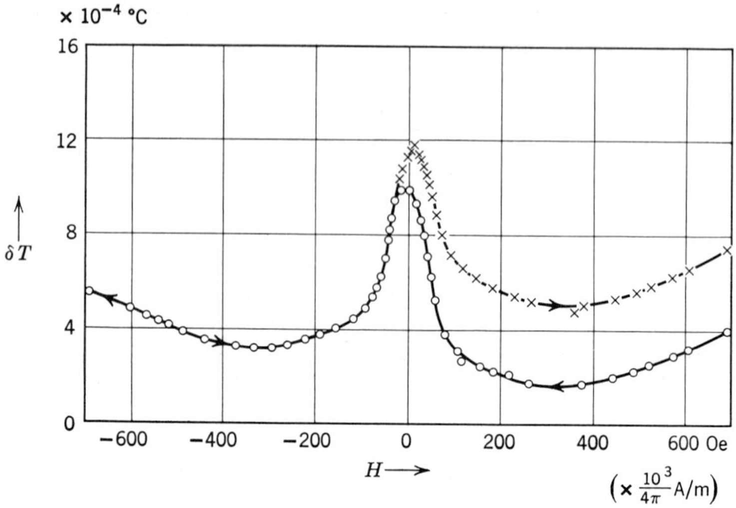

Fig. 19.10. Temperature change of Fe caused by its magnetization as a function of the applied field (Okamura,[38] from Becker and Döring[G,3]).

changes of temperature in the high field regions. Figure 19.10 shows the measurement made by Okamura[38] for soft iron; we see the absorption of heat caused by magnetization in a weak field region. This change is thought to be due to a reversible rotation of spontaneous magnetization against the anisotropy energy. The work done by the magnetic field is mainly used to raise the anisotropy energy, and a rotation of magnetization results in a decrease of the anisotropy field, which, like a decrease of the molecular field, opens up the angular distribution of spins, increasing the entropy of the system and resulting in an absorption of heat. Since the anisotropy field is changed by $\Delta H = -10K_1/3I_s$ (cf. 7.40) during the rotation of magnetization from $\langle 100\rangle$ to $\langle 111\rangle$, the heat generated during this process is calculated from (19.17) to be

$$\delta Q = -\frac{10\Theta L'(\alpha)}{T - 3\Theta L'(\alpha)} K_1. \tag{19.19}$$

If $K_1 < 0$, the rotation takes place from $\langle 111\rangle$ to $\langle 100\rangle$, and δQ is still negative. For iron, $K_1 = 4.2 \times 10^4$, $\Theta = 1043°\mathrm{K}$, $L'(\alpha) \approx 0.0025$ at $T = 300°\mathrm{K}$, and $C = 3.5 \times 10^6$ J/deg m³. The temperature change is calculated to be

$$\delta T = \frac{\delta Q}{C} = -\frac{10 \times 1043 \times 0.0025}{300 - 3 \times 1043 \times 0.0025} \times \frac{4.2 \times 10^4}{3.5 \times 10^6}$$

$$= -1.08 \times 10^{-3}\,°\mathrm{C}, \tag{19.20}$$

which is in good agreement with the magnitude of heat absorption seen in Fig. 19.10. Experimental separation of the reversible temperature change from the irreversible one was tried by Bates and Sherry.[42]

19.3 Magnetomechanical Effect

By magnetomechanical effect is meant the phenomenon in which the mechanical deformations or stresses are interrelated with the change in magnetization. We begin with the magnetostriction observed as a function of the intensity of the magnetic field. If a uniform magnetic field is applied, the magnetic body elongates or contracts as shown in Fig. 19.11*a* or 19.11*b*, depending on the sign of the magnetostriction constant λ. The deformation is caused by rotation magnetization or a displacement of 90° walls, but no deformation can be induced by 180° wall displacement. The deformation caused by rotation magnetization is given by, from (8.41) and (13.10),

$$\Delta\left(\frac{\delta l}{l}\right) = -3\lambda \sin\theta_0 \cos\theta_0 \,\Delta\theta = \frac{3\lambda I_s}{2K_u} \sin^2\theta_0 \cos\theta_0 \,\Delta H. \tag{19.21}$$

If the easy axes of the uniaxial anisotropy are distributed, it turns out that $\overline{\sin^2\theta_0 \cos\theta_0} = \frac{1}{4}$, so that (19.21) becomes

$$\Delta\left(\frac{\delta l}{l}\right) = \frac{3\lambda I_s}{8K_u} \Delta H. \tag{19.22}$$

If the 90° wall is effective in giving rise to the magnetostriction, the elongation observed along the direction of the magnetic field (cf. Fig. 13.7) is

$$\Delta\left(\frac{\delta l}{l}\right) = \frac{3}{2}\lambda_{100}\left[\cos^2\left(\theta_0 - \frac{\pi}{4}\right) - \cos^2\left(\theta_0 + \frac{\pi}{4}\right)\right]Ss = 3\lambda_{100} \sin\theta_0 \cos\theta_0 Ss. \tag{19.23}$$

Since displacement s is given by

$$s = \frac{\sqrt{2}\, I_s \cos\theta_0}{\alpha} \Delta H$$

for a 90° wall, (19.23) becomes

$$\Delta\left(\frac{\delta l}{l}\right) = \frac{3\sqrt{2}\,\lambda_{100} I_s S}{\alpha} \sin\theta_0 \cos^2\theta_0 \,\Delta H. \tag{19.24}$$

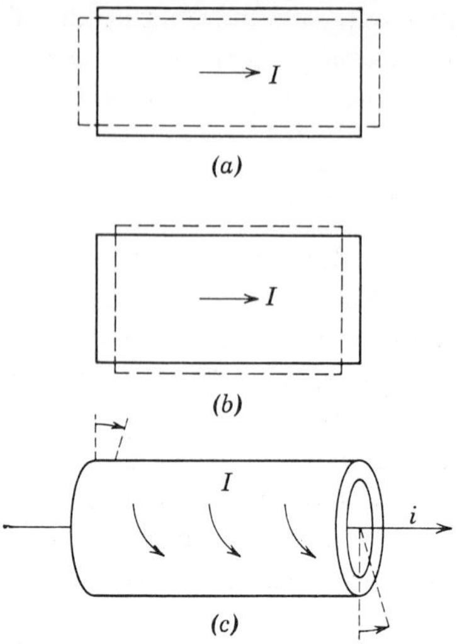

Fig. 19.11. Deformation of ferromagnetic bodies caused by their magnetization.

For a polycrystalline material, $\overline{\sin\theta_0 \cos^2\theta_0} = \pi/16$, so that

$$\Delta\left(\frac{\delta l}{l}\right) = \frac{3\sqrt{2}\,\pi\lambda_{100} I_s S}{16\alpha}\,\Delta H. \qquad (19.25)$$

It appears in (19.22) and (19.25) that large deformation is expected for a material having large magnetostriction. If, however, the magnetostriction is large enough to govern the ease of magnetization, it turns out that $K_u = \frac{3}{2}\lambda\sigma$ for rotation magnetization and $\alpha = \pi\lambda_{100}\sigma_i S/2$ for 90° wall displacement, as known from (13.47); hence (19.22) and (19.25) can be expressed as

$$\Delta\left(\frac{\delta l}{l}\right) = \frac{I_s}{4\sigma_i}\,\Delta H \qquad (19.26)$$

for rotation magnetization and as

$$\Delta\left(\frac{\delta l}{l}\right) = \frac{3\sqrt{2}\,I_s}{8\sigma_i}\,\Delta H \qquad (19.27)$$

for 90° wall displacement. It is interesting to note that both results are independent of λ. Since, of course, the deformation is proportional to the magnetostriction constant as long as the permeability is determined by other factors, equations (19.26) and (19.27) provide the ultimate value of deformation for infinitely large magnetostriction.

In Fig. 19.11*c*, the magnetic field is produced by an electric current flowing parallel to the axis of the cylindrical specimen as well as the uniform magnetic field parallel to the axis. The resultant direction of the magnetic field is helical as shown in the figure. The magnetization of the cylinder by this field gives rise to a torsion of the cylinder. This phenomenon was observed by Wiedemann[43] in 1862, and it is called the Wiedemann effect.

Next we discuss the inverse effect of magnetostriction, that is, the magnetization change caused by an application of mechanical stress. The domain magnetization tends to align itself parallel to the applied tension, provided $\lambda > 0$, whereas it tends to point perpendicular to the applied tension for $\lambda < 0$ (Fig. 19.12*a* and 19.12*b*). If the cylinder is twisted, the effective tension makes an angle of 45° from the axis, so that the magnetization lines up in a helical way. If the ac magnetic field is applied parallel to the axis, it induces the magnetization change along the helical line, which can be detected by a secondary coil wound around the cylinder wall as shown in Fig. 19.12*c*. In the absence of torsion, the magnetization changes along the axis of the cylinder, and no signal can be induced in the secondary coil. Thus the ac signal picked up by the secondary coil is a good measure of the torsional force. This phenomenon is called the inverse Wiedemann effect.

In principle, the magnetization caused by the tension σ can be determined by minimizing the sum of the magnetostrictive energy (8.79) and

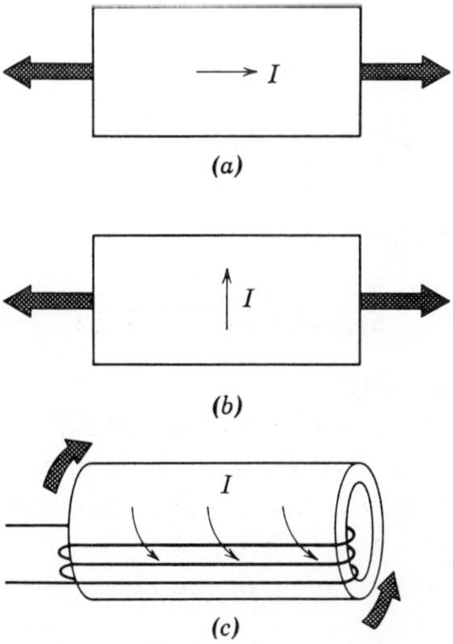

Fig. 19.12. Magnetization of ferromagnetic bodies caused by their mechanical deformation.

the potential energy of domain magnetizations. This, however, can be solved from a thermodynamical relation, provided that the deformation is known as a function of the applied field. If the internal energy is a function of magnetization I and the elongation $\delta l/l$, the equilibrium condition gives

$$\frac{\partial E}{\partial I} = H \tag{19.28}$$

and

$$\frac{\partial E}{\partial(\delta l/l)} = \sigma. \tag{19.29}$$

Since

$$\frac{\partial^2 E}{\partial(\delta l/l)\,\partial I} = \frac{\partial^2 E}{\partial I\,\partial(\delta l/l)}, \tag{19.30}$$

we have

$$\frac{\partial H}{\partial(\delta l/l)} = \frac{\partial \sigma}{\partial I}. \tag{19.31}$$

From (19.21), therefore, we have for rotation magnetization

$$\Delta I = \frac{3\lambda I_s}{2K_u} \sin^2\theta_0 \cos\theta_0\,\Delta\sigma, \tag{19.32}$$

and from (19.24) we have for the 90° wall displacement

$$\Delta I = \frac{3\sqrt{2}\,\lambda_{100} I_s S}{\alpha} \sin\theta_0 \cos^2\theta_0\,\Delta\sigma. \tag{19.33}$$

For a polycrystalline material we have

$$\Delta I = \frac{3\lambda I_s}{8K_u}\Delta\sigma \tag{19.34}$$

for rotation magnetization, and

$$\Delta I = \frac{3\sqrt{2}\,\pi\lambda_{100} I_s S}{16\alpha}\Delta\sigma \tag{19.35}$$

for displacement of 90° walls.

It should be noted here that equation (19.30) is valid only when the energy is a unique function of I and $\delta l/l$. In the presence of 180° walls the magnetostrictive energy is not a unique function of magnetization I, and (19.30) is no longer valid. This is the reason why the magnetization is insensitive to the applied stress in the demagnetized state, in contrast to (19.32) or (19.33).

Now we discuss the phenomenon that the applied stress causes a deformation through the change in domain magnetization in addition to the elastic deformation. This phenomenon is called the ΔE effect, because the effective Young's modulus is changed by the additional deformation. Figure 19.13 is a schematic illustration of the stress-strain curves of ferromagnetic and non-magnetic substances. For rotation magnetization,

$$\Delta I = -I_s \sin\theta_0\,\Delta\theta, \tag{19.36}$$

and we have, from (19.21) and (19.32),

$$\begin{aligned}\Delta\left(\frac{\delta l}{l}\right) &= -3\lambda \sin\theta_0 \cos\theta_0\,\Delta\theta \\ &= \frac{3\lambda}{I_s}\cos\theta_0\,\Delta I \\ &= \frac{9\lambda^2}{2K_u}\sin^2\theta_0\cos^2\theta_0\,\Delta\sigma.\end{aligned} \tag{19.37}$$

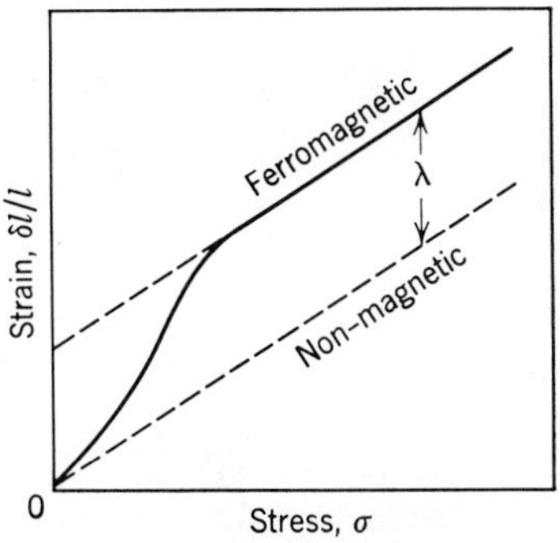

Fig. 19.13. A stress-strain curve of a ferromagnetic substance and its comparison with that of a non-ferromagnetic substance.

For polycrystalline materials,

$$\overline{\sin^2\theta_0\cos^2\theta_0} = 2/15,$$

so that

$$\Delta\left(\frac{\delta l}{l}\right) = \frac{3\lambda^2}{5K_u}\Delta\sigma. \tag{19.38}$$

Since this gives the additional elongation,

$$\frac{3\lambda^2}{5K_u} = \frac{1}{E_{\text{mag}}} - \frac{1}{E_0} \fallingdotseq -\frac{\Delta E}{E_0{}^2},$$

or

$$\frac{\Delta E}{E} = -\frac{3\lambda^2}{5K_u}E_0. \tag{19.39}$$

If $K_u = \frac{3}{2}\lambda\sigma_i$, $\Delta E/E$ becomes

$$\frac{\Delta E}{E} = -\frac{2\lambda}{5\sigma_i}E_0. \tag{19.40}$$

For 90° walls, by using $\Delta I = \sqrt{2}\,I_s \cos\theta_0 Ss$, we have from (19.23) and (19.33)

$$\begin{aligned}\Delta\left(\frac{\delta l}{l}\right) &= 3\lambda_{100}\sin\theta_0\cos\theta_0 Ss \\ &= \frac{3\lambda_{100}}{\sqrt{2}\,I_s}\sin\theta_0\,\Delta I \\ &= \frac{9\lambda_{100}^2 S}{\alpha}\sin^2\theta_0\cos^2\theta_0\,\Delta\sigma.\end{aligned} \tag{19.41}$$

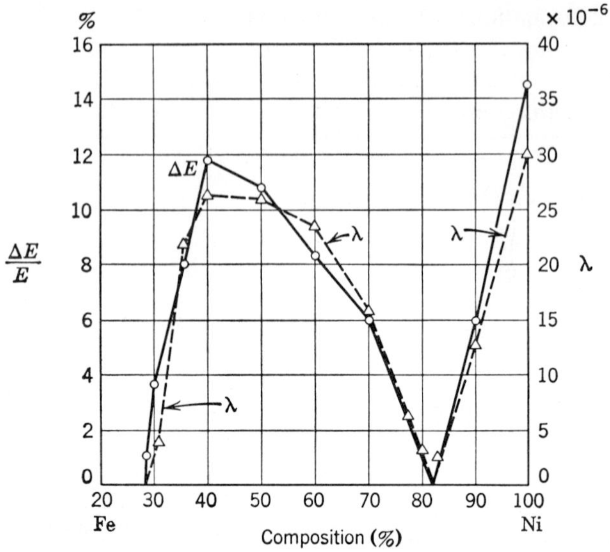

Fig. 19.14. ΔE effect of Fe-Ni alloys and its comparison with the compositional dependence of λ (Köster,[44] from Bozorth[G10]).

For polycrystalline materials (19.41) becomes

$$\Delta\left(\frac{\delta l}{l}\right) = \frac{6\lambda_{100}^2 S}{5\alpha}\Delta\sigma, \tag{19.42}$$

from which we have

$$\frac{\Delta E}{E} = -\frac{6\lambda_{100}^2 S}{5\alpha}E_0. \tag{19.43}$$

If α is governed by $\lambda\sigma_i$, $\Delta E/E$ becomes

$$\frac{\Delta E}{E} = -\frac{12\lambda_{100}}{5\pi\sigma_i}E_0, \tag{19.44}$$

which gives a value almost equal to that in (19.40). Thus the contribution of 90° walls is almost equal to that of rotation magnetization for highly stressed materials. These results show that $\Delta E/E$ is proportional to λ, being in good agreement with experiment (Fig. 19.14).

The elastic modulus is also influenced by spontaneous magnetization through volume magnetostriction. The increase of spontaneous magnetization caused by stress is, from (19.31),

$$\frac{\partial I_s}{\partial\sigma} = \frac{\partial}{\partial H}\left(\frac{\delta l}{l}\right) = \frac{1}{3}\frac{\partial\omega}{\partial H}, \tag{19.45}$$

where ω is a fractional volume change (cf. 8.58). The change in magnetization then causes a change in length:

$$\Delta\left(\frac{\delta l}{l}\right) = \Delta(\tfrac{1}{3}\omega) = \frac{\dfrac{1}{3}\dfrac{\partial \omega}{\partial H}}{\dfrac{\partial I_s}{\partial H}} \Delta I_s = \frac{\left(\dfrac{1}{3}\dfrac{\partial \omega}{\partial H}\right)^2}{\chi_0} \Delta\sigma, \tag{19.46}$$

where χ_0 is given by (13.85). By a procedure similar to that in (19.38) and (19.39), we have from (19.46)

$$\frac{\Delta E}{E} = -\frac{\left(\dfrac{1}{3}\dfrac{\partial \omega}{\partial H}\right)^2}{\chi_0} E_0. \tag{19.47}$$

In general, the elastic modulus decreases slightly with an elevation of temperature, whereas $\Delta E/E$ increases (its absolute value decreases) with the disappearance of ferromagnetism. In most ferromagnetic substances the effect of technical magnetization, (19.39) or (19.43), is much greater than the effect of volume magnetization (19.47). In exceptional ferromagnetic substances such as 30 ~ 45% Ni-Fe alloys, however, the latter effect is large enough to compensate the temperature dependence of elastic modulus.[45] Elinvar[46] (the alloy containing 50% Fe, 35% Ni, and 9.1% Cr) has a temperature coefficient of Young's modulus as small as 0.7×10^{-5}, and Vibralloy,[47] Coelinvar[48] and Velinvar[49] show temperature coefficients less than 10^{-6}.

In a magnetic substance, mechanical vibration is accompanied by motion of domain magnetizations, the energy of which can be absorbed more easily than the normal vibration of the crystal lattice. In other words, the internal friction of a ferromagnetic substance is much larger than that of a non-magnetic substance. For this reason magnetic 13 Cr stainless steel is sometimes used instead of non-magnetic 18-8 stainless steel for turbine blades in order to suppress the mechanical vibration.[50,51]

References

19.1. W. Thomson: *Proc. Roy. Soc. (London)* **8,** 546 (1851); *Phil. Mag.* IV, **15,** 469 (1858).
19.2. N. F. Mott: *Proc. Roy. Soc. (London)* **156,** 368 (1936).
19.3. T. Kasuya: *Progr. Theoret. Phys. (Kyoto)* **16,** 58 (1956).
19.4. R. V. Colvin, S. Legvold, and F. H. Spedding: *Phys. Rev.* **120,** 741 (1960).
19.5. E. Englert: *Ann. Physik.* **14,** 589 (1932).
19.6. W. Döring: *Ann. Physik.* **32,** 259 (1938).

19.7. T. Hirone and N. Hori: *Sci. Rept. Tohoku Imp. Univ.* **30,** 125 (1942).
19.8. W. L. Webster: *Proc. Roy. Soc.* (*London*) **113,** 196 (1926); **114A,** 611 (1927); *Proc. Phys. Soc.* (*London*) **42,** 431 (1930).
19.9. S. Kaya: *Sci. Repts. Tohoku Imp. Univ.* **17,** 1027 (1928).
19.10. Y. Shirakawa: *Sci. Rept. Tohoku Imp. Univ.* **29,** 132, 152 (1940).
19.11. S. Chikazumi: *J. Phys. Soc. Japan* **5,** 327, 333 (1950).
19.12. Y. Gondo and Z. Funatogawa: *J. Phys. Soc. Japan* **7,** 41 (1952).
19.13. E. Tatsumoto and K. Kuwahara: *J. Sci. Hiroshima Univ.* **19,** 385, 396 (1955).
19.14. E. Tatsumoto, K. Kuwahara, and H. Kimura: *J. Sci. Hiroshima Univ.* **24,** 359 (1960)
19.15. J. Kondo: *Progr. Theoret. Phys.* (*Kyoto*) **27,** 772 (1962).
19.16. J. Smit: Thesis (Rijksuniversiteit of Leiden, Netherland, 1956); *Physica* **17,** 612 (1951).
19.17. J. P. Jan: *Solid State Physics* (Academic Press, New York, 1957), Vol. 5, p. 73.
19.18. See Ref. G.10, Fig. 10-18, p. 453.
19.19. T. Wakiyawa and S. Chikazumi: *J. Phys. Soc. Japan* **15,** 1975 (1960).
19.20. Lord Kelvin: *Papers* Vol. II (1884), p. 267.
19.21. K. E. Grew: *Phys. Rev.* **41,** 356 (1932).
19.22. Z. Funatogawa: *J. Phys. Soc. Japan* **5,** 311 (1950); **6,** 256 (1951); Y. Gondo and Z. Funatogawa: *ibid.*, **7,** 589 (1952); N. Miyata and Z. Funatogawa: *ibid.*, **9,** 967 (1954).
19.23. A. Kundt: *Wied. Ann.* **49,** 257 (1893).
19.24. A. W. Smith and R. W. Sears: *Phys. Rev.* **34,** 1466 (1929).
19.25. E. M. Pugh: *Phys. Rev.* **36,** 1503 (1930).
19.26. A. W. Smith: *Phys. Rev.* **30,** 1 (1910).
19.27. E. M. Pugh, N. Rostoker, and A. I. Schindler: *Phys. Rev.* **80,** 688 (1950).
19.28. J. P. Jan and H. M. Gijsman: *Physica* **5,** 277 (1952).
19.29. E. M. Pugh and N. Rostoker: *Rev. Mod. Phys.* **25,** 151 (1953).
19.30. A. I. Schindler and E. M. Pugh: *Phys. Rev.* **89,** 295 (1953).
19.31. S. Foner and E. M. Pugh: *Phys. Rev.* **91,** 20 (1953).
19.32. W. L. Webster: *Proc. Cambridge Phil. Soc.* **23,** 800 (1925).
19.33. R. Kaplus and J. M. Luttinger: *Phys. Rev.* **95,** 1154 (1954).
19.34. See Ref. G.3.
19.35. P. Weiss and R. Forrer: *Ann. Physique* **10,** 153 (1926).
19.36. E. B. Ellwood: *Nature* **123,** 797 (1929); *Phys. Rev.* **36,** 1066 (1930).
19.37. A. Townsend: *Phys. Rev.* **47,** 306 (1935).
19.38. T. Okamura: *Sci. Rept. Tohoku Univ.* **24,** 745 (1935).
19.39. L. F. Bates and J. C. Weston: *Proc. Phys. Soc.* (*London*) **53,** 5 (1941); L. F. Bates and D. R. Healey: *ibid.*, **55,** 188 (1943); L. F. Bates and A. S. Edmondson: *ibid.*, **59,** 329 (1947); L. F. Bates and E. G. Harrison: *ibid.*, **60,** 213 (1948); L. F. Bates: *J. Phys. Radium* **10,** 353 (1949); **12,** 459 (1951); L. F. Bates and G. Marshall: *Rev. Mod. Phys.* **25,** 17 (1953).
19.40. L. F. Bates and N. P. R. Sherry: *Proc. Phys. Soc.* (*London*) **68B,** 304 (1955).
19.41. L. F. Bates and A. W. Simpson: *Proc. Phys. Soc.* (*London*) **68B,** 849 (1955).
19.42. L. F. Bates and N. P. R. Sherry: *Proc. Phys. Soc.* (*London*) **68B,** 642 (1955).
19.43. G. Wiedemann: *Pogg. Ann.* **117,** 631 (1862).
19.44. W. Köster: *Z. Metallk.* **35,** 194 (1943).
19.45. G.3, p. 356.
19.46. C. E. Guillaume: *Proc. Phys. Soc.* (*London*) **32,** 374 (1920).
19.47. M. E. Fine and W. C. Ellis: *J. Metals* **2,** 1120 (1950); **3,** 761 (1951).

19.48. H. Masumoto, H. Saito, and T. Kono: *Sci. Rept. Res. Inst. Tohoku Univ.* **A6,** 529 (1954); H. Masumoto, H. Saito, and T. Sugai: *ibid.*, **A7,** 533 (1955).
19.49. H. Masumoto, H. Saito, and T. Kobayashi: *Sci. Rept. Res. Inst. Tohoku Univ.* **A4,** 255 (1952). H. Masumoto, H. Saito and K. Goto: *ibid.*, **A9,** 159 (1957).
19.50. L. Lazan and L. J. Demer: *Proc. ASTM* **51,** 611 (1951).
19.51. A. W. Cochardt: *J. Appl. Mech.* **20,** 196 (1953).

20

Special Spin Configurations and Rare Earth Metals

20.1 Triangular and Helical Spin Configurations

A triangular arrangement of spins was first treated theoretically by Yafet and Kittel[1] to explain an anomalously small value of spontaneous magnetization in certain ferrites. An example is shown in Fig. 5.10, where we see that the magnetic moment of mixed Zn ferrites drops off with addition of a large amount of Zn and deviates from the theoretical curves which are calculated by assuming a combination of parallel and antiparallel moments. Since pure Zn ferrite is known to be antiferromagnetic, it is expected that the B-B interaction will be antiferromagnetic even in the mixed ferrimagnetic Zn ferrite. The effect of this interaction is usually masked by the strong A-B interaction which causes the spins in the B sites to be aligned parallel to each other. If, however, the A-B interaction is weakened by, say, the introduction of non-magnetic Zn ions into A sites, the tendency toward an antiferromagnetic arrangement in the B sites will be increased. To treat this problem we divide the A and B sublattices into four sublattices: A_1, A_2 and B_1, B_2. Let $\boldsymbol{I}_{A1}$, $\boldsymbol{I}_{A2}$, $\boldsymbol{I}_{B1}$, and $\boldsymbol{I}_{B2}$ denote the magnetization vectors of each sublattice, where we assume that $|\boldsymbol{I}_{A1}| = |\boldsymbol{I}_{A2}|$ and $|\boldsymbol{I}_{B1}| = |\boldsymbol{I}_{B2}|$. Then the molecular field acting on each sublattice is given by

$$\boldsymbol{H}_{mA1} = -w(\alpha\boldsymbol{I}_{A1} + \alpha'\boldsymbol{I}_{A2} + \boldsymbol{I}_B) = -w[(\alpha - \alpha')\boldsymbol{I}_{A1} + \alpha'\boldsymbol{I}_A + \boldsymbol{I}_B],$$

$$\boldsymbol{H}_{mA2} = -w(\alpha'\boldsymbol{I}_{A1} + \alpha\boldsymbol{I}_{A2} + \boldsymbol{I}_B) = -w[(\alpha - \alpha')\boldsymbol{I}_{A2} + \alpha'\boldsymbol{I}_A + \boldsymbol{I}_B]. \qquad (20.1)$$

Similarly, for the B_1 and B_2 sublattices, we have

$$\begin{aligned} \boldsymbol{H}_{mB1} &= -w(\boldsymbol{I}_A + \beta \boldsymbol{I}_{B1} + \beta' \boldsymbol{I}_{B2}) = -w[(\beta - \beta')\boldsymbol{I}_{B1} + \beta' \boldsymbol{I}_B + \boldsymbol{I}_A], \\ \boldsymbol{H}_{mB2} &= -w(\boldsymbol{I}_A + \beta' \boldsymbol{I}_{B1} + \beta \boldsymbol{I}_{B2}) = -w[(\beta - \beta')\boldsymbol{I}_{B2} + \beta' \boldsymbol{I}_B + \boldsymbol{I}_A], \end{aligned} \tag{20.2}$$

where $\boldsymbol{I}_A$ and $\boldsymbol{I}_B$ are the resultant magnetizations of the A and B sublattices, respectively, or

$$\begin{aligned} \boldsymbol{I}_A &= \boldsymbol{I}_{A1} + \boldsymbol{I}_{A2}, \\ \boldsymbol{I}_B &= \boldsymbol{I}_{B1} + \boldsymbol{I}_{B2}. \end{aligned} \tag{20.3}$$

Since the magnetization must be lined up parallel to the molecular field at each site, the first equation in (20.1) tells us that $\alpha' \boldsymbol{I}_A + \boldsymbol{I}_B$ must be parallel to $\boldsymbol{I}_{A1}$. Similarly, from the second equation of (20.1), it follows that $\alpha' \boldsymbol{I}_A + \boldsymbol{I}_B$ must be parallel to $\boldsymbol{I}_{A2}$. If, therefore, $\boldsymbol{I}_{A1}$ and $\boldsymbol{I}_{A2}$ make some angle other than 0° or 180° with each other, it necessarily follows that

$$\alpha' \boldsymbol{I}_A + \boldsymbol{I}_B = 0. \tag{20.4}$$

By the same reasoning, if angle formation occurs in the B sublattice, we have

$$\beta' \boldsymbol{I}_B + \boldsymbol{I}_A = 0. \tag{20.5}$$

Since simultaneous occurrence of (20.4) and (20.5) necessarily leads to $I_A = I_B = 0$, provided $\alpha'\beta' \neq 1$, the angle formation is expected to occur in only one of the two original sublattices.

Now we suppose that the angle formation occurs only in the B sublattice. Then it follows from (20.5) that $\boldsymbol{I}_B = -\boldsymbol{I}_A/\beta'$, so that

$$\boldsymbol{H}_{mA} = -w(\alpha_0 \boldsymbol{I}_A + \boldsymbol{I}_B) = w\left(\frac{1}{\beta'} - \alpha_0\right)\boldsymbol{I}_A, \tag{20.6}$$

where $\alpha_0 = (\alpha + \alpha')/2$, and

$$\begin{aligned} \boldsymbol{H}_{mB1} &= w(\beta' - \beta)\boldsymbol{I}_{B1}, \\ \boldsymbol{H}_{mB2} &= w(\beta' - \beta)\boldsymbol{I}_{B2}. \end{aligned} \tag{20.7}$$

Since these equations are identical with the expression for the Weiss molecular field (cf. 4.22), the temperature dependence of each sublattice magnetization is expected to be identical with that of a normal ferromagnetic substance. If, therefore, $I_{B1} = I_{B2}$ at $T = 0$, it follows that $I_{B1}(T) = I_{B2}(T)$ at any T. Then the angle ϕ between $\boldsymbol{I}_A$ and $\boldsymbol{I}_{B1}$ or $\boldsymbol{I}_{B2}$ is, from (20.5),

$$\cos\phi = \frac{I_{A1}}{\beta' I_{B1}}. \tag{20.8}$$

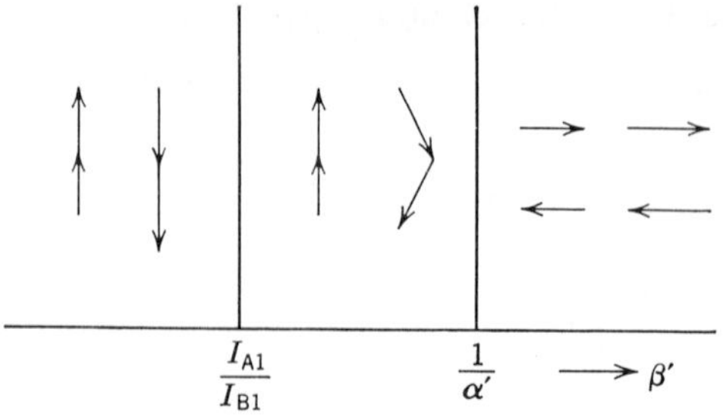

Fig. 20.1. Change in spin configuration due to a change in β' or in the interaction between B_1 and B_2 sites.

According to this equation, angle formation in B sites is possible only when $\beta' > I_{A1}/I_{B1}$. If the value of β' is changed from a small value to a large value while α, I_A, and I_B remain constant, the stable spin configuration is changed as shown in Fig. 20.1. For $\beta' < I_{A1}/I_{B1}$, no angle formation occurs in B sites; hence we expect a normal antiparallel alignment of A and B spins. For $\beta' > I_{A1}/I_{B1}$, angle formation on B sites occurs; and the angle between A and B spins is given by (20.8). If β' is increased further and becomes equal to $1/\alpha'$, angle formations are expected to occur for both A and B sites. For larger β', the sublattice spins $\boldsymbol{I}_{A1}$ and $\boldsymbol{I}_{A2}$ and $\boldsymbol{I}_{B1}$ and $\boldsymbol{I}_{B2}$ are expected to be antiparallel to each other, respectively. It is easily seen that the triangular arrangement cannot exist for very large values of β', because the molecular field at the A site given by (20.6) becomes antiparallel to its magnetization. For further detailed description of this phenomenon and a discussion of the transformations between various configurations during the course of heating, the reader may refer to the paper by Lotgering,[2] who discussed this problem in detail in relation to the magnetism of various kinds of oxide and sulfide spinels.

Direct experimental verification of triangular arrangement was first attempted by Prince[3] for $CuCr_2O_4$ by means of neutron diffraction. He concluded that the diffraction pattern does not conflict with the triangular model. Pickart and Nathans,[4] however, did not find any evidence of triangular arrangement by same means for $MnFe_{2-x}Cr_xO_4$, while Edwards[5] showed that the magnetic properties and X-ray diffraction pattern of $MnCr_{2-x}Al_xO_4$ can be explained in terms of the Yafet-Kittel-Lotgering model. Jacobs[6] examined the triangular arrangement by magnetizing specimens in a very high magnetic field, strong enough to rotate the canted spins toward the direction of the magnetic field. Figure 20.2 compares the magnetization curves of Fe_3O_4 and Mn_3O_4. The latter compound has a saturation magnetization lower than the value expected

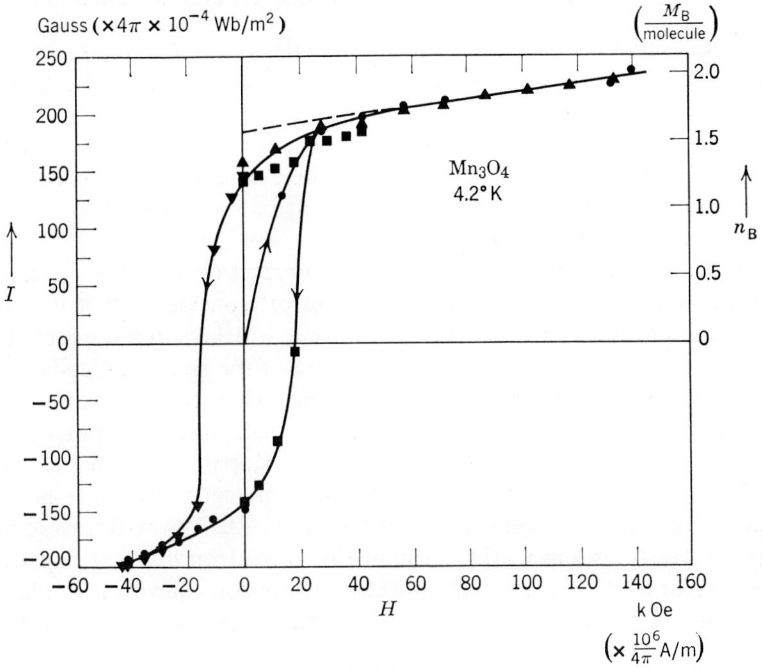

(a)

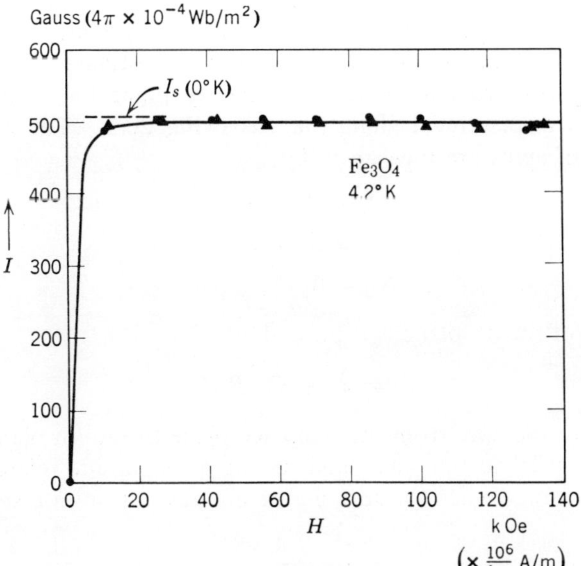

(b)

Fig. 20.2. Magnetization curves for Fe_3O_4 and Mn_3O_4 (after Jacobs[6]).

from the Néel model. The saturation magnetization of Mn_3O_4 is increased with an increase of the intensity of the magnetic field, while that of magnetite remains almost constant.

A triangular arrangement was also discussed by Hirone and Adachi[7] for the nickel-arsenide type of crystals. They explained the composition dependence of the magnetic transition point of the MnSb–CrSb system by this model (cf. Fig. 5.24).

There are some other possibilities for interpreting the low saturation moment of chromites. Baltzer and Wojtowicz[8] considered that the Cr^{3+} ion, which normally has the total spin $S = \frac{3}{2}$, will take the doublet state with $S = \frac{1}{2}$ as the ground state, if the B site of the spinel lattice is given a sufficient tetragonal or trigonal distortion. They tried to explain low saturation moments of various chromites in terms of a low spin state of Cr^{3+} ion. Another proposal was made by Kaplan and others,[9] who pointed out the possibility of having a ferrimagnetic spiral (or helical) structure in spinels for some range of the ratio of B-B to A-B interaction. Actually Hastings and Corliss[10] found that a neutron diffraction pattern observed for $MnCr_2O_4$ is semiquantitatively in agreement with a calculation based on the ferrimagnetic spiral structure. As for mixed Zn ferrites, there is also a possibility of accounting for the low saturation moments in terms of superparamagnetism caused by local magnetizations which are isolated by non-magnetic Zn ions, as shown by Ishikawa.[11]

A helical spin configuration was proposed independently by Yoshimori[12] in relation to the antiferromagnetic spin structure of MnO_2, by Villain[13] in relation to $MnAu_2$, and by Kaplan.[14] First we postulate that spins are all aligned parallel to one another in each xy plane but change their direction with displacement along the z axis (Fig. 20.3). The x, y, and z components of spins are then given by

$$\begin{aligned} S_{ix} &= S\cos\phi_i, \\ S_{iy} &= S\sin\phi_i, \\ S_{iz} &= 0. \end{aligned} \tag{20.9}$$

The angle ϕ_i of the ith plane is expressed as

$$\phi_i = \phi_0 + \boldsymbol{Q}\cdot\boldsymbol{R}_i, \tag{20.10}$$

where $\boldsymbol{R}_i$ is the distance from the reference plane to the ith plane and $\boldsymbol{Q}$ is a vector parallel to the z axis and having a magnitude related to the pitch of the screw. When the spin makes one revolution in n steps, Q is given by

$$Q = \frac{2\pi}{nc}, \tag{20.11}$$

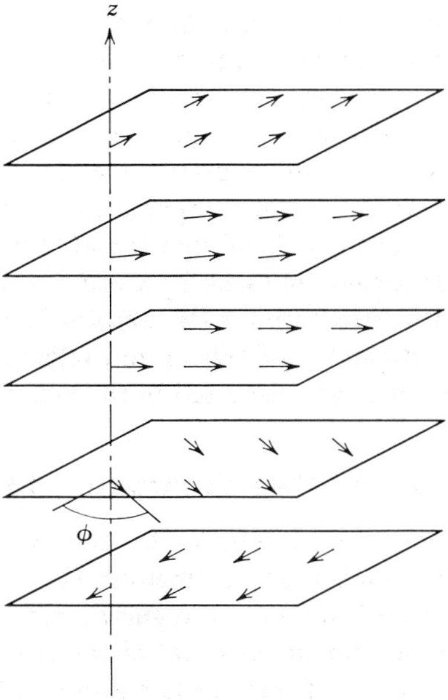

Fig. 20.3. A helical spin configuration.

where c is the separation of the xy lattice planes. For $n = 1$ or ∞, the spin arrangement is ferromagnetic; for $n = 2$, the arrangement is antiferromagnetic. In the general case, the configuration becomes helical or screw-type, with n not necessarily an integer. Now we assume that the exchange interaction is ferromagnetic between the nearest neighbor planes, while it is antiferromagnetic between the second nearest neighbor planes. Then the exchange energy stored in a unit volume is given by

$$E_{ex} = -2NS^2(J_1 \cos Qc + J_2 \cos 2Qc), \tag{20.12}$$

where N is the number of magnetic atoms in a unit volume, and J_1 and J_2 are the exchange integrals for the interactions between one atom and all atoms in the first and second neighbor planes, respectively ($J_1 > 0$ and $J_2 < 0$). The stable configuration is found by minimizing (20.12); then we have

$$\cos Qc = -\frac{J_1}{4J_2}. \tag{20.13}$$

According to (20.13), the necessary condition for a helical spin configuration is $|J_2| > J_1/4$.

Such a configuration was first investigated by Yoshimori[12] for a rutile type of structure. He compared the screw-type configuration with an antiferromagnetic configuration and also with the more complicated antiferromagnetic configuration which was proposed by Bizette,[15] Erickson,[16] and Yosida[17] and concluded that the screw type is energetically stable for some values of J_2/J_1. He calculated susceptibility, spin waves, antiferromagnetic resonance, and expected lines of neutron diffraction on the basis of this spin structure and compared them with experiment.

A helical spin structure was also found for $MnAu_2$ by Herpin, Meriel, and Villain,[18] and for chromium metal by Corliss, Hastings, and Weiss.[19] Enz[20] also reported this structure for $(Ba,Sr)_2Zn_2Fe_{12}O_{22}$. Recently there have been many investigations of helical and related spin structures for rare earth metals, and they are discussed in the next section.

20.2 Electron Structure and Magnetic Properties of Rare Earth Metals

The rare earth metals are the fifteen elements which range from lanthanum (La), atomic number 57, to lutetium (Lu), atomic number 71. They are usually arranged outside the regular array of the atomic table, because they all exhibit similar chemical characteristics. The reason is that the electronic structure of these metals is normally given by

$$(4f)^n(5s)^2(5p)^6(5d)^1(6s)^2,$$

where n increases from 0 to 14 as the atomic number increases from 57 to 71, and thus the outer shell electron structure, which determines the chemical properties of elements, is the same for all rare earth metals. Normally three electrons of the open outer shells $(5d)^1$ and $(6s)^2$ are easily removed from these atoms, giving trivalent ions. Because of this similarity in chemical character, it has been difficult to separate these elements from each other and to obtain them in pure form. The development of the ion exchange method, however, made possible the separation of these elements[21] and the preparation of pure rare earth metals.[22] Scandium, with atomic number 21, and yttrium, with number 39, are also usually included with the rare earth metals, because they have similar external electronic structures.

Various physical constants of the rare earth metals are listed in Table 20.1. As seen in this table, most of the rare earth elements crystallize in the hexagonal closed-packed structure, although there are a few exceptions. The atomic radius remains almost the same for all rare earth elements, decreasing slightly from about 1.8 Å to about 1.74 Å as the atomic number increases from 57 (La) to 71 (Lu).[23] Exceptional are Eu and Yb, which are believed to be divalent in the metallic state.

The incomplete $4f$ shell has a close relation to the magnetic properties of this series of elements, as the $3d$ shell does for the iron-group transition elements. The only difference is that in rare earth metals the $4f$ shell is enclosed by an outer electron shell composed of $(5s)^2(5p)^6$, even if the atoms are ionized by the loss of three electrons. Therefore the orbital angular momentum of $4f$ electrons remains unquenched by the crystalline field of neighboring ions. The $4f$ shell has seven orbitals with orbital quantum number $l = 3$, so that the magnetic quantum numbers of the orbitals are $-3, -2, -1, 0, 1, 2$, and 3. According to the Pauli principle, each orbit can have two electrons with spins $\frac{1}{2}$ and $-\frac{1}{2}$. All the possible electronic states can therefore be shown as the 14 boxes of Fig. 20.4. According to Hund's rule,[24] in the most stable state (i) the spin arrangement should have the maximum total spin angular momentum and (ii) the orbital arrangement should have the maximum orbital angular momentum within the restriction of rule (i) and of the Pauli principle (cf. Section 3.4). Thus the electrons are expected to fill up the boxes in Fig. 20.4 in the order indicated by the numbers in the circles. The resultant values of L and S are given in Table 20.1 for all trivalent ions.

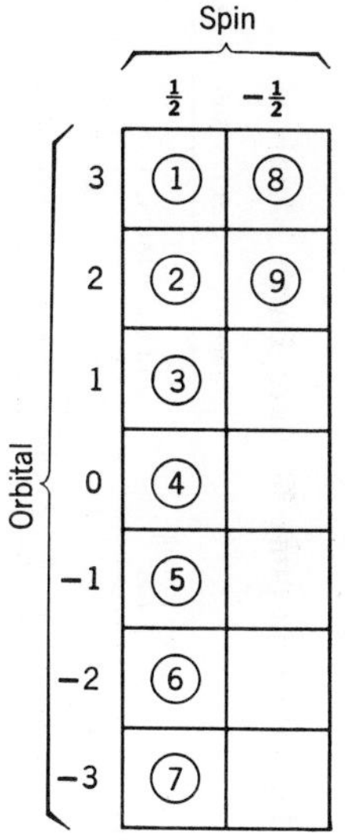

Fig. 20.4. Spin and orbital states of electrons in the $4f$ shell.

The values of J in the table are calculated by rule (iii), that $J = L - S$ for a less than half-filled shell ($n \leq 7$) and $J = L + S$ for a more than half-filled shell ($n \geq 7$). The values of S, L, and J are plotted in Fig. 20.5 as a function of the number of $4f$ electrons, n. It is seen that J takes two maxima, small and large, at $n = 3$ and 10, respectively.

In order to calculate the magnetic moment of atoms, it is necessary to calculate the g value first. It can be done by using the Landé formula (3.52), and the values of S, L, and J in Table 20.1. The effective magnetic moment and the saturation moment are calculated by (4.17) and (4.13), respectively, using the values of g and J, and are given in Table 20.1. The effective moment is also shown in Fig. 20.6 (the solid line) as a function of the number of $4f$ electrons. The experimental points[25] indicated by circles were determined from the temperature dependence of the susceptibility of rare earth salts, as solids or in solution.

The agreement between theory and experiment is excellent except for Eu^{3+} and Sm^{3+}. Van Vleck[26] interpreted this discrepancy in terms of the fact that the energy separation between the ground state and the first excited state (multiplet interval) is fairly small for these two elements where L and S nearly compensate each other. Since such an excited state

Table 20.1. Various Physical and Magnetic Constants of Rare Earth Metals

Element	Atomic No., Z	Mol. Wt.	Density, g/cc	Cryst. Form at Room Temp.	Transition Point, °C	M.P., °C	Curie Point, θ_f, °K	Néel Point, θ_N, °K	Asymptotic Curie Point, θ_a, °K	Electronic State of 3+ Ion: Number of 4f Electrons	S	L	J	Magnetic Moment in Bohr Magnetons: Effective Moment: Theory: Hund	V.V.-F.	Experiment: 3+ Ion	Metal	Saturation Moment: Theory gJ	Obs.	Refs.
Sc	21	44.96	2.992	h.c.p.	1335	1539														
Y	39	88.92	4.478	h.c.p.	1459	1509														
La	57	138.92	6.174	h.c.p.	310	920				0	0	0	0	0	0	0	0			
			6.186	f.c.c.	868															
Ce	58	140.13	6.771	f.c.c.	725	795		12.5	−46	1	$\frac{1}{2}$	3	$2\frac{1}{2}$	2.54	2.56	2.52	2.51	2.14		28
Pr	59	140.92	6.782	hex.	798	935			−21	2	1	5	4	3.58	3.62	3.60	2.56	3.20		29
Nd	60	144.27	7.004	hex.	862	1024		7.5	−16	3	$\frac{3}{2}$	6	$4\frac{1}{2}$	3.62	3.68	3.50	3.3–3.71	3.27		30, 31
Pm	61	(147)	—	—		1035				4	2	6	4	3.68	2.83	—	—	2.40		
Sm	62	150.35	7.536	rhomb.	917	1072		14.8		5	$\frac{5}{2}$	5	$2\frac{1}{2}$	0.85	1.55	—	1.74	0.72		28
Eu	63	152.0	5.259	b.c.c.		826		(90)	15	6	3	3	0	0.00	3.40	—	8.3	0.0		23, 36, 44
Gd	64	157.26	7.895	h.c.p.	1264	1312	289		310	7	$\frac{7}{2}$	0	$3\frac{1}{2}$	7.94	7.94	7.80	7.93	7.0	7.12	32, 33, 40
Tb	65	158.93	8.272	h.c.p.	1317	1356	218	230	236	8	3	3	6	9.72	9.70	9.74	9.62	9.0	9.25	32, 41, 47
Dy	66	162.51	8.536	h.c.p.		1407	90	179	151	9	$\frac{5}{2}$	5	$7\frac{1}{2}$	10.64	10.6	10.5	10.67	10.0	10.2	27, 32, 41, 44
Ho	67	164.94	8.803	h.c.p.		1461	20	133	87	10	2	6	8	10·60	10.6	10.6	10.9	10.0	9.7[a]	27, 34, 43, 45
Er	68	167.27	9.051	h.c.p.		1497	20	80(53)	41.6	11	$\frac{3}{2}$	6	$7\frac{1}{2}$	9.58	9.6	9.6	10.0	9.0	8.3[a]	27, 46, 47
Tm	69	168.94	9.332	h.c.p.		1545	22	53	20	12	1	5	6	7.56	7.6	7.1	7.56	7.0		35, 47
Yb	70	173.04	6.977	f.c.c.	798	824				13	$\frac{1}{2}$	3	$3\frac{1}{2}$	4.53	4.5	4.4	0.0	4.0		
Lu	71	174.99	9.842	h.c.p.		1652				14	0	0	0	0	0	0	0	0		
Ref.			(70)		(70)	(70)														

[a] Determined by neutron diffraction.

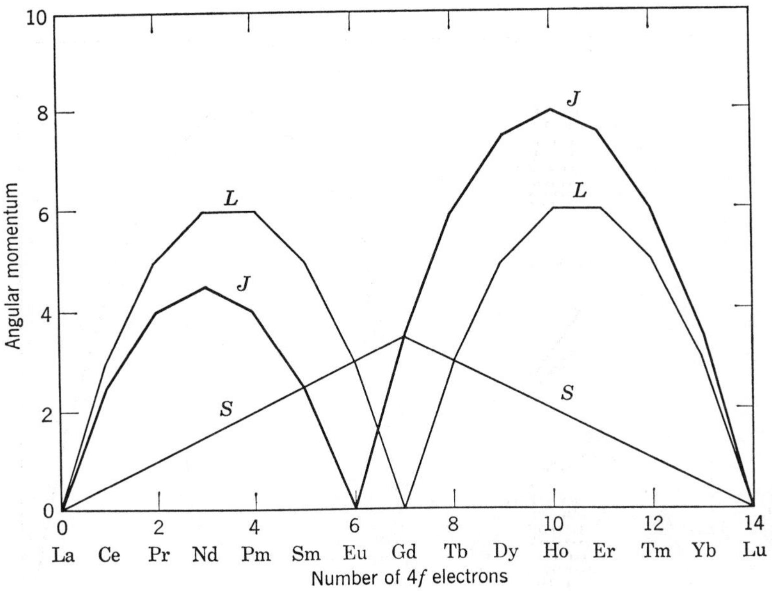

Fig. 20.5. Spin S, orbital L, and resultant angular momentum J as functions of the number of $4f$ electrons of trivalent rare earth ions.

has a larger value of J than the ground state for the less than half filled $4f$ shell, thermal excitation is expected to increase the magnetic moment above that calculated according to the Hund rule. The values of magnetic moment calculated in this way by Van Vleck and Frank[26] are listed in Table 20.1 and plotted in Fig. 20.6 as a broken line, which shows excellent agreement with experiment.

The experimental points indicated by crosses in Fig. 20.6 are observed for rare earth elements in the metallic state.[27–36] These points are also in fair agreement with the theory for trivalent ions except for Eu and Yb, both of which are supposed to have "divalent ion cores." The anomalously large ionic radii of these two elements can be interpreted along the same lines.[23] If we suppose that an extra electron goes into the $4f$ shell, the inner core structure of Eu^{2+} and Yb^{2+} should be just the same as that of Gd^{3+} and Lu^{3+}, respectively. Actually the magnetic moments of these metals can be adequately explained by this picture.

The similarity of the magnetic moments of the rare earths in the ionic and metallic states indicates that, even in the metallic state, $4f$ electrons are well localized in an inner core of each atom. Nevertheless, some of the rare earth metals exhibit reasonably strong interactions between atoms, giving rise to ferromagnetism, antiferromagnetism, or, sometimes, helical

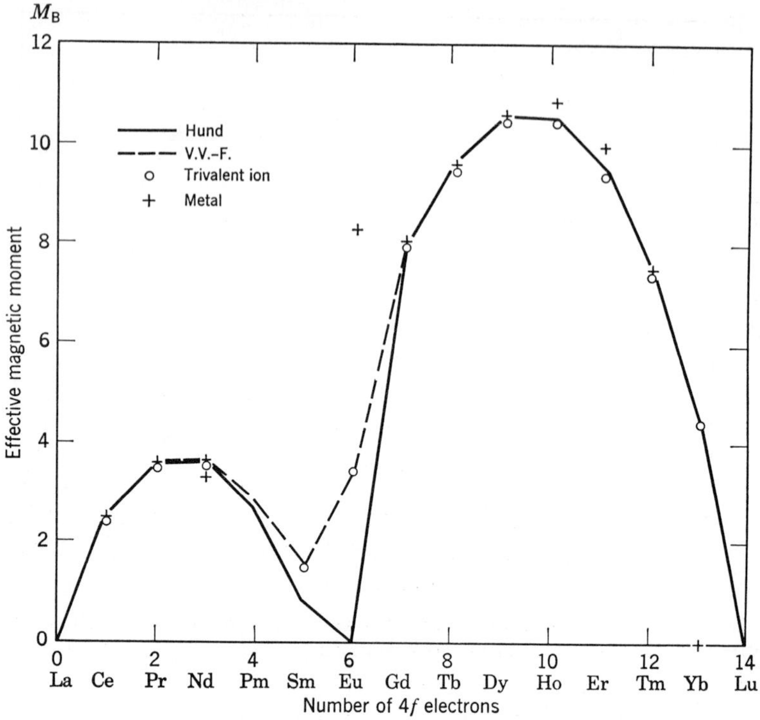

Fig. 20.6. Atomic magnetic moment as a function of the number of 4*f* electrons measured for trivalent rare earth ions in compounds and for rare earth metals, and comparison with the Hund and Van Vleck-Frank theories.

spin structures. In Fig. 20.7 the reader may see that the asymptotic Curie point increases monotonically with a decrease in the number of 4*f* electrons from Tm to Gd. This tendency is quite similar to the behavior of S shown in Fig. 20.5. Since the coupling between neighboring atomic magnetic moments is thought to occur through exchange coupling between 4*f* and conduction electrons,[37–39] and such an exchange coupling is expected to be proportional to S^2 (not to J^2), we can reasonably understand this tendency. This picture, however, cannot explain the actual behavior of θ_a in less than half-filled rare earth metals, for which θ_a is very small, and is even negative for some metals.

Most of the rare earth metals with a more than half-filled 4*f* shell exhibit ferromagnetism at low temperature. Gadolinium shows ferromagnetism below 289°K, and its temperature dependence obeys the $T^{3/2}$ law.[33,40] The absolute saturation moment also agrees well with the theoretical saturation moment gJ calculated for the trivalent ion. This is also true for Tb.[41] For Dy it is fairly hard to saturate the magnetization at low temperature, because of its large crystal anisotropy. Saturation magnetization measured along the easy direction [11$\bar{2}$0] by using a single crystal gives a value[42] which agrees with gJ fairly well, as seen in Table 20.1. Ho also has very large crystal anisotropy, so that its apparent saturation magnetization extrapolated to 0°K gives a saturation moment of $8.55M_B$, which is definitely less than the theoretical value $10.0M_B$. Neutron diffraction experiments, however, revealed that magnetic moments of about $9.7M_B$ are arranged on a conical surface about the c axis.[43] The large magnitude of magnetocrystalline anisotropy of these metals is thought to originate in their large orbital moment, which interacts electrically with the crystalline field produced by the surrounding atoms.

The most striking feature of the magnetism of rare earth metals with a

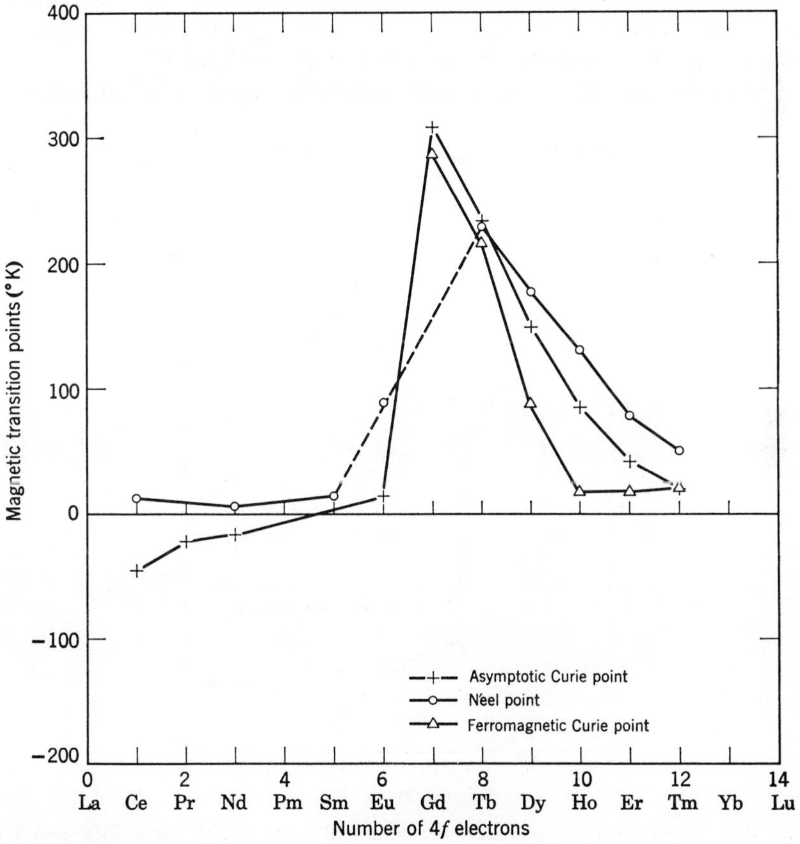

Fig. 20.7. Asymptotic Curie point, Néel point and ferromagnetic Curie point of rare earth metals as a function of the number of 4f electrons.

more than half-filled $4f$ shell is that above the Curie point most of these metals show a helical arrangement of spins which is again destroyed at the Néel point. These temperatures are also given in Table 20.1. These spin structures have been investigated for various rare earth metals by Koehler, Wilkinson, and others by means of neutron diffraction.[43–49] Figure 20.8 shows part of the diffraction pattern and also the intensity of the (0002) diffraction line observed for dysprosium[44] as a function of temperature. Below 90°K the intensity of the (0002) line increases with decreasing temperature, indicating an increase in the ferromagnetic moment. Above 90°K the intensity of the (0002) line decreases to a value which can be attributed to a nuclear reflection (cf. Chapter 21). Between 90° and 179°K, however, satellite reflections appear on each side of the (0002) nuclear reflection. These reflections indicate the presence of helical spin structure with moments perpendicular to the hexagonal c axis. The period n or the interlayer turn angle Qc is observed to change with temperature as shown in the upper graph of Fig. 20.8.

Holmium appears to be antiferromagnetic between 20° and 133°K,

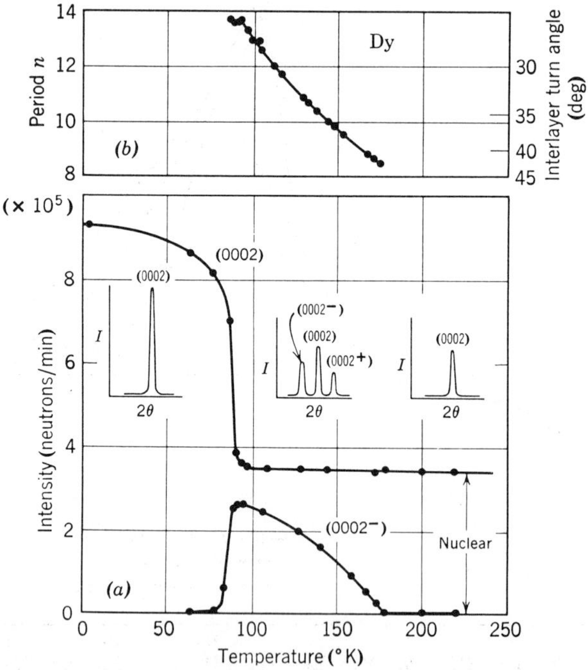

Fig. 20.8. Temperature dependence of the (0002) and ($000\bar{2}$) reflections and the interlayer turn angle or the period n (number of atomic layers in one revolution) of helical structures for dysprosium (after Wilkinson et al.[44,49]).

and ferromagnetic below 20°K.[34] In the antiferromagnetic region it exhibits a helical spin structure[43,45] with its moment parallel or nearly parallel to the *c* plane. The turn angle varies from 50° to 36° as the temperature decreases from 133° to 35°K. Even in the ferromagnetic region there remains a helical spin structure of turn angle 36° with a small component of moment parallel to the *c* axis ($2M_B$).

Erbium[46,47] is antiferromagnetic below 80°K and ferromagnetic below 20°K. The antiferromagnetic region is divided into two subregions. In the upper region, between 80° and 52°K, the spins are oscillating parallel to the *c* axis with a period of 7 because of a strong anisotropy with the easy axis parallel to the *c* axis. Between 52° and 20°K there appears a component of moment parallel to the *c* plane which forms a helical spin structure. Below 20°K diffraction lines are explained by a model wherein the spins have a ferromagnetic component parallel to the *c* axis but show a helical spin arrangement in the *c* plane.

Thulium[47] is antiferromagnetic below 53°K where an oscillation of the z component of spins occurs with a period of 7. Terbium[47] is also found to exhibit a helical spin structure similar to that of Dy, but its temperature range is very narrow, 218° to 230°K.

All these spin configurations are explained by Miwa and Yosida[50] in terms of a helical spin structure modified by the presence of a magneto-crystalline anisotropy. Similar calculations have been made by Kaplan[51] and Elliot.[52]

As discussed in the preceding section, the period of screw rotation is related to the ratio of the first nearest neighbor interaction J_1 to the second nearest neighbor interaction J_2 (cf. 20.13). Experiment shows that for most of the heavy rare earth metals the turn angle decreases with decreasing temperature from the Néel point and stays at constant value over some range of low temperatures. In Fig. 20.9 the experimental values of the reciprocal of the period n or the interlayer rotational angle Qc at the Néel points and low temperatures measured in units of 2π are plotted as a function of the number of $4f$ electrons. In this graph we see that the first neighbor interaction J_1 relative to the second neighbor interaction J_2 becomes stronger and stronger as the number of $4f$ electrons is decreased from 10 (corresponding to Ho) and that the relation (20.13) is no longer valid at Gd. Actually no antiferromagnetism or helical spin structure has been found in gadolinium metal. This behavior was interpreted by Yosida and Watabe[53] in terms of the s-f exchange interaction.

In rare earth metals with a less than half-filled $4f$ shell no ferromagnetism appears, for reasons which remain unclear. Cerium shows paramagnetism above 12.5°K,[48] where a thermal hysteresis has been found. This is caused by a phase transition from f.c.c. to condensed f.c.c. which is

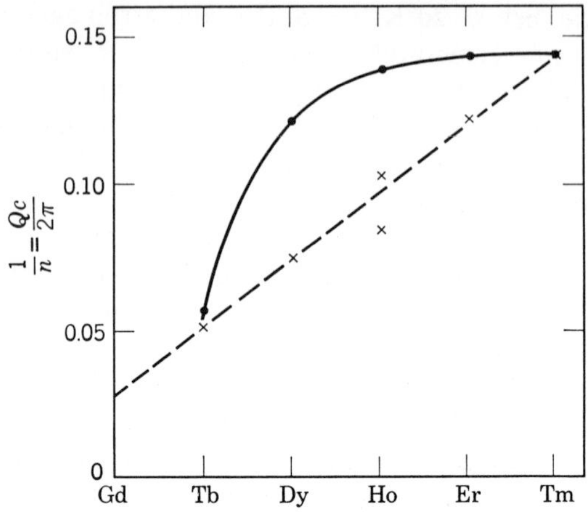

Fig. 20.9. Interlayer fractional revolution of the helical spin configuration observed for several rare earth metals at the Néel temperature (dots) and low temperatures (crosses) as functions of the number of $4f$ electrons (after Yosida and Watabe;[53] experimental points are due to Koehler et al.[47]).

thought to be caused by a transfer of a $4f$ electron to the $5d6s$ band. The same type of transition was found to result from the application of high pressure.[54] An odd fact is that the thermal hysteresis becomes smaller and smaller during repeated heating and cooling cycles.[28] The transfer of a $4f$ electron at low temperature has been confirmed by diffuse neutron scattering.[49] The hexagonal close packed phase was also found to exist in cerium, and it becomes antiferromagnetic below 12.5°K.[48,49]

20.3 Magnetic Alloys and Compounds Containing Rare Earth Elements

A number of investigations have been made on the general properties of alloys and compounds containing rare earth elements, for which the reader may refer to the book by Gschneidner.[55] In this section we confine ourselves to alloys and compounds which exhibit strong magnetism or interesting magnetic behavior.

The magnetic properties of La-Ce alloys were investigated by Roberts and Lock.[56] These alloys exhibit only weak magnetism: from temperature dependence these workers found that the effective magnetic moment per one Ce atom increases from 2.5 to 4.0 as the La content increases from 0% to 95%. They interpreted this fact in terms of a transfer of electrons from the $5d6s$ band to the $4f$ shell. The alloy system La-Nd was also investigated by Lock.[57] He found that the effective magnetic moment per Nd atom, determined from the temperature dependence of the susceptibility, stays almost constant in the range 0% to 80% La. Lock found also that ferromagnetism occurs below about 10°K for 17% to 80% La. Saturation moments determined from ferromagnetic saturation were found to be 0.8 to $1.2M_B$, which are lower than the theoretically expected value. The systems Gd-La and Gd-Y were investigated by Thoburn, Legvold, and Spedding.[58] For alloys of high Gd content, paramagnetic-to-ferromagnetic transitions are observed with decreasing temperature; and, for smaller Gd content, ferromagnetic ordering is replaced by antiferromagnetic ordering. The magnetic moment per Gd atom determined in the ferromagnetic and antiferromagnetic ranges was found to be somewhat larger than the theoretical values.

The alloys of Gd and 3*d* transition metals have been investigated by many workers.[59–62] There are numerous papers on Gd-Co and Gd-Fe alloys, in which the intermetallic compounds $GdCo_5$, $GdCo_2$, $GdFe_5$, and $GdFe_2$ are found to exist. According to measurements on single crystals, the magnetic moment of Co in $GdCo_5$ is aligned exactly antiparallel to that of Gd, while for $GdFe_5$ some of the Fe moments are antiparallel, and the others parallel, to the Gd moment.[62] Systematic studies of RCo_5-type compounds (R: rare earth element) were made by two groups.[63,64] Almost complete antiparallel alignment of the moments of R and Co was observed for R = Y, Sm, Ce, Tm, Gd, Tb, Er, Dy, and Ho. For Nd and Pr the saturation moments were considerably higher than those expected for antiparallel alignment.

The intermetallic compounds between rare earth metals and 4*d* or 5*d* transition metals have also been investigated. Bozorth et al.[65] measured the Curie points and the saturation moments for RIr_2, ROs_2, and RRu_2. The Curie point or exchange coupling is highest (80°K) for R = Gd and decreases on both sides of Gd when it is plotted as a function of the number of 4*f* electrons in R. The saturation moments are lower than gJ but higher than $2S$. This is interpreted in terms of partial quenching of the orbital magnetic moment. Compton et al.[66] also found that $PrPt_2$, $NdPt_2$, and $GdPt_2$ are ferromagnetic below 7.9°, 6.7°, and 77°K, respectively.

The RAl_2-type compounds were also found to be ferromagnetic for all rare earths.[67] The maximum Curie point of 175°K occurs also for R = Gd, with the Curie point decreasing with an increase or a decrease in the number of 4*f* electrons. Here again the saturation moments are found to be smaller than gJ; this is partially due to a lack of saturation by the applied field.

Compounds between rare earths and the elements of the Vb column, that is, N, P, As, and Sb, have been examined by the Oak Ridge group.[49,68] They found that TbN, DyN, HoN, and ErN are ferromagnetic, with

Curie points 42°, 26°, 18°, and 5°K, respectively. HoP is also ferromagnetic below 5.5°K, but all others, that is, TbP, TbAs, TbSb, HoSb, ErP, and ErSb are antiferromagnetic, with Néel points of 9°, 12°, 14°, 9°, 3.1°, and 3.7° K, respectively.

For oxides Pauthenet[69] determined the saturation magnetic moment of ferromagnetic garnets and found that the orbital moments of the rare earth elements in these compounds seem to be partially quenched below about 100°K. Remarkable magnetic properties were found by Matthias, Bozorth, and Van Vleck[70] for EuO. Below 77°K this oxide exhibits strong ferromagnetism in which the magnetic moments of Eu^{2+} ions are aligned parallel. At low temperatures the saturation magnetization is as high as $I_s = 2.41$ Wb/m^2 (= 1917 gauss) which may be compared with iron, gadolinium, or Fe–Co alloy. Many other oxides of rare earths have been investigated, but most of them show only weak magnetism. For further details the reader is referred to books by Gschneidner[55] and by Spedding and Daane.[71]

References

20.1. Y. Yafet and C. Kittel: *Phys. Rev.* **87,** 290 (1952).
20.2. F. K. Lotgering: *Philips Res. Repts.* **11,** 190, 337 (1956).
20.3. E. Prince: *Acta Cryst.* **10,** 554 (1957).
20.4. S. J. Pickart and R. Nathans: *Phys. Rev.* **116,** 317 (1959).
20.5. P. L. Edwards: *Phys. Rev.* **116,** 294 (1959).
20.6. I. S. Jacobs: *Phys. Chem. Solids* **11,** 1 (1959).
20.7. T. Hirone and K. Adachi: *J. Phys. Soc. Japan* **12,** 156 (1957).
20.8. P. K. Baltzer and P. J. Wojtowicz: *J. Appl. Phys.* **30,** 27S (1959).
20.9. T. A. Kaplan, K. Dwight, D. Lyons, and N. Menyuk: *J. Appl. Phys.* **32,** 13S (1961).
20.10. J. M. Hastings and L. M. Corliss: *J. Phys. Soc. Japan* **17,** Suppl. BIII, 43 (1962).
20.11. Y. Ishikawa: *J. Phys. Soc. Japan* **17,** 1877 (1962).
20.12. A. Yoshimori: *J. Phys. Soc. Japan* **14,** 807 (1959).
20.13. J. Villain: *Chem. Phys. Solids* **11,** 303 (1959).
20.14. T. A. Kaplan: *Phys. Rev.* **116,** 888 (1959).
20.15. H. Bizette: *J. Phys. Radium* **12,** 161 (1951).
20.16. R. A. Erickson: *Phys. Rev.* **85,** 745 (1952).
20.17. K. Yosida: *Progr. Theoret. Phys.* **8,** 259 (1952).
20.18. A. Herpin, P. Meriel, and J. Villain: *Compt. Rend.* **249,** 1334 (1959).
20.19. L. M. Corliss, J. M. Hastings, and R. J. Weiss: *Phys. Rev. Letters,* **3,** 211 (1959).
20.20. U. Enz: *J. Appl. Phys.* **32,** 22S (1961).
20.21. F. H. Spedding, A. F. Voight, E. M. Gladrow, and N. R. Sleight: *J. Am. Chem. Soc.* **69,** 2777 (1947); F. H. Spedding and J. E. Powell: *J. Metals* **6,** 1131 (1954).
20.22. F. H. Spedding and A. H. Daane: *J. Metals* **6,** 504 (1954).
20.23. W. Klemm and H. Bommer: *Z. Anorg. u. Allegem. Chem.* **231,** 138 (1937); **241,** 264 (1939).

20.24. F. Hund: *Z. Physik.* **33,** 855 (1925).
20.25. P. W. Selwood: *Magnetochemistry* (Interscience, New York, 1943).
20.26. Van Vleck: G. 1, p. 245.
20.27. F. H. Spedding, S. Legvold, A. D. Daane, and L. D. Jennings: *Progr. Low Temp. Phys.* **2,** 368 (North-Holland Pub. Co., Amsterdam, 1957).
20.28. J. M. Lock: *Proc. Phys. Soc. (London)* **B70,** 566 (1957).
20.29. C. H. La Blanchetais: *Comp. Rend.* **234,** 1353 (1952).
20.30. J. F. Elliot, S. Legvold, and F. H. Spedding: *Phys. Rev.* **94,** 50 (1954).
20.31. D. R. Behrendt, S. Legvold, and F. H. Spedding: *Phys. Rev.* **106,** 723 (1957).
20.32. Sigurds Arajs and R. V. Colvin: *J. Appl. Phys.* **32,** 336S (1961).
20.33. W. E. Henry: *J. Appl. Phys.* **29,** 524 (1958).
20.34. B. L. Rhodes, S. Legvold, and F. H. Spedding: *Phys. Rev.* **109,** 1547 (1958).
20.35. D. D. Davis and R. M. Bozorth: *Phys. Rev.* **118,** 1543 (1960).
20.36. R. M. Bozorth and J. H. Van Vleck: *Phys. Rev.* **118,** 1493 (1960).
20.37. C. Zener: *Phys. Rev.* **81,** 446 (1951); **82,** 403 (1951); **83,** 299 (1951); **85,** 324 (1951).
20.38. L. Pauling: *Proc. Natl. Acd. Sci. USA* **39,** 551 (1953).
20.39. T. Kasuya: *Progr. Theoret. Phys.* **16,** 45 (1956).
20.40. W. C. Thoburn, S. Legvold, and F. H. Spedding: *Phys. Rev.* **110,** 1298 (1958).
20.41. W. C. Thoburn, A. Legvold, and F. H. Spedding: *Phys. Rev.* **112,** 56 (1958).
20.42. D. R. Behrendt, S. Legvold, and F. H. Spedding: *Phys. Rev.* **109,** 1544 (1958).
20.43. W. C. Koehler: *J. Appl. Phys.* **32,** 20S (1961).
20.44. M. K. Wilkinson, W. C. Koehler, E. O. Wollan, and J. W. Cable: *J. Appl. Phys.* **32,** 48S (1961).
20.45. W. C. Koehler, J. W. Cable, E. O. Wollan, and M. K. Wilkinson: *Bull. Am. Phys. Soc. Ser. II* **5,** 459 (1960).
20.46. J. W. Cable, E. O. Wollan, W. C. Koehler, and M. K. Wilkinson: *J. Appl. Phys.* **32,** 49S (1961).
20.47. W. C. Koehler, J. W. Cable, E. O. Wollan, and M. K. Wilkinson: *J. Phys. Soc. Japan* **17,** Suppl. BIII, 32 (1962).
20.48. M. K. Wilkinson, H. R. Child, C. J. McHargue, W. C. Koehler, and E. O. Wollan: *Phys. Rev.* **122,** 1409 (1961).
20.49. M. K. Wilkinson, H. R. Child, W. C. Koehler, J. W. Cable, and E. O. Wollan: *J. Phys. Soc. Japan* **17,** Suppl. BIII, 27 (1962).
20.50. H. Miwa and K. Yosida: *Progr. Theoret. Phys.* **26,** 693 (1961).
20.51. T. A. Kaplan: *Phys. Rev.* **124,** 329 (1961).
20.52. R. J. Elliot: *Phys. Rev.* **124,** 340 (1961).
20.53. K. Yosida and A. Watabe: *Progr. Theoret. Phys.* **28,** 361 (1962).
20.54. A. W. Lawson and Ting-Youn Tang: *Phys. Rev.* **76,** 301 (1949).
20.55. K. Gschneidner, Jr.: *Rare Earth Alloys* (D. Van Nostrand Co., Princeton, N.J., 1961).
20.56. L. M. Roberts and J. M. Lock: *Phil. Mag.* **2,** 811 (1957).
20.57. J. M. Lock: *Phil. Mag.* **2,** 726 (1957).
20.58. W. C. Thoburn, S. Legvold, and F. H. Spedding: *Phys. Rev.* **110,** 1298 (1958).
20.59. E. A. Nesbitt, J. H. Wernick, and E. Corenzwit: *J. Appl. Phys.* **30,** 365 (1959).
20.60. R. C. Vickery, W. C. Sexton, V. Novy, and E. V. Kleber: *J. Appl. Phys.* **31,** 366 (1960).
20.61. W. M. Hubbard, E. Adams, and J. V. Gilfrich: *J. Appl. Phys.* **31,** 368S (1960).
20.62. W. M. Hubbard and E. Adams: G. 37, p. 143.
20.63. K. Nassau, L. V. Cherry, W. E. Wallace: *Phys. Chem. Solids* **16,** 131 (1960).

20.64. E. A. Nesbitt, H. J. Williams, J. H. Wernick, and R. C. Sherwood: *J. Appl. Phys.* **32,** 342S (1961).

20.65. R. M. Bozorth, B. T. Matthias, H. Suhl. E. Corenzwit, and D. D. Davis: *Phys. Rev.* **115,** 1595 (1959).

20.66. V. B. Compton and B. T. Matthias: *Acta Cryst.* **12,** 651 (1959).

20.67. H. J. Williams, J. H. Wernick, E. A. Nesbitt, and R. C. Sherwood: G. 37, p. 91.

20.68. M. K. Wilkinson, H. R. Child, J. W. Cable, E. O. Wollan, and W. C. Koehler: *J. Appl. Phys.* **31,** 358S (1960).

20.69. R. Pauthenet: *Comp. Rend.* **243,** 1737 (1956).

20.70. B. T. Matthias, R. M. Bozorth, and J. H. Van Vleck: *Phys. Rev. Letters* **7,** 160 (1961).

20.71. F. H. Spedding and A. H. Daane: *The Rare Earths* (John Wiley & Sons, Inc., New York, 1961).

21

Other Techniques for Investigating Internal Magnetic Structure

21.1 Neutron Diffraction

There is no doubt that neutron diffraction is the best means for investigating internal magnetic structure. We have already quoted several experimental results of neutron diffraction in relation to various magnetic metals and compounds. In this section we describe the basic concepts of neutron diffraction, its experimental procedures, and important experimental results.

The neutron is an elementary particle which has no electric charge but has a magnetic moment of -1.913 in units of the nuclear magneton which is equal to 1/1840 of Bohr's magneton. It is, therefore, expected that a neutron can pass through matter without being disturbed by the presence of electric charges, but that it can be scattered by magnetic dipoles. If we regard the neutron beam as a de Broglie wave, we expect that its wavelength $\lambda = h/mv$ is 0.9×10^{-12} cm for an energy of 10 MeV, which is common for neutrons produced by nuclear fission. This wavelength is too small compared to the lattice constants of solids. If, however, the neutron is slowed down by collision with atoms in the material used as moderator, the wavelength becomes longer and thus makes the diffraction from crystal lattices observable. This idea was first proposed by Elsasser[1] and experimentally confirmed by Halban[2] and others.[3] Since, at that time, only the very weak neutron beam which is emitted from a radioactive substance was available, it was hard to do a precise experiment. Recently, intense beams of neutrons have been available from thermal nuclear reactors. It was fortunate, in particular for solid state physicists, that its most abundantly available neutrons have a wavelength of the same magnitude as atomic dimensions (~1 Å). This has favored accomplishment of neutron diffraction experiments, from which much precise knowledge of magnetic structures as well as of crystal structures has been obtained.

There are two ways in which neutrons interact with matter. One is an interaction with the nucleus, and the other is an interaction with atomic magnetic moments. In contrast to X-ray scattering, the former interaction does change in a quite irregular way with an increase of atomic number. This is partly because of the existence of strong resonances associated with the scattering process, which affect not only the magnitude of the cross section but also the phase changes between the incident and the scattered neutron waves. If the nucleus has a spin, the scattering cross section depends also on the sign of nuclear spin relative to the neutron spin. In contrast to X-ray scattering or to magnetic scattering of neutrons, the scattering of neutrons from nuclei bound to atoms in a condensed system is isotropic with the scattering angle, because the nucleus is essentially a point, when it is measured by the scale of the wavelength of thermal neutrons.

Magnetic interaction depends on the spin and orbital magnetic moment of atoms. The magnetic scattering of neutrons was first discussed by Bloch[4] and Schwinger[5] and later treated in detail by Halpern and Johnson.[6] According to their calculations, the magnetic scattering from paramagnetic ions is described by the differential scattering cross section into a unit solid angle,

$$\frac{\partial \sigma}{\partial \Omega} = \tfrac{2}{3} S(S+1)\left(\frac{\mu_0{}^2 e^2 \gamma}{m}\right)^2 f^2, \tag{21.1}$$

where S is the total spin of the paramagnetic ion, γ the neutron moment in the unit of nuclear magnetons (cf. 21.11), and f the amplitude form factor of the ion. For ferromagnetic or antiferromagnetic materials, in which atomic magnetic moments are arranged regularly, the differential scattering cross section into a unit solid angle is given by

$$\frac{\partial \sigma}{\partial \Omega} = D^2(\boldsymbol{q} \cdot \boldsymbol{\lambda})^2 \tag{21.2}$$

if we ignore the nuclear scattering. In this equation

$$D = \frac{\mu_0{}^2 e^2 \gamma S}{m} f. \tag{21.3}$$

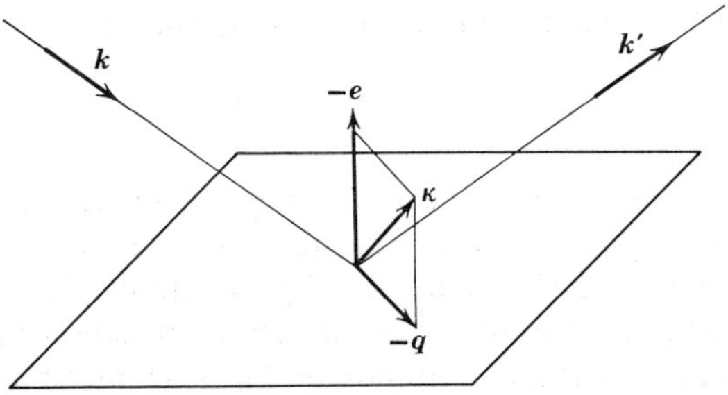

Fig. 21.1. Relation between various vectors involved in neutron diffraction: $\boldsymbol{k}$, $\boldsymbol{k}'$ the incident and the scattered wave vectors, $\boldsymbol{e}$ the scattering vector, $\boldsymbol{\varkappa}$ the unit vector parallel to the magnetic moment, and $\boldsymbol{q}$ the vector given by (21.4).

The vector $\boldsymbol{\lambda}$ in (21.2) is the unit vector parallel to the polarization of the neutron, and $\boldsymbol{q}$ is the vector given by

$$\boldsymbol{q} = \boldsymbol{e}(\boldsymbol{e} \cdot \boldsymbol{\varkappa}) - \boldsymbol{\varkappa}, \tag{21.4}$$

where the scattering vector $\boldsymbol{e}$ is defined as

$$\boldsymbol{e} = \frac{\boldsymbol{k} - \boldsymbol{k}'}{|\boldsymbol{k} - \boldsymbol{k}'|}. \tag{21.5}$$

In this formula $\boldsymbol{k}$ and $\boldsymbol{k}'$ are the incident and the scattered wave vectors of the neutron (Fig. 21.1). In other words, the scattering vector $\boldsymbol{e}$ is a unit vector parallel to the bisector between $\boldsymbol{k}$ and $\boldsymbol{k}'$. The vector $\boldsymbol{\varkappa}$ is a unit vector parallel to the magnetic moment of atom. The magnitude of the vector $\boldsymbol{q}$ is equal to the projection of the vector $\boldsymbol{\varkappa}$ onto the plane perpendicular to the vector $\boldsymbol{e}$ and the direction of $\boldsymbol{q}$ is parallel to this plane. If the polarization of the incident neutrons, $\boldsymbol{\lambda}$, is parallel to the vector $\boldsymbol{q}$, the magnetic scattering of neutrons is maximum, while, if the vector $\boldsymbol{\lambda}$ is perpendicular to $\boldsymbol{q}$, no magnetic scattering should occur. If the magnetic moment $\boldsymbol{\varkappa}$ is parallel to the scattering vector $\boldsymbol{e}$, no scattering of neutrons should occur in the direction $\boldsymbol{k}'$, no matter how the neutron polarization $\boldsymbol{\lambda}$ is oriented.

If we take the nuclear scattering into consideration, the differential scattering cross section for scattering into a unit solid angle is given by

$$\frac{\partial\sigma}{\partial\Omega} = [C + D(\boldsymbol{q} \cdot \boldsymbol{\lambda})]^2 = C^2 + 2CD\boldsymbol{q} \cdot \boldsymbol{\lambda} + D^2q^2, \tag{21.6}$$

where C is the nuclear scattering amplitude. When we use an unpolarized neutron beam, we must average (21.6) for all possible orientations of $\boldsymbol{\lambda}$, so that $\boldsymbol{q} \cdot \boldsymbol{\lambda}$ averages to zero. Then the differential cross section is given by the sum of the nuclear and magnetic contributions:

$$\frac{\partial\sigma}{\partial\Omega} = C^2 + D^2q^2. \tag{21.7}$$

The contributions of the nuclear and magnetic scattering are sometimes comparable in magnitude; then the magnetic scattering is fairly easily separated by changing the value of q by the application of a magnetic field or by raising temperature to the magnetic transition point. If, however, the magnetic scattering amplitude Dq is smaller than the nuclear scattering amplitude C, the separation of the two contributions by this means becomes fairly difficult. If we assume that Dq is 10% of C, it follows from (21.7) that the magnetic contribution is only 1% of the nuclear contribution. If we use a polarized neutron beam, which will be discussed in detail later, we can alter the second term in (21.6) by switching the direction of polarization of the neutron ($\lambda = \pm 1$). The change caused by this procedure becomes about 40% of the nuclear term, provided that the spin orientation of atoms is favorable for the polarization of neutrons. Values of C and D for various magnetic atoms and ions are given in Table 21.1. Knowing the value of D, we can determine the magnetic moment of atoms from (21.3)

Table 21.1. Differential Cross Sections for Nuclear and Magnetic Scattering (after Bacon[7])

Atom or Ion	Nuclear Scattering Amplitude C, 10^{-12} cm	Magnetic Scattering Amplitude D, 10^{-12} cm	
		For $\theta = 0$	For $\sin\theta/\lambda = 0.25$
Cr^{2+}	0.35	1.08	0.45
Mn^{2+}	−0.37	1.35	0.57
Fe (metal)	0.96	0.60	0.35
Fe^{2+}		1.08	0.45
Fe^{3+}		1.35	0.57
Co (metal)	0.28	0.47	0.27
Co^{2+}		1.21	0.51
Ni (metal)	1.03	0.16	0.10
Ni^{2+}		0.54	0.23

The diffraction of neutrons from crystals is quite similar to that of X-rays. Nuclear scattering from each atom gives rise to diffraction lines which are determined by the atomic arrangement in the crystal. Magnetic scattering occurs, however, only from magnetic atoms, and the scattering

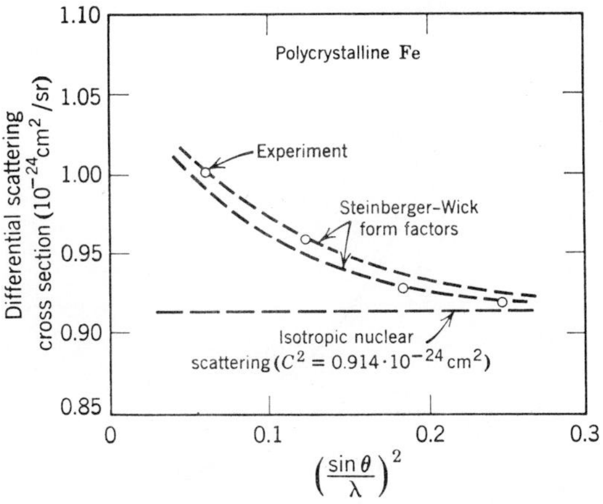

Fig. 21.2. Differential scattering cross sections as a function of scattering angle for magnetic and nuclear scattering of Fe (Shull, Wollan, and Koehler[8]).

amplitude depends on the orientation of the magnetic moments of these atoms, so that the diffraction lines are determined entirely by magnetic structures, whose unit cell is, in general, different from the crystal unit cell. Moreover, the angular variation of scattering amplitude is different for nuclear and magnetic scatterings. As mentioned before, nuclear scattering is isotropic with scattering angle, whereas magnetic scattering decreases fairly rapidly with an increase of the scattering angle. Figure 21.2 shows the angular dependence of differential scattering cross sections as determined for polycrystalline iron by Shull, Wollan, and Koehler.[8] It turns out that each diffraction line should thus contain the nuclear and the magnetic parts in different proportions depending on the crystal and magnetic structure factors and also on the diffraction angle. Incidentally, the angular dependence of atomic form factors is sharper for neutron magnetic scattering than for X-ray diffraction as shown in Fig. 21.3, because the spin magnetic moments which contribute to the magnetic scattering of neutrons are located outside the average position of electrons which contribute to X-ray scattering.

Let us look now at the apparatus used for neutron diffraction. Figure 21.4 shows a cross-sectional view of the atomic pile Cp-5, JRR-2,[10] which has been installed at the Japan Atomic Energy Research Institute. At the center of this pile there are a number of fuel rods made from 90% condensed uranium plates which emit neutrons by nuclear fission. The emitted neutrons are slowed down by heavy water, and finally their average energy is reduced by a factor 10^{-8} to the order of kT. We call these neutrons thermal neutrons. The thermal neutrons go through a

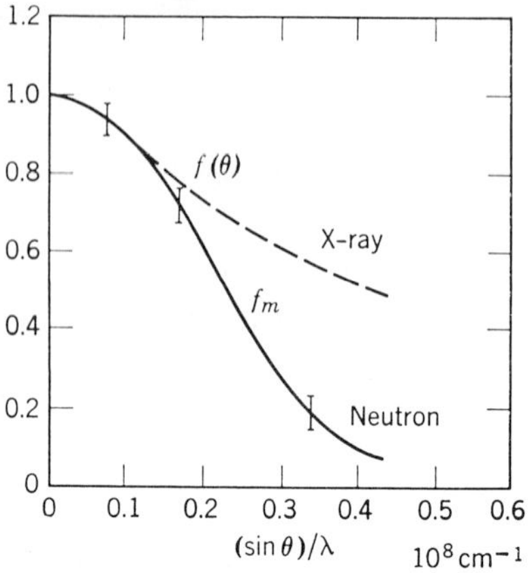

Fig. 21.3. Comparison of atomic form factors by X-ray and by neutrons (Bacon[7]). The curve for neutrons is due to Shull, Strauser, and Wollan.[9]

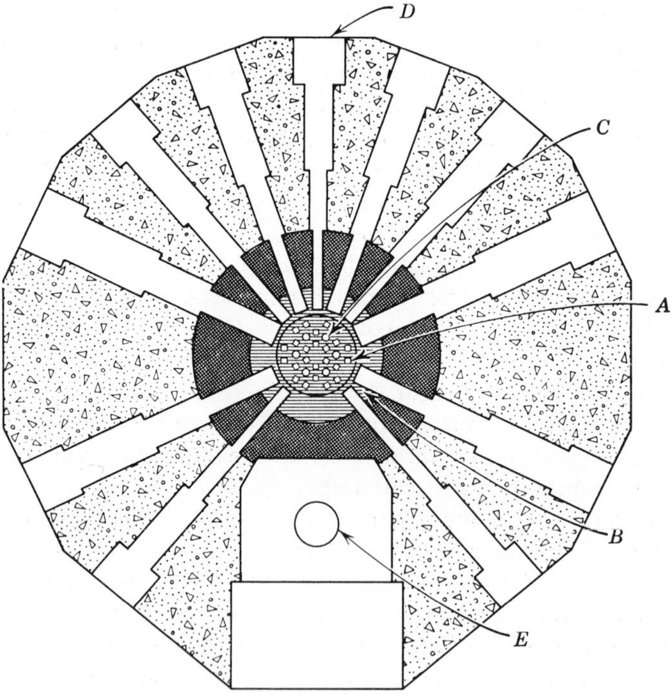

Fig. 21.4. Cross-sectional view of the CP5 atomic pile (JRR-2): *A*, Atomic fuel; *B*, heavy water; *C*, controlling rod; *D*, collimator; *E*, vertical hole for irradiation.

collimator which is fixed in the heavy concrete shields surrounding the center part of the pile and can be used for neutron diffraction. Figure 21.5 shows the top and side view of the neutron diffraction apparatus[11] which was installed in JRR-2. Graphite blocks at the top of the collimator serve to protect the collimator from emerging fast neutrons. In this particular apparatus the neutron beam is separated into two parts; one is diffracted by a monochromator crystal **A** and goes to the **A** goniometer, and the other goes through crystal **B** and is diffracted to goniometer **B**. For goniometer **A** the direction of the incident beam is fixed at 42° from the original beam, which is suitable for (311) reflection of a lead single crystal. The monochromatic beam thus produced is 1.08 Å in wavelength. This goniometer is the double-axis type, which permits the energy analysis of scattered neutrons. At the same time goniometer **B** can be rotated around the rotation axis of crystal table **B**, so that one can select various wavelengths. Monochromator crystal **B** can be magnetized so as to permit an experiment using a polarized neutron beam. The flux density of neutrons is 10^{12} to 10^{14} neutrons/cm^2 sec at the center of the pile; then it decreases to $10^6 \sim 10^8$ after passing through a collimator which leads neutrons from the pile to the monochromator crystal. Reflection of a neutron beam by a crystal usually reduces the intensity to $10^4 \sim 10^6$.

The first idea for production of a polarized neutron beam was proposed by Shull.[12] Later this idea was actually reduced to the practical apparatus by Nathans and Shull and their co-workers.[13,14] Figure 21.6 is a schematic diagram of a polarized neutron spectrometer. First the unpolarized neutron is diffracted by a crystal which is magnetized in the vertical direction or parallel to the reflecting atomic plane. That is, the vector $\boldsymbol{\varkappa}$ is perpendicular to $\boldsymbol{e}$ in (21.4), and hence

$$\boldsymbol{q} = -\boldsymbol{\varkappa}. \tag{21.8}$$

Then the cross section (21.6) becomes

$$\frac{\partial\sigma}{\partial\Omega} = (C - D)^2, \tag{21.9}$$

provided that the neutron is polarized parallel to the magnetization of the diffracting crystal; it becomes

$$\frac{\partial\sigma}{\partial\Omega} = (C + D)^2 \tag{21.10}$$

if the polarization of neutrons is opposite to the magnetization of the crystal. If we choose a crystal for which the nuclear magnetic amplitude C is just equal to the magnetic scattering D, the cross section of a positively polarized neutron (21.9) vanishes, and we have only a negatively polarized

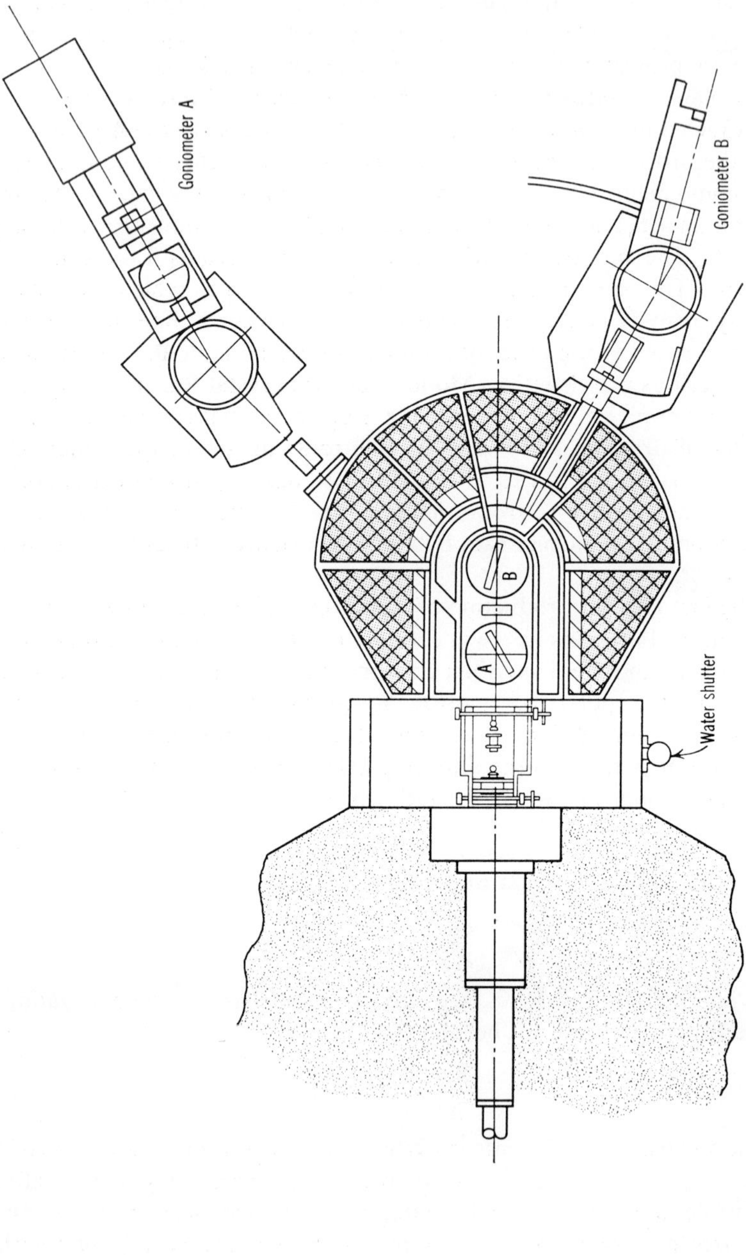
Goniometer A
Goniometer B
Water shutter
A
B

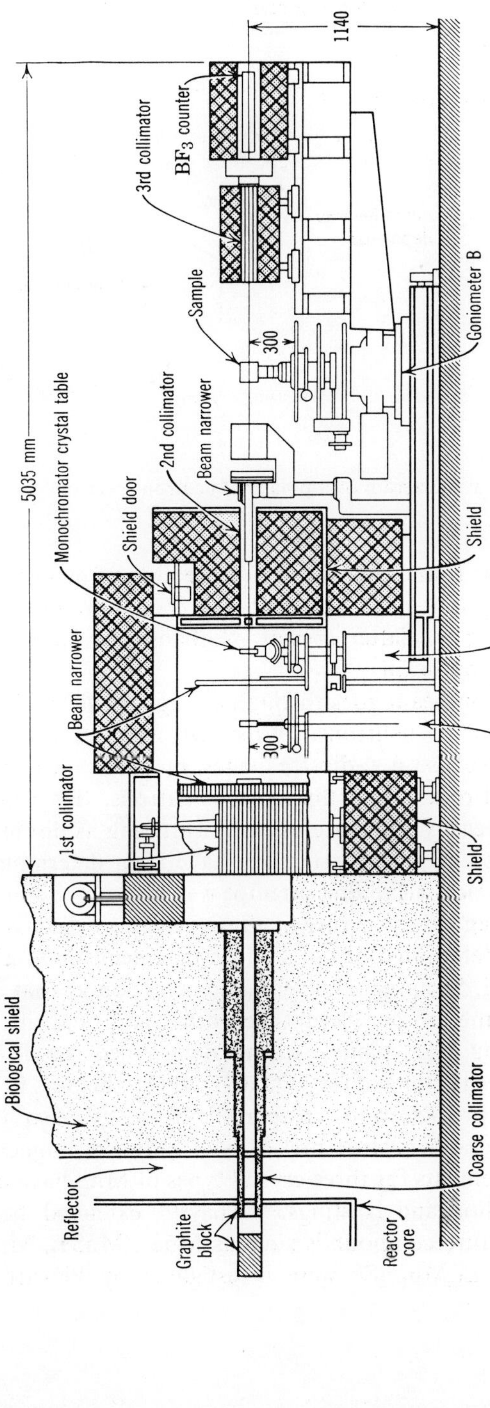

Fig. 21.5. Neutron diffraction apparatus.[11]

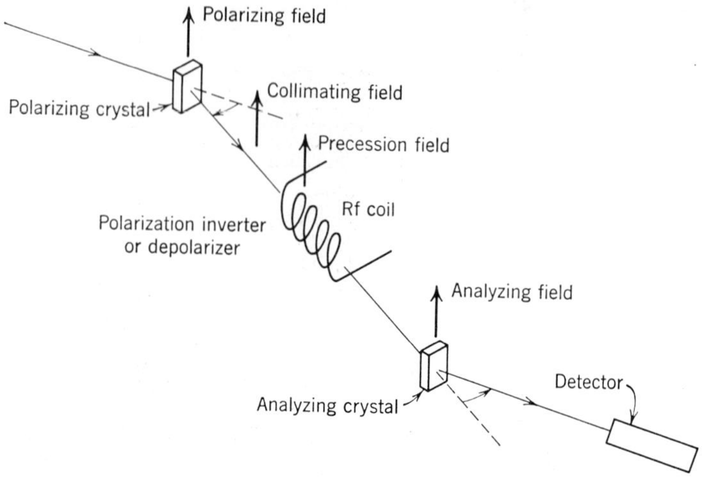

Fig. 21.6. Schematic diagram of a polarized neutron spectrometer (after Nathans and others[14]).

neutron. It was found that the (220) reflection of magnetite and the (111) and (200) reflections of f.c.c. cobalt-iron crystal are both suitable for this purpose. Using the latter crystal, Nathans and Shull obtained a 95% polarized neutron beam.

The diffraction beam goes through a magnetic field which is parallel to the polarization of neutrons. When we want to reverse the sense of polarization, we use a radio-frequency magnetic field which induces a precession and results in a flipping of neutrons. By using both neutron beams we can easily eliminate nuclear scattering as discussed above.

A number of magnetic structures have been determined by means of neutron diffraction. The first attempt was made by Shull and Smart[15] on the antiferromagnetic structure of MnO, which we discussed in Chapter 5. More precise determinations of spin arrangement for MnO and FeO were made by Shull, Strauser, and Wollan,[9] who found that all spins in the same (111) plane are parallel to each other and opposite to the spins on the neighboring (111) plane. Roth[16] determined the directions of spins to be all in the (111) plane for NiO and MnO; this was in agreement with a theoretical prediction by Kaplan.[17] Roth[18,19] also investigated the antiferromagnetic twin structures of NiO and their magnetization process.

Spin arrangements for three crystal types of MnS have been determined by Corliss, Elliot, and Hastings.[20] They[21,22] extended their investigations to pyrite structure compounds such as MnS_2, $MnSe_2$, $MnTe_2$. Magnetic transitions of $Li_xMn_{1-x}Se$ were investigated by Pickart, Nathans, and Shirane.[23]

Antiferromagnetic structures of various kinds of halides have also been investigated: MnF_2, FeF_2, CoF_2, and NiF_2 were studied by Erickson,[24] $FeCl_2$ by Herpin,[25] $MnBr_2$ by Wollan, Koehler, and Wilkinson,[26] $FeBr_2$, $CoBr_2$, $FeCl_2$, and $CoCl_2$ by Wilkinson, Cable, Wollan, and Koehler,[27] CrF_2, and $CrCl_2$ by Cable, Wilkinson, and Wollan,[28] CoF_3, FeF_3, CoF_3, and MnF_3 by Wollan, Child, and Koehler,[29] MoF_3 by Wilkinson et al.,[30] and $KMeF_3$ (Me = Cr, Mn, Fe, Co, Ni, and Cu) by Scatturin et al.[31]

Corundum-type compounds such as Cr_2O_3,[32] α-Fe_2O_3,[11] $FeTiO_3$ and its solid solutions with α-Fe_2O_3,[33] $MnTiO_3$,[34] and $NiTiO_3$[34] were also investigated by neutron diffraction. Perovskite structures such as (La, Ca)MnO_3,[35] La(Mn, Cr)O_3,[36] $LaFeO_3$,[37] $NdFeO_3$,[38] $HoFeO_3$,[38] and $ErFeO_3$[38] were also examined. Some of these compounds have been discussed in Chapter 5.

Magnetic structures of ferrites have also been fully investigated by a number of researchers. Nickel ferrite,[38] copper ferrites,[39] and cobalt-ferrites[40] were all found to be inverse spinels, having the magnetic structure expected by Néel. The magnesium ferrite[41,42] turned out to be only an incompletely inverse spinel; this finding accounts for the magnitude of the molar magnetic moment. Manganese ferrite[43] was found to be 81% normal and 19% inverse spinel. The zinc ferrite is a normal spinel, but it was found to show a complicated antiferromagnetic spin arrangement below 9°K.[44] The ion arrangement of Verwey order observed in magnetite at low temperatures was also confirmed by neutron diffraction[45] (cf. Chapter 17).

NiAs-type compounds have also been investigated by neutron diffraction. For instance, CrSb was found, by Snow,[46] to have its spin axis parallel to the *c* axis, while FeS_{1+x} was found, by Sidhu,[47] to have the spins perpendicular to the *c* axis. MnAs was studied by Bacon and Street.[48]

Ferrimagnetism in Mn_4N and Fe_4N was also analyzed by neutron diffraction[49,50] and explained in terms of the difference in the magnetic moment of Mn ions in the two different lattice sites.

Magnetic structures and atomic moments in ferromagnetic metals and alloys are also investigated by using neutron techniques as partly described in Section 4.3. Shull and others[8,51] examined the possibility of a Zener structure in body-centered cubic iron and concluded that the local difference in the magnetic moment of iron atoms is less than 0.4 to $0.6 M_B$. Face-centered cubic iron, which was prepared in precipitate form in a copper matrix, was found by Abrahams et al.[52] to be antiferromagnetic below 8°K with a sublattice moment of $0.7 M_B$. A number of alloys which include $3d$ transition metals were investigated by Shull and Wilkinson,[53] who showed that, according to one of the two possible analyses, the magnetic moment of each kind of magnetic atom changes rather slowly with composition compared with the variation of average magnetic moment. Further investigations on iron-nickel[54,55] and iron-cobalt[56] alloys have confirmed this general character. Low and Collins[55] examined the localization of magnetic moment on impurity atoms such as Fe, Co, Cr, and V in the matrix of Ni by using scattering of slow neutrons, and they found that the magnetic moment is well localized on Fe and Co atoms, whereas Cr and V atoms disturb the magnetic moment of the neighboring Ni atoms (cf. Section 4.3). Nathans and others[13] studied the alloy Fe_3Al by means of a polarized neutron beam and found that Fe atoms have two magnetic moments, $1.46M_B$ and $2.14M_B$, depending on the lattice sites.

Cr and α-Mn were studied by Shull and Wilkinson[51] and were found to exhibit antiferromagnetic diffraction below 470°K and 100°K, respectively. The magnetic moment of the Cr atom was found to be less than $0.4M_B$. Detailed investigation[57] of α-Mn shows that the magnetic moments of manganese atoms are quite different in different lattice sites. Similar investigations were also made on Mn-Cr,[58] Mn-Cu,[59,60] and AuMn.[61]

Spatial distribution of electron spins around each nucleus can be determined from the magnetic form factor. The first trial was made by Hughes et al.[62] for iron by means of transmission measurements. Shull et al.[8] determined the magnetic form factor of iron from the ferromagnetic contributions to the coherent reflections in the iron powder pattern. Brockhouse et al.[63] also determined the magnetic form factor of Fe^{3+} ion from paramagnetic scattering studies on zinc ferrite. These results are shown in Fig. 21.7. The experimental points are in reasonably good accord with the results of Hartree-type calculations made by Steinberger and Wick.[65]

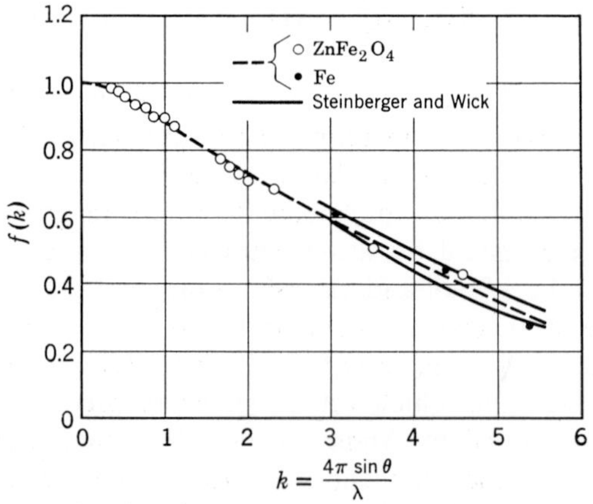

Fig. 21.7. Comparison of observed and calculated magnetic form factors for iron (after Shull and Wollan[64]).

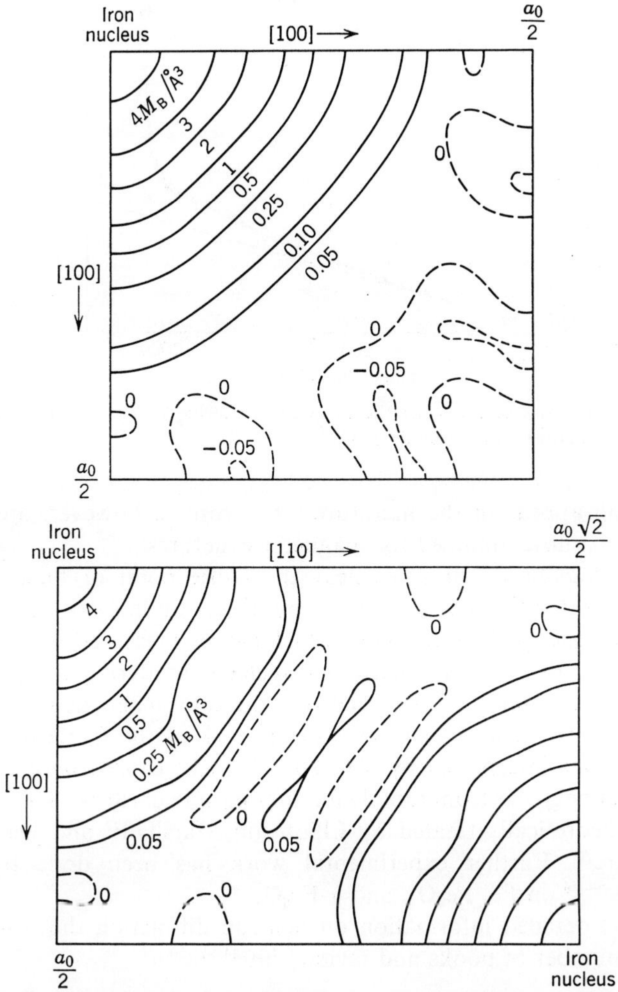

Fig. 21.8. Spin map on the (001) and the (110) planes of body-centered cubic iron (after Shull and Yamada[71]).

A number of investigations have been made on the magnetic form factors of various magnetic ions, such as manganese,[9,20,22,51] cobalt,[66] nickel,[67] erbium, and neodymium.[68,69] Three-dimensional spin maps were obtained in a similar way by Pickart and Nathans[70] for Fe_3Al, and by Shull and Yamada[71] for iron. Figure 21.8 is the spin map of iron in the (100) and (110) planes; we can see spin density spreading along the [100] direction more than along the [110] direction.

Some investigations[72,73] of magnetic domain structures by neutron

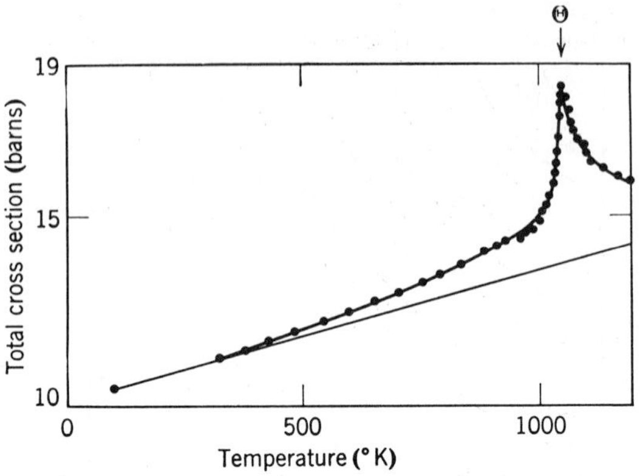

Fig. 21.9. Total cross sections for the inelastic scattering of neutron from iron as a function of temperature (after Squires[75]).

transmission appear in the literature. The results, however, are not so conclusive as those obtained for magnetic structures.

Thermal fluctuation of spins near the Curie point or fluctuation in domain magnetization can be studied by observing inelastic scattering of neutrons the energy distribution of which can be measured by the double spectrometer. Brockhouse[74] examined inelastic scattering by spin waves from magnetite. Squires[75] observed that the total magnetic cross section of iron for slow neutrons shows a sharp peak at the Curie point (Fig. 21.9); this was interpreted by Van Hove[76] to be caused by an inelastic scattering of neutrons from rapidly fluctuating spin clusters. This problem was also theoretically treated by Elliot and Marshall[77] and De Gennes and Villain.[78] Further experimental work has been done by many researchers[79–86] on Fe, Fe_3O_4, and α-Fe_2O_3.

For more detailed information on neutron diffraction the reader may refer to a number of books and reviews.[7,10,64,87–90]

21.2 Measurement of Internal Field at Nucleus by Nuclear Magnetic Resonance and Mössbauer Effect

In general, nuclei have spins and their associated magnetic moments which are measured by the unit of the nuclear magneton. The values of nuclear magnetic moment are given in Table 21.2 for various isotopes. The magnitude of the nuclear magneton is given by

$$M_N = \frac{\mu_0 e\hbar}{2m_p} = 6.33 \times 10^{-33} \text{ Wb-m}, \qquad (21.11)$$

Table 21.2. Nuclear Spin, Magnetic Moment, and Quadrupole Moment of Various Isotopes (after Tomita and Itoh[91,92])

Element	Mass Number, A	Spin, $I/\hbar$	Magnetic Moment, M/M_N	Quadrupole Moment, $Q/10^{-24}$ cm^2
H	1	$\frac{1}{2}$	2.79245	—
	2	1	0.85732	2.77×10^{-3}
He	3	$\frac{1}{2}$	−2.218	—
Li	6	1	0.8221	4.6×10^{-4}
	7	$\frac{3}{2}$	3.257	2×10^{-2}
Be	9	$\frac{3}{2}$	−1.178	2×10^{-2}
B	10	3	1.810	6×10^{-2}
	11	$\frac{3}{2}$	2.689	3×10^{-2}
C	13	$\frac{1}{2}$	0.7023	—
N	14	1	0.4037	2×10^{-2}
	15	$\frac{1}{2}$	−0.2831	—
O	17	$\frac{5}{2}$	−1.893	-5×10^{-3}
F	19	$\frac{1}{2}$	2.628	—
Ne	21	$\frac{3}{2}$	−0.662	
Na	23	$\frac{3}{2}$	2.217	0.1
Mg	25	$\frac{5}{2}$	−0.8547	
Al	27	$\frac{5}{2}$	3.641	0.149
Si	29	$\frac{1}{2}$	−0.555	—
P	31	$\frac{1}{2}$	1.131	—
S	33	$\frac{3}{2}$	0.6429	-8×10^{-2}
Cl	35	$\frac{3}{2}$	0.8210	-8.0×10^{-2}
	37	$\frac{3}{2}$	0.6835	-6.2×10^{-2}
K	39	$\frac{3}{2}$	0.3910	0.1
	41	$\frac{3}{2}$	0.2148	-2×10^{-2}
Sc	45	$\frac{7}{2}$	4.749	
Ti	47	$\frac{5}{2}$	−0.7873	
	49	$\frac{7}{2}$	−1.102	
V	51	$\frac{7}{2}$	5.145	
Cr	53	$\frac{3}{2}$	0.58	
Mn	55	$\frac{5}{2}$	3.462	
Fe	57	$\frac{1}{2}$	0.0903	
Co	59	$\frac{7}{2}$	4.460	
Cu	63	$\frac{3}{2}$	2.226	−0.15
	65	$\frac{3}{2}$	2.386	−0.14
Zn	67	$\frac{5}{2}$	0.874	
Ga	69	$\frac{3}{2}$	2.012	0.2318
	71	$\frac{3}{2}$	2.556	0.1461
Ge	73	$\frac{9}{2}$		−0.2
As	75	$\frac{3}{2}$	1.435	0.3

Table 21.2 (Continued)

Element	Mass Number, A	Spin, $I/\hbar$	Magnetic Moment, M/M_N	Quadrupole Moment, $Q/10^{-24}\,cm^2$
Se	77	$\frac{1}{2}$	0.5333	
Br	79	$\frac{3}{2}$	2.106	0.35
	81	$\frac{3}{2}$	2.269	0.28
Kr	83	$\frac{9}{2}$	−0.968	0.15
Rb	85	$\frac{5}{2}$	1.349	0.24
	87	$\frac{3}{2}$	2.749	0.12
Sr	87	$\frac{9}{2}$	−1.1	
Y	89	$\frac{1}{2}$	−0.14	—
Zr	91	$\frac{5}{2}$		
Nb	93	$\frac{9}{2}$	6.167	
Mo	95	$\frac{5}{2}$	−0.18	
	97	$\frac{5}{2}$	−0.19	
Ru	99	$\frac{5}{2}$		
	101	$\frac{5}{2}$		
Rh	103	$\frac{1}{2}$	−0.10	
Pd	105	$\frac{5}{2}$	−0.57	
Ag	107	$\frac{1}{2}$	−0.086	—
	109	$\frac{1}{2}$	−0.159	—
Cd	111	$\frac{1}{2}$	−0.5923	—
	113	$\frac{1}{2}$	−0.6196	—
In	113	$\frac{9}{2}$	5.497	1.144
	115	$\frac{9}{2}$	5.509	1.161
Sn	115	$\frac{1}{2}$	−0.9134	—
	117	$\frac{1}{2}$	−0.9951	—
	119	$\frac{1}{2}$	−1.0411	—
Sb	121	$\frac{5}{2}$	3.343	−1.3
	123	$\frac{7}{2}$	2.534	−1.7
Te	123	$\frac{1}{2}$	−0.732	
	125	$\frac{1}{2}$	−0.882	
I	127	$\frac{5}{2}$	2.809	−0.59
Xe	129	$\frac{1}{2}$	−0.7726	—
	131	$\frac{3}{2}$	0.8	−0.15
Cs	133	$\frac{7}{2}$	2.579	0.003
Ba	135	$\frac{3}{2}$	0.837	
	137	$\frac{3}{2}$	0.936	
La	139	$\frac{7}{2}$	2.778	−0.2
Pr	141	$\frac{5}{2}$	3.9	
Nd	143	$\frac{7}{2}$	−1.0	$\lesssim$1.2
	145	$\frac{7}{2}$	−0.62	$\lesssim$1.2

Table 21.2 (Continued)

Element	Mass Number, A	Spin, $I/\hbar$	Magnetic Moment, M/M_N	Quadrupole Moment, $Q/10^{-24}$ cm^2
Sm	147	$\frac{7}{2}$	−0.30	0.72
	149	$\frac{7}{2}$	−0.25	0.72
Eu	151	$\frac{5}{2}$	3.4	~1.2
	153	$\frac{5}{2}$	1.5	~2.5
Tb	159	$\frac{3}{2}$		
Ho	165	$\frac{7}{2}$		
Tm	169	$\frac{1}{2}$		
Yb	171	$\frac{1}{2}$	0.45	—
	173	$\frac{5}{2}$	−0.65	3.9
Lu	175	$\frac{7}{2}$	2.6	5.9
Ta	181	$\frac{7}{2}$	2.1	6.5
W	183	$\frac{1}{2}$		
Re	185	$\frac{5}{2}$	3.143	2.8
	187	$\frac{5}{2}$	3.175	2.6
Os	189	$\frac{3}{2}$	0.70	2
Ir	191	$\frac{3}{2}$	0.16	~1.2
	193	$\frac{3}{2}$	0.17	~1.0
Pt	195	$\frac{1}{2}$	0.6005	—
Au	197	$\frac{3}{2}$	0.136	—
Hg	199	$\frac{1}{2}$	0.4994	—
	201	$\frac{3}{2}$	−0.607	0.5
Tl	203	$\frac{1}{2}$	1.611	—
	205	$\frac{1}{2}$	1.627	—
Pb	207	$\frac{1}{2}$	0.5837	—
Bi	209	$\frac{9}{2}$	4.040	−0.4
U	235	$\frac{5}{2}$		

where m_p is the mass of the proton. The spin angular momentum of the nucleus is still measured by the unit $\hbar$, as is the electron's, and is usually denoted by I. As seen in Table 21.2, the magnetic moment of the nucleus is not necessarily proportional to the angular momentum. Therefore the nuclear g factor defined by

$$g = \frac{M/M_N}{I} \tag{21.12}$$

scatters over a fairly wide range, from about 0.1 to 5.6. When the nucleus is placed in a magnetic field, it exhibits precession motion, which can be detected by means of a resonance technique. The resonance frequency is, however, lower by a factor 1/1836 than that of electron spin resonance, because of the smallness of the nuclear magnetic moment. That is, if we use a magnetic field of 0.8 MA/m† (= 10,000 Oe), the resonance frequency,

$$\omega = g \frac{M_N}{\hbar} H, \tag{21.13}$$

is calculated as

$$f = \frac{(6.33 \times 10^{-33}) \times (0.8 \times 10^6)}{2\pi \times (1.054 \times 10^{-34})} = 7.7 \text{ Mc} \tag{21.14}$$

for $g = 1$. Thus the frequency for most nuclear magnetic resonances is in the range of radio wave frequencies. Since the g factor varies over a wide range, the resonance frequency of nuclear magnetic resonance depends strongly on the sort of nucleus observed.

One more thing which must be taken into consideration for nuclear magnetic resonance is the presence of the electric quadrupole moment of the nucleus which gives rise to a shift of the resonance line in the presence of a gradient of electric field. The quadrupole moment is defined by

$$Q = \frac{1}{e} \int \rho(3z_n^{\ 2} - r_n^{\ 2})\, dv, \tag{21.15}$$

where ρ is the charge density in the nucleus, r_n is the radial coordinate, and z_n is the coordinate parallel to the nuclear spin axis. If $Q > 0$, the shape of the nucleus is prolate; if $Q < 0$, the shape of the nucleus is oblate. Suppose that the nucleus is in a place where the electric field is changing in intensity along the z_0 axis. In other words, the second derivative of the electric potential,

$$\frac{\partial^2 \phi}{\partial z_0^{\ 2}} = eq, \tag{21.16}$$

is assumed to be non-zero at the nucleus. If the shape of nucleus is prolate, its long axis tends to be aligned parallel to the direction of the field gradient, provided $q < 0$. Such an electrostatic interaction can exert a torque on the nucleus, causing a shift of resonance lines. The order of energy involved in this interaction is roughly expressed as e^2qQ. This gradient of electric field is thought to be produced mainly by a distortion of the outer electron cloud and partly by a displacement of neighboring atoms due to a lattice distortion. The value of Q is also given in Table 21.2 for various isotopes.

The first studies of nuclear magnetic resonance in an antiferromagnetic substance were those of Poulis and Hardeman,[93,94] who studied the proton resonance in $CuCl_2 \cdot 2H_2O$ in the liquid helium range. They observed a

† Megaampere/meter.

splitting of the proton resonance at temperatures below the Néel point for the antiferromagnetic arrangement of spins of Cu atoms. Upon the application of a magnetic field stronger than a critical field of 0.58 MA/m (= 7200 Oe) parallel to the spin axis, the splitting suddenly collapses into a pair of lines because of a flopping of the spin axis into a direction normal to the field.

Jaccarino and Shulman[95] observed the nuclear magnetic resonance of F^{19} in MnF_2 which is antiferromagnetic below 72°K. They found that the internal field acting on the nucleus of F^{19} is 3.2 MA/m (= 40 kOe), part of which arises from a dipolar field of neighboring Mn ions and the other part from a partial 2*s* hole on the F^{19} ion as the result of covalency. F^{19} resonance was also studied by Shulman[96] for NiF_2, which shows a parasitic ferromagnetism caused by a slight canting of the antiferromagnetic spin arrangement. The nuclear magnetic resonance of F^{19} and Co^{59} was observed by Jaccarino et al.[97] for CoF_2.

For ferromagnetic substances it has been found that the nuclear magnetic resonance is enhanced by a factor of 10^5 by the displacement of domain walls driven by the external r-f field. This phenomenon was first observed by Portis and Gossard[98] for cobalt fine particles 1 to 5μ in diameter which have a multidomain structure. Suppose that a domain wall of width δ is oscillating back and forth with an amplitude x. For a spherical particle the amplitude x is mainly determined by an increase of demagnetizing field, so that

$$H_{ex} \doteq \frac{I_s}{3\mu_0} \frac{x}{d}, \tag{21.17}$$

where d is the diameter of the particle. A spin in the middle of the oscillating wall rotates by the angle $\theta = \pi x/\delta$, so that the internal field H_n which acts on the nucleus of the atom with the oscillating spin should produce the enhanced r-f field:

$$H_e = H_n\theta = \frac{H_n \pi x}{\delta} = \frac{3\pi\mu_0 H_n H_{ex}}{I_s}\frac{d}{\delta}. \tag{21.18}$$

Since $H_n \approx 10^7$ A/m, $I_s \approx 1$ Wb/m², and $d/\delta \approx 10$, (21.18) gives

$$\frac{H_e}{H_{ex}} \approx 10^3. \tag{21.19}$$

Considering that the signal obtained should be proportional to the square of the amplitude of the field and proportional to the fractional number of oscillating atoms in the particle (given by δ/d), we can conclude that the signal is enhanced by a factor of $(10^3)^2 \times 10^{-1} = 10^5$, which is just equal to the experimental value.

Portis and Gossard[98] proved this by the following experiments. The intensity of the signal was reduced by the application of a static magnetic field, and it finally dropped practically to the zero level when the intensity of the static field reached the demagnetizing field $I_s/3\mu_0$ under which no domain walls can be present. During this process the resonance frequency stayed constant, because walls are located at the places where $H = 0$. In contrast to this, the signal observed for single domain particles was weaker by a factor 10^4 than that of multidomain particles. (Even for single domain particles an enhancement of about 100 times can be expected from a rotation of magnetization.) Moreover, the resonance frequency for single domain particles was slightly different from that of multidomain particles, because the demagnetizing field $I_s/3\mu_0$ is acting on the resonating nucleus for single domain particles, while no field is present for multidomain particles. Shulman[96] observed an enhancement of 10^6 even for weakly magnetic NiF_2. The nuclear magnetic resonance of Fe^{57} in b.c.c.[99,100] Fe, of Co^{59} in f.c.c.[101] and h.c.p.[102] Co, and of Ni^{61} in f.c.c.[103] Ni was also observed by a number of investigators. The experiments were extended to various magnetic alloys and compounds. The results will be discussed later together with the results of the Mössbauer effect.

The internal field which acts on a nucleus can also be measured by means of the Mössbauer effect for some kinds of nucleus. This effect utilizes the fact that a γ-ray can be emitted from a nucleus without recoil if the nucleus is bound into a solid. Thus the emitted γ-ray has an energy equal to the energy separation between the two states of the nucleus before and after the emission, so that this γ-ray can be again absorbed by another nucleus of the same species by selective absorption. This fact was first pointed out by Lamb[104] and experimentally verified by Mössbauer[105] by using the isotope Ir^{191}. The recoil energy of the free nucleus associated with the emission of the γ-ray is given by $E_0^2/2m_nc^2$, where E_0 is the nuclear transition energy, c the light velocity, and m_n the mass of the nucleus. The tightness of binding of the nucleus in the solid can be measured by the Debye temperature. In order to obtain the Mössbauer effect, it is desirable to select an isotope which has a small recoil energy and to use a solid with a large Debye temperature. The Mössbauer effect has been observed on Co^{57} which is disintegrated to Fe^{57},[106–110] and on these disintegrations: Co^{61} to Ni^{61};[111] Sn^{119} to Sn^{119};[112] Cu^{67} to Zn^{67};[113] Tb^{161} to Dy^{161};[114] Gd^{153} to Eu^{153};[115] and Pt^{197} to Au^{197}.[116] Of these combinations, Co^{57} to Fe^{57} is particularly interesting for researchers on magnetism, because it allows the detailed measurement of the internal field which acts on the nucleus of Fe^{57} in magnetic materials.

The energy scheme of the nuclear transition of Co^{57} to Fe^{57} is shown in Fig. 21.10. Co^{57}, which can be made by irradiation of Fe with 4 MeV

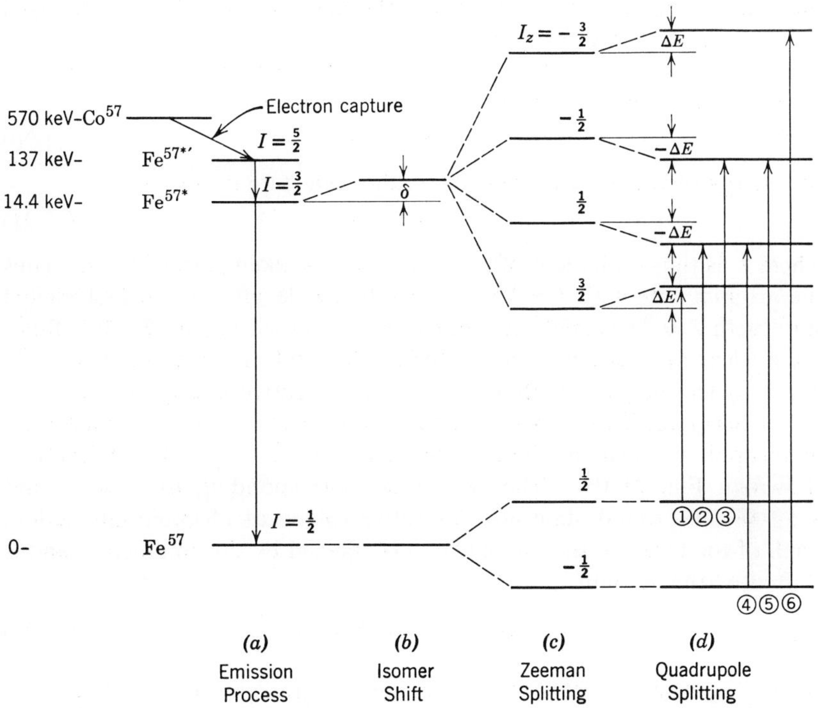

Fig. 21.10. Energy scheme (*a*) of the nuclear transition of Co^{57} to Fe^{57} and energy levels (*b*, *c*, *d*) of the ground and excited states of Fe^{57}.

deuterons, is decayed by electron capture with a half-life of 270 days to the second excited state of Fe^{57} with spin $I = 5/2$; then it makes a γ-transition to the first excited state with $I = 3/2$ and finally transfers with a period of 10^{-7} sec to the ground state with $I = 1/2$. The energy separation between the last two levels is 14.4 keV.

Now, if a nucleus of Fe^{57} exists in a magnetic material, the energy levels of the ground and the first excited states are split into six levels, as shown in Fig. 21.10(*b*–*d*) by various effects. One is the isomer shift, which is a shift of the excited level upwards by δ (Fig. 21.10*b*). One of the reasons for this is explained in terms of the difference in the size of the nucleus between the ground and the excited states, which gives rise to the difference in Coulomb interaction between the nucleus and surrounding electrons. If the electron density about the nucleus is different between emitter and absorber for some reason or other, such difference in Coulomb interaction should result in a shift of energy levels. The isomer shift is also caused by a temperature difference or a difference in Debye temperature between the emitter and the absorber because the mass of the nucleus is changed during emission or absorption of the γ-ray.

The second effect is the internal magnetic field H_i which acts on the nucleus. Since the nucleus has a magnetic moment

$$M = gM_N I, \tag{21.20}$$

the energy of the nucleus depends on the spin state in this way:

$$E = -gM_N I_z H_i, \tag{21.21}$$

where I_z is the component of I along the z axis taken parallel to H_i. Thus the ground state with $I = \frac{1}{2}$ is split into two levels, and the first excited state with $I = \frac{3}{2}$ is split into four levels as shown in Fig. 21.10*c*. Since the nuclear magnetic moment of the excited state is negative, $I_z = -\frac{3}{2}$ has a higher energy level than $I_z = \frac{3}{2}$ in a negative internal field.

Furthermore, if there is a gradient of an electric field at the nucleus, a quadrupole moment of the nucleus gives rise to a splitting of levels as shown in Fig. 21.10*d*. The two states corresponding to $I_z = \frac{1}{2}$ and $-\frac{1}{2}$ of the ground state are insensitive to the quadrupole interaction. Each of the levels of the excited state is affected by this interaction and is shifted by the amount

$$\Delta E = \tfrac{1}{4}e^2 qQ \frac{3\cos^2\theta - 1}{2}, \tag{21.22}$$

where θ is the angle between the spin and the gradient of the electric field. The sense of the shift is the same for $I_z = \frac{3}{2}$ and $-\frac{3}{2}$ or $I_z = \frac{1}{2}$ and $-\frac{1}{2}$, because the quadrupole moment is insensitive to the sign of I.

Now let us describe the apparatus for the measurement of the Mössbauer effect (Fig. 21.11). Experiments always involve a source which contains the radioactive species, an absorber which contains the stable isotope, and a detector of the γ-rays which penetrate through the absorber. The substance under study may be used as either the source or the absorber. When Co^{57} is used as the source, it is diffused into a stainless steel or

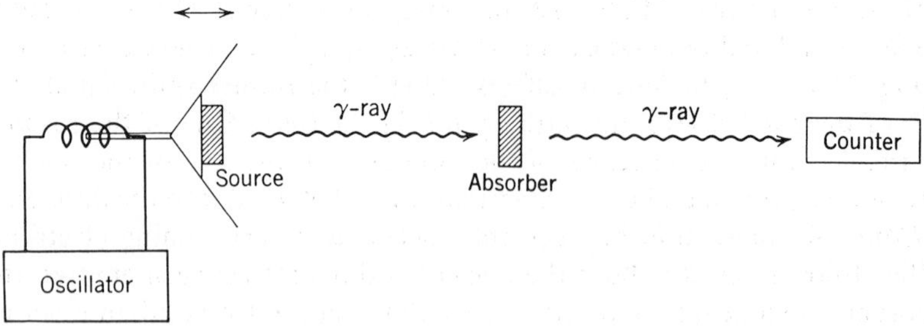

Fig. 21.11. Schematic illustration of experimental arrangement for measurement of the Mössbauer effect.

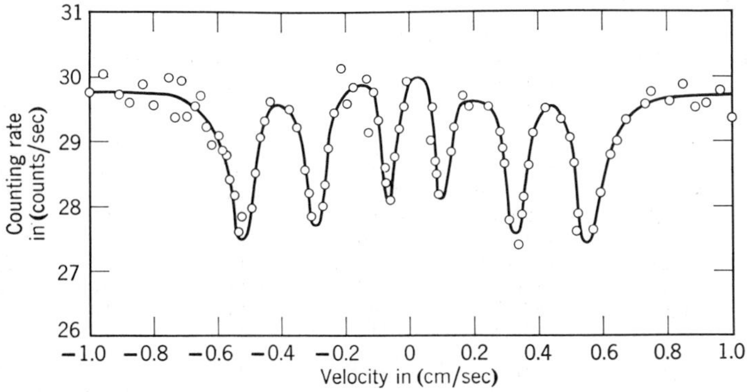

Fig. 21.12. The hyperfine spectrum of Fe^{57} in iron metal, produced by the Mössbauer effect with a stainless steel source and a natural iron absorber 0.001 in. thick (after Wertheim[110]).

chromium foil, which does not give rise to a hyperfine interaction with the radioactive cobalt. The source is set on a movable part of a dynamic speaker which drives the source back and forth with a velocity ranging from 0 to 1 cm/sec. The wavelength of the γ-ray which is thus emitted from the source is modulated by a Doppler shift so as to cover the energy range of the six possible absorption lines shown in Fig. 21.10*d*. The specimen used as the absorber should contain at least 0.1 mg/cm^2 of Fe^{57}. Since the natural abundance of Fe^{57} is only 2.14% of the total amount of Fe, the use of enriched samples may be desirable if the total amount of Fe is small. The absorption of the γ-ray can be measured by means of a scintillation counter and of a multichannel pulse height analyzer as a function of instantaneous velocity of the source.[117]

Figure 21.12 shows an example of the Mössbauer spectrum observed for Fe^{57} in iron metal. The six absorption lines correspond to the six transitions shown in Fig. 21.10*d*. From the relative positions of the six lines, we can find the isomer shift δ, the internal field H_i, and the quadrupole coupling e^2qQ.

As we have already pointed out, the isomer shift δ depends on the density of electrons at the nucleus. For Fe^{57}, the isomer shift measured by using stainless steel source is largest for the divalent state, next for the trivalent state, and smallest for a metallic component. Walker, Wertheim, and Jaccarino[118] calculated the shifts for various numbers of 3*d* and 4*s* electrons and compared them with experiment.

A quadrupole shift also provides some significant information about magnetic states of the material. Since the quadrupole shift ΔE depends on the direction of nuclear spin relative to the direction of the electric

Table 21.3. The Internal Magnetic Field of Several Isotopes Included in Various Magnetic Materials (after Watson and Freeman[120] and Ishikawa[121])

Nucleus	Host	Internal Field MA/m	Internal Field kOe	Temperature, °K	Method	Refs.
^{57}Fe	Fe	−27.3	−342	0	M	108
	Fe	|27.1|	|339|	0	NMR	99, 100
	Co	−24.9 ± 0.4	−312 ± 5	0	M	108
	Ni	−22.3 ± 0.4	−280 ± 5	0	M	108
	Fe_3Al	−22.3	−280	78	M	119
		−17.6	−220			
	Fe_2Zr	−15.2 ± 0.8	−190 ± 10	room	M	122
	Fe_2Ti	<0.8	<10	room	M	122
	Fe_3N	−27.5	−345	room	M	
		−17.2	−215			
$^{57}Fe^{3+}$	YIG (tetra)	−36.7	−460	78	M	123
		|37.3|	|468|	78	NMR	124–126
	YIG (octa)	−43.1	−540	78	M	123
		|43.9|	|550|	78	NMR	124–126
	DyIG (tetra)	−36.7	−460	78	M	123
	DyIG (octa)	−43.1	−540	78	M	123
	α-Fe_2O_3	−41.1	−515	room	M	127
	γ-Fe_2O_3	−41.1 ± 1.6	−515 ± 20	85	M	128
	$NiFe_2O_4$	−40.7 ± 1.6	−510 ± 20	room	M	128
	$Li_{0.5}Fe_{2.5}O_4$ (order)	−40.5 ± 1.6	−508 ± 20	room	M	129
	$Li_{0.5}Fe_{2.5}O_4$ (disorder)	−40.7 ± 1.6	−510 ± 20	room	M	129
	MgO	−43.9	−550	1.3	ESR	130
	ZnO	−39.1	−490	290.4	ESR	131
	CoO	−44.4 ± 0.5	−557 ± 6	78	M	132
	Fe_3O_4 (tetra)	−40.7 ± 1.6	−510 ± 20	50	M	119
		−40.7	−510	78	NMR	124
$\frac{1}{2}(^{57}Fe^{3+} + {}^{57}Fe^{2+})$	Fe_3O_4	−35.9	−450	300	M	119, 128
		−36.9	−462	302	NMR	124
$^{57}Fe^{2+}$	Fe_3O_4 (octa)	−37.1 ± 1.6	−465 ± 20	50	M	119
	CoO	−16.0 ± 0.8	−200 ± 10	169	M	132
	FeF_2	−27.1	−340	0	M	133
	$FeS_{1.00}$	−25.5	−320	300	M	119
	$FeTiO_3$	−5.6	−70	0	M	134
^{59}Co	Co (f.c.c.)	−17.36	−217.5	0	NMR	98
	Co (h.c.p.)	−18.2	−228	0	NMR	102
	Fe	−23.1	−289	0	NMR	100
	Ni	−6.4	−80	0	Cv	135
^{61}Ni	Ni	−13.6	−170	room	NMR	103
	Ni	−13.6	−170	room	M	111
^{119}Sn	Fe	−6.4 ± 0.3	−81 ± 4	100	M	136
	Co	−1.64 ± 0.12	−20.5 ± 1.5	100	M	136
	Ni	+1.48 ± 0.08	+18.5 ± 1.0	100	M	136
	Mn_4Sn	−3.6	−45	0	M	112
	Mn_2Sn	+16.0	+200	0	M	112
^{65}Cu	Fe	|17.0|	|212.7|	273	NMR	137
	Co	|12.6|	|157.5|	283	NMR	137
^{197}Au	Fe	|22.5|	|282|		M	138
	Co	|0.97|	|12.2|		M	138
^{161}Dy	DyIG	+279 ± 44	+3500 ± 550	85	M	114
^{159}Tb	Tb	+335 ± 80	+4200 ± 1000	0	NMR	139

field gradient as shown in (21.22), it is possible to determine the direction of nuclear spin and, accordingly, the direction of the atomic magnetic moment of Fe atoms, provided that the direction of the electric field is known. Actually Ōno, Ishikawa, Ito, and Hirahara[119] observed for α-Fe_2O_3 that ΔE changes from positive to negative at 250°K (Morrin temperature), where the spin axis changes from the basal plane to the c axis. The absolute value of the change is explained by (21.22).

Next we discuss the internal magnetic field at the nucleus as determined by the Mössbauer effect, together with the data of nuclear magnetic resonance. The experimental values of the internal field of several isotopes are given in Table 21.3. The most striking feature of the results is that the internal field in almost all cases is negative, or opposite to the magnetic polarization of the atom. Moreover, the intensity of the internal field is an order of magnitude larger than that expected from a dipolar field of unpaired $3d$ electrons.

The origin of the internal field was first investigated by Marshall[140] and has been calculated in greater detail by Watson and Freeman.[120,141] According to them, the main negative internal field arises from the polarization of $1s$ and $2s$ electron cores, which have high electron densities near the nucleus and thus give rise to a large internal field through a Fermi contact[142] to the nucleus (exchange polarization). Since a $3s$ electron is expected to produce a positive internal field, the actual internal field is composed of the difference between these negative and positive contributions unless the other mechanisms are in operation. According to the calculation by Watson and Freeman on the Mn^{2+} ion which has no orbital moment, the contributions of the $1s$, $2s$, and $3s$ cores are -2.4, -112, and $+59$ MA/m (-30, -1400, and $+740$ kOe), respectively, resulting in a total of -55 MA/m (-690 kOe), which is in good agreement with the experimental value of -52 MA/m (-650 kOe).

The internal field of the Fe^{3+} ion listed in Table 21.3 is in the range of -40.7 to -43.9 MA/m (-510 to -550 kOe). The calculation based on the exchange polarization gives -50.3 MA/m (-630 kOe). Watson and Freeman[120] explained the difference between the two in terms of the contraction of the $3d$ shell of Fe^{3+} compared with that of Mn^{2+}. The average intensity of the internal field of Fe^{2+} listed in Table 21.3 is much smaller than $|-43.9|$ MA/m ($|-550|$ kOe), which is calculated on the basis of exchange polarization. The difference is supposed to be caused by the internal field produced by an unquenched orbital magnetic moment. The large positive internal magnetic field of ^{161}Dy listed in Table 21.3 is explained in terms of its large orbital magnetic moment (cf. Section 21.1). The contribution of the orbital magnetic moment was calculated by Kondo[143] for various rare earth metals.

It is seen in Table 21.3 that the internal field of ^{57}Fe in magnetic metals is fairly low compared with that in magnetic compounds. This is partly because of the smallness of the magnetic moment of the Fe atom (that is, $2.2M_B$ in iron, which is small compared with $5M_B$ for Fe^{3+}), but this is still inadequate to account for the results of the experiment. Another possible source of the internal field is the polarization of 4*s* electrons caused by the polarization of 3*d* electrons of the Fe^{57} atom and of the neighboring atoms. From the fact that a non-magnetic Sn atom shows a fairly large internal field as shown in Table 21.3, we suppose that the internal field of any nucleus in ferromagnetic metals and alloys can be affected by the polarization of neighboring atoms. This occurs partly through the polarization of 4*s* electrons, the exchange polarization of the inner *s* shells, or the admixture of the unpaired 3*d* spin into the closed shell of neighboring atoms.[120] The expansion of the 3*d* shell in metals may also be one of the causes of the change in the internal field.

Although too many factors have been considered to be origins of the internal field, there still is no doubt that measurement of the internal field will provide much fruitful information about the magnetic moment of individual atoms in ferromagnetic materials. The reader may refer to the review by Portis[144] on nuclear magnetic resonance, that of Wertheim[110] on the Mössbauer experiment, and a paper by Watson and Freeman[120] on the interpretation of internal field.

References

21.1. W. M. Elsasser: *Compt. Rend.* **202,** 1029 (1936).
21.2. H. Halban and P. Preiswerk: *Compt. Rend.* **203,** 73 (1936).
21.3. D. P. Mitchell and P. N. Powers: *Phys. Rev.* **50,** 486 (1936).
21.4. F. Bloch: *Phys. Rev.* **50,** 259 (1936); **51,** 994 (1937).
21.5. J. Schwinger: *Phys. Rev.* **51,** 544 (1937).
21.6. O. Halpern and M. H. Johnson: *Phys. Rev.* **55,** 898 (1939).
21.7. G. E. Bacon: *Neutron Diffraction* (Oxford, Clarendon Press, 1955).
21.8. C. G. Shull, E. O. Wollan, and W. C. Koehler: *Phys. Rev.* **84,** 912 (1951).
21.9. C. G. Shull, W. A. Strauser, and E. O. Wollan: *Phys. Rev.* **83,** 333 (1951).
21.10. S. Hoshino: *Neutron Diffraction* (in Japanese) (Maki Publishing Co., Tokyo, 1961).
21.11. S. Miyake, S. Hoshino, K. Suguki, H. Katsuragi, S. Hagiwara, T. Yoshie, and K. Miyashita: *J. Phys. Soc. Japan* **17,** Suppl. B-II, 358 (1962).
21.12. C. G. Shull: *Phys. Rev.* **81,** 626 (1951).
21.13. R. Nathans, M. T. Pigott, and C. G. Shull: *Phys. Chem. Solids* **6,** 38 (1958).
21.14. R. Nathans, C. G. Shull, G. Shirane, and A. Anderson: *Phys. Chem. Solids* **10,** 138 (1959).
21.15. C. G. Shull and J. S. Smart: *Phys. Rev.* **76,** 1256 (1949).
21.16. W. L. Roth: *Phys. Rev.* **110,** 1333 (1958).

21.17. J. J. Kaplan: *J. Chem. Phys.* **22,** 1709 (1954).
21.18. W. L. Roth and G. A. Slack: *J. Appl. Phys.* **31,** 352S (1960).
21.19. W. L. Roth: *J. Appl. Phys.* **31,** 2000 (1960).
21.20. L. M. Corliss, N. Elliott, and J. M. Hastings: *Phys. Rev.* **104,** 924 (1956).
21.21. L. M. Corliss, N. Elliott, and J. M. Hastings: *J. Appl. Phys.* **29,** 391 (1958).
21.22. J. M. Hastings, N. Elliott, and L. M. Corliss: *Phys. Rev.* **115,** 13 (1959).
21.23. S. J. Pickart, R. Nathans, and G. Shirane: *Phys. Rev.* **121,** 707 (1961).
21.24. R. A. Erickson: *Phys. Rev.* **90,** 779 (1953).
21.25. A. Herpin and P. Merial: *Compt. Rend.* **245,** 650 (1957).
21.26. E. O. Wollan, W. C. Koehler, and M. K. Wilkinson: *Phys. Rev.* **110,** 638 (1958).
21.27. M. K. Wilkinson, J. W. Cable, E. O. Wollan, and W. C. Koehler: *Phys. Rev.* **113,** 497 (1959).
21.28. J. W. Cable, M. K. Wilkinson, and E. O. Wollan: *Phys. Rev.* **118,** 950 (1959).
21.29. E. O. Wollan, H. R. Child, W. C. Koehler, and M. K. Wilkinson: *Phys. Rev.* **112,** 1132 (1958).
21.30. M. K. Wilkinson, E. O. Wollan, H. R. Child, and J. W. Cable: *Phys. Rev.* **121,** 74 (1961).
21.31. V. Scatturin, L. Corliss, N. Elliott, and J. Hastings: *Acta Cryst.* **14,** 19 (1961).
21.32. B. N. Brockhouse: *J. Chem. Phys.* **21,** 961 (1953).
21.33. G. Shirane, S, J. Pickart, R. Nathans, and Y. Ishikawa: *Phys. Chem. Solids* **10,** 35 (1959).
21.34. G. Shirane, S. J. Pickart, and Y. Ishikawa: *J. Phys. Soc. Japan* **14,** 1352 (1959).
21.35. E. O. Wollan and W. C. Koehler: *Phys. Rev.* **100,** 545 (1955).
21.36. V. H. Bents: *Phys. Rev.* **106,** 225 (1957).
21.37. W. C. Koehler, and E. O. Wollan: *Phys. Chem. Solids* **2,** 100 (1957); W. C. Koehler, E. O. Wollan, and M. K. Wilkinson: *Phys. Rev.* **118,** 58 (1960).
21.38. J. M. Hastings and L. M. Corliss: *Rev. Mod. Phys.* **25,** 114 (1953).
21.39. E. Prince and R. G. Treuting: *Acta Cryst.* **9,** 1025 (1956).
21.40. E. Prince: *Phys. Rev.* **102,** 674 (1956).
21.41. G. E. Bacon and F. F. Robert: *Acta Cryst.* **6,** 57 (1953).
21.42. L. M. Corliss, J. M. Hastings, and F. G. Brockman: *Phys. Rev.* **90,** 1013 (1953).
21.43. J. M. Hastings and L. M. Corliss: *Phys. Rev.* **104,** 328 (1956).
21.44. J. M. Hastings and L. M. Corliss: *Phys. Rev.* **102,** 1460 (1956).
21.45. W. C. Hamilton: *Phys. Rev.* **110,** 1050 (1958).
21.46. A. I. Snow: *Rev. Mod. Phys.* **25,** 127 (1953).
21.47. S. S. Sidhu: Pittsburgh Diffraction Conference (Abstract), 1954.
21.48. G. E. Bacon and R. Street: *Nature* **175,** 518 (1955).
21.49. B. C. Frazer: *Phys. Rev.* **112,** 751 (1958).
21.50. W. J. Takei, G. Shirane, and B. C. Frazer: *Phys. Rev.* **119,** 122 (1960).
21.51. C. G. Shull and M. K. Wilkinson: *Rev. Mod. Phys.* **25,** 100 (1953).
21.52. S. C. Abrahams, L. Guttman, and J. S. Kasper: *Phys. Rev.* **127,** 2052 (1962).
21.53. C. G. Shull and M. K. Wilkinson: *Phys. Rev.* **97,** 304 (1955).
21.54. M. F. Collins, R. V. Jones, and R. D. Lowde: *J. Phys. Soc. Japan* **17,** Suppl. B-III, 19 (1962).
21.55. G. G. E. Low and M. F. Collins: *J. Appl. Phys.* **34,** 1195 (1963).
21.56. W. M. Lomer: *J. Phys. Radium* **23,** 716 (1962).
21.57. J. S. Kasper and B. W. Roberts: *Phys. Rev.* **101,** 537 (1956).
21.58. J. S. Kasper and R. M. Waterstrat: *Phys. Rev.* **109,** 1551 (1957).
21.59. D. Meneghetti and S. S. Sidhu: *Phys. Rev.* **105,** 130 (1957).

21.60. G. E. Bacon, I. W. Dunmur, J. H. Smith, and R. Street: *Proc. Roy. Soc. (London)* **A241,** 223 (1957).
21.61. G. E. Bacon and R. Street: *Proc. Roy. Soc. (London)* **72,** 470 (1958).
21.62. D. J. Hughes, J. R. Wallace, and R. H. Holtzman: *Phys. Rev.* **73,** 1277 (1948).
21.63. B. N. Brockhouse, L. M. Corliss, and J. M. Hastings: *Phys. Rev.* **98,** 1721 (1955).
21.64. C. G. Shull and E. O. Wollan: "Application of Neutron Diffraction to the Solid State Problem," *Solid State Physics*, Vol. 2 (Academic Press, New York, 1956).
21.65. J. Steinberger and G. C. Wick: *Phys. Rev.* **76,** 994 (1949).
21.66. R. Nathans and A. Paoletti: *Phys. Rev. Letters* **2,** 254 (1959).
21.67. H. A. Alperin: *Phys. Rev. Letters* **6,** 55 (1961).
21.68. G. T. Trammel: *Phys. Rev.* **92,** 1387 (1953).
21.69. W. H. Kleiner: *Phys. Rev.* **90,** 168 (1953).
21.70. S. J. Pickart and R. Nathans: *Phys. Rev.* **123,** 1163 (1961).
21.71. C. G. Shull and Y. Yamada: *J. Phys. Soc. Japan* **17,** Suppl. B-III, 1 (1962).
21.72. M. Burgy, D. J. Hughes, J. R. Wallance, R. B. Heller, and W. E. Woolf: *Phys. Rev.* **80,** 953 (1950).
21.73. O. Halpern and T. Holstein: *Phys. Rev.* **59,** 1960 (1941).
21.74. B. N. Brockhouse: *Phys. Rev.* **106,** 859 (1957); **111,** 1273 (1958).
21.75. G. L. Squires: *Proc. Phys. Soc. (London)* **A67,** 248 (1954).
21.76. L. Van Hove: *Phys. Rev.* **93,** 268 (1954); **95,** 249 (1954); **95,** 1374 (1954).
21.77. R. J. Elliott and W. Marshall: *Rev. Mod. Phys.* **30,** 75 (1958).
21.78. P. G. De Gennes and J. Villain: *Phys. Chem. Solids* **13,** 10 (1960).
21.79. A. M. McReynolds and T. Riste: *Phys. Rev.* **95,** 1161 (1954).
21.80. M. K. Wilkinson and C. G. Shull: *Phys. Rev.* **103,** 516 (1956).
21.81. H. A. Gersch, C. G. Shull, and M. K. Wilkinson: *Phys. Rev.* **103,** 525 (1956).
21.82. R. D. Lowde: *Rev. Mod. Phys.* **30,** 69 (1958).
21.83. M. Ericson and B. Jacrot: *Phys. Chem. Solids* **13,** 235 (1960).
21.84. T. Riste, K. Blinowski, and J. Janik: *Phys. Chem. Solids* **9,** 153 (1959).
21.85. T. Riste: *Phys. Chem. Solids* **17,** 308 (1961).
21.86. T. Riste and A. Wanic: *Phys. Chem. Solids* **17,** 318 (1961).
21.87. L. S. Kothari and K. S. Singwi: "Interaction of Thermal Neutrons with Solids," *Solid State Physics*, Vol. 8 (Academic Press, New York, 1959).
21.88. D. J. Hughes: *Neutron Optics* (Interscience Publishers, Inc., New York, 1954).
21.89. G. R. Ringo: "Neutron Diffraction and Interference, *Handbook der Physik*, XXXII (Springer, 1957).
21.90. D. J. Hughes: *Pile Neutron Research* (Addison-Wesley, 1953).
21.91. K. Tomita and J. Itoh: *Bussei-Butsurigaku Koza* (Kyoritsu Publishing Co., Tokyo), Vol. 7, p. 4 (in Japanese).
21.92. For further details refer to: O. Strominger, J. M. Hollander, and G. T. Seaborg: *Rev. Mod. Phys.* **30,** 585 (1958).
21.93. N. J. Poulis and G. E. Hardeman: *Physica* **18,** 201, 315, 429 (1952); **19,** 391 (1953).
21.94. G. E. G. Hardeman, N. J. Poulis, W. van der Lugt, and W. P. A. Hass: *Physica* **22,** 48 (1956); **23,** 907 (1957); **24,** 280 (1958).
21.95. V. Jaccarino and R. G. Shulman: *Phys. Rev.* **107,** 1196 (1957).
21.96. R. G. Shulman: *Phys. Rev.* **121,** 125 (1961).
21.97. V. Jaccarino, R. G. Shulman, J. L. Davis, and J. W. Stout: *Bull. Am. Phys. Soc. Ser. II* **3,** 41 (1958).

21.98. A. C. Gossard and A. M. Portis: *Phys. Rev. Letters* **3**, 164 (1959); A. M. Portis, and A. C. Gossard: *J. Appl. Phys.* **31**, 205S (1960).
21.99. C. Robert and J. M. Winter: *Comp. Rend.* **250**, 3831 (1960).
21.100. Y. Koi, A. Tsujimura, T. Hihara, and T. Kushida: *J. Phys. Soc. Japan* **16**, 1040 (1961).
21.101. V. Jaccarino: *Bull. Am. Phys. Soc. Ser. II* **4**, 461 (1959).
21.102. Y. Koi, A. Tsujimura, and T. Kushida: *J. Phys. Soc. Japan* **15**, 1342, 2100 (1960).
21.103. L. J. Bruner, J. I. Budnick, R. J. Blume, and E. L. Boyd: *Bull. Am. Phys. Soc. Ser II* **5**, 491 (1960); *Phys. Rev.* **121**, 83 (1961).
21.104. W. E. Lamb: *Phys, Rev.* **55**, 676 (1960).
21.105. R. L. Mössbauer: *Z. Physik.* **151**, 124 (1958), *Z. Naturforsch*, **149**, 211 (1959).
21.106. R. V. Pound and G. A. Rebka, Jr.; *Phys. Rev. Letters* **3**, 439 (1960).
21.107. J. P. Schiffer and W. Marshall: *Phys. Rev. Letters*, **3**, 556 (1959).
21.108. S. S. Hanna, J. Heberle, C. Littlejohn, G. J. Perlow, R. S. Preston, and D. H. Vincent: *Phys. Rev. Letters* **4**, 28, 177, 513 (1960).
21.109. G. De Pasquali, H. Frauenfelder, S. Margulies, and R. N. Peacock: *Phys. Rev. Letters* **4**, 71 (1960).
21.110. G. K. Wertheim: *J. Appl. Phys.* **32**, 110S (1961).
21.111. F. E. Obenhain and H. H. F. Wegener: *Phys. Rev.* **121**, 1344 (1961).
21.112. L. M. Schützmeister, R. S. Preston, and S. S. Hanna: *Phys. Rev.* **122**, 1717 (1962).
21.113. P. P. Craig, D. E. Nagle, and D. R. F. Cochran: *Phys. Rev. Letters* **4**, 561 (1960).
21.114. R. Brauminger, S. G. Cohen, A. Marinon, and S. Ofer: *Phys. Rev. Letters* **6**, 467 (1961).
21.115. A. Bussière de Nercy, M. Langevin, and M. Spighel: *J. Phys. Radium* **21**, 288 (1960).
21.116. N. D. Nagle: *Phys, Rev. Letters* **4**, 237 (1960).
21.117. S. L. Ruby, L. M. Epstein, and K. H. Sun: *Rev. Sci. Instr.* **31**, 580 (1960).
21.118. L. R. Walker, G. K. Wertheim, and V. Jaccarino: *Phys. Rev. Letters* **6**, 98 (1961).
21.119. K. Ōno, Y. Ishikawa, A. Ito, and E. Hirahara: *J. Phys. Soc. Japan* **17**, Suppl. B-I, 125 (1962); K. Ōno and A. Ito: *ibid.*, **17**, 1012 (1962).
21.120. R. E. Watson and A. J. Freeman: *Phys. Rev.* **123**, 2027 (1961).
21.121. Y. Ishikawa: *Metal Phys.* **8**, 21, 65 (1962).
21.122. C. W. Kocher and P. J. Brown: *J. Appl. Phys.* **33**, 1091 (1962).
21.123. R. Bauminger, S. G. Cohen, A. Marinov, and S. Ofer: *Phys. Rev.* **112**, 743 (1961).
21.124. S. Ogawa and S. Morimoto: *J. Phys. Soc. Japan* **17**, 654 (1962).
21.125. C. Robert: *Compt. Rend.* **251**, 2684 (1960); **252**, 1442 (1961).
21.126. E. L. Boyd, L. J. Bruner, J. I. Budnick, and R. J. Blume: *Bull. Am. Phys. Soc. Ser. II* **6**, 159 (1961).
21.127. O. C. Kistner and A. W. Sunyan: *Phys. Rev. Letters* **4**, 412 (1960).
21.128. R. Bauminger, S. G. Cohen, A. Marinov, S. Ofer, and E. Segal: *Phys. Rev.* **122**, 1447 (1961).
21.129. W. H. Kelley, V. J. Folen, M. Hass, W. N. Schreiner, and G. B. Beard: *Phys. Rev.* **124**, 80 (1961).
21.130. E. S. Rosenvasser: *Bull. Am. Phys. Soc.* **6**, 117 (1961).
21.131. W. M. Walsh: *Bull. Am. Phys. Soc.* **6**, 117 (1961).

21.132. G. K. Wertheim: *Phys. Rev.* **124,** 764 (1961).
21.133. G. K. Wertheim: *Phys. Rev.* **121,** 63 (1961).
21.134. G. Shirane and S. L. Ruby: *J. Phys. Soc. Japan* **17,** Suppl. B-I, 133 (1962).
21.135. V. Arp. D. Edmunds and R. Petersen: *Phys. Rev. Letters* **3,** 212 (1959).
21.136. A. J. F. Boyle, D. St. D. Bunburg, and C. Edward: *Phys. Rev. Letters* **5,** 553 (1960).
21.137. Y. Koi, A. Tsujimura, T. Hirahara, and T. Kushida: *J. Phys. Soc. Japan* **17,** Suppl. B-I, 96 (1962).
21.138. D. A. Shirely and M. Kaplan: *Phys. Rev.* **123,** 816 (1961).
21.139. L. Herve and P. Veillet: *Compt. Rend.* **252,** 99 (1961).
21.140. W. Marshall: *Phys. Rev.* **110,** 1280 (1958).
21.141. R. E. Watson and A. J. Freeman: *J. Appl. Phys.* **32,** 118S (1961).
21.142. E. Fermi: *Z. Physik.* **60,** 320 (1930); E. Fermi and E. Segrè: *Z. Physik.* **82,** 729 (1933).
21.143. J. Kondo: *J. Phys. Soc. Japan* **16,** 1690 (1961).
21.144. A. M. Portis: *Magnetism* Vol. II, Academic Press, New York, 1964.

part 6

ENGINEERING APPLICATIONS OF MAGNETIC MATERIALS

22

Kinds of Magnetic Materials

22.1 Metallic Core Materials

The largest use of magnetic materials is for magnetic cores of transformers, motors, inductors, and generators. Desirable properties of core materials are high permeability, low magnetic loss, and low coercivity. In addition, high induction and low cost are important factors, particularly for large electrical equipment. Unalloyed iron and silicon-iron and aluminum-iron alloys are now being used for high power machines. However, for some critical applications, more expensive materials, such as Permalloy, Supermalloy, and other high quality alloys, with superior magnetic properties are often used.

Iron. Unalloyed iron is the simplest and the cheapest magnetic material. It usually contains appreciable amounts of impurities such as carbon, manganese, silicon, phosphor, or sulfur, unless it is prepared from electrolytic iron or heat-treated with special precautions. These impurities are generally harmful to the magnetic properties. For example, if a commercial iron containing 0.2% of impurities is heat-treated at 950°C, it exhibits $\bar{\mu}_a \simeq 150$, $\bar{\mu}_{\mathrm{max}} \simeq 5000$, and $H_c \simeq 80$ A/m (= 1.0 Oe). If, however, the impurities are reduced to 0.05% by annealing at 1480°C in hydrogen, the magnetic properties are greatly improved: $\bar{\mu}_a \cong 10{,}000$, $\bar{\mu}_{\mathrm{max}} \cong 200{,}000$, and $H_c = 4$ A/m (≒ 0.05 Oe).[1] Such treatment is too expensive for practical purposes. Moreover, for unalloyed iron there is the phenomenon that the permeability decreases on annealing at 130°C. This phenomenon is called magnetic aging, and it is thought to be due to the precipitation of a small amount of iron nitride or carbide at this temperature.[2] A small addition of Al, Ti, or V is found to be effective in suppressing this effect. The low resistivity of unalloyed iron [$\rho = 10 \times 10^{-8}$ Ω-m (= 10 μΩ-cm)] is also undesirable if this material is to be used for ac machines, because of the eddy current loss. It is reported that, when unalloyed iron is annealed above 900°C, the cooling rate in the temperature range 900° to 800°C, where the crystal transforms from face-centered cubic to body-centered cubic, should be no faster than 5°/min. In spite of this fact, annealing above 1000°C generally results in better maximum permeability than annealing below the transformation point.

Silicon-iron alloys. The addition of a small amount of silicon results in an improvement of the maximum permeability, in a reduction of magnetic aging and also in an increase of resistivity. On the other hand, the silicon causes a decrease in saturation magnetization and an increase in brittleness. Silicon is soluble in iron up to 15% Si. The Curie point decreases from 770°C to about 500°C with increase of silicon content from 0% to 15%. In the range 15% to 33% Si, α, and ϵ (FeSi) phases coexist, but, since the ϵ phase is non-magnetic, only one Curie point is observed. Above 33% Si the alloy is non-magnetic. The temperature range of the γ-phase of pure iron, 900° to 1400°C, is diminished by the addition of silicon up to 2.5% Si, beyond which the γ-phase disappears. Single crystals of 3% silicon-iron alloy are often used for experiments, because this alloy has no lattice transformation during cooling and therefore single crystals grown from melt can be safely cooled to room temperature.

Various magnetic constants are changed by addition of Si to Fe. The magnetocrystalline anisotropy K_1 decreases almost linearly with addition of Si. The data is available up to about 7%, and extrapolation indicate that K_1 should become zero at about 11% Si (= 20 at% Si) (cf. Fig. 7.8). The magnetostriction constants λ_{100} and λ_{111} go through zero at about 5% ~ 6% Si (10 ~ 11 at% Si) (cf. Fig. 8.12). In contrast to this, the saturation magnetization I_s decreases with addition of Si as if the magnetic moment of iron is simply diluted by non-magnetic silicon atoms. In other words, I_s vs composition extrapolates to zero at 100% Si, at least in the low concentration range of Si. As discussed in Part 4, the magnetic properties of soft magnetic materials are improved by reducing K_1 and λ, provided that I_s remains constant. In this sense the addition of Si to Fe should improve the soft magnetic properties. Actually the hysteresis loss W_h and the coercive force H_c decrease, and the maximum permeability increases with the addition of Si, as shown in Fig. 22.1, although the reason is partly that the annealing temperature is higher for larger silicon content. Because of the brittleness of the high silicon alloys, the

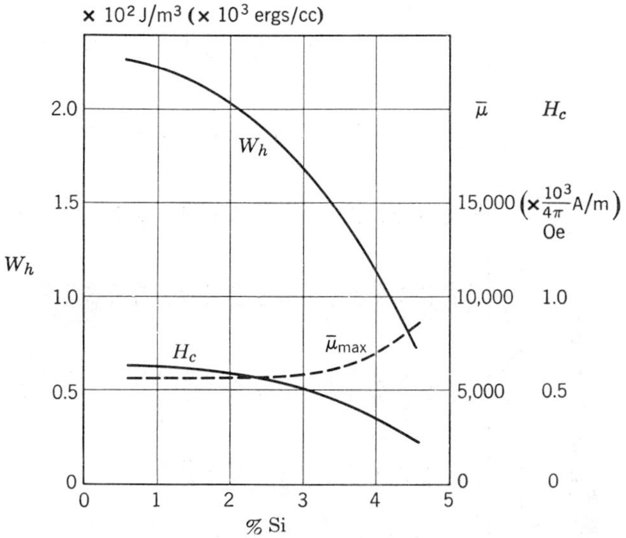

Fig. 22.1. Variation of magnetic properties of Si-Fe alloys with addition of silicon (Bozorth[3]).

3% ~ 4.5% Si-Fe alloy is used for transformers, whereas 1% ~ 3% Si-Fe is used for rotating machines such as motors and generators.

In most high power machines, iron cores are magnetized as high as 70% ~ 80% of the saturation magnetization; hence it is desirable that rotation magnetization take place as easily as possible. One of the possible means to realize this is to reduce the value of K_1. However, if we reduce K_1 by increasing the silicon content, the alloy becomes increasingly brittle. Another possibility is to orient the crystal grains so as to align one of the easy directions in each grain parallel to the direction of magnetization. This was accomplished by Goss[4] in 1934, and the material thus manufactured is called Goss steel or oriented silicon steel and is extensively used in transformers. The process of manufacture consists of a proper sequence of hot rolling and annealing, followed by severe cold rolling and recrystallization annealing. The recrystallization texture is (110) [001]. The orientation of crystallines can be as high as 90% perfect; this material has magnetic properties much superior to those of non-oriented silicon steel (cf. Table 22.1).

Recently it has been demonstrated that the so-called cubic texture with (001) [100] crystal orientation can be manufactured by secondary grain growth[5] or by rolling and recrystallizing an ingot with columnar grains.[6] The cubic texture is thought to be more useful than the unoriented texture, because it contains two easy directions in the plane of the sheet,

Table 22.1. Magnetic Properties of Soft Magnetic Alloys (Mainly from Bozorth[G.10])

Material	Composition	Heat Treatment,[a] °C	$\bar{\mu}_a$	$\bar{\mu}_{max}$	H_c		I_s		θ, °C	ρ, $\times 10^{-8}\,\Omega m$	σ, g/cc
					A/m	Oe	Wb/m²	Gauss			
Iron	0.2 (imp.)	950	150	5,000	80	1.0	2.15	1,710	770	10	7.88
Purified iron	0.05 (imp.)	1,480 (H_2); 880	10,000	200,000	4	0.05	2.15	1,710	770	10	7.88
Silicon-iron	4Si	800	500	7,000	40	0.5	1.97	1,570	690	60	7.65
Silicon-iron (oriented)	3Si	800	1,500	40,000	8	0.1	2.00	1,590	740	47	7.67
Silicon-iron (cubic)	3Si	—	—	116,000	5.6	0.07	2.00	1,590	740	47	7.67
Silicon-iron (cubic)	3Si	—	—	65,000	6.4	0.08	2.00	1,590	740	47	7.67
Aluminum iron	3.5Al	1,100	500	19,000	24	0.3	1.90	1,510	750	47	7.5
Alfer	13Al	—	700	3,700	53	0.66	1.20	955	—	90	6.7
Alperm	16Al	600 Q	3,000	55,000	3.2	0.04	0.80	637	400	140	6.5
16 Alfenol	16Al	600 Q	600 ~ 4,100	4,000 ~ 90,000	—	—	0.80	637	—	153	6.5
78 Permalloy	78.5Ni	1,050; 600 Q	8,000	100,000	4	0.05	1.08	860	600	16	8.6
Supermalloy	5Mo, 79Ni	1,300 C	100,000	1,000,000	0.16	0.002	0.79	629	400	60	8.77
Cr Permalloy	3.8Cr, 78Ni	1,000	12,000	62,000	4	0.05	0.80	637	420	65	8.5
Mumetal	5Cu, 2Cr, 77Ni	1,175 (H_2)	20,000	100,000	4	0.05	0.65	517	—	62	8.58
Hipernik	50Ni	1,200 (H_2)	4,000	70,000	4	0.05	1.60	1,270	500	45	8.25
50 Isoperm	50Ni	1,100 CR	90	100	480	6	1.60	1,270	500	40	8.25
Deltamax	50Ni	1,075	500	200,000	—	—	1.55	1,230	—	45	8.25
Thermoperm	30Ni	1,000	—	—	—	—	0.20	159	—	—	—
Permendur	50Co	800	800	5,000	160	2.0	2.45	1,950	980	7	8.3
45–25 Perminvar	25Co, 45Ni	1,000; 400	400	2,000	95	1.2	1.55	1,230	715	19	—
7–70 Perminvar	7Co, 70Ni	1,000; 425	850	4,000	48	0.6	1.25	995	650	16	8.6

[a] Q, quenched; C, controlled cooling rate; CR, severely cold rolled; H_2, annealed in pure hydrogen.

so that magnetic flux can easily change its path in the plane of the sheet.[7] This property may be useful for rectangular iron cores and for rotors of rotating machines.

Iron-aluminum alloys. Aluminum is soluble in iron up to 32 wt% Al; in which range it forms the superlattices Fe_3Al at 13.9% Al and FeAl at 32.6% Al. Saturation magnetization decreases with addition of aluminum as if Al simply dilutes the magnetic ions of Fe, and the value of I_s becomes 1.4 Wb/m^2 (= 1100 gauss) at Fe_3Al. Further addition of aluminum causes a more rapid decrease of I_s, and I_s becomes zero at 18% Al. The magnetocrystalline anisotropy constant K_1 decreases rapidly with addition of Al as in the silicon-iron alloys and becomes negative near Fe_3Al. On the other hand, the magnetostriction constant increases with addition of Al and reaches a large value ($\lambda = 37 \times 10^{-6}$) at Fe_3Al. This is the reason why 13% Al-Fe (Alfer) is used as a magnetostriction material. It is expected from the compositional dependence of I_s and K_1 that the soft magnetic properties should be improved with addition of aluminum. The situation is quite similar to that for silicon-iron, except that λ increases. Since, however, a large value of λ is harmful only when an internal stress σ is present in the material, high permeability is expected as long as the material is prepared without internal stress. Actually it is reported that iron-aluminum exhibits a sharp maximum in permeability at 13% Al when it is prepared by melting in a high vacuum and by cooling slowly to room temperature.[8,9] It was also reported by Masumoto and Saito[10] that high permeability is obtained for 16% Al-Fe when it is quenched from 600°C. They called this alloy Alperm. It is called Alfenol[11] in the United States. A detailed investigation of its workability has been reported. Because of their poor workability and their hard oxide which frequently damages the rolling machine, iron-aluminum alloys have not been extensively used in the past. Recently, however, with progress in high vacuum melting technique and in rolling technique, these alloys have been more or less available as iron core materials. The high resistivity of iron-aluminium alloys makes them advantageous for use in magnetic sheet steel.

Iron-nickel alloys. This alloy system is composed of two regions: the "irreversible" alloys from 5% to 30% Ni-Fe, and the "reversible" alloys from 30% to 100% Ni-Fe. Irreversible alloys transform from f.c.c. to b.c.c. during cooling from high temperature. This transformation, which is called a martensite transformation or diffusionless transformation, occurs over a certain temperature range, and the average transformation temperature differs considerably between cooling and heating. Because of this phenomenon, various magnetic properties exhibit thermal hysteresis, and this is the reason why these alloys are called irreversible alloys.

The reversible alloys are single phase f.c.c. solid solutions. The saturation magnetization at absolute zero temperature shows a simple variation along the Slater-Pauling curve except near 30% Ni-Fe, where both the saturation magnetization and the Curie point of the f.c.c. phase become very small and presumably drop off to zero.[12] This anomalous decrease of the Curie point has been a factor in the development of various alloys. For example, Thermoperm[13] is the 30% Ni-Fe alloy whose saturation magnetization or permeability decreases rapidly with increasing temperature. A piece of this alloy is used as a shunt in a magnetic circuit to compensate for the temperature change of the magnetic flux in the circuit. Alloys of this kind are called magnetic compensating alloys. Some of the Ni-Cu alloys which have Curie points just above room temperature are also used for this purpose. Invar is the 36% Ni-Fe alloy whose thermal expansion coefficient is very small, because the usual thermal expansion is compensated by a temperature decrease of volume magnetostriction. Superinvar[14] (31% Ni, 4% ~ 6% Co) is an alloy of this kind with an even smaller thermal expansion coefficient. Elinvar[15] (36% Ni, 12% Co) and others[16] are alloys with temperature independent elastic moduli attained by compensation of the usual temperature dependence of the elastic modulus by that due to volume magnetostriction (cf. Section 19.3).

Both the magnetocrystalline anisotropy and magnetostriction go through zero at 18% to 25% Fe-Ni, as shown in Figs. 7.9, and 8.11. As expected, the permeability is very high around this composition (Fig. 13.3). Permalloy with 21.5% Fe is one of the representative soft magnetic materials. It attains very high permeability when quenched from 600°C (cf. Section 17.1). The addition of a small amount of Mo, Cr, or Cu to this alloy eliminates the necessity for the quenching process, improves the permeability, and increases the resistivity (cf. Table 20.1). It is believed that the high permeability in this case is attained by simultaneous realization of zero K_1 and zero λ at a certain composition by adding the third element.

It is reported by Yensen[17] that very high permeability is attained also for 45% ~ 50% Fe-Ni by annealing at 1200°C in a dry hydrogen atmosphere. The reason for the high permeability in this case is thought to be the easy displacement of domain walls in the impurity-free material. Because of the fairly large values of K_1 and λ in this alloy (cf. Figs. 7.9 and 8.11), rotation magnetization begins at a relatively small value of I/I_s, but the maximum permeability is still fairly large because the saturation magnetization is fairly high. A relatively low nickel content also reduces the cost of the alloy.

It has been found that the 50% Fe-Ni alloy develops (001) [100] recrystallization texture by annealing at 1100°C subsequent to severe cold rolling and therefore exhibits an excellent rectangular hysteresis loop. Deltamax,[18] a recrystallized sheet of this kind, is used in applications which make use of its non-linear hysteresis curve (cf. Section 23.1).

Isoperm, which has constant permeability over a wide range of magnetic induction, is manufactured by rolling the recrystallized 50% Fe-Ni alloy down to 50% reduction of the thickness. The roll magnetic anisotropy thus induced results in a linear magnetization curve (cf. Section 17.2), which is useful for iron cores of circuit transformers and choke coils.

Iron-cobalt alloys. Cobalt is soluble in iron up to 75% Co-Fe. The alloys are body-centered cubic and form a superlattice of the FeCo type which has a transition point at 730°C. As seen in the Slater-Pauling curve (Fig. 4.9), the saturation magnetization at 0°K has a maximum at about 35% Co. The maximum value at room temperature is attained at about 40% Co, where K_1 also goes through zero, so that we may expect high permeability. Actually, sharp maxima in μ_a and $\mu_{\max}$ are observed to occur at about 50% Co-Fe. The composition is made commercially under the name of Permendur. Sometimes small amounts of V, Mo, W, or Ti are added to the alloy for the purpose of improving the workability. Its high flux density is useful when it is used as the vibrating plate of a telephone receiver, because it retains a reasonably high permeability for ac magnetization even if it is dc magnetized by a permanent manget.

Iron-cobalt-nickel alloys. The best-known magnetic alloy which belongs to this system is Perminvar: 25% Co, 45% Ni, and 30% Fe. As reported by Elmen[19] and Masumoto,[20] the feature of this alloy is its constant permeability over a wide range of magnetic induction. In order to attain this characteristic, the alloy must be annealed for a fairly long time (for instance, for 24 hours) at 400° $\sim$ 450°C. The necessity for this treatment was interpreted by Taniguchi[21] as follows: All the spins in the domain walls are stabilized by a uniaxial anisotropy induced by directional ordering during the anneal, which keeps the domain walls in their original positions. Since the energy of the domain wall is expressed as a quadratic function of its displacement (cf. 15.44), the domain wall displacement and, accordingly, the induced magnetization should be proportional to the intensity of the applied field. If the alloy is magnetized up to a reasonably strong magnetization, the domain walls may escape from their stabilized poistions and thus become incapable of giving constant permeability. Then the hysteresis loop develops a wasp-waisted shape as shown in Fig. 22.2. The reason for this shape was also explained by Taniguchi. It is also possible to make the hysteresis loop rectangular, because this alloy responds sensitively to magnetic annealing (cf. Section 17.1).

Nickel-cobalt alloys. This alloy system forms f.c.c. solid solutions from 0% to 70% Co. In this range the saturation magnetization I_s and the Curie point Θ change monotonically. In spite of a fairly complicated dependence of K_1 on the alloy composition, various magnetic constants such as $\bar{\mu}_a$, $\bar{\mu}_{\max}$, and H_c also change monotonically with composition. In the range 70% to 100% Co the alloys are h.c.p. at room temperature, transforming to f.c.c. at high temperatures. A special kind of magnetic annealing is effective in this range, as mentioned in Section 17.3.

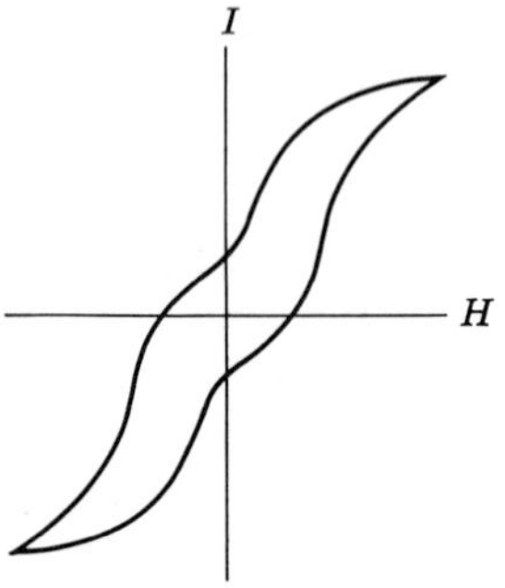

Fig. 22.2. Wasp-waisted hysteresis loop.

22.2 Pressed Powder and Ferrite Cores

For high frequency applications, reduction of the various losses accompanying high frequency magnetization is more important than static magnetic characteristics. The eddy current loss is of primary importance, and, in order to reduce it, division of magnetic metals and alloys into sheets or fine particles is effective. Ferrites and other magnetic compounds are particularly good for high frequency applications because of their high resistivity (cf. Table 22.2).

Table 22.2. Magnetic Properties of Pressed Powder and Ferrite Cores

Material	Composition	Treatment	$\bar{\mu}_a$	H_c A/m	H_c Oe	I_s Wb/m²	I_s Gauss	Θ, °C	ρ, Ω-m
Carbonyl Iron powder	100Fe	Press	20	1200	15	1.56	1240	770	10^8
Mo Permalloy powder	2Mo, 81Ni	Press, 650	125	—	—	0.70	560	480	10^4
Sendust powder	5Al, 10Si	Press, 800	80	100	1.25	0.45	360	500	
Mn-Zn ferrite	50Mn, 50Zn	1150	2000	8	0.1	0.25	200	110	1
Cu-Zn ferrite	40Cu, 60Zn	1000	1100	40	0.5	—	—	90	10^3
Cu-Mn ferrite	40Cu, 60Mn	1250	—	80	1.0	0.29	230	—	
Ni-Zn ferrite	30Ni, 70Zn	1050	80	—	—	0.40	320	130	10^{10}
Mg-Zn ferrite	50Mg, 50Zn	—	500	80	1.0	0.26	207	120	
Mg-Mn ferrite	50Mg, 50Mn	1400	—	40	0.5	0.27	215	130	

Carbonyl iron. Carbonyl iron $Fe(CO)_5$ can be made by reacting CO gas with iron under high pressures at high temperatures. By decomposing it at 350°C, we get spherical iron particles of uniform size ranging in diameter from 3 to 20 μ. By pressing these particles with an insulating binder, we have magnetic pressed cores with low high-frequency losses.

Permalloy dust core. The addition of several per cent sulfur to Permalloy or Mo Permalloy makes these alloys brittle at room temperature. After hot rolling into thin sheets, these alloys are ground into small particles with an average diameter of about 40 μ. The powders are heat-treated, reground, compressed into final shape with insulating material, and re-annealed.

Sendust. The alloy containing 5% Al, 10% Si, and 85% Fe exhibits extremely high permeabilities $\bar{\mu}_a = 30{,}000$ and $\bar{\mu}_{\max} = 120{,}000$ as cast from the melt because this composition just meets the conditions $K_1 = 0$ and $\lambda = 0$.[22] Since this alloy is very brittle, it is usually ground into small particles about 10 μ in diameter, which, after annealing, are compressed into final shapes with insulating material.

These metal powder cores have been mostly replaced by insulating magnetic materials such as ferrites and garnets. The fundamental structures and magnetic properties of these materials have already been mentioned in Sections 5.2 and 5.3. Here we discuss the magnetic properties of ferrites and other compounds which have been used for practical purposes. Snoek[22] found that the addition of Zn to various simple ferrites causes a decrease of the Curie point to relatively low temperatures, which is effective in producing high permeabilities at room temperature on account of the Hopkinson effect (cf. Section 13.2). Values of the magnetocrystalline anisotropy and the magnetostriction of various mixed ferrites are sensitive to the content of excess iron. Disaccommodation is also sensitive to the content of Fe^{2+} and of vacancies (cf. Section 15.3). Thus the magnetic properties of mixed ferrites are very sensitive to the content of constituent metal ions, as for magnetic alloys, and also to the program of heat treatment and to the atmosphere existing during this treatment.

Mn-Zn ferrites. As seen in Table 5.2 and Fig. 5.9, Mn ferrite has the highest saturation magnetic moment of any of the simple ferrites. Mn-Zn ferrite is a good soft magnetic core material in this sense, but it is not very suitable for extremely high frequency purposes because of its relatively low resistivity ($\rho = 0.1 \sim 1\ \Omega$m). Maximum permeability occurs at $50 \sim 75$ mol % $MnFe_2O_4$, $50 \sim 25$ mol % $ZnFe_2O_4$. After being sintered at 1400°C, Mn-Zn ferrite is usually quenched from 800°C or slowly cooled in nitrogen atmosphere to prevent segregation of αFe_2O_3.

Cu-Zn ferrites. This ferrite was discovered by Kato and Takei[23] in 1932 long before the recent development of ferrites, and it was named Oxide Core. The permeability is maximum at 40 mol % $CuFe_2O_4$ and 60 mol % $ZnFe_2O_3$. The resistivity is fairly high ($\rho = 10^3\ \Omega$m), a property which is advantageous for high frequency applications. Since Cu^{2+} ions tend to be reduced to Cu^{1+} at high temperatures, this ferrite is usually sintered at 1000°C.

Cu-Mn ferrites. A rectangular hysteresis loop is observed at about 40 mol % $CuFe_2O_4$ and 60 mol % $MnFe_2O_4$. Sintering at 1250°C is most effective for producing this characteristic.

Ni-Zn ferrites. Permeability is maximum at about 30 mol % $NiFe_2O_4$ and 60 mol % $ZnFe_2O_4$. Resistivity is as high as 10^3 to 10^7 Ωm. The high frequency characteristics depend on the Zn content, as seen in Fig. 16.7. This ferrite is used as a magnetostriction material because of its high magnetostriction. Since only the doubly charged state is stable for nickel ions, this ferrite can be sintered at high temperatures.

Mg-Zn, Mg-Mn ferrites. Permeability is fairly high at 50 mol % $MgFe_2O_4$ and 50 mol % $ZnFe_2O_4$. Mg-Mn ferrite is used as a rectangular hysteresis loop material and also as a gyrator material (cf. Section 23.2).

Ferroxplana. As already discussed in Section 5.3, this magnetoplumbite-type iron oxide has a uniaxial magnetocrystalline anisotropy with the easy plane parallel to the c plane, and its high frequency usefulness extends to several hundred megacycles (cf. Section 16.2).

22.3 Permanent Magnets

Another important engineering application of magnetic materials is as permanent magnets. An ideal permanent magnet should maintain its magnetization constant when it is subjected to internal or external demagnetizing fields or to changes in temperature. Let us consider a magnetic circuit which contains a permanent magnet as shown in Fig. 22.3. In this figure, P is a permanent magnet which magnetizes the iron yokes Y and thus produces a magnetic field at the air gap G. The magnetostatic energy stored in this system is given by (10.24) as

$$U = \tfrac{1}{2}\iiint (\boldsymbol{B}, \boldsymbol{H})\, dv, \tag{22.1}$$

where $\boldsymbol{B}$ is the magnetic flux density which does not include the permanent magnetization, as already remarked in Section 10.1. The integrand in (22.1), therefore, becomes H^2 inside the permanent magnet. This part of the energy is never useful for the original purpose of the magnetic circuit. In order to calculate a useful energy produced by the permanent magnet, integration should be made over G and Y, or

$$U = \tfrac{1}{2}\int_{G\&Y} BSH_s\, ds = \tfrac{1}{2}\Phi \int_{G\&Y} H_s\, ds \tag{22.2}$$

where S is the cross-sectional area of the magnetic circuit ($dv = S\, ds$). Although the main purpose of the magnetic circuit is to produce a magnetic field at the air gap G, the energy stored in Y is also included in (22.2), because this part of the energy is also considered to be supplied by the

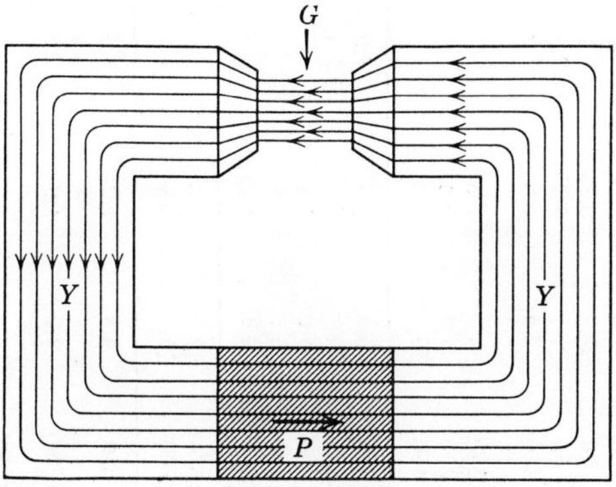

Fig. 22.3. Magnetic circuit containing a permanent magnet.

magnet (similarly to the line loss of an electric circuit). Considering that

$$\oint H_s\, ds = 0 \tag{22.3}$$

for a complete circuit, we have

$$\int_{G\&Y} H_s\, ds = -\int_P H_s\, ds. \tag{22.4}$$

Therefore (22.2) becomes

$$U = -\tfrac{1}{2}\Phi \int_P H_s\, ds = -\tfrac{1}{2}\int_P (\boldsymbol{B}, \boldsymbol{H})\, dv, \tag{22.5}$$

where the integration should be made over the volume of the permanent magnet. This $\boldsymbol{B}$ does include the permanent magnetization. Thus the quality of a permanent magnet can be measured by the value of the product $-(\boldsymbol{B}, \boldsymbol{H})$. (Note that $H < 0$ inside the magnet.) The product (BH) depends not only on the magnet material but also on the working point on the magnetization curve. For instance, Fig. 22.4 shows the demagnetizing part of hysteresis curve, where point A' has large B but small $|H|$ while point A'' has large $|H|$ but small B. It is, therefore, most effective to select the working point at A, where (BH) takes the maximum value $(BH)_{\max}$.†

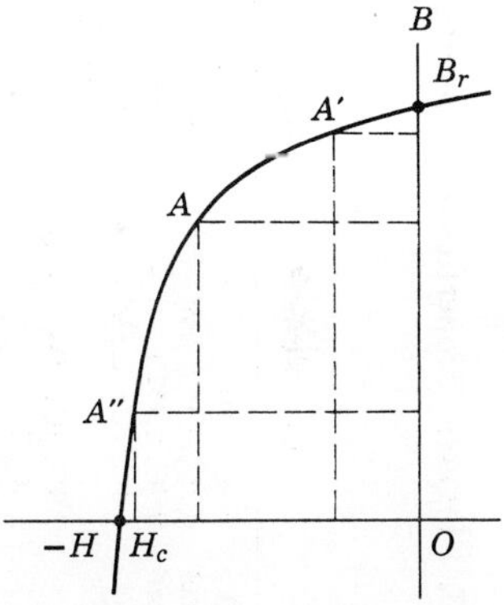

Fig. 22.4. Demagnetizing curve of a permanent magnet.

† $(BH)_{\max}$ does not necessarily correspond to V_m (cf. 2.42). For instance, if μ_r is small, V_m is usually large, but in this case the reluctance of the magnet R_m becomes large also, resulting in no large flux.

Table 22.3. Magnetic Properties of Permanent Magnetic Materials (Partly from Bozorth[G.10])

Type	Material	Composition	Heat Treatment[a]	B_r Wb/m²	H_c A/m × 10²	H_c Oe	$\frac{1}{2}(BH)_{max}$[b] J/m³ × 10³	$\frac{1}{2}(BH)_{max}$[b] Gauss · Oe[c] × 10⁶	Density g/cc
Lattice transformation hardening	Carbon steel	0.9C, 1Mn	800 Q	1.0	40	50	0.8	0.2	7.8
	Tungsten steel	0.7C, 0.3Cr, 6W	850 Q	1.0	56	70	1.2	0.31	8.1
	3.5% Cr Steel	0.9C, 0.35Cr	830 Q	0.98	48	60	1.1	0.27	7.7
	15% Co Steel	1.0C, 7Cr, 0.5Mo, 15Co	1150 AQ, 780 FC, 1000 Q	0.82	143	180	2.4	0.6	7.9
	KS Steel	0.9C, 3Cr, 4W, 35Co		0.90	200	250	4.0	1.0	
	MT Steel	2.0C, 8.0Al		0.60	160	200	1.8	0.45	
	Vicalloy	52Co, 14V	600 B	1.0	360	450	12	3.0	
Precipitation hardening	MK Steel	16Ni, 10Al, 12Co, 6Cu		0.8	446	560	6.4	1.6	7.0
	Alnico 5	14Ni, 24Co, 8Al, 3Cu	1300 AF, 600 B	1.2	438	550	20	5.0	7.3
	Cunife	60Cu, 20Ni		0.54	438	550	6.0	1.5	
	Cunico	50Cu, 21Ni, 29Co		0.34	525	660	3.6	0.9	
	Silmanal	9Mn, 4Al, 87Ag	250 B	0.055	4770	${}_IH_c = 6000$	0.3	0.08	
Superlattice	Pt-Fe	78Pt		0.58	1250	1570	12	3.0	10
	Pt-Co	23Co, 77Pt		0.45	2070	2600	15	3.8	11
Fine particle	Co ferrite (O.P. Magnet)	$(3CoO + FeO)Fe_2O_3$		0.25	518	650	4.8	1.2	3.5
	Ba ferrite (Ferroxdure)	$BaO{\cdot}6Fe_2O_3$		0.20	1200	1500	4.0	1.0	4.5
	MnBi	20Mn, 80Bi		0.42	2630	3300	16.7	4.2	
	Iron powder	100Fe		0.40	400	500	3.0	0.75	
	Elongated	100Fe		0.57	613	770	6.4	1.6	
	FeCo powder	55Fe, 45Co		1.02	630	790	18	4.5	

[a] Q, quenched; AQ, quenched in air; FC, furnace cooled; AF, cooled in magnetic field; B, baked.
[b] Energy per unit volume.
[c] This column shows $(BH)_{max}$ in cgs. The energy per unit volume is $1/(8\pi)$ of the values in this column.

In order to attain a large value of $(BH)_{max}$, three factors are important: a large value of B_r ($= I_r$), a large value of H_c, and a rectangular hysteresis loop. The first requirement is fulfilled by increasing the value of I_s, and also by making the ratio I_r/I_s as close to unity as possible. The latter requirement is closely related to the realization of a rectangular hysteresis loop. The actual methods for achieving this are magnetic annealing, orientation of crystallites by means of magnetic pressing, or special casting. In order to increase H_c, various methods as discussed in Chapter 14 are useful: One is to increase internal stress by utilizing lattice transformation, precipitation hardening, or superlattice formation. The other is to utilize fine particles small enough to have a single domain structure. It should be noticed here that the coercive force important for permanent magnets is the point where $B = 0$ (not $I = 0$). We normally distinguish this coercive force by denoting by ${}_BH_c$, and the normal coercive force by ${}_IH_c$. The value of ${}_BH_c$ is related to B_r by

$$|{}_BH_c| = \frac{I_c}{\mu_0} < \frac{I_r}{\mu_0} = \frac{B_r}{\mu_0}, \tag{22.6}$$

where I_c is the value of I at the point $B = 0$. From this relation we find that ${}_BH_c$ is also limited by the value of B_r. In this sense it can be said that essential improvement of permanent magnets can be made only by improving the value of B_r or, accordingly, the value of I_s.

In the early history of permanent magnets, various iron-base magnets such as carbon steel, tungsten steel, and chromium steel were developed and then replaced by alloy magnets of high quality such as KS steel, MK steel, and Alnico V. More recently, various oxide or compound magnets, such as OP magnet, Ba ferrite, and MnBi, have been developed. The values of ${}_BH_c$ of these magnets have been increased to the limit given by (22.6), so that further development in quality must come from an increase in B_r. The recent development of elongated fine particle magnets of Fe and Fe-Co is worth noticing because of the high remanence of these materials (cf. Table 22.3).

Lattice transformation hardening magnets. When iron-rich alloys are cooled from the γ-phase (f.c.c.), the crystal normally has a martensitic transformation to the α-phase (b.c.c.), in which case the crystal lattice is distorted by strong internal stresses. This structure is also magnetically hard because of its internal stress and its magnetic heterogeneity. Tungsten steel, chromium steel, and cobalt steel all utilize this phenomenon. Both KS steel, which was invented by Honda,[24] and MT magnet steel, which was invented by Mishima and Makino,[25] belong to this category. The former is composed of iron, cobalt, and tungsten, while the latter is composed of iron, carbon, and aluminum. Vicalloy[26] is composed of iron, cobalt, and vanadium. The real mechanism of its high coercivity has not been clarified, but it is supposed to be related to the α-γ and order-disorder transformations. Because of its good workability for drawing and rolling, this alloy can be used as a magnetic recording material.

Precipitation hardening magnets. The first magnet of this type was MK steel, which was invented by Mishima.[27] Alnico 5 has almost the same composition. This type of magnet is composed of Fe, Ni, Co, Al, and Cu. The fundamental composition is approximately expressed by the formula Fe_2NiAl. Above 1300°C this alloy has an ordered b.c.c. structure, which segregates into an iron-rich b.c.c. phase and another b.c.c. phase during the process of slow cooling. During this precipitation process, large internal stresses are produced and at the same time the structure is finely divided. It was first thought that the origin of its high coercivity lay in its strong internal stresses, but later Nesbitt[28] found that a high coercivity is also observed for an alloy of the same kind which has no magnetostriction; accordingly he suggested that a single domain structure of fine precipitated particles must be the origin of the high coercivity. Cooling this alloy from 1300°C in a magnetic field results in a rectangular hysteresis loop and accordingly a large $(BH)_{\rm max}$. The origin of this magnetic annealing has been considered to be the reorientation of elongated precipitated particles.

Cunife[29] and Cunico[30] are precipitation hardening alloys composed of Cu-Ni-Fe and Cu-Ni-Co, respectively. These are mechanically soft and machinable and are used as magnetic recording materials. Silmanal[31] (composition $MnAg_5Al$), is probably a precipitation alloy. Although the saturation magnetization is only $I_s = 0.088$ Wb/m^2 (= 64 gauss), its coercive force is as high as 480 kA/m† (= 6000 Oe). This material is also mechanically soft and easily drawn into wire.

Superlattice magnets. Typical magnets which belong to this type are Fe-Pt and Co-Pt. These alloys form superlattices of FePt and CoPt types which are tetragonal and exhibit very high coercive force.

Fine particle magnets. OP (oxide powder) magnet, which was invented by Kato and Takei[32] in 1933, is a ferrite magnet of the composition $3CoFe_2O_4 + Fe_3O_4$. Since cobalt ferrite has a large magnetocrystalline anisotropy and also a large magnetostriction (cf. Tables 7.5 and 8.2), it is expected that, when this material is divided into single domain fine particles, it should exhibit high coercivity. Even in polycrystalline materials in which crystal grains are in contact with each other, it is possible that each grain can become a single domain below some critical

† Kiloampere/meter.

size (cf. Section 11.3) and thus attain a high coercivity. If, however, crystallites grow too large because of overfiring, the coercive force should be reduced. This material also responds to magnetic annealing (cf. Section 17.1), which is effective in increasing $(BH)_{max}$. Ba ferrite, which is a magnetic oxide of the magnetoplumbite type, was invented at the Philips laboratory[33] in 1951. The composition is $BaO{\cdot}6Fe_2O_3$ or $BaFe_{12}O_{19}$. Because of its large magnetocrystalline anisotropy, this material can have single domain structure with high coercive force when it is fired into a fine grained polycrystal. It has been reported that magnetic pressing is effective in aligning the easy axes of the particles and thus increases $(BH)_{max}$ by a factor of 3 as compared to the non-oriented materials.[34]

Manganese bismuthide has been known to be ferromagnetic since its discovery by Heusler[35] in 1904. Later Guillaud[36] found that it exhibits a large coercive force $H_c = 96$ kA/m (= 1200 Oe) in powdered form. It is prepared by heating a mixture of Mn and Bi in a helium atmosphere at 700°C, and after annealing for a long time at 440°C it is ground into fine powder by a high speed hammer mill. The average diameter of particles is about 8 μ. The remanent magnetization and the coercive force are fairly large compared to those of ferrite magnets. However, its susceptibility to corrosion in air at room temperature has so far prevented its commercial use. Manganese-aluminum has been found to be ferromagnetic by Kono,[37] Koch et al.,[38] and Köster and Wachtel.[39] The coercive force is as large as 80 kA/m (= 1000 Oe) for the alloy as cast from the melt, but it increases to 367 kA/m (= 4600 Oe) when it is subjected to severe machining such as swaging.

Metal powder magnets can be prepared either by decomposing a chemical compound in powder form or by depositing the material electrolytically on a liquid metal electrode such as mercury. It is possible by the latter method to make elongated fine particles which exhibit a large coercive force because of their large shape anisotropy and single domain structure. Iron and iron-cobalt particle magnets have been developed.[40]

References

22.1. R. M. Bozorth: G. 10, p. 61.
22.2. M. Asanuma and S. Ogawa: *J. Phys. Soc. Japan* **12,** 955 (1957).
22.3. R. M. Bozorth: G. 10, p. 80.
22.4. N. P. Goss: U.S. Pat. 1,965,559 (Appl. 8/7/33); *Trans. Am. Soc. Metals* **23,** 515 (1935); U. S. Pat. 2,084,336-7 (Appl. 1/30/34 and 12/1/34).
22.5. F. Assumus et al.: *Z. Metallk.* **48,** 341, 344 (1957).
22.6. W. R. Hibbard and J. L. Walter: Japan Pat. Syo 33-7509 (1958).

22.7. J. L. Walter, W. R. Hibbard, H. C. Fielder, H. E. Grenoble, R. H. Pry, and P. G. Frischmann: *J. Appl. Phys.* **29,** 363 (1958).
22.8. J. F. Nachman and W. J. Buehler: Naval Ord. Report 4130 (1955).
22.9. M. Sugihara: *J. Phys. Soc. Japan* **15,** 1456 (1960); *Rep. Elec. Commum. Lab. Nippon Tel. Tel. Co.* **7,** 333 (1959).
22.10. H. Masumoto and H. Saito: *Sci. Rept. Res. Inst. Tohoku Univ.* **3,** 523 (1951); **4,** 321 (1952); **6,** 338 (1954).
22.11. J. F. Nachman and W. J. Buehler: *J. Appl. Phys.* **25,** 307 (1954).
22.12. J. S. Kouvel and R. H. Wilson: *J. Appl. Phys.* **32,** 435 (1961).
22.13. F. Stablein: *Tech. Mitt. Krupp* **2,** 127 (1934).
22.14. H. Masumoto: *Sci. Rept. Tohoku Imp. Univ.* **20,** 101 (1931).
22.15. C. E. Guillaume: *Proc. Roy. Soc. (London)* **32,** 374 (1920).
22.16. G. E. Shubrooks: *Metal Progr.* **21,** 58 (1932).
22.17. T. D. Yensen: *Elec. J. (London)* **28,** 286 (1931); *Trans. Am. Soc. Metals* **27,** 797 (1939).
22.18. J. H. Crede and J. P. Martin: *J. Appl. Phys.* **20,** 966 (1949).
22.19. G. W. Elmen: *J. Franklin Inst.* **206,** 317 (1928); **207,** 583 (1929).
22.20. H. Masumoto: *Sci. Rept. Tohoku Imp. Univ.* **18,** 195 (1929).
22.21. S. Taniguchi: *Sci. Repts. Tohoku Univ.* **8A,** 173 (1956).
22.22. J. L. Snoek: G. 7.
22.23. Y. Kato and T. Takei: Japan Pat. 98844 (1932); T. Takei: *Elec. Chem.* **5,** 411 (1939).
22.24. K. Honda and S. Saito: *Sci. Rept. Tohoku Imp. Univ.* **9,** 417 (1920).
22.25. T. Mishima and N. Makino: *Iron & Steel* **43,** 557, 647, 726 (1956).
22.26. E. A. Nesbitt: *Metals Tech.* **13,** No. 1973, 1 (1946); *Trans. Am. Inst. Min. Met. Engrs.* **166,** 415 (1946).
22.27. T. Mishima: *Ohm* **19,** 353 (1932).
22.28. E. A. Nesbitt: *J. Appl. Phys.* **21,** 879 (1950).
22.29. H. Neumann, A. Büchner, and H. Reinboth: *Z. Metallk.* **29,** 173 (1937).
22.30. W. Darmöl and H. Neumann: *Z. Metallk.* **30,** 217 (1938).
22.31. H. H. Potter: *Phil. Mag.* **12,** 255 (1931).
22.32. K. Kato and T. Takei: *J. Japan Elec. Eng.* **53,** 408 (1933).
22.33. J. J. Went, G. W. Rathenau, E. W. Gorter, and G. W. von Oosterhout: *Philips Tech. Rev.* **13,** 194 (1952).
22.34. G. W. Rathenau, J. Smit, and A. L. Stuyts: *Z. Physik.* **133,** 250 (1952).
22.35. F. Heusler: *Z. Angew. Chem.* **17,** 260 (1904).
22.36. C. Guillaud: Thesis (Strasbourg, 1943).
22.37. H. Kono: *J. Phys. Soc. Japan* **13,** 1444 (1958).
22.38. A. J. J. Koch, P. Hokkeling, M. G. v. d. Steeg, and K. J. de Vos: *J. Appl. Phys.* **31,** 75S (1960).
22.39. W. Köster and E. Wachtel: *Z. Metallk.* **51,** 271 (1960).
22.40. T. O. Paine, L. I. Mendelsohn, and F. E. Luborsky: *Phys. Rev.* **100,** 1055 (1955); *J. Appl. Phys.* **26,** 1274 (1955); *Elec. Eng.* **76,** 851 (1957); *J. Appl. Phys.* **28,** 344, 485 (1958).

23

Special Applications of Magnetic Materials

23.1 Applications of Non-linear Characteristics

Non-linearity of the magnetization curve has been utilized for various purposes: magnetic memories, dc converters, and magnetic amplifiers. The simplest and best-known application of this kind is the utilization of remanent states for memory devices. Magnetic materials with rectangular hysteresis loops have two possible remanent states, positive and negative (Fig. 23.1). These two states can be used to represent "0" and "1" of the binary number system. The most common arrangement of memory cores is shown in Fig. 23.2. A number of wires parallel to the x and y axes pass through magnetic cores which are arranged in a regular array. If all the cores are originally in state "zero," and if a current pulse which produces a magnetic field of about half the coercive force is sent to one of the x wires and also to one of the y wires simultaneously, the magnetic core which is at the intersection of the two wires is switched, while the magnetic states of all other cores remain unchanged. Thus we can write "1" in any desired core. In order to read out this memory, all we can do is to send a pulse of the

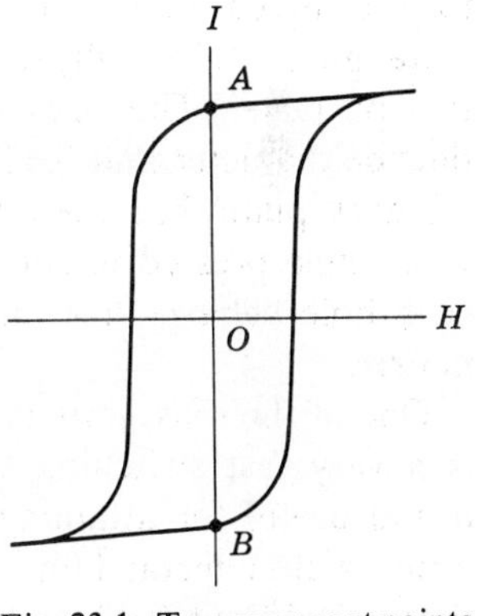

Fig. 23.1. Two remanent points on a rectangular hysteresis loop.

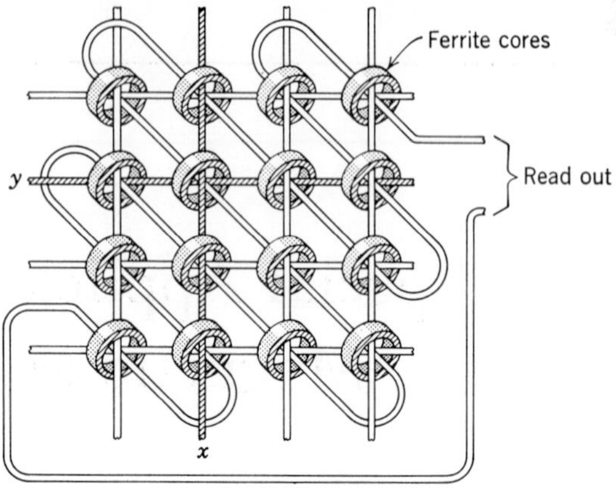

Fig. 23.2. Magnetic memory matrix using ferrite cores.

opposite sign to each core successively, and to observe whether a signal is produced in the read-out wire which goes through all cores. A signal will be produced only when the state "1" is switched back to "0." Because "read out" procedure is destructive, if we want to restore the original memory we must write in the same information again. A number of non-destructive read-out methods have been developed to avoid this complicated process.[1]

The required characteristic of the magnetic cores is that the reversible susceptibility in the remanent state be sufficiently small not to give rise to too much noise. In other words, the hysteresis loop should be rectangular. Normally Mn-Mg mixed ferrite with $\lambda_{111} = 0$ is used for this purpose. The sizes of memory cores range from 1.25 to 2.0 mm outside diameter and 0.32 to 0.52 mm inside diameter. A more compact memory plate[2] has been developed which consists of a number of holes in a single pressed ferrite core. A small part of the ferrite plate around each hole behaves like an individual memory core in the usual memory matrix.

One of the most important requirements of a magnetic memory device is a very fast switching time. Thin films of magnetic alloy have been found useful for attaining this. It has been reported that the switching time of thin metal films is as small as 10^{-8} sec, which is close to the theoretical limit 10^{-9} sec (cf. Section 16.3) as long as the available field is at most a few oersteds.

The most common arrangement of a magnetic thin film memory is a regular array of metal film spots 2 to 5 mm in diameter on a glass plate, as shown in Fig. 23.3. The thickness of the film is about 1000 Å. The method of deposition may be vacuum evaporation, electroplating, or sputtering. By applying a magnetic field during deposition it is possible to produce uniaxial anisotropy in the film with the easy axis parallel to the applied magnetic field. The films thus treated have, therefore, two remanent states with the magnetization either in the positive or the negative direction along the easy axis. In order to switch the magnetization, we simply pass a pulse current through copper ribbons which are stretched on the array of film spots parallel to the x and y axes: As discussed in Section 18.1, the film spot which exists at the intersection of the current-carrying x and y ribbons can easily reverse its magnetization, while all the others remain unchanged.

There are a number of magnetic film memories employing other geometries. The Twistor[3] was originally an assembly of twisted thin wires in each of which the easy axis is established 45° from the long axis of the wire by magnetostrictive anisotropy. The magnetization of the wire can be switched either by a pulse current passing through the wire itself or by a magnetic field applied parallel to the length of the wire. A similar element can be constructed by electroplating magnetic films onto a copper wire.[4] The magnetic rod,[5] cylinder,[6–8] and Tensor[9] are all similar devices.

Magnetic recording is also an application of the non-linearity of magnetization curves. The magnetic medium must retain residual magnetization proportional to the applied signal field. If we apply a signal field OB to a magnetic medium by a recording head, the tape will be magnetized to point A on its magnetization curve (cf. Fig. 23.4), and then return to residual magnetization A'. Figure 23.5 shows residual magnetization OA' as a function of the signal field OB. As in this graph,

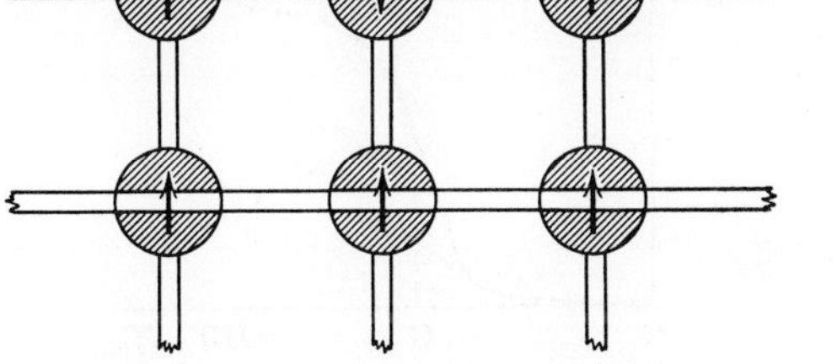

Fig. 23.3. Magnetic memory matrix using thin film spots.

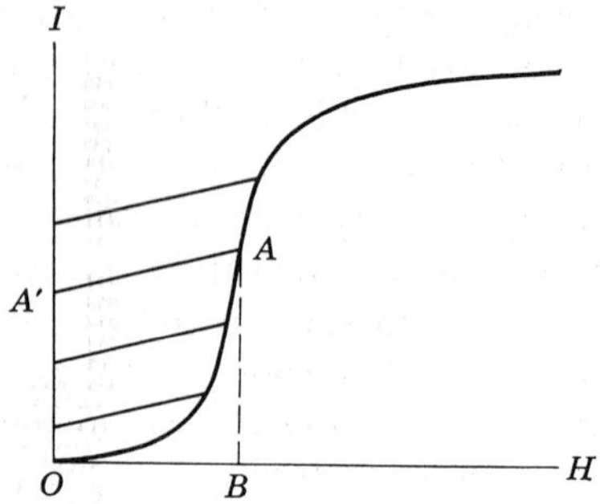

Fig. 23.4. Magnetization curve of magnetic recording tape.

the residual magnetization must be very small until the signal field becomes larger than a certain value. In magnetic tapes this is important, because otherwise the tape will be easily magnetized by an external disturbing field or a stray field from the neighboring layer of the wound tape. In recording, a high frequency field is added as a bias to the signal field in order to secure a proportionality between the signal field and the residual magnetization of the recording tape. In order to secure this characteristic, the hysteresis loop must be rectangular and must have a fairly large coercive force. In the most common magnetic tape, elongated fine particles of γFe_2O_3 are aligned on the tape to obtain this characteristic.

Rectangular hysteresis loops have also been utilized for a contact dc converter.[10] If we insert in an ac circuit an inductor made of a magnetic core with a rectangular hysteresis loop (Fig. 23.6), the time change of the current becomes very small when the instantaneous current changes its sign (Fig. 23.7), because of the large flux change of the core which is equivalent to the insertion of a large inductance. This waveform is useful for switching the direction of the current to produce direct current, since, at the instant of switching, $i = 0$ and $di/dt = 0$, both of which eliminate the possibility of arcing at the switch contacts.

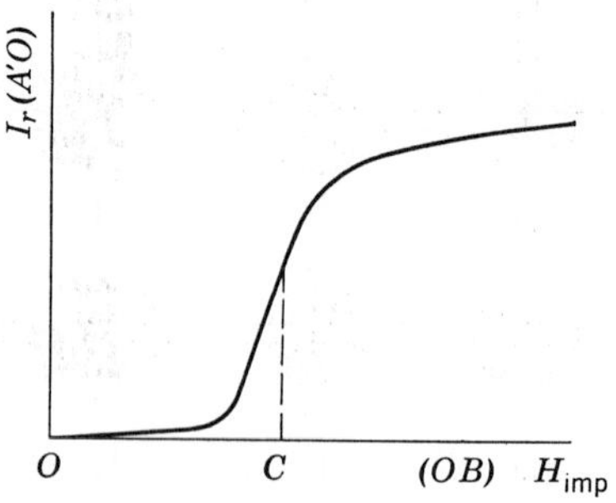

Fig. 23.5. Residual magnetization as a function of signal field.

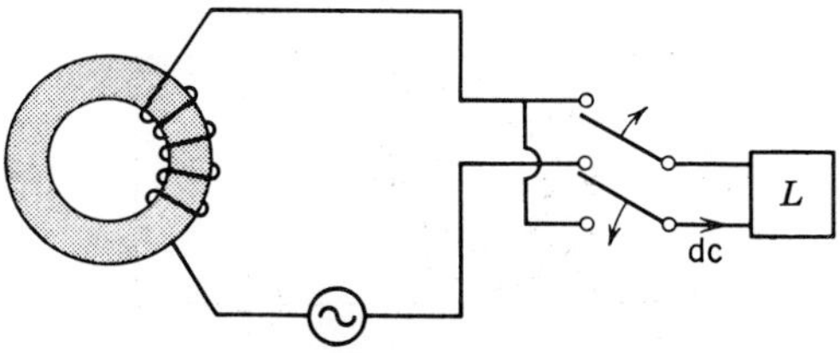

Fig. 23.6. Schematic diagram of arcless ac-dc contact converter.

Magnetic amplifiers[11] also utilize a non-linear magnetization curve. This kind of amplifier is used to control high power currents. Figure 23.8 shows the principle of the magnetic amplifier. The core is made of a magnetic material with a rectangular hysteresis loop. The impedance of the coil is large enough to suppress the ac current which goes through the load L (curve A in Fig. 23.9). If, however, the core is magnetized by a dc signal current through the left-hand coil, the hysteresis loop is shifted as shown in Fig. 23.10 and the impedance becomes practically zero in a part of the cycle where the core is saturated (curve B in Fig. 23.9). In practice, a pair of cores is used instead of a single core, and rectifiers are combined with magnetic cores. As a result, the output current becomes the alternating current.

The magnetic amplifier discussed above is a power amplifier. There is also a magnetic voltage amplifier. Figure 23.11 shows one example of this kind of amplifier, where two similar magnetic cores with rectangular hysteresis loops are magnetized by 180° out-of-phase alternating currents. The voltage which is induced in a secondary coil A by a flux reversal of the upper core is compensated by the voltage induced in coil B, so that we have no output signal. If, however, a signal of any waveform is sent to the secondary coil, it causes shifts of the hysteresis loops of the two cores in opposite directions and thus gives rise to uncompensated pulses

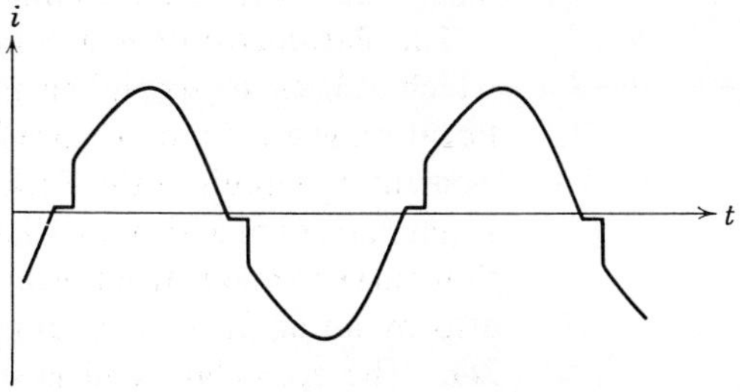

Fig. 23.7. Waveform which passes through an inductor with a rectangular hysteresis core.

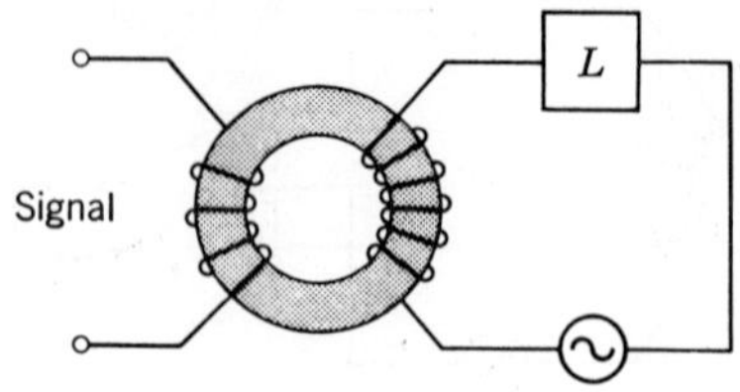

Fig. 23.8. Principle of the magnetic amplifier.

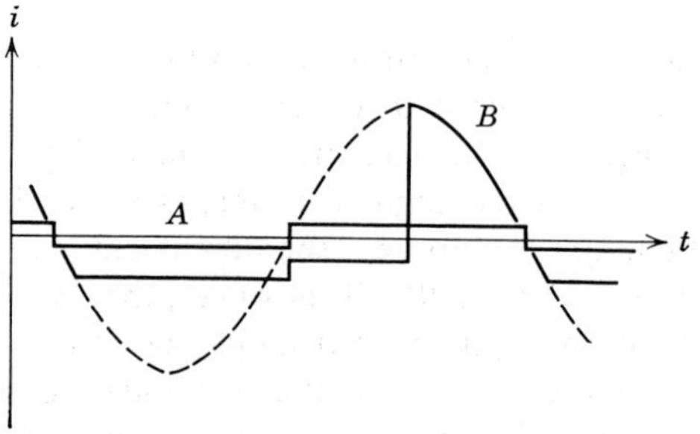

Fig. 23.9. Waveform of the ac current which passes through a magnetic amplifier.

the heights of which are proportional to the shifts of the hysteresis loop and accordingly to the signal (Fig. 23.12). Instead of a dc signal, we can use an external magnetic field by changing the shape of the magnetic cores to a convenient form, and thus we can detect a very weak magnetic field.

Next we discuss the utilization of the higher harmonics caused by non-linearity of magnetization curves. As discussed in Section 14.3, the third harmonic can be induced by cycling the Rayleigh loop. If we magnetize the magnetic core with a dc field and superimpose an ac current, we obtain the second harmonic. These phenomena can be utilized for the production of the second and third harmonics of the ac current.

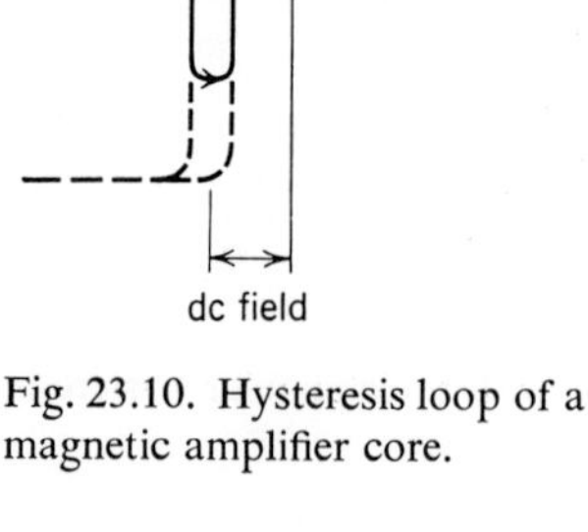

Fig. 23.10. Hysteresis loop of a magnetic amplifier core.

The Parametron[12] is a computer element which utilizes the second harmonic of a non-linear magnetic core. Figure 23.13 shows an elementary circuit of the Parametron, where a magnetic core with a non-linear magnetization curve is magnetized by a dc current and also by an ac current of angular frequency 2ω. The secondary coil is connected to a capacitor so as to resonate at angular frequency ω. Since the current of angular

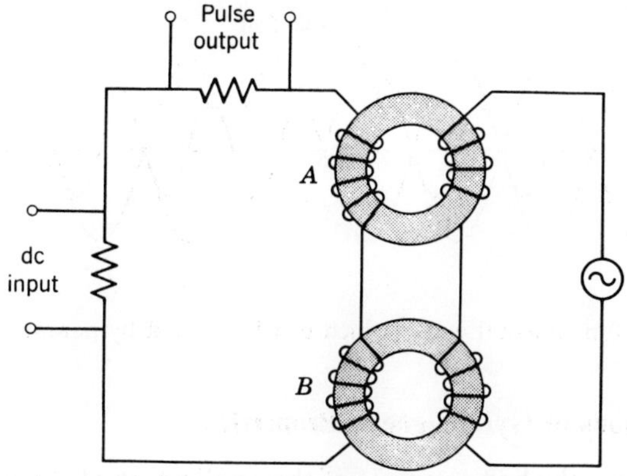

Fig. 23.11. Magnetic voltage amplifier.

frequency ω is necessarily accompanied by the second harmonic of frequency 2ω, the current of frequency 2ω in the primary coil will supply energy to the resonant circuit. Now there are two kinds of fundamental waves which can be excited by the second harmonic, as shown by curves A and B in Fig. 23.14. If we apply a small signal of frequency ω of either 0° or 180° phase, in advance of the build-up of the resonant current, a current having the same phase as that of the signal will be built up and will memorize the phase of the signal. This circuit has been used for computer elements, memories, and amplifiers.

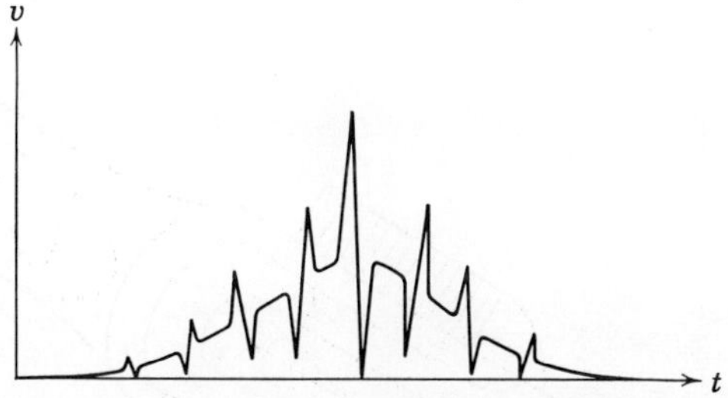

Fig. 23.12. Waveform of the output voltage of a magnetic voltage amplifier.

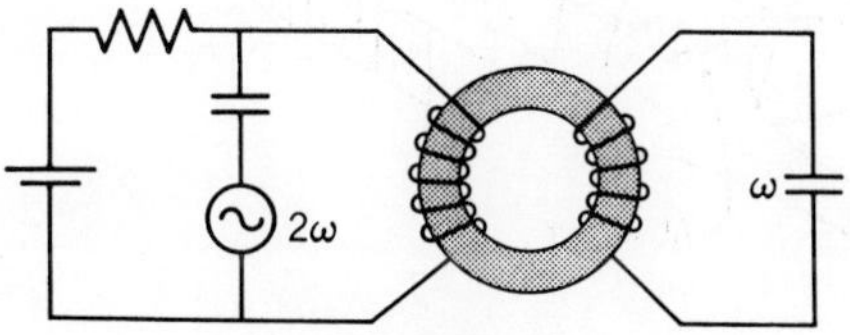

Fig. 23.13. Elementary circuit of the Parametron.

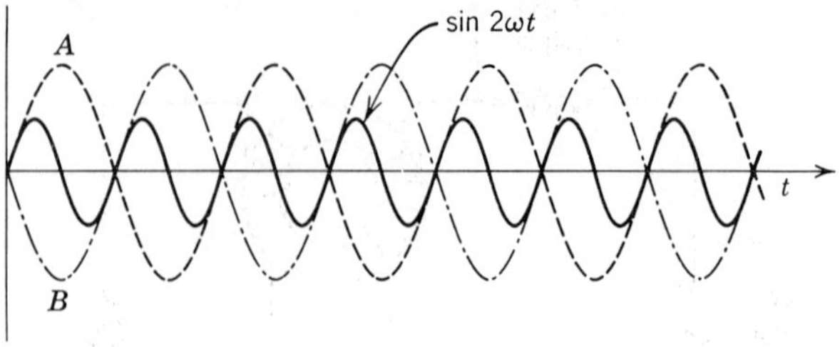

Fig. 23.14. Two fundamental waves which can be excited by parametric excitation.

23.2 Applications of Gyromagnetic Properties

The gyrator[13] is the best example of the application of the gyromagnetic properties of magnetic materials. A rod of ferrite is magnetized along its long axis in a microwave guide (Fig. 23.15). The spins inside the ferrite are ready to precess about the long axis. A linearly polarized incident microwave can be decomposed into two circularly polarized waves whose magnetic vectors are rotating in opposite directions. Since the circularly polarized wave which has the same sense of rotation as the precession of the spins in the ferrite will be partly absorbed by excitation of spin precession in the ferrite, the polarization of the output wave will rotate by some angle θ. If part of the output wave is reflected back into this system by some means, the polarization will rotate once again by θ in the same direction, making a total angle 2θ with that of the incident wave. Such a wave is easily eliminated by an absorber which is so placed as not to

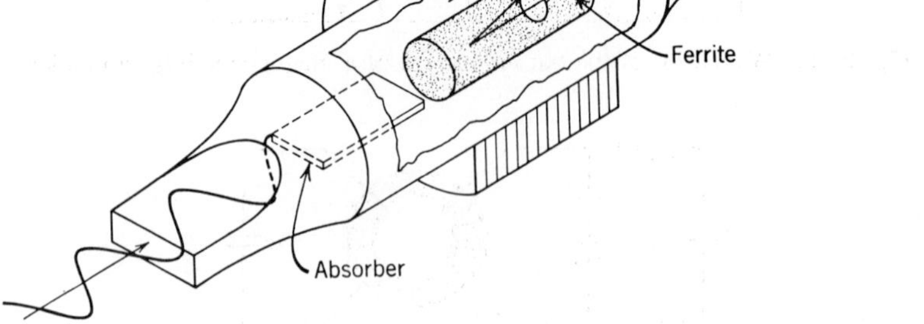

Fig. 23.15. Gyrator.

affect the incident beam. Such an isolator can protect the waveguide from disturbance by reflected waves.

The solid state "maser" is a microwave amplifier which utilizes the gyromagnetic property of paramagnetic substances.[14] The principle of the paramagnetic maser is that spins in the system are first excited by a pumping microwave to a high energy level; then the excited spins will emit a microwave of the same frequency by stimulation by a small signal microwave. The term maser was coined by Gordon, Zeiger, and Townes,[15] who first operated an ammonia gas maser (the name stands for Microwave Amplification by Stimulated Emission of Radiation). A comprehensive explanation[16] of the mechanism of the paramagnetic maser is as follows: When a spin is placed in a magnetic field, it will precess and gradually change its average direction toward the direction of the field, provided that some damping acts on the precessional motion. This motion is described by (16.46). If, however, the spin is first excited to $\theta_0 = 0$, the spin should stay forever in the same orientation as we see in (16.46). The situation is the same even if we increase the intensity of the magnetic field. When, however, a weak microwave of the same frequency as that of the precession of the spin arrives at the excited spin, the precession motion will easily be excited and will emit a stronger microwave. In this process the energy of the emission is supplied from the magnetic energy of the system. The maser is also used as a microwave oscillator.

References

23.1. Q. W. Simkins: *J. Appl. Phys.* **33S,** 1020 (1962).
23.2. J. A. Rajchman: *Proc. Inst. Radio Engrs.* **45,** 325 (1957).
23.3. A. H. Bobeck: *J. Appl. Phys.* **29,** 485 (1958).
23.4. S. J. Schwartz and J. S. Sallo: *IRE Trans. Elec. Computers* **EC8,** 465 (1959).
23.5. D. A. Meier: *Proc. Electron Computer Conf.* (1960), p. 22.
23.6. T. R. Long: *J. Appl. Phys.* **31,** 123 (1960).
23.7. G. R. Hoffman, J. A. Turner, and T. Kilburn: *J. Brit. Inst. Radio Engrs.* **20,** 31 (1960).
23.8. G. Rostky: *Electric Design* **9,** 26 (1961).
23.9. U. F. Gianola: *J. Appl. Phys.* **29,** 849 (1958).
23.10. F. Koppelmann: *Electro-technische Z.* **62,** 5, 3 (1941).
23.11. H. F. Storm: *Magnetic Amplifier* (McGraw-Hill Book Co., 1955).
23.12. Parametron Laboratory: *Studies on the Parametron* (Kyoritsu Publishing Co., Tokyo, 1959) (in Japanese).
23.13. C. L. Hogan: *Rev. Mod. Phys.* **25,** 253 (1953).
23.14. J. R. Singer: "Masers" (John Wiley & Sons, Inc., New York, 1959).
23.15. J. P. Gordon, H. J. Zeiger, and C. H. Townes: *Phys. Rev.* **95,** 282 (1954); **99,** 1264 (1955).
23.16. H. Takahashi: *Metal Phys.* **4,** 151 (1958).

General References

G.1. J. H. Van Vleck: *Theory of Electric and Magnetic Susceptibilities* (Clarendon Press, Oxford, 1932).
G.2. E. C. Stoner: *Magnetism and Matter* (Methuen and Co., London, 1934).
G.3. R. Becker and W. Döring: *Ferromagnetismus* (Springer, Berlin, 1939).
G.4. F. Brailsford: *Magnetic Materials* (Methuen & Co., London, 1948).
G.5. *Symposium on Magnetic Testing* (American Society for Testing Materials, 1948).
G.6. C. Kittel: *Physical Theory of Ferromagnetic Domains, Rev. Mod. Phys.* **21,** 541 (1949).
G.7. J. L. Snoek: *New Developments in Ferromagnetic Materials* (Elsevier, Amsterdam, 1949).
G.8. P. Weiss and G. Foëx: *Le Magnétisme* (Librairie Armond Colin, Paris, 1951).
G.9. Colloque International de Ferromagnétisme at d'Antiferromagnétisme de Grénoble, *J. Phys. Radium* **12,** No. 3 (1951).
G.10. R. M. Bozorth: *Ferromagnetism* (D. Van Nostrand Co., Princeton, N.J., 1951).
G.11. W. Jellinghaus: *Magnetische Messungen an ferromagnetischen Stoffen* (Walter de Gruyter & Co., Berlin, 1952).
G.12. K. Hoselitz: *Ferromagnetic Properties of Metals and Alloys* (Oxford, 1952).
G.13. F. Pawleck: *Magnetische Werkstoffe* (Springer, Berlin, 1952).
G.14. L. F. Bates: *Modern Magnetism* (Cambridge University Press, Oxford, 1953).
G.15. C. Kittel: *Introduction to Solid State Physics* (John Wiley & Sons, Inc., New York, second edition, 1956).
G.16. T. Nagata: *Rock-magnetism* (Maruzen Co., Tokyo, 1953).
G.17. Washington Conference on Magnetism, *Rev. Mod. Phys.* **25,** No. 1 (Jan. 1953).
G.18. M. G. Say: *Magnetic Alloys and Ferrites* (George Newnes, London, 1954).
G.19. K. H. Stewart: *Ferromagnetic Domains* (Cambridge University Press, 1954).
G.20. First Conference on Magnetism and Magnetic Materials, Pittsburgh (American Institute of Electrical Engineers, New York, 1955).
G.21. E. W. Gorter: Some Properties of Ferrites in Connection with Their Chemistry, *Proc. IRE* **43,** 245 (1955).

G.22. W. Köster: *Beiträge zur Theorie des Ferromagnetismus und der Magnetisierungs Kurve* (Springer, Berlin, 1956).
G.23. C. Kittel and J. K. Galt: Ferromagnetic Domain Theory (Solid State Physics, Vol. III, pp. 439–564, Academic Press, New York, 1956).
G.24. Ferrite Issue, *Proc. IRE* **44,** No. 10, 1229–1516 (1956).
G.25. Second Conference on Magnetism and Magnetic Materials, Boston (American Institute of Electrical Engineers, New York, 1957).
G.26. Convention on Ferrite, *Proc. IRE* **104B,** 127–265 (1957).
G.27. P. A. Miles, W. B. Westphal, and A. von Hippel: Dielectric Spectroscopy of Ferromagnetic Semiconductors, *Rev. Mod. Phys.* **29,** No. 3, 279–307 (1957).
G.28. Varenna Lectures on Magnetism, *Nuovo Cimento* Suppl. **6,** 895 (1957).
G.29. Third Conference on Magnetism and Magnetic Materials, Washington D.C., *J. Appl. Phys.* **29,** No. 3 (1958).
G.30. Fourth Conference on Magnetism and Magnetic Materials, Philadelphia, *J. Appl. Phys.* **30,** Suppl. to No. 4 (1959).
G.31. Colloque International de Magnétism de Grénoble, *J. Phys. Radium* **20,** No. 2–3 (1959).
G.32. J. Smit and H. P. J. Wijn: Ferrites (John Wiley and Sons, Inc., New York, 1959).
G.33. R. M. Bozorth et al.: *Magnetic Properties of Metals and Alloys* (American Society for Metals, Ohio, 1959).
G.34. Fifth Conference on Magnetism and Magnetic Materials, Detroit, *J. Appl. Phys.* **31,** No. 5S (1960).
G.35. Sixth Conference on Magnetism and Magnetic Materials, New York, *J. Appl. Phys.* **32,** No. 3S (1961).
G.36. Seventh Conference on Magnetism and Magnetic Materials, Phoenix, *J. Appl. Phys.* **33,** No. 3S (1962).
G.37. International Conference on Magnetism and Crystallography, Kyoto, *J. Phys. Soc. Japan* **17,** Suppl. B-I (1962).
G.38. E. Kneller: *Ferromagnetismus, mit einem Beitrag Quanten-theorie u. Electronen-Theorie des Ferromagnetismus* (Springer-Verlag, Berlin, 1962).
G.39. J. B. Goodenough: *Magnetism and the Chemical Bond* (Interscience Division of John Wiley & Sons. Inc., New York, 1963).
G.40. Eighth Conference on Magnetism and Magnetic Materials, Pittsburgh *J. Appl. Phys.* **34,** No. 4, Part 2 (1963).

Solutions of Problems

2.1. 5.62×10^5 A/m ($= 7.05 \times 10^3$ Oe), 8.82×10^6 A/m^2 ($= 1.11 \times 10^3$ Oe/cm).
2.2. $\bar{\mu}/[1 + 12(\bar{\mu} - 1)/e^8]$.
2.3. 9.61×10^4 A/m ($= 1210$ Oe).
3.1. 2.21×10^7 rad/sec.
3.2. 4.14×10^5 A/m ($= 5200$ Oe).
3.3. 9.0×10^{-6} kg-m^2/s.
3.4. $S = 2$, $L = 2$, $J = 4$, $^5D_4$.
4.1. 2.35×10^{-7}.
4.2. $\partial H^2/\partial x = 2.13 \times 10^{14}$ A^2/m^3 ($= 3.35 \times 10^8$ Oe^2/cm).
4.3. 42 at % Fe-Cr, 60 at % Ni-Co, 80 at % Fe-Ni.
5.1. χ_0, χ_0, $\frac{2}{3}\chi_0$.
5.2. $I_a = 2NM$, $I_b = -2NM$; $\Theta_f/\Theta_a = -1.02$; no compensation point.
7.1. Isotropic.
7.2. 4.58×10^5 A/m ($= 5760$ Oe).
7.3. 4.39×10^4 J/m^3.
7.4. $E_a = K_d \cos \phi$.
8.1. $\delta l/l = \frac{3}{2}\lambda_{100}\, I/I_s$.
8.2. $(\delta l/l)_{[100]} = h_1 \cos^2\theta + h_3 \sin^2\theta \cos^2\theta$; $(\delta l/l)_{[001]} = h_3 \sin^2\theta \cos^2\theta$.
8.3. $E_{\sigma[100]} = -\frac{3}{28}\lambda_{100}\sigma$, $E_{\sigma[010]} = -\frac{3}{7}\lambda_{100}\sigma$, $E_{\sigma[001]} = -\frac{27}{28}\lambda_{100}\sigma$.
9.1. $\gamma = \sqrt{24A\lambda\sigma}$.
9.2. $\gamma_{100} = 0$, $\gamma_{110} = \frac{2}{3}\sqrt{AK_2}$.
9.3. 1.22×10^{-3} J/m^3.
10.1. $\pi R^2 I_s^2/2\mu_0$.
10.2. $I_s^2 d/8\pi\mu_0$.
10.3. $\epsilon_m = 5.40 \times 10^4 I_s^2 d(1 - \sin 2\phi)/(1 + I_s^2/4\mu_0 K_1)$.

11.1. (i) $d \propto K_u^{1/4}$, (ii) $d \propto K_u^{-1/4}$.
11.2. 0.331 mm.
11.3. 0.01 μ.
12.1. $2/\pi$ (= 0.636).
12.2. The following domains are reversed from plus to minus directions from the remanent state: (i) from $\theta = -30°$ to $+30°$; (ii) from 24.3° to 65.7° and from −24.3° to −65.7°.
12.3. The following domains are reversed from minus to plus directions from the demagnetized state: (i) inside the cone with the vertical half-angle 45°; (ii) between the cones with the vertical half-angles 30° and 60°.
13.1. (i) $\beta\mu_0$; (ii) $2\beta\mu_0$.
13.2. $\bar{\chi}_{a\ 30°} = 16.25$, $\bar{\chi}_{a\ 45°} = 12.5$, $\bar{\chi}_{a\ 60°} = 8.75$, $\bar{\chi}_{a\ \mathrm{poly}} = 10$.
13.3. $I = I_s[1 - (4K_u^2/15I_s^2)(1/H^2) - \cdots]$.
14.1. 45° from the x direction toward the $-y$ direction, $I_r/I_s = 2/\pi$.
14.2. (i) $2\gamma_{180}/I_s a$ for 180° wall between the x and $\bar{x}$ domains; (ii) $2\sqrt{2}\,\gamma_{90}/I_s a$ for 90° wall between x and $\bar{y}$ domains; (iii) $4\sqrt{2}\,\gamma_{90}/I_s a$ for 90° wall between x and $\bar{z}$ domains.
14.3. $-0.4\eta H_1^2$.
15.1. $10^2:1:5.0 \times 10^{-7}$.
15.2. $E_a = -(N_c l_c^2/3kT)[\alpha_1^2 + (\alpha_2^2 - \alpha_1^2)e^{-t/\tau}]$.
15.3. 25 times.
15.4. For $H \parallel [010]$, $\chi_a = \chi_{a0}[1 - (N_c l_c^2/3K_1kT)(1 - 2e^{-t/\tau})]$; for $H \parallel [001]$, $\chi_a = \chi_{a0}[1 - (N_c l_c^2/3K_1kT)(1 - e^{-t/\tau})]$.
16.1. $P = (4I_s^2v^2/\rho d)z(d - z)$, where z is the distance between the top surface and the wall.
16.2. 6.93×10^{-6} sec; 24.5 turns.
16.3. 4.02×10^3 A/m (= 50.5 Oe).
16.4. $s = mv/\beta = v/4\pi\lambda = 8.0 \times 10^{-9}$ m = 80 Å.
17.1. $K_{u\ 001}:K_{u\ 110}:K_{u\ 111} = 0:1:2$.
17.2. s.c., 0:1; b.c.c., 2:1; f.c.c., 4:3.
17.3. $E_a = -\frac{1}{24}Nl_0pS^2s\cos^2\theta$, where θ is the angle between magnetization and $[1\bar{1}0]$.

appendix I

Symbols Used in the Text

A	exchange energy constant
B	magnetic flux density
B_1, B_2	magnetoelastic constants
C	Curie constant; specific heat
E	energy density; electric field; Young's modulus
E_H	magnetic field energy density
E_a	magnetic anisotropy energy density
E_{el}	elastic energy density
E_{mag}	magnetostatic energy density
E_{magel}	magnetoelastic energy density
E_σ	magnetostrictive anisotropy energy density
F	force
H	magnetic field
H_m	molecular field
H_0	critical field
H_c	coercive force
H_a	anisotropy field
I	intensity of magnetization
I_s	saturation magnetization
I_0	saturation magnetization at 0°K
I_r	residual magnetization
J	exchange integral; total angular momentum
K	magnetic anisotropy constant
K_u	uniaxial anisotropy constant
K_d	unidirectional anisotropy constant
L	torque (density); orbital angular momentum

M	magnetic moment
M_B	Bohr magneton
N	the number of atoms (or spins) in a unit volume; demagnetizing factor; total turns
N_c	the number of carbon atoms in a unit volume
P	angular momentum; power
Q	activation energy
R	Hall coefficient; radius
R_m	reluctance
S	long range order parameter; area; spin angular momentum
T	temperature (degrees Kelvin)
U	energy
V	thermoelectrical power
V_m	magnetomotive force
W	work
W_h	hysteresis loss
W_r	rotational hysteresis loss
Z	atomic number
a	lattice constant; coefficient of the shape effect (magnetostriction)
b	coefficient in law of approach to saturation
c	elastic modulus; compressibility; light velocity
c_{11}, c_{12}, c_{44}	elastic moduli
d	width of domains
e	strain; electronic charge
f	frequency
g	g factor; anisotropy energy function; exchange term
h	Planck's constant; magnetostriction constants
i	current; current density
i_{ma}	macroscopic eddy current density
i_{mi}	microscopic eddy current density
k	Klirr factor; dimensional ratio; Boltzmann factor
l	coefficient of dipole-dipole interaction; length; thickness
m	magnetic pole; mass
n	number of turns per unit length
l, m, n	direction cosines
p	pressure; probability
q	coefficient of quadrupole interaction
r	radius; distance; roll reduction
s	displacement; skin depth; slip density
s_{11}, s_{12}, s_{44}	elastic constants
t	time
v	velocity; volume
w	molecular field coefficient; pair energy
x, y, z	Cartesian coordinates
z	number of nearest neighbors

α	curvature of potential valley; variable of Langevin and Brillouin functions; thermal expansion coefficient; damping factor
$(\alpha_1, \alpha_2, \alpha_3)$	direction cosines of magnetization
α, β	molecular field coefficients
β	damping coefficient of domain wall; packing factor
$(\beta_1, \beta_2, \beta_3)$	direction cosines of observation direction and the annealing field
γ	surface density of domain wall energy
$(\gamma_1, \gamma_2, \gamma_3)$	direction cosines of atomic pair
δ	thickness of domain wall; loss angle
ϵ	surface density of energy; ratio of orbit to spin
ϵ_m	magnetostatic energy per unit area
ϵ_w	wall energy per unit area
ζ	ratio of delayed to instantaneous magnetization
η	Rayleigh constant
θ	angle (particularly between magnetization and the field)
λ	magnetostriction constant; relaxation frequency; spin-orbit parameter
λ, μ	relative number of Fe^{3+} ions in A and B sites
μ	permeability
$\bar{\mu}$	relative permeability
μ_0	permeability of vacuum
μ_a	initial permeability
μ_{max}	maximum permeability
ν	gyromagnetic constant
ρ	resistivity
ρ_m	volume density of magnetic pole
σ	electrical conductivity; tension; short range order parameter
τ	relaxation time
ϕ	angle (azimuthal); magnetic potential
χ	magnetic susceptibility
$\bar{\chi}$	relative magnetic susceptibility
ψ	angle
Θ	Curie point
Θ_f	ferromagnetic Curie point
Θ_N	Néel point
Θ_a	asymptotic Curie point
Φ	magnetic flux

appendix 2

Fundamental Equations for Magnetism

mksa	cgs
$B = I + \mu_0 H$	$B = 4\pi I + H$
$I = \chi H = \mu_0 \bar{\chi} H$	$I = \chi H$
$B = \mu H = \mu_0 \bar{\mu} H$	$B = \mu H$
$\mu = \chi + \mu_0$	$\mu = 4\pi\chi + 1$
$\bar{\mu} = \bar{\chi} + 1$	
$H_d = -\dfrac{1}{\mu_0} NI$	$H_d = -NI$
$B = \mu_a H + \frac{1}{2}\eta H^2$	$B = \mu_a H + \frac{1}{2}\eta H^2$
$I = \chi_a H + \frac{1}{2}\eta H^2$	$I = \chi_a H + \dfrac{1}{8\pi}\eta H^2$
$E_m = \frac{1}{2}(BH)_{\max}$	$E_m = \dfrac{1}{8\pi}(BH)_{\max}$

appendix 3

Conversion Table of Magnetic Quantities—mksa and cgs Systems[a]

Quantities	Symbols	mksa Unit	Conversion Ratio: mksa value / cgs value	Conversion Ratio: cgs value / mksa value	cgs Unit
Magnetic pole	m	Wb	1.257×10^{-7}	7.96×10^{6}	
Magnetic flux	Φ	Wb	10^{-8}	10^{8}	Mx
Magnetic moment	M	Wb-m	1.257×10^{-9}	7.96×10^{8}	
Magnetization	I	Wb/m^2	1.257×10^{-3}	7.96×10^{2}	G
Magnetic flux density	B	Wb/m^2	10^{-4}	10^{4}	G
Magnetic field	H	A/m	7.96×10	1.257×10^{-2}	Oe
Magnetic potential	ϕ_m	AT	7.96×10^{-1}	1.257	gilbert
Magnetomotive force	V_m				
Magnetic susceptibility	χ	H/m	1.579×10^{-5}	6.33×10^{4}	
Relative susceptibility	$\bar{\chi}$		equal to $4\pi\chi$ in CGS		
Permeability	μ	H/m	1.257×10^{-6}	7.96×10^{5}	
Relative permeability	$\bar{\mu}$		equal to μ in CGS		
Permeability of vacuum	$\mu_0 = 4\pi \times 10^{-7}$	H/m			$= 1$
Demagnetizing factor	N		7.96×10^{-1}	1.257×10	
Rayleigh constant	η	H/A	1.579×10^{8}	6.33×10^{7}	Oe^{-1}
Reluctance	R_m	AT/Wb	7.96×10^{7}	1.257×10^{-8}	gilbert/Mx
Inductance	L	H	10^{-9}	10^{9}	abhenry
Anisotropy constant	K				
Magnetostatic energy density	E_m				
Energy product	$\frac{1}{2}(BH)_{\max}$ $\left(\frac{1}{8\pi}(BH)_{\max}\right)$	J/m^3	10^{-1}	10	erg/cc
Magnetostriction constant	λ		1	1	
Ordinary Hall Coefficient	R	Ωm^2/A	1.257×10^{-4}	7.96×10^{3}	Ωcm/Oe

[a] Wb (weber), A (ampere), AT (ampere-turn), H (henry), J (joule), Ω (ohm), Mx (maxwell), G (gauss), Oe (oersted). $1.257 = 4\pi \cdot 10$, $7.96 = 10^2/4\pi$, $1.579 = (4\pi)^2/10^2$, $6.33 = 10^3/(4\pi)^2$.

appendix 4

Conversion Table for Various Units of Energy†

eV	cm^{-1}	°K	J	cal
1	= 0.80657×10^{4}	= 1.16049×10^{4}	= 1.60210×10^{-19}	= 3.8291×10^{-20}
1.23981×10^{-4} =	1	= 1.43879	= 1.98630×10^{-23}	= 4.7474×10^{-24}
0.86170×10^{-4} =	0.69511	= 1	= 1.38053×10^{-23}	= 3.2995×10^{-24}
0.62418×10^{19}	= 0.50344×10^{23}	= 0.72435×10^{23} =	1	= 2.3900×10^{-1}
2.61158×10^{19}	= 2.10642×10^{23}	= 3.03071×10^{23} =	4.1840	= 1

† The values in this table are based on the recommendations of the NAS-NRC, U.S.A. (cf. *Physics Today*, p. 48, February, 1964).

appendix 5

Value of Some Constants†

Light velocity	$c = 2.997925 \times 10^8$ m/sec
Acceleration of gravity	$g = 9.80665$ m/sec^2
Universal gravitation constant	$G = 6.670 \times 10^{-11}$ N-m^2/kg^2
Planck's constant	$h = 6.6256 \times 10^{-34}$ J-sec
	$\hbar = h/2\pi = 1.05450 \times 10^{-34}$ J-sec
Mechanical equivalent of heat	$J = 4.1840$ J/15° cal
Boltzmann's constant	$k = 1.38054 \times 10^{-23}$ J/deg
Value of kT at 0°C	$kT_0 = 3.771 \times 10^{-21}$ J
Avogadro's number	$N = 6.02252 \times 10^{23}$/mol
Electronic mass	$m = 9.1091 \times 10^{-31}$ kg
Electronic charge	$e = 1.60210 \times 10^{-19}$ C
Ratio charge/mass	$e/m = 1.758796 \times 10^{11}$ C/kg
Faraday constant	$F = Ne = 9.64870 \times 10^4$ C/mole
Bohr magneton	$M_B = 1.16530 \times 10^{-29}$ Wb-m
Gyromagnetic constant	$\gamma = 1.1051 \times 10^5 g$ m/A sec

$\pi = 3.1415926535$, $e = 2.7182818285$, $\log_e 10 = 2.302685$, $\log_{10} e = 0.43429$, $\log_e 2 = 0.69315$, $\log_{10} 2 = 0.30103$, $\log_e \pi = 1.14473$, $\log_{10} \pi = 0.49715$, $\sqrt{2} = 1.41421$, $\sqrt{3} = 1.73205$, $\sqrt{5} = 2.23606$, $\sqrt{10} = 3.16227$, 1 rad $= 57.295°$, $1° = 0.01745$ rad.

† The constants in this table are based on the recommendations of the NAS-NRC, U.S.A. (cf. *Physics Today*, p. 48, February, 1964).

appendix 6

Periodic Table of the Elements

Each block in the table gives Atomic Number, Symbol, Name, Electronic Structure, Spectroscopic Ground Term, Atomic Weight, and Melting Point.

(After IUPAC, 1957; Smithsonian Physical Tables, 1959; and other sources.)

Group \ Period		2 He Core ($1s^2$)		
Ia	3 Li	Lithium 6.939	$2s^1\ ^2S_{1/2}$ 179°C
IIa	4 Be	Beryllium 9.0122	$2s^2\ ^1S_0$ 1285°C

1 H	Hydrogen 1.00797	$1s^1\ ^2S_{1/2}$ −259.19°C

IIIb	5 B	Boron 10.811	$2s^22p^1\ ^2P_{1/2}$ (2300°C)
IVb	6 C	Carbon 12.01115	$2s^22p^2\ ^3P_0$ (3700°C)
Vb	7 N	Nitrogen 14.0067	$2s^22p^3\ ^4S_{1\frac{1}{2}}$ −209.97°C
VIb	8 O	Oxygen 15.9994	$2s^22p^4\ ^3P_2$ −218.79°C
VIIb	9 F	Fluorine 18.9984	$2s^22p^5\ ^2P_{1\frac{1}{2}}$ −219.61°C
2 He Helium 4.0026 $1s^2\ ^1S_0$	10 Ne	Neon 20.183	$2s^22p^6\ ^1S_0$ −248.59°C

Group	3 Ne Core ($2s^22p^2$)			Group	4 Ar Core ($3s^23p^6$)		
Ia	11 Na	Sodium 22.9898	$3s^1\ ^2S_{1/2}$ 97.82°C		19 K	Potassium 39.102	$3d^04s^1\ ^2S_{1/2}$ 63.2°C
IIa	12 Mg	Magnesium 24.312	$3s^2\ ^1S_0$ 650°C		20 Ca	Calcium 40.08	$3d^04s^2\ ^1S_0$ 851°C
				IIIa	21 Sc	Scandium 44.956	$3d^14s^2\ ^2D_{1\frac{1}{2}}$ 1540°C
				IVa	22 Ti	Titanium 47.90	$3d^24s^2\ ^3F_2$ 1680°C
				Va	23 V	Vanadium 50.942	$3d^34s^2\ ^4F_{1\frac{1}{2}}$ 1847°C
				VIa	24 Cr	Chromium 51.996	$3d^54s^1\ ^7S_3$ 1903°C
				VIIa	25 Mn	Manganese 54.9381	$3d^54s^2\ ^6S_{2\frac{1}{2}}$ 1244°C
				VIII	26 Fe	Iron 55.847	$3d^64s^2\ ^5D_4$ 1539°C
					27 Co	Cobalt 58.9332	$3d^74s^2\ ^7F_{4\frac{1}{2}}$ 1492°C
					28 Ni	Nickel 58.71	$3d^84s^2\ ^3F_4$ 1453°C
				Ib	29 Cu	Copper 63.54	$3d^{10}4s^1\ ^2S_{1/2}$ 1083°C
				IIb	30 Zn	Zinc 65.37	$3d^{10}4s^2\ ^1S_0$ 419.5°C
IIIb	13 Al	Aluminum 26.9815	$3s^23p^1\ ^2P_{1/2}$ 659°C		31 Ga	Gallium 69.72	$4s^24p^1\ ^2P_{1/2}$ 29.75°C
IVb	14 Si	Silicon 28.086	$3s^23p^2\ ^3P_0$ 1420°C		32 Ge	Germanium 72.59	$4s^24p^2\ ^3P_0$ 936°C
Vb	15 P	Phosphorus 30.9738	$3s^23p^3\ ^4S_{1\frac{1}{2}}$ 44.2°C		33 As	Arsenic 74.9216	$4s^24p^3\ ^4S_{1\frac{1}{2}}$ 814°C
VIb	16 S	Sulfur 32.064	$3s^23p^4\ ^3P_2$ 119°C		34 Se	Selenium 78.96	$4s^24p^4\ ^3P_2$ 217.4°C
VIIb	17 Cl	Chlorine 35.453	$3s^23p^5\ ^2P_{1\frac{1}{2}}$ −100.99°C		35 Br	Bromine 79.909	$4s^24p^5\ ^2P_{1\frac{1}{2}}$ −7.2°C
0	18 Ar	Argon 39.948	$3s^23p^6\ ^1S_0$ −189.37°C		36 Kr	Krypton 83.80	$4s^24p^6\ ^1S_0$ −157.3°C

	5 Kr Core ($3d^{10}4s^24p^6$)			6 Xe Core ($4d^{10}5s^25p^6$)		
Ia	37 Rb	Rubidium 85.47	$4d^05s^1\,{}^2S_{1/2}$ 39°C	55 Cs	Cesium 132.905	$5d^06s^1\,{}^2S_{1/2}$ 28.5°C
IIa	38 Sr	Strontium 87.62	$4d^05s^2\,{}^1S_0$ 770°C	56 Ba	Barium 137.34	$5d^06s^2\,{}^1S_0$ 710°C
IIIa	39 Y	Yttrium 88.905	$4d^15s^2\,{}^2D_{1\frac{1}{2}}$ 1509°C	57–71 (Lanthanides)		
IVa	40 Zr	Zirconium 91.22	$4d^25s^2\,{}^3F_2$ 1857°C	72 Hf	Hafnium 178.49	$5d^26s^2\,{}^3F_2$ 2220 ± 30°C
Va	41 Nb	Niobium 92.906	$4d^45s^1\,{}^6D_{1/2}$ 2470°C	73 Ta	Tantalum 180.948	$5d^36s^2\,{}^4F_{1\frac{1}{2}}$ 2996°C
VIa	42 Mo	Molybdenum 95.94	$4d^55s^1\,{}^7S_3$ 2610°C	74 W	Tungsten 183.85	$5d^46s^2\,{}^5D_0$ 3380°C
VIIa	43 Tc	Technetium (99)	$4d^55s^2$	75 Re	Rhenium 186.2	$5d^56s^2\,{}^6S_{1\frac{1}{2}}$ 3180°C
	44 Ru	Ruthenium 101.07	$4d^75s^1\,{}^5F_5$ 2250°C	76 Os	Osmium 190.2	$5d^66s^2\,{}^5D_4$ 3000°C
VIII	45 Rh	Rhodium 102.905	$4d^85s^1\,{}^4F_{4\frac{1}{2}}$ 1960°C	77 Ir	Iridium 192.2	$5d^76s^2\,{}^4F_{1\frac{1}{2}}$ 2443°C
	46 Pd	Palladium 106.4	$4d^{10}5s^0\,{}^1S_0$ 1552°C	78 Pt	Platinum 195.09	$5d^96s^1\,{}^3D_3$ 1769°C
Ib	47 Ag	Silver 107.870	$4d^{10}5s^1\,{}^2S_{1/2}$ 960.8°C	79 Au	Gold 196.967	$5d^{10}6s^1\,{}^2S_{1/2}$ 1063°C
IIb	48 Cd	Cadmium 112.40	$4d^{10}5s^2\,{}^1S_0$ 320.9°C	80 Hg	Mercury 200.59	$5d^{10}6s^2\,{}^1S_0$ −38.87°C
IIIb	49 In	Indium 114.82	$5s^25p^1\,{}^2P_{1/2}$ 156.6°C	81 Tl	Thallium 204.37	$6s^26p^1\,{}^2P_{1/2}$ 303.6°C
IVb	50 Sn	Tin 118.69	$5s^25p^2\,{}^3P_0$ 231.91°C	82 Pb	Lead 207.19	$6s^26p^2\,{}^3P_0$ 327.3°C
Vb	51 Sb	Antimony 121.75	$5s^25p^3\,{}^4S_{1\frac{1}{2}}$ 630.5°C	83 Bi	Bismuth 208.980	$6s^26p^3\,{}^4S_{1\frac{1}{2}}$ 271°C
VIb	52 Te	Tellurium 127.60	$5s^25p^4\,{}^3P_2$ 450°C	84 Po	Polonium (210)	$6s^26p^4\,({}^3P_2)$ 254°C
VIIb	53 I	Iodine 126.9044	$5s^25p^5\,{}^2P_{1\frac{1}{2}}$ 113.6°C	85 At	Astatine (210)	$6s^26p^5$
0	54 Xe	Xenon 131.30	$5s^25p^6\,{}^1S_0$ −112.5°C	86 Rn	Radon (222)	$6s^26p^6\,{}^1S_0$ −71°C

	7 Rn Core ($4f^{14}5d^{10}6s^{2}6p^{6}$)					
Ia	87 Fr	Francium (223)	$6d^{0}7s^{1}$			
IIa	88 Ra	Radium (226)	$6d^{0}7s^{2}\ ^{1}S_{0}$ 700°C	Lanthanides (Rare Earth) Xe Core ($4d^{10}5s^{2}5p^{6}$)		
Actinides	89 Ac	Actinium (227)	$6d^{1}7s^{2}\ ^{2}D1\frac{1}{2}$ 1047°C	57 La	Lanthanum 138.91	$4f^{0}5d^{1}6s^{2}\ ^{2}D1\frac{1}{2}$ 920°C
	90 Th	Thorium 232.038	$6d^{2}7s^{2}\ ^{3}F_{2}$ 1707°C	58 Ce	Cerium 140.12	$4f^{1}5d^{1}6s^{2}\ ^{3}H_{4}$ 795°C
	91 Pa	Protactinium (231)	$6d^{3}7s^{2}\ (^{4}F1\frac{1}{2})$ 1427°C	59 Pr	Praseodymium 140.907	$4f^{3}5d^{0}6s^{2}\ ^{4}I4\frac{1}{2}$ 935°C
	92 U	Uranium 238.03	$5f^{3}6d^{1}7s^{2}\ (^{5}L_{6})$ 1133°C	60 Nd	Neodymium 144.24	$4f^{4}5d^{0}6s^{2}\ ^{5}I_{4}$ 1024°C
	93 Np	Neptunium (237)	$5f^{(4)}6d^{(1)}7s^{2}$ 640°C	61 Pm	Promethium (147)	$4f^{5}5d^{0}6s^{2}$ 1297°C
	94 Pu	Plutonium (242)	$5f^{(5)}6d^{(1)}7s^{2}$ 639.5 ± 2°C	62 Sm	Samarium 150.35	$4f^{6}5d^{0}6s^{2}\ ^{7}F_{0}$ 1072°C
	95 Am	Americium (243)	$5f^{7}6d^{0}7s^{2}$	63 Eu	Europium 151.96	$4f^{7}5d^{0}6s^{2}\ ^{8}S3\frac{1}{2}$ 826°C
	96 Cm	Curium (247)	$5f^{(7)}6d^{(1)}7s^{2}$	64 Gd	Gadolinium 157.25	$4f^{7}5d^{1}6s^{2}\ ^{9}D_{2}$ 1312°C
	97 Bk	Berkelium (249)	$5f^{(8)}6d^{(1)}7s^{2}$	65 Tb	Terbium 158.924	$4f^{8}5d^{1}6s^{2}$ 1356°C
	98 Cf	Californium (251)	$5f^{(9)}6d^{(1)}7s^{2}$	66 Dy	Dysprosium 162.50	$4f^{9}5d^{1}6s^{2}$ 1407°C
	99 Es	Einsteinium (254)	$5f^{(10)}6d^{(1)}7s^{2}$	67 Ho	Holmium 164.930	$4f^{10}5d^{1}6s^{2}$ 1461°C
	100 Fm	Fermium (253)	$5f^{(11)}6d^{(1)}7s^{2}$	68 Er	Erbium 167.26	$4f^{11}5d^{1}6s^{2}$ 1497°C
	101 Md	Mendelevium (256)	$5f^{(12)}6d^{(1)}7s^{2}$	69 Tm	Thulium 168.934	$4f^{13}5d^{0}6s^{2}\ ^{2}F3\frac{1}{2}$ 1545°C
	102 No	Nobelium	$5f^{(13)}6d^{(1)}7s^{2}$	70 Yb	Ytterbium 173.04	$4f^{14}5d^{0}6s^{2}\ ^{1}S_{0}$ 824°C
				71 Lu	Lutetium 174.97	$5d^{1}6s^{2}\ ^{2}D2\frac{1}{2}$ 1652°C

Author Index

Italic page numbers indicate the places where the authors' names are referred to in the text. Page numbers followed by "f" and "t" indicate figures and tables, respectively.

Abrahams, S. C., 396, *469,* 485
Adachi, K., 109, *444,* 456
Adams, E., 457
Aiyama, Y., *369,* 370f, 395
Akimoto, S., 109
Alperin, H. A., 486
Anderson, A., 484
Anderson, P. W., *81, 82,* 108, *345,* 355
Andrä, W., 417
Aoyagi, K., 369t, 369, 395
Arajs, Sigurds, 457
Arnold, H. D., 395
Arnold, R. D., *343,* 355
Arp, V., 488
Arrott, A., *77,* 78
Asamuma, M., 505
Assumus, F., 505
Azumi, K., 173t, 184

Bacon, G. E., 462t, 464f, *469,* 484, 485, 486
Bader, F., 78
Baker, W. O., *220,* 241
Baldwin, M. E., 355
Baltzer, P. K., *444,* 456
Barkhausen, H., *114,* 127
Barnett, S. J., *45,* 59
Bates, L. F., 35, *59, 233,* 233f, 241, *429, 431,* 438, 517
Bauminger, R., 487
Bean, C. P., 14, *224,* 225f, 241, *314,* 320, *391,* 393f, *394,* 396, *398,* 400f, *411, 413,* 417, 418
Beard, G. B., 487
Becker, J. J., 395, *413,* 418
Becker, R., 69f, *115,* 166f, 167f, 172f, 180f, 181f, *249,* 259, *266, 277, 279,* 280, 308f, 420f, 428f, 430f, 517
Becquerel, J., *11,* 13
Behrendt, D. R., 457
Benard, J., 13
Bents, V. H., 485
Bertaut, F., *98,* 104f, 105t, *107,* 109
Bickford, L. R., Jr., 139f, 140t, 142t, 159, 160, 175t, *370,* 371f, *388,* 388f, 390f, 395, 396
Bijvoet, J., *295,* 302

Bindels, J., *295,* 302
Bitter, F., *115,* 127, *219,* 241
Bittrich, G., 241
Bizette, H., *446,* 456
Blinowski, K., 486
Bloch, F., *71,* 78, *115,* 127, *186,* 204, *460,* 484
Bloembergen, N., 355
Blois, M. S., 417
Blume, R. J., 487
Bobeck, A. H., 515
Boersch, H., *405,* 408f, 417
Bommer, H., 456
Boyd, E. L., *406,* 417, 487
Boyle, A. J. F., 488
Bozorth, R. M., 13, 19t, 22t, *30,* 35, 49f, 59, 73f, 109, *117,* 127, *134,* 140t, 142t, 143t, 145f, 159, 160, 174f, 175t, 185, *205,* 218, *219, 222,* 241, *252, 253,* 259, *264,* 264f, 280, *301,* 302, *359, 360, 389,* 394, 396, 418, 429f, 436f, *455, 456,* 457, 458, 493f, 494t, 502t, 505, 517, 518
Brailsford, F., 517
Braun, P. B., 109, 332f, *355*
Brissonneau, P., *295,* 302, 320
Brockhouse, B. N., *470, 472,* 485, 486
Brockman, F. G., *328,* 329f, 354, 485
Brown, P. J., 487
Brown, W. F., *277,* 280
Brownlow, J. M., 160, 395
Bruner, L. J., 487
Büchner, A., 506
Budnick, J. I., 487
Buehler, W. J., 506
Bunburg, D. St. D., 488
Bunge, H. J., *387,* 396
Burgers, W. G., *373,* 395
Burgy, M., 486
Bussiere de Nercy, A., 487

Cable, J. W., 457, 458, *469,* 485
Calhoun, B. A., *389,* 396
Callen, H. B., *345,* 355
Carr, W. J., Jr., 78, *150, 151,* 151f, 152f, 160, 174f, 184
Chandrasekhar, B. S., 159
Chattock, A. P., 35
Cherry, L. V., 457
Chikazumi, S., 73f, 78, *116,* 127, 132f, 159, 192f, 202f, 204, 218, 231f, 237f, 241, 280, 360f, *361, 362,* 363f, 367f, 368f, 369t, 374f, *375, 387,* 394, 395, 396, 418, 422f, 438
Child, H. R., 457, 458, *469,* 485
Cioffi, P. P., 35
Clogston, A. M., *345,* 355
Cochardt, A. W., 439
Cochran, D. R. F., 487
Cohen, M. S., *406,* 417
Cohen, S. G., 487
Coleman, R. V., *415,* 418
Collins, M. F., *76,* 78, *470,* 485
Colvin, R. V., *420,* 437, 457
Compton, V. B., *455,* 458
Conger, R. L., *410,* 418
Conradt, H. W., *373,* 395
Corenzwit, E., 457, 458
Corliss, L. M., *91,* 109, *444, 446,* 456, *468,* 485, 486
Craig, P. P., 487
Craik, D. J., *220,* 241
Crede, J. H., 506
Creveaux, H., 109, 396
Crittenden, E. C., Jr., 417
Cumming, U. K., 355
Curie, P., *69*
Czerlinsky, E., *277,* 278f, 280

Daane, A. H., *456,* 456, 457, 458
Dahl, O., *361,* 373, 395
Darmöl, W., 506
Davis, D. D., 457, 458
Davis, J. L., 486
De Blois, R. W., *203,* 204, *415,* 416f, *417,* 418
de Freudenreich, J., *299,* 302
De Gennes, P. G., *472,* 486
de Haas, W. J., 59
Dehlinger, U., 78
Demer, L. J., 439
De Pasquali, G., 487
de Vos, K. J., 506
De Vries, G., *295,* 302, *311,* 320
Dietrich, W., 355

Dillinger, J. F., *359,* 394
Dillon, J. F., Jr., 109, *224,* 241, *342,* 355
Domenicali, C. A., 396
Dorfman, J., *240,* 241
Döring, W., 69f, *115,* 166f, 167f, 172f, 180f, 181f, *249,* 259, *266, 277, 279,* 280, 308f, *348,* 355, *397,* 417, 420f, *422,* 428f, 430f, 437, 517
Dorothy, E. W., 109
Dowling, P. H., 354
Dunmur, I. W., 486
Dwight, K., 396, 456
Dzialoshinski, I., *12,* 14

Edmondson, A. S., 438
Edmunds, D., 488
Edward, C., 488
Edwards, P. L., *442,* 456
Eich, A. M., 417
Einstein, A., 59
Elbinger, G., 143t, 144t, 160
Elliot, J. F., 457
Elliot, R. J., *453,* 457, *472,* 486
Elliott, N., *468,* 485
Ellis, W. C., 438
Ellwood, E. B., *429,* 438
Elmen, G. W., *361,* 395, *497,* 506
Elmore, W. C., *116,* 127, *219,* 241
Elsasser, W. M., *459,* 484
Emerson, W. H., 160
Englert, E., 421f, 437
Enz, U., 141t, 159, *318,* 320, *446,* 456
Epstein, L. M., 487
Erickson, R. A., *446,* 456, *469,* 485
Ericson, M., 486
Esveld, C. J., 280, 302
Ewing, J. A., *291, 309,* 320

Fallot, M., 13
Ferguson, E. T., *369,* 395
Fermi, E., 488
Fielder, H. C., 506
Fine, M. E., 438
Foëx, G., 35, 517
Folen, V. J., 160, *345,* 355, 487
Foner, S., *31,* 35, 438
Forrat, F., 109
Forrer, R., *387,* 396, 428f, *428,* 429f, 438
Frank, A., 449
Frauenfelder, H., 487
Frazer, B. C., 485
Freedman, J. P., *406,* 417
Freeman, A. J., 482t, *483, 484,* 487, 488
Frenkel, J., *240,* 241
Friedel, J., *75, 76,* 78
Frischmann, P. G., 506
Fuller, H. W., *226,* 241, *405,* 417
Funatogawa, Z., 141t, 159, 160, 175t, 185, 417, *423,* 438

Galt, J. K., 141t, 143t, 160, *344, 352,* 352f, *354,* 355, 518
Ganzhorn, K., 78
Geller, S., 109
Gersch, H. A., 486
Gersdorf, R., *311,* 320
Geschwind, S., *345, 348,* 355
Gianola, U. F., 515
Gijsman, H. M., 438
Gilbert, T. L., *334,* 355
Gilbert, W., *12*
Gilfrich, J. V., 457
Gilleo, M. A., 109
Gladrow, E. M., 456
Goertz, M., *389,* 396
Goldmans, P., 280, 302
Gondo, Y., *397,* 417, *423,* 438
Goodenough, J. B., *59,* 78, *102,* 106t, 109, *211,* 218, *399, 402,* 404f, 417, 518
Gordon, J. P., *515,* 515
Gorter, C. J., 108
Gorter, E. W., *92,* 93f, *98,* 99f, 100t, 109, 159, 280, 302, 506, 517
Goss, N. P., *493,* 505
Gossard, A. C., *477, 478,* 487
Goto, K., 439
Graham, C. D., Jr., *77,* 78, 159, *197, 203,* 204, *373, 391,* 395, 396, *415,* 416f, 418
Gran, B. E., 418
Grenoble, H. E., 506
Grew, K. E., *423,* 438
Griffith, P. M., *220,* 241

Griffiths, J. H. E., *49,* 59
Gschneidner, K., Jr., *454, 456,* 457
Guillaud, C., *92,* 109, 140t, 159, 280, 396
Guillaud, P. C., 13
Guillaume, C. E., 438, 506
Guttman, L., 485

Haayman, P. W., 396
Hagiwara, S., 484
Halban, H., *459,* 484
Hale, M. E., *226,* 241, *405,* 417
Hall, R. C., 174f, 184
Halpern, O., *460,* 484, 486
Hamilton, W. C., *387, 389,* 396, 485
Hamos, L. V., *115,* 127, *219,* 241
Hanna, S. S., 487
Haraldsen, H., *106,* 109
Hardeman, G. E., *476,* 486
Harrison, E. G., 438
Harrison, F. W., *92,* 109
Hass, M., 487
Hass, W. P. A., 486
Hastings, J. M., *91,* 109, *444, 446,* 456, *468,* 485, 486
Healey, D. R., 438
Heberle, J., 487
Heidenreich, R. D., *134,* 159, *372,* 395
Heisenberg, W., *73,* 78
Heller, R. B., 486
Henry, W. E., 64f, 78, 457
Herpin, A., *446,* 456, *469,* 485
Herve, L., 488
Heusler, F., 506
Hibbard, W. R., 505
Hidnert, P., 178f, 185
Hihara, T., 487
Hirahara, E., *483,* 487
Hirahara, T., 488
Hirone, T., 78, 106f, 107f, 109, *422,* 438, *444,* 456
Hirota, E., 355
Hoffman, G. R., 515
Hoffman, R. W., *397, 405,* 417
Hogan, C. L., 515
Hokkeling, P., 506
Holeman, J. O., 418
Hollander, J. M., 486
Holstein, T., 486
Holtzman, R. H., 486
Honda, K., *114,* 127, 136f, 159, *503,* 506
Hori, N., *422,* 438
Hoselitz, K., 517
Hoshino, S., 484
Hubbard, W. M., 457
Huber, E. E., *402,* 404f, 417
Hughes, D. J., *470,* 486
Hund, F., 59, 457

Ichinokawa, T., *405,* 407f, 417
Iida, S., *369,* 370f, 395
Inoue, T., 395
Ishikawa, Y., 101f, 109, 144t, 160, 176t, 185, *444,* 456, 482t, *483,* 485, 487
Ito, A., *483,* 487
Ito, K., 160
Itoh, J., 473
Iwata, H., 132f, 395, 396
Iwata, T., *369,* 374f, *375, 387,* 395

Jaccarino, V., *477, 481,* 486
Jacobs, I. S., *77,* 78, *88,* 108, *413, 415,* 415f, 418, *442,* 443f, 456
Jacrot, B., 486
Jan, J. P., 425f, 438
Janik, J., 486
Jellinghaus, W., 517
Jennings, L. D., 457
Johnson, M. H., *460,* 484
Jones, R. V., 78, 485
Jonker, G. H., 103f, 109, 332f, 355

Kaplan, J. J., 485
Kaplan, M., 488
Kaplan, T. A., *444, 453,* 456, 557
Karplus, R., *426,* 438
Kasper, J. S., 485
Kasuya, T., *420,* 437, 457
Kato, Y., 395, *499, 504,* 506
Katsuragi, H., 484
Kaufman, A. R., 396
Kaufman, W., *405,* 417
Kaya, S., *114, 116,* 127, 136f, 159, 166f, 167f, 184, *252,* 253f, 259, *361,* 395, *422,* 438
Keffer, F., *153,* 160

Kelley, J. M., 355
Kelley, W. H., 487
Kelsall, G. A., *359,* 394
Kelvin, Lord, *see* Thomson, W.
Kersten, M., *267, 271,* 272f, *273, 274,* 274f, 280, *287,* 287f, *288,* 302
Kikuchi, R., *334, 338,* 355
Kilburn, T., 515
Kimura, H., 424f, 438
Kip, A. F., *343,* 355
Kistner, O. C., 487
Kittel, C., 14, 18f, 47t, *50,* 51t, *52,* 59, 78, *197,* 204, *211,* 218, *240,* 241, *323, 333, 343, 345, 351,* 354, 355, *372,* 395, *397,* 398f, 399f, 414, 417, 418, *440,* 456, 517, 518
Kleber, E. V., 457
Klein, M. J., *397,* 417
Kleiner, W. H., 486
Klemm, W., 456
Kneller, E., *411,* 412f, *413,* 413f, 418, 518
Knorr, J. C., *405,* 417
Kobayashi, T., 160, 439
Koch, A. J. J., 506
Kocher, C. W., 487
Koehler, W. C., 78, *452,* 454f, 457, 458, *463,* 463f, *469,* 484, 485
Koi, Y., 487, 488
Kojima, Y., 143t, 160
Kondo, J., *423,* 438, *483,* 488
Kondorsky, E., *76,* 78, *267,* 280
Konno, H., 417
Kono, H., 506
Kono, T., *389,* 392f, 396, 409f, 418, 439
Kono, Y., 73f, 78
Kooi, C. F., 355
Koppelmann, F., 515
Korestoff, E., 418
Kornetzki, M., 181f, 185
Koster, G. F., *74,* 75f, 78
Köster, W., 436f, 438, 506, 518
Kothari, L. S., 486
Kouvel, J. S., 73f, *76, 77,* 78, 396, 506
Kramers, H. A., *71,* 78, *81,* 108
Krupicka, S., 320
Krutter, H., *74,* 78
Kubo, R., *81, 88,* 108, 160, 355
Kundt, A., *425,* 438
Kushida, T., 487, 488
Kuwahara, K., *423,* 424f, 438

La Blanchetais, C. H., 457
Lamb, W. E., *478,* 487
Lambeck, M., *405,* 408f, 417
Landau, L., *117,* 127, *205,* 218, 355
Landé, A., *58,* 59
Langevin, M., 487
Lawson, A. W., 457
Lazan, L., 439
Lee, E. W., 173t, 178f, 184, *233,* 241
Legvold, S., *420,* 437, *455,* 457
Li, C. H., *387,* 396
Li, Y. Y., 14
Lichtenberger, F., 172f, 184
Lifshitz, E., *117,* 127, *205, 214,* 218, 355
Lin, S. T., 396
Littlejohn, C., 487
Lock, J. M., *454, 455,* 457
Lomer, W. M., 485
Lommel, J. M., 418
Long, T. R., 515
Lotgering, F. K., 160, *442,* 456
Low, G. G. W., *76,* 78, *470,* 485
Lowde, R. D., 78, 485, 486
Luborsky, F. E., 415, 415f, 418, 506
Luttinger, J. M., *426,* 438
Lyons, D., 456

Maeda, S., 109
Makino, N., *503,* 506
Málek, Z., 417
Margulies, S., 487
Marinov, A., 487
Marshall, G., 438
Marshall, W., *472, 483,* 486, 487, 488
Martin, J. P., 506
Masumoto, H., 439, *495, 497,* 506
Matcovich, T., 418
Matthias, B. T., 13, 160, *456,* 458
Mayer, H., *410,* 418
Mayer, W., *405,* 417
Maxwell, J. L., *44,* 59
McGuire, T. R., *348,* 355
McHargue, C. J., 457
McKeehan, L. W., 145f, 160

McReynolds, A. M., 486
Mee, C. D., *233,* 233f, 241
Meier, D. A., 515
Meiklejohn, W. H., *391,* 393f, *394,* 396
Mendelsohn, L. I., 506
Meneghetti, D., 485
Menyuk, N., 396, 456
Mercereau, J. E., *342,* 355
Meriel, P., *446,* 456, 485
Merritt, F. R., 160, *344,* 355
Michel, A., 13
Middelhoek, S., 417
Miles, P. A., 518
Miller, R. W., 396
Mishima, T., *503, 504,* 506
Mitchell, D. P., 484
Mitchell, E. N., 418
Mitui, T., *373,* 395
Miyahara, S., *318,* 320, *373,* 395
Miyake, S., 484
Miyashita, K., 484
Miyata, N., 141t, 143t, 144t, 159, 160, 175t, 185, *423,* 438
Miwa, H., *453,* 457
Mizuta, H., 395
Morimoto, S., 487
Moriya, T., *12,* 14
Moser, P., *295,* 302
Mössbauer, R. L., *478,* 487
Mott, N. F., 78, *419, 420, 423,* 437
Müller, H. G., 396

Nachman, J. F., 506
Nagamiya, T., *88, 108,* 109, 160, *345,* 355
Nagasaki, S., 109
Nagata, T., 517
Nagle, D. E., 487
Nakagawa, Y., 396
Nakamichi, T., 174f, 184
Nakamura, Y., 174f, 184
Nassau, K., 457
Nathans, R., 109, *442,* 456, *465, 468,* 468f, *470, 471,* 484, 485, 486
Néel, L., *9,* 13, 14, *82, 88,* 91f, 96f, *104,* 108, 109, *117, 120,* 127, *167,* 184, *205, 214,* 218, *232,* 232f, *240,* 241, *271, 277,* 280, *295, 300, 301,* 302, *311, 313, 314, 315,* 320, *362, 365, 372, 373,* 395, *399,* 401f, *410, 414,* 417, 418, *469*
Nesbitt, E. A., *134,* 159, *372,* 395, 458, *504,* 506
Neugebauer, C. A., *397,* 417
Neumann, H., 506
Neurath, P. W., *197,* 204
Nilan, T. G., 241
Novy, V., 457

Obenhain, F. E., 487
Ofer, S., 487
Ogawa, S., 409f, 418, 487, 505
Ohta, K., 141t, 143t, 160, 175t, 185, *318, 319,* 320
Ohyama, T., *414,* 418
Okamura, T., 143t, 160, 396, *429, 430,* 430f, 438
Oliver, D. A., 395
Olson, C. D., *410,* 411f, 418
Ōno, K., *483,* 487
Onsager, L., *71,* 78
Oomura, T., *362,* 395
Osborn, J. A., 34
Osmond, W. P., 109

Paine, T. O., 506
Palmer, W., 141t, 159
Paoletti, A., 486
Pappis, J., 184
Pauli, W., 13, 59
Pauling, L., 78, 457
Paulis, N. J., *476,* 486
Pauthenet, R., *98,* 104f, 105t, 109, *456,* 458
Pawleck, F., 517
Paxton, W. S., 241
Peacock, R. N., 487
Pearce, R. R., 34
Pearson, R. F., 141t, 142t, 144t, 159, 160
Penoyer, R. F., 140t, 146t, 159, 160, *370, 371,* 395
Perlow, G. J., 487
Petersen, R., 488
Pickart, S. J., 109, *442,* 456, *468, 471,* 485, 486
Pigott, M. T., 484
Pittelli, E., *345,* 355

Pohm, A. V., *410,* 411f, 418
Portis, A. M., *477, 478, 484,* 487, 488
Potter, H. H., 506
Poulis, J. A., 108
Poulis, N. J., *88,* 108
Pound, R. V., 487
Powell, J. E., 456
Powers, P. N., 484
Preisach, F., *299,* 302
Preiswerk, P., 484
Preston, R. S., 487
Prince, E., *442,* 456, 485
Proebster, W. E., 355
Prosen, R. J., 418
Pry, R. H., 506
Pryce, M. H. L., *158,* 160
Pugh, E. M., *425, 426,* 426f, 438
Pugh, E. W., *406,* 417

Rado, C. T., 144t, *332, 343,* 355
Rajchman, J. A., 515
Rayleigh, Lord, *296,* 302, *309,* 320
Rathenau, G. W., 109, 159, *295,* 302, *311,* 320, *373,* 395, *414,* 414f, 418, 506
Rebka, G. A., Jr., 487
Reinboth, H., 506
Remeika, J. P., 160, 241
Rhodes, B. L., 457
Richter, G., *309, 310,* 320
Ringo, G. R., 486
Riste, T., 486
Robert, F. F., 485
Roberts, B. W., *224,* 225f, 241, *398,* 400f, 417, 485
Roberts, C., 487
Roberts, L. M., *454,* 457
Rodbell, D. A., 14
Romeijn, F. C., 396
Rosenvasser, E. S., 487
Rostky, G., 515
Rostoker, N., 426f, 438
Roth, W. L., *468,* 484, 485
Ruby, S. L., 487, 488
Russell, H. N., 59

Sage, M., 109
Saito, H., 439, *495,* 506
Saito, S., 506
Sallo, J. S., 515
Sato, H., *77,* 78, 159
Saunders, F. A., 59
Say, M. G., 517
Scatturin, V., *469,* 485
Schiffer, J. P., 487
Schindler, A. I., 438
Schmeckenbecher, A., 418
Schnitzler, A., *345,* 355
Schreiner, W. N., 487
Schüppel, W., 417
Schützmeister, L. M., 487
Schwarts, S. J., 515
Schwinger, J., *460,* 484
Scott, G. G., *47,* 59, *415,* 418
Seaborg, G. T., 486
Sears, R. W., *425,* 438
Seavey, M. H., Jr., *343,* 355
Sedov, V. L., *76,* 78
Segal, E., 487
Segrè, E., 488
Sekiya, J., 127
Sekizawa, H., *369,* 370f, 395
Selwood, P. W., 457
Sexton, W. C., 457
Shafer, M. W., 140t, 146f, 159
Shanabarger, M. R., 355
Shedden, J. W., 395
Shenker, H., 142t, 143t, 146f, 160
Sherry, N. P. R., *431,* 438
Sherwood, R. C., *224,* 241, *399, 401,* 401f, 402f, 403f, *405,* 417, 458
Shih, J. W., 145f, 160
Shimizu, K., 237
Shirakawa, Y., *422,* 438
Shirane, G., *101,* 109, *468,* 484, 485, 488
Shirely, D. A., 488
Shockley, W., *117,* 127, *205, 214,* 218, *219, 222,* 241, *323,* 354, 372, 395
Shubina, L. A., *372,* 395
Shubrooks, G. E., 506
Shull, C. G., *77,* 78, 80f, *81,* 108, *463,* 463f, 464f, *465, 468, 469, 470,* 470f, *471,* 471f, 484, 485, 486
Shulman, R. G., *477, 478,* 486
Shur, J. S., *372,* 395
Sidhu, S. S., *469,* 485
Simkins, Q. W., 515

Simoisaka, J., 396
Simpson, A. W., 438
Singer, J. R., 515
Singwi, K. S., 486
Six, W., *373,* 395
Sixtus, K. J., *114, 115,* 127, *373,* 395
Slack, G. A., 485
Slater, J. C., *74,* 78
Sleight, N. R., 456
Slonczewski, J. C., *156,* 157f, 160, *370, 371,* 395
Smart, J. S., 80f, *81,* 108, *468,* 484
Smit, J., 140t, 142t, 159, 160, 355, *423,* 425f, 438, 506, 518
Smith, A. W., *425,* 438
Smith, D. O., *402,* 404f, *405, 406, 410,* 417, 418
Smith, J. H., 486
Smith, P. S., *397,* 417
Smoluchowski, R., 174f, 184
Snoek, J. L., *306, 310, 316,* 316f, *318,* 320, *329,* 355, *373,* 395, *499,* 506, 517
Snow, A. I., *469,* 485
Solt, I. H., *342,* 355
Soohoo, R. F., 418
Spedding, F. H., *420,* 437, *455, 456,* 456, 457, 458
Spighel, M., 487
Squires, G. L., *472,* 472f, 486
Stablein, F., 506
Stapels, R. P., 160
Steeg, M. G. v. d., 506
Steinberger, J., *470,* 486
Stemme, O., 417
Steneck, W. G., 354
Stephan, W., 127, *203,* 204
Stevens, K. W. H., 78
Stewart, K. H., 517
Stominger, O., 486
Stoner, E. C., 34, *74,* 78, *240,* 241, 517
Storm, H. F., 515
Stout, J. W., 486
Strauser, W. A., 464f, *468,* 484
Street, R., *309,* 309f, 320, *469,* 485, 486
Stuijts, A. L., 355
Stull, J. L., 184
Sturkin, D. A., 174f, 184
Stuyts, A. L., 506
Sucksmith, W., 34, 140t, 159
Sugai, T., 439
Sugihara, M., 506
Suguki, K., 484
Suhl, H., *345,* 355, 458
Sun, K. H., 487
Sunyan, A. W., 487
Suzuki, K., *116,* 127, *202,* 204, 218, 231f, 237f, 369t, 374f, *375, 387,* 395, 396
Syono, Y., 144t, 160, 176t, 185

Tachiki, M., *157, 158,* 158f, 160
Takahashi, H., 515
Takahashi, M., *389,* 392f, 396, *408,* 409f, 418
Takaki, H., 166f, 167f, 174f, 184
Takata, Y., 228f, *229,* 241
Takei, T., 395, *499, 504,* 506
Takei, W. J., 485
Tamagawa, K., 396
Tang, Ting-Youn, 457
Taniguchi, S., *295,* 296f, 302, *362, 373,* 395, *497,* 506
Tannenwald, P. E., *343,* 355
Tarasov, L. P., 145f, 160
Tatsumoto, E., 174f, 184, *423,* 424f, 438
Teale, R. W., 109
Thiessen, P. A., *115,* 127, *219,* 241
Thoburn, W. C., *455,* 457
Thompson, J. E., 140t, 159
Thomson, W., *419, 423,* 437, 438
Tilden, E. F., 160, 185
Tomita, K., 473t, 486
Tomono, Y., 304f, *305,* 306f, 320, 361, 395
Tonks, L., *114, 115,* 127
Townes, C. H., *515,* 515
Townsend, A., *429,* 429f, 438
Trammel, G. T., 486
Treuting, R. G., 485
Tsubokawa, I., 106t, 109
Tsujimura, A., 487, 488
Tsuya, N., 106t, 109, 355
Turner, J. A., 515

Ubbink, J., 108
Usami, S., 159

Valenta, L., *397*, 417
Van den Handel, J., *11*, 13, 108
Van der Heide, H., *319*, 320, *328*, 354
Van der Lugt, W., 486
Van Geest, D. W., *311*, 320
Van Hove, L., *472*, 486
Van Kranendonk, J., 78
Van Oosterhout, G. W., 109, 159, 506
Van Santen, J. H., 109
Van Vleck, J. H., 13, *59*, *71*, 78, *82*, 108, *147*, *149*, *153*, 160, *447*, *449*, *456*, 457, 458, 517
Veillet, P., 488
Verwey, E. J. W., *344*, 355, *387*, 396
Vickery, R. C., 457
Villain, J., *444*, *446*, 456, 486
Vincent, D. H., 487
Vogt, E., 14
Voight, A. F., 456
Von Hippel, A., 518

Wachtel, E., 506
Wakiyama, T., 395, 438
Walker, J. G., 145f, 160, 174f, 184
Walker, L. R., *342*, 342f, 343f, *345*, *348*, 355, *481*, 487
Wallace, W. E., 457
Wallance, J. R., 486
Walsh, W. M., 487
Walter, J. L., 505, 506
Wangsness, R. K., *348*, 355
Wanic, A., 486
Wannier, G. H., *71*, 78
Watabe, A., *77*, 78, *453*, 454f, 457
Watanabe, D., 409f, 418
Watanabe, H., 78, 109
Waterstrat, R. M., 485
Watson, R. E., 482t, *483*, *484*, 487, 488
Weber, G. H., *255*, 259
Webster, W. L., 165f, 184, *422*, *426*, 438
Weertman, J. R., *343*, 355
Wegener, H. H. F., 487
Weiss, G. P., *406*, 417
Weiss, P., *29*, 35, *65*, 78, *106*, 109, *113*, 127, *299*, 302, *387*, 396, *428*, 428f, 429f, 438, 517
Weiss, P. R., *70*, *71*, 78, *186*, 204
Weiss, R. J., 446
Went, J. J., 109, 140t, 159, 506
Wernick, J. H., 458
Wertheim, G. K., *481*, 481f, *484*, 487, 488
Weston, J. C., 438
Westphal, W. B., 518
White, R. L., *342*, 355
Wick, G. C., *470*, 486
Wiedemann, G., *433*, 438
Wigen, P. E., 355
Wijn, H. P. J., 109, *140t*, 159, 280, 302, *319*, 320, *328*, 332f, 354, 355, 518
Wilkinson, M. K., 78, *452*, 452f, 457, 458, *469*, *470*, 485, 486
Williams, A. J., 109, 160, 185
Williams, C., 178f, 185
Williams, H. J., *30*, 35, *117*, *120*, 127, *134*, 159, *205*, 218, *219*, *222*, *224*, 241, *266*, 280, *323*, *324*, 324f, 354, *389*, 396, *399*, *401*, 401f, 402f, 403f, *405*, 417, 458
Wilson, R. H., 73f, *76*, 78, 506
Winslow, F. H., *220*, 241
Winter, J. M., 487
Wohlfarth, E. P., 78, *240*, 241
Wojtowicz, P. J., *444*, 456
Wolf, P., *338*, 355
Wolf, W. P., *159*, 160
Wollan, E. O., 78, 457, 458, *463*, 463f, 464f, *468*, *469*, 470f, 484, 485, 486
Wooley, J. C., *309*, 309f, 320
Woolf, W. E., 486

Yafet, Y., *440*, 456
Yager, W. A., 49f, 59, 160, *344*, 355
Yamada, Y., *471*, 471f, 486
Yamadaya, T., *318*, *319*, 320
Yamamoto, M., 174f, 184, *362*, *373*, 395
Yanase, A., *319*, 320
Yensen, T. D., *496*, 506
Yoshie, T., 484
Yoshimori, A., *444*, *445*, 456
Yosida, K., *77*, 78, *81*, *88*, 108, *157*, *158*, 158f, 160, *347*, 355, *446*, *453*, 454f, 456, 457
Yuzuri, M., 106t, 107f, *108*, 109

Zeiger, H. J., *515*, 515
Zener, C., *77*, 78, *102*, 109, *150*, 160, 457

Subject Index

Italic page numbers indicate the places where definitions or detailed explanations are given. Page numbers followed by "f" and "t" indicate figures and tables, respectively.

Absolute saturation magnetization, 67
Ag_3Mn, 394
Al-Fe alloys, 174f, 424f, 494t
  13% Al-Fe, 495
  16% Al-Fe, 495
$Al_xFe^{2+}_yFe^{3+}_zO_4$, 144t
Alfenol, 495
  16 Alfenol, 494t
Alfer, 494t
Alnico 2, 372
Alnico 5, 18f, 19t, 65, 134, 309, 309f, 372, 373, 502t, 503, *504*
Alperm, 494t, 495
(Alpha)α-Fe_2O_3, 11, 86t, 100, 469, 472, 482t, 483
(Alpha)α-iron, 310, 310f, 469
(Alpha)α-manganese, 77, 86t, 470
Al-Si-Fe alloy, *see* Sendust
Ampère's theorem, 27
Ampere-turn, 27
Anisotropic magnetostriction, 165
Anisotropic exchange, 148
Anisotropy field, 138, 330, 332t
Anomalous specific heat, 71, 178
Anomalous thermal expansion, 178f
Antiferromagnetic resonance, *345,* 347f
Antiferromagnetic substances, 86t, 87f, 468, 469
Antiferromagnetism, 8, 77, *79,* 82f, 85f, 449
  spin arrangement, 9f, 80f, 87f
Approach to saturation magnetization, 246, *274,* 278f
Atomic form factors, 463, 463f, 464f
Atomic magnetic moment, 58f, 450f
Atomic number, 54
Atom pairs, 148f, 167f, 362, 363
Atomic pile, 463, 464f
Au-Mn alloy, 77, 470

$BaCo_{1.85}Fe_{16.2}O_{27}$, 140t
$Ba_2Co_2Fe_{12}O_{22}$, 140t
$Ba_3Co_2Fe_{24}O_{41}$, 140t

Ba-ferrite, 101, 240, 285, 414, 414f, 502t, 503, 505
Ballistic galvanometer, 31, 32f
$Ba^{2+}Mn^{4+}O_3$, 102
$BaO \cdot 6Fe_2O_3$ or $BaFe_{12}O_{19}$, 140t, 505
$BaO \cdot 2FeO \cdot 8Fe_2O_3$ or $BaFe_{18}O_{27}$, 140t
$BaO \cdot 2MO \cdot 8Fe_2O_3$, 101
$2BaO \cdot 2MO \cdot 6Fe_2O_3$, 101
$3BaO \cdot 2MO \cdot 12Fe_2O_3$, 101
Barkhausen effect, 114f, 292
Barnett method, 46, 46f
$(Ba,Sr)_2Zn_2Fe_{12}O_{22}$, 446
$Ba_2Zn_{1.5}Fe_{12.5}O_{22}$, 140t
$(BH)_{max}$, 501, 502t
Bitter figure technique, *see* Powder pattern method
Blade-shaped domain, 119, 120f, 212, 212f
Bloch theory of spin wave, 70, 397
Bloch-type wall, *399,* 401, 401f, 403f
Bloch wall, 115, 186
Bohr magneton, *40*
Boltzmann constant, 61
Boltzmann factor, 61, 65, 311, 364
Brillouin function, 63
Bulk modulus, 177

$Ca^2Mn^{4+}O_3$, 102
Carbon steel, 304f, 306f, 502t, 503
Carbonyl iron, 308, 498, 498t
Cerium, 453
Chain structure, 406
Checkerboard pattern, 211, 211f, 227, 228
Chromites, 444; *see also* $MnCr_2O_4$
Chromium, 77, 86t, 446, 470
Cioffi's recording fluxmeter, 32, 33f
Circular pattern, 211, 211f, 228
Closure domains, 119, 229, 229f, 230, 231f, 398
$Co_2Ba_3Fe_{24}O_{41}$, 332t
Cobalt, 19t, 47t, 51t, 140t, 152f, 228f, 229f, 248, 261t, 274f, 301t, 391, 426, 429, 477, 478, 482t
$CoBr_2$, 469
$CoCl_2$, 86t, 469
Co-CoO system, 391, 393f
  spin arrangement, 393f
Co-Cr alloys, 73f
Co-Cu alloys, 414
  2% Co-Cu, 373
Coelinvar, 437
Coercive force, *16,* 18f, 19t, 287f, 292, 293, 294, 410, 414, 414f, 493f, 494t, 498t, 502t
CoF_2, 86t, 469, 477
CoF_3, 469
Co-Fe alloys, 145f
  12.5% Co-Fe alloy, 369t
  20% Co-Fe alloy, 369t
  40 at% Co-Fe, 173t
  50 at% Co-Fe, 173t
  70 at% Co-Fe, 173t
$CoFe_2O_4$ or Co ferrite, 91f, 99, 100t, 134, 156, 369, 370f, 371f, 469, 502t, 504
$Co_{0.01}Fe_{2.99}O_4$, 157f
$Co_xFe_yO_4$, 142t
Coil constant, 32
Collective electron model, 74
Co-Mn alloy, 73f, 394
Compensation temperature, *98,* 105, 105t
$Co(NH_4)_2(SiO_4)_2 \cdot 6H_2O$, 86t
Co-Ni alloy, 145f, 174f, 369, 391, 392f
Co-Ni-Fe alloy, *see* Fe-Co-Ni alloy
$Co_xNi_yFe_zO_4$, 143t
Contact dc converter, 507, *510,* 511f
CoO, 86t, 106t, 108, 391, 482t
CoP, 106t
Coplanar strips, 209f, 211
Copper, 74
CoPt, 504
Corundum-type oxides, 100
CoS, 106t
CoSb, 106t
CoSe, 106t
Co steel, 503
  15% Co steel, 502t
CoTe, 106t
Coulomb interaction, *3,* 13, 56, 205
Co_2Z, 101, 332, 332f
$Co_xZn_yFe_zO_4$, 143t, 175t
Cr alloy, 76
CrAs, 106t
$CrCl_2$, 86t, 469
CrF_2, 469
Critical damping, 336, 337

Cr-Ni alloy, 73f
CrO_2, 10
Cr_2O_3, 86t, 469
Cross-tie walls, *402,* 404f, 405f, 408f
CrP, 106t
Cr Permalloy, 494t
CrS, 86t, 106t, 108
$Cr_{1.17}S$, 107f, 108
CrSb, 86t, 106t, 469
CrSe, 106t
Cr steel, 503
 3.5% Cr steel, 502t
CrTe, 106t
Crystal effect, 179, 180, 180f
Crystal type, of antiferromagnetics, 86t
 of Ni-As type compounds, 106t, 106f
 of rare earth metals, 448t
Cubic texture, 493
$CuBr_2$, 86t
$CuCl_2$, 86t
$CuCl_2 \cdot 2H_2O$, 86t, 88, 347, 476
$CuCr_2O_4$, 442
Cu-Fe-Ni alloy, 413, 413f
$CuFe_2O_4$ or Cu ferrite, 91t, 100t, 389, 469
$Cu_xFe_yO_4$, 143t
Cu_3Mn, 394
Cu-Mn ferrite, 498t, *500*
Cu-Ni alloys, 145f, 174f, 426
Cunico, 502t, 504
Cunife, 502t, 504
Curie constant, 86t
Curie law, 8, 62
Curie temperature or Curie point, 9, 67, 69, 186
 asymptotic, 70, 85, 86t, 95, 448t, 450, 451f
 ferrimagnetic, 95, 100t, 105t, 498t
 ferromagnetic, 70, 448t, 451t, 494t, 498t
Curie-Weiss law, 10, 69
Cu-Zn ferrite, 318, 498t, *499*

Damping coefficient, 349
Damping term, 333, 351
Debye temperature, 478
Degeneracies, 155
(Delta)ΔE effect, *435,* 435f
Deltamax, 253, 494t, 497
Demagnetization, *257*
 ac, 257, 258f
 thermal, 257
Demagnetized state, 15
Demagnetizing curve, 16, 28f
Demagnetizing factor, 19, *20,* 179, 180, 208, 372
 for cylinder, 22
 for ellipsoid of rotation, 21, 181
 for flat ellipsoid, 21
 for oblate ellipsoid, 21, 22t
 for plate, 20, 22
 for prolate ellipsoid, 22t
 for rod, 22t
 for sphere, 22
Demagnetizing field, 19, 50, 478
Density, of ferrites, 100t
 of garnets, 105t
 of permanent magnet, 502t
 of rare earth metals, 448t
 of soft magnetic materials, 494t
d(epsilon)ϵ orbits, 153, 154f, 156f
d(gamma)γ orbits, 153, 154f, 156f
Diamagnetism, *7,* 7f
Differential scattering cross section, *see* Neutron scattering cross section
Differential susceptibility, 16
Diffusion aftereffect, *see* Magnetic aftereffect
Diffusionless transformation, 495
Dimensional resonance, 329, 329f
Dipolar narrowing, 345
Dipole-dipole interaction, 147, 149, 168; *see also* Pseudodipolar interaction
Directional order, 264, 295, *362,* 375, *377,* 497
Direction of easy magnetization, *see* Easy direction
Disaccommodation, *316,* 316f, 499
Dislocation, 277, 288, 378
Domain magnetization, distribution at coercive force point, *254,* 257f
 distribution at remanence, *248,* 250f, 257f
 distribution on magnetization curves, 249f, 255f
Domain pattern, 192f, *219,* 221f, 223f, 228f, 237f, 375f, 378f, 379f, 400f, 401f, 407f, 416f

Domain structure, 118f–123f, 125f, *219,* 226f, 361f, 374f, 416f; *see also* Magnetic domain
of non-uniform substance, 233
of polycrystals, 233
of stressed crystals, 237
of thin films, 397, 399f, 406f
of uniform substance, 226
Domain wall, 115, 123, *186,* 477
configuration, 193f, 194f, 196f, 197f, 199f, 203f, 324f
constraints, 120, 238, 288
90° domain wall, 192, *198,* 269f
180° domain wall, 191f, *192,* 272f
displacement, 124, 124f, 125f, 265f, 269f, 272f
energy, 188, 192, 195, 196f, 200f, 227, 273, 286f, 350, 401f
expansion, 272, 288, 289
equation of motion, 348
fractional volume, 353
structure, 187f, 189f, 191f, 198f, 214f, 349f, 399
thickness, 188, 191, 286
velocity, 324, 351, 417
Doublet, 154, 155
Double wall, 401, 402f, 403f
Dust core, 328
DyIG, 482; *see also* Ferrimagnetic garnet
DyN, 455
Dysprosium, 450, 452, 452f

Easy direction or easy axis, *114,* 131
Eddy current, 325f, 337, 344; *see also* Macroscopic eddy current; Microscopic eddy current
Eddy current anomaly, 325
Eddy current loss, 321, 327
Effective atomic moment, 63, 448t
Effective field, *22,* 33
Einstein-de Hass method, 46, 46f
Elastic constants, 182
Elastic energy, *169*
Elastic moduli, 169, 173
Electrolytic polishing, 219, *221,* 222f
Electromagnetic mode, 343
Electron diffusion, 319, 328, 344, 352, 387
Electronic configuration, 55t, 446, 448t
Electron microscopy, 405, 406f, 407f
Electron shell, 54f
Elinvar, 437, 496
Erbium, 453
$ErFeO_3$, 469
ErN, 455
ErP, 456
ErSb, 456
Euler equation, 190
EuO, 10, 456
Europium, 446, 447, 449
Exchange energy, 186, 445
Exchange integral, 71, 73, 186, 445
Exchange interaction, 147
Exchange polarization, 483
Extraordinary Hall effect, 425, 426f

Faraday effect or Faraday rotation, 224, 399, 405, 408f
Fe-Al alloys, 369, 394, *495*
FeAl, 495
Fe_3Al, 237, 237f, 387, 470, 471, 482t, 495
$Fe_{3.35}Al$, 369t
$Fe_{4.7}Al$, 369t
FeAs, 106t
$FeBr_2$, 469
$FeCl_2$, 11, 86t, 469
$FeCl_3$, 86t
$FeCO_3$, 86t
$FeCO_3 \cdot 2MgCO_3$, 86t
Fe-Co alloy, 73f, 369, 470, *497*
Fe-Co-Ni alloy, *497*
FeCo powder magnet, 415, 502t, 503, 505
Fe-Cr alloy, 73f
FeF_2, 86t, 469, 482t
FeF_3, 469
Fe-FeO system, 394
Fe_3N, 482t
Fe_4N, 469
Fe_2NiAl, 134, 504
Fe-Ni alloys, 73f, 75, 145f, 172f, 174f, 263, 264f, 288, 359, 361, 362, 363f, 366, 369, 373, 406, 408, 409f, 436f, 470, *495; see also* Ni-Fe alloys; Permalloy
15% Fe-Ni, 171

Fe-Ni alloys, 19% Fe-Ni, 263
21% Fe-Ni, 252, 263, 264
21.5% Fe-Ni, 359, 360f, 361, 422, 422f, 496
24% Fe-Ni, 263
40% Fe-Ni, 171
50% Fe-Ni, 247, 262, 373, 496
FeO, 86t, 106t, 108, 468
Fe_3O_4, 12, 47t, 51t, 88, 91f, 100t, 139, 157, 157f, 158, 220f, 328, 354, 390f, 442, 443f, 468, 469, 472, 482t
FeP, 106t
FePt, 18f
Fermi contact, 483
Ferrimagnetic garnets, 98, 102, 104f, 105t, 224, 348, 456, 482t
Ferrimagnetic oxide, *99*
Ferrimagnetic resonance, 347
Ferrimagnetism, 9, 10f, *79, 88,* 95f
spin arrangement, 10f, 90f, 442f
Ferrites, *88,* 91f, 98f, *100,* 100t, 270
anisotropy constants, 140t–144t
cores, 295, 498, 498t
magnetic annealing, 369
magnetostriction, 175t
molecular weight, 100t
resistivity, 328
Ferromagnetic resonance, *48,* 49f, 51t, 137, 338f, 388, 390f
Ferromagnetism, 10, 10f, *60,* 64, 449, 450
Ferroxdure, 502
Ferroxplana, 43, *331,* 332t, 332f, 500
$Fe_{0.875}S$, 106
$FeS_{1.00}$, 86t, 106t, 107, 482t
FeS_{1+x}, 469
FeSb, 106t
FeSe, 106t
FeTe, 86t, 106t
Fe_2Ti, 482t
$FeTiO_3$, 100, 469, 482t
$FeTiO_3$-αFe_2O_3 system, *100,* 469
Fe-V alloy, 73f
Fe_2Zr, 482t
Fine particle, 397, *411,* 415
Flopping of the spin axis, 88, 477
Fluxmeter, 32
Forced magnetostriction, 178, 180f
Form effect, 176, 179, 180, 180f
Form factor, 460
4*f* electron, 448t, 449f, 450f, 451f
4*f* shell, 446, 447
Fourier analysis, 132, 133t

Gadolinium, 450, 453
$Ga_xFe_y^{2+}Fe_z^{3+}O_4$, 144t
Galvanomagnetic effect, *419*
Garnets, *see* Ferrimagnetic garnets
Gauss' theorem, *20,* 212
(Gamma)γFe_2O_3, 100, 100t, 482t, 510
(Gamma)γ-iron, 77, 469
$GdCo_2$, 455
$GdCo_5$, 455
Gd-Co alloy, 455
$GdFe_2$, 455
$GdFe_5$, 455
Gd-Fe alloy, 455
Gd-Fe garnet, 348
Gd-La alloy, 455
$GdPt_2$, 455
Gd-Y alloy, 455
g factor, 41, 47t, 51t, 58, 447
nuclear, 475
Goss steel, 493
Grain-oriented silicon steel, 253
Ground states, 155t
Gyrator, 500, *514,* 514f
Gyromagnetic constant, 41, 329
Gyromagnetic effect, 41
Gyromagnetic experiment, 47t
Gyromagnetic properties, 514
Gyromagnetic ratio, 41

Half-value width, 342
Hall effect, *423; see also* Extraordinary Hall effect; Ordinary Hall effect
temperature dependence, 426f
Hard magnetic materials, 18, 500
Harmonics, second, 512
third, 297
Helical spin configuration, 440, *444,* 445f, 449, *452,* 452f, 453
pitch of screw or turn angle, 444, 452, 452f, 453, 454f
Hematite, 100; *see also* (Alpha)α-Fe_2O_3

Heusler alloy, 47t, 51t
Hipernik, 494t
$HoFeO_3$, 469
Holmium, 452
HoN, 455
HoP, 456
Hopkinson effect, 247, 273, 499
Hopping motion of electrons, *see* Electron diffusion
Horizontal magnetic pendulum, 29, 30f
HoSb, 456
Hund's rule, 56, 90, 154, 447, 449
Hypersil, 18f
Hysteresis, 11, 16, 17f, 246, 512f
  constricted or wasp-waisted, 295, 296f, 497
  rectangular, *500,* 503, 507f, 510
  shifted, 299f
Hysteresis loss, 17, 292, 293, 294, 297, 301t, 327, 493f

Ilmenite, 100, 101f
Incoherent rotation, *see* Rotation magnetization
Incremental susceptibility, 16
Induced magnetic anisotropy, 319, *359,* 363f, 367f, 368f, 387, 392f, 405, 408, 409f
Inelastic scattering of neutrons, 472, 472f
Inertia of the domain wall, 348
Inertia term, 333
Initial permeability or initial susceptibility, 16, 19t, 260–274, 264f, 271f, 296, 301t, 330f, 332t, 494t, 498t; *see also* Magnetic susceptibility
Intensity of magnetization, *6*
Internal field, 472, 477, *480,* 482t, *483,* 484
Internal stresses, 202f, 286, 287f
Intra-atomic dipole-dipole interaction, 155t, 157, 158
Intrinsic magnetization, 72
Invar, 496
Inverse magnetostrictive effect, *182,* 433
Inverse spinel, 90, 90f
Inverse Wiedemann effect, 433
Ion exchange method, 446
Ionic migration, 319
Iron, 18f, 47t, 51t, 77, 136f, 151f, 165f, 166f, 167f, 173t, 181f, 253f, 261t, 274f, 278f, 288, 295, 301t, 343, 405, 423, 424f, 426, 428, 429, 430f, 463, 463f, 470, 470f, 471, 471f, 472, 472f, 478, 481, 481f, 482t, 491, 494t
Iron powder, 301t, 415, 502t, 503, 505
Irreversible aftereffect, *see* Magnetic aftereffect
Irreversible alloys, 495
Irreversible magnetization, 246, *281*
Irreversible rotation magnetization, *281*
  critical field, 283, 283f, 285
Irreversible susceptibility, 16
Irreversible wall displacement, *285*
  critical field, 288, 290f, 324
Isolator, 515
Isomer shift, *479,* 481
Isoperm, 247, 253, 262, 373, 374f, *497*
  50 Isoperm, 494t
Isotropic magnetostriction, 163, 171
Itinerant electron model, 74

Jordan-type aftereffect, *see* Magnetic aftereffect

Kaya's law, 253
Kirchhoff's law, 27, 28, 29
Kittel mode, *339,* 342
Klirr factor, 298
$KMeF_3$ (Me: Cr, Mn, Fe, Co, Ni, and Cu), 469
KS steel, 18f, 502t, 503

$(La,Ca)MnO_3$, 103f, 469
La-Ce alloy, 454
$La^{3+}Co^{3+}O_3$, 102
$LaFeO_3$, 469
$La(Mn,Cr)O_3$, 469
$La^{3+}Mn^{3+}O_3$, 102
La-Nd alloy, 454
Landau-Lifshitz equation, *333,* 339, 351
Landé formula, *58,* 447
Langevin function, 62, 62f, 66, 83, 412
Langevin theory, *60*

Laplace equation, 24, 206, 209, 210
Large Barkhausen effect, 115, 115f
(La,Sr)MnO_3, 103f
Lattice constant, of ferrites, 100t
  of garnets, 105t
Lattice sites, 8*a* sites, 89
  16*a* site, 104
  24*c* site, 104
  16*d* sites, 90
  24*d* site, 104
  96*h* site, 104
  interstitial, 310f
Lattice transformation hardening magnets, 502t, *503*
Lattice vacancy, 319
Law of approach to saturation, *see* Approach to saturation magnetization
Law of electromagnetic induction, 321
Legendre polynomials, 147, 151
Li-Cr ferrite, 98, 99f
$Li_xFe_y^{2+}Fe_z^{3+}O_4$, 144t
$Li_{0.5}Fe_{2.5}O_4$, 100t, 482t
$Li_xMn_{1-x}Se$, 468
Line width mechanism, 344
lmn law, 253
Localized model, 74
Long range order-Fine slip type (L. F. type), *382*, 383t, 384, 385
Lorentz field, 24, 65, 68
Loss, 321, *327*, 353
Loss factor, 298, 305, 327
  temperature dependence, 306f
Low temperature transition of magnetite, *387*, 388f

Macroscopic eddy current, 326, 326f
Maghemite, *see* (Gamma)γFe_2O_3
Magnetic absorption peaks, 49f, 342f, 344f
Magnetic aftereffect, 248, *303*, 304f, 309f, 328
  diffusion, 313, 315
  irreversible, 316
  Jordan type, 309
  mechanism, *309*
  phenomenology, 303
  reversible, 316
  Richter-type, 308f, 309
Magnetic aftereffect, superposition principle, 316
  thermal fluctuation, 313, 315
Magnetic aging, 491
Magnetic amplifiers, 507, *511*, 512f, 513f
Magnetic anisotropy, *128*, 147, 155, 409, 451, 453; *see also* Induced magnetic anisotropy; Magnetostrictive anisotropy; Oblique-incidence anisotropy; One-ion anisotropy; Perpendicular anisotropy; Roll magnetic anisotropy; Rotatable anisotropy; Surface anisotropy; Uniaxial anisotropy; Unidirectional anisotropy
Magnetic anisotropy constants, of alloys, 145f, 492, 495
  of cubic crystals, 130
  of ferrite, 140t–144t, 146f, 389, 390f
  of uniaxial crystals, 129, 140
  temperature dependence, 150, 151f, 152f, 157f, 158f
Magnetic anisotropy energy, 114, 128, 134, 311
Magnetic annealing effect, 134, 156, 295, 319, *359*, 497, 503, 505
  on ferrites, 369, 370f
  on precipitation alloys, 371
  on solid solutions, 362, 363f, 367f
  on thin films, 408
Magnetic balance, 29, 30f
Magnetic circuits, *24*, 26f, 501f
Magnetic colloid, 219
Magnetic compensating alloys, 496
Magnetic cooling, 387
Magnetic cylinder, 509
Magnetic dipolar interaction, *5*, 147, 157, 319
Magnetic dipole, *5*
Magnetic domain, 10, *113*, 117, 162f; *see also* Domain structure
  energy, 227, 228, 230, 239
  size, 227, 228, 230, 239
Magnetic energy, 71, 71f, 88, 137
Magnetic field, *3*
Magnetic flux density, *6*, 26
Magnetic form factor, 470f
Magnetic granulometry, 413

Magnetic induction, *6*
Magnetic Kerr effect, 224, 225f, 398
Magnetic materials, 18f, 19t, *491*
Magnetic moment, *4*, 447
Magnetic poles, 3
  distribution, 209f, 211f, 212f, 215f
Magnetic potentiometer, 33, 33f
Magnetic recording, 509
Magnetic recording tape, 283, 291, 510f
Magnetic resistance, *26*
Magnetic resonance, *see* Antiferromagnetic resonance; Dimensional resonance; Ferrimagnetic resonance; Ferromagnetic resonance; Natural resonance; Nuclear magnetic resonance; Spin wave resonance
Magnetic rod, 509
Magnetic scattering amplitude or cross section, 460, 462, 462t, 463
Magnetic susceptibility, *6*, 15f, 18f, 316; *see also* Differential susceptibility; Incremental susceptibility; Initial susceptibility; Irreversible susceptibility; Maximum susceptibility; Relative susceptibility; Reversible susceptibility; Total susceptibility
  frequency dependence, 330f, 332f, 340f
  temperature dependence, 8–12, 70f, 82f, 85f, 95f, 107f
Magnetic torque curve, 131, 132f
Magnetic viscosity, 248
Magnetite, *see* Fe_3O_4
Magnetization by stress, 433f
Magnetization curve, 11f, *15*, 15f, 64f, 126f, 136f, *245*, 245f, 262f, 284f, 290f, *291*, 292f, 293f, 294f, 360f, 374f, 393f, 412f, 443f, 510
Magnetization reversal, 415f
Magnetocaloric effect, 427
Magnetocrystalline anisotropy or crystal magnetic anisotropy, *see* Magnetic anisotropy
Magnetoelastic energy, 169
Magnetomechanical effect, *431*
Magnetomechanical factor, 52
Magnetometer, 31, 31f
Magnetomotive force, *27*
Magnetoplumbite, 399, 401f
Magnetoplumbite structure oxide, 101, 102f
  M type, 101
  W type, 101
  Y type, 101
  Z type, 101
Magnetoresistance effect, 32, 360, *419*, 421f, 422f, 424f, 425f
Magnetostatic energy, 179, *205*, 226, 260, 500
  formulas, 205, 207, 208, 209
  for surface distribution of poles, 211
  of inclusions, 238
  of poly-cubic crystal, 236
  of poly-uniaxial crystal, 234
Magnetostriction, 114, 161, 432f, 436f
  field dependence, 161f, 431
  magnetization dependence, 164f, 165f, 166f, 167f
Magnetostriction constants, 170, 171, 172, 366
  of ferrites, 175t, 500
  of metals and alloys, 172f, 173f, 174f, 492, 495
Magnetostrictive anisotropy, 128
Magnetothermal effect, *427*, 428f, 429f, 430f
Magnetothermoelectric effect, *423*
Martensite transformation, 495
Maser, 515
Mass of domain wall, 349, 351
Mass of proton, 475
Maximum susceptibility, *16*, 19t, 292, 293, 294, 493f, 494t
Maxwell experiment, 44, 45f
Maze domain, 116, 116f, 117, 203
$MCoO_3$, 102
Measurement, of magnetic field, *29*, 32
  of magnetization, *29*
  of magnetostriction, 173
  of torque curve, 130f, 131
Memory core or memory device, 410, 507, 508f, 509f
Metal powder magnet, 285
Metamagnetism, 11, 11f
$MFeO_3$, 102
$Mg_2Ba_2Fe_{12}O_{22}$, 332t
$MgFe_2O_4$ or Mg ferrite, 100t, 226, 469

$Mg_xFe_yO_4$, 144t
Mg-Mn ferrite, 498t, *500*, 508
Mg_2Y, 101
Mg-Zn ferrite, 318, *500*
Microscopic eddy current, 326, 326f
Microwave oscillator, 515
Mild steel, 19t
Minor loop, 16
MK steel, 18f, 19f, 372, 502t, 503, 504
$MMnO_3$, 102
MnAl, 505
MnAs, 11, 86t, 106t, 469
$MnAu_2$, 11, 444, 446
MnBi, 11, 86t, 106t, 140t, 224, 398, 399, 400f, 502t, 503, 505
$MnBr_2\cdot 4H_2O$, 86t, 469
$MnCl_2\cdot 4H_2O$, 86t
$Mn_xCo_yFe_zO_4$, 142t, 175t
Mn-Cr alloy, 77, 470
$MnCr_{2-x}Al_xO_4$, 442
$MnCr_2O_4$, 444
Mn-Cu alloy, 77, 470
MnF_2, 86t, 88, 469, 477
MnF_3, 469
Mn-Fe-Cr ferrites, 99
$MnFe_{2-x}Cr_xO_4$, 442
$MnFe_2O_4$ or Mn ferrite, 91, 91f, 100t, 157, 158, 158f, 318, 342f, 469
$Mn_xFe_yO_4$, 140t, 146f, 175t
Mn_4N, 469
Mn-Ni alloy, 412
$Mn_xNi_yFe_zO_4$, 141t
MnO, 79, 80f, 86t, 106t, 108, 468
MnO_2, 86t, 444
Mn_3O_4, 442, 443f, 444
MnP, 106t
MnS, 106t, 468
MnS_2, 468
MnSb, 106t
Mn_2Sb, 140t
MnSb-CrSb system, 107f, 108, 444
MnSe, 86t, 106t
$MnSe_2$, 468
Mn_2Sn, 482t
Mn_4Sn, 482t
MnTe, 86t, 106t
$MnTe_2$, 468
$Mn_xTi_yFe_zO_4$, 142t
$MnTiO_3$, 469
$Mn_xZn_yFe_zO_4$ or Mn-Zn ferrite, 99, 141t, 175t, 316, 316f, 318, 319, 328, 329, 329f, 342, 498t, *499*
MoF_3, 469
$MO\cdot Fe_2O_3$, 88
$MO\cdot 6Fe_2O_3$, 101
$3M_2O_3\cdot 5Fe_2O_3$, 102, 105t
Molecular field, 65, 67, 83, 94, 440
Molecular field factor, 67
Mo Permalloy, 4-79 Mo Permalloy, 301t
  Mo Permalloy film, 401, 402f, 403f
  Mo Permalloy powder, 498t, 499
  2-81 Mo Permalloy powder, 301t
Morrin temperature, 483
Mössbauer effect, *478*
  apparatus, 480, 480f
  spectrum, 481, 481f
$MTiO_3$-Fe_2O_3, 101f
MT magnet, 502t, *503*
(Mu)μ^* correction, *214*
Mumetal, 494t

Natural resonance, 247, 329, 330
$NdFeO_3$, 469
$NdPt_2$, 455
Néel temperature, 8, *82*, 83, 86t, 391, 448t, 451t, 452
Néel-type wall, *399*, 401f
Neutron, 459
  flipping, 468
Neutron diffraction, 74, 77, 79, 80f, 442, 444, 446, 452, *459*, 461f
  apparatus, 463, 467f
Neutron moment, 460
Neutron scattering, 76, 77, 454, 461, 462t, 463f, 464f
Neutron scattering cross section, 460, 461, 462t, 463f, 465, 472, 472f
Ni-Al ferrites, 99
NiAs-type compounds, 105, 106t, 106f, 444, 469
NiBi, 106t
Nickel, 18f, 19t, 47t, 51t, 74, 86t, 152f, 173t, 261t, 274f, 278f, 287, 287f, 301t, 405, 408, 419, 420f, 423, 426, 426f, 428f, 429f, 478, 482t
$NiCl_2$, 86t
Ni-Co alloy, 73f, 76, 387, 426, *498*

Ni-Co ferrites, 146f
Ni-Cr alloy, 73f
Ni-Cu alloy, 73f, 423, 496
NiF_2, 469, 477, 478
Ni-Fe alloys, 30 ~ 45% Ni-Fe, 76, 437, 496
36% Ni-Fe, 496
54% Ni-Fe, 369t
83% Ni-Fe, 369t
Ni_3Fe, 362, 366, 367f, 369t, 375, 375f, 376f, 377f, 378f, 379f, 380f
$NiFe_2O_4$ or Ni ferrite, 91f, 100t, 157, 158, 344, 352, 371, 389, 429, 469, 482t
$Ni_xFe_yO_4$, 143t
Ni-Mn alloy, 73f, 394
Ni-Mn-Ti ferrites, 99
NiO, 86t, 106t, 108, 468
NiP, 106t
NiS, 106t
NiSb, 106t
NiSe, 106t
NiTe, 106t
$NiTiO_3$, 469
Ni-V alloy, 73f, 76
Ni-Zn alloy, 73f
$Ni_xZn_yFe_zO_4$ or Ni-Zn ferrite, 99, 143t, 175t, 318, 319, 329, 332, 498t, *500*
Non-linear characteristics, 507
Normal spinel, 90
Nuclear magnetic moment, 472, 473t
Nuclear magnetic resonance, *472*
Nuclear magneton, 459, 472
Nuclear scattering amplitude or cross section, 462, 462t, 463
Nuclear spin, 472, 473t, 475
Nuclear transition, 479f

Oblique-incidence anisotropy, 406
Octahedral site, 90, 154f, 155t, 156f
One-ion anisotropy, 345
One-ion model, 153
OP (oxide powder) magnet, 285, 369, 502t, 503, 504
Orbital angular momentum quantum number, 53, 449f
Orbital magnetic moment, 39, 51, 57, 91, 423, 447, 483
Orbital state, 153, 447f
Ordinary Hall effect, *425*
Oriented silicon steel, 19t, 493
Oscillograph hysteresis tracer, 32
Oxide core, 499

Pair energy, 147, 167, 168
Pair model, 147
Palladium, 419, 420f
Paramagnetism, *8,* 8f, *60,* 64f
Parametric excitation, 514f
Parametron, 512, 513f
Parasitic ferromagnetism, 11, 12f, 100, 477
Pauli paramagnetism, 8
Pauli principle, 54, 73, 447
$PbFe_{11}Al_1O_{19}$, 101, 399
P domains, 232, 232f, 233
Perfect order, 361
Permalloy, 19f, 121, 192f, 238, 247, 262, 327, 361, 496; *see also* Fe-Ni alloys
45-Permalloy, 301t
76-24 Permalloy, 261t
78-Permalloy, 18f, 19t, 47t, 51t, 494t
80-20 Permalloy, 404f
Permalloy dust core, 499
Permalloy thin film, 405, 407f, 408
Permalloy problem, 361
Permanent magnets, 429, *500,* 501f, 502t
Permeability, *see* Magnetic susceptibility
of vacuum, *3*
Permendur, 494t, 497
Perminvar, 295, 497
7-70 Perminvar, 494t
45-25 Perminvar, 301t, 494t
Perovskite-type oxides, 102, 102f
Perpendicular anisotropy, 410
Picture-frame single crystals, 266f, 324f, 352f
Planck's constant, *40*
Poisson equation, 206
Polarized neutron, 462, 465, 468f
Powder pattern method, 219, 222f, 223f
Precession motion, 42, 42f, 334
Precipitation hardening magnets, 504
Pressed powder core, 498, 498t

Pressure acting on the wall, *272,* 285, 286, 349
Principal quantum number, *see* Total quantum number
Proton resonance, 476
$PrPt_2$, 455
Pseudodipolar interaction, 148, 311, 345, 362
Pseudoquadrupolar interaction, 148
Pt-Co magnet, 502t, 504
Pt-Fe magnet, 502t, 504
Pyrrotite, 106

Q domains, 232, 232f, 233f
Quadrupole moment, 473t, 476, *480*
Quadrupole shift, 481

RAl_2 (R: rare earth), 455
Rare earth alloys and compounds, *454*
Rare earth metals, 11, 420, *446,* 448t, 449f, 450f, 451f, 483
Rayleigh constant, 296, 301t
Rayleigh loop, 296f, 297, 297f, 298f, 301t
RCo_5 (R: rare earth), 455
Recrystallization texture, 373, 493
Rectangular hysteresis loop, *500,* 503, 507f, 510
Relative permeability, *6*
Relative susceptibility, *6*
Relaxation frequency, 333, 337f, 338, 342, 351
Reluctance, *26*
Remanence, 16, 246, *248,* 253f, 254f, 291f, 292, 293, 294, *327,* 502t, 510f
Residual magnetization, *see* Remanence
Resistivity, 328, 494t, 498t, 500
  field dependence, 421f
  magnetization dependence, 422f
  temperature dependence, 419, 420f
Resonance field, 139f
Resonance frequency, 50, 138, 331, 332t, 336, 475, 476
Restoring coefficient, 349
Resultant orbital vector, 55
Resultant spin vector, 55
Resultant total angular momentum, 55, 449f
Reverse domain, 228
Reversible aftereffect, *see* Magnetic aftereffect
Reversible alloy, 495, 496
Reversible magnetization process, *260*
Reversible rotation magnetization, *260*
Reversible susceptibility, *16*
Reversible wall displacement, *264*
Richter-type aftereffect, *see* Magnetic aftereffect
RIr_2 (R: rare earth), 455
Rock magnetism, 301, 517
Roll magnetic anisotropy, 363f, *373,* 376f, 377f
ROs_2 (R: rare earth), 455
Rotatable anisotropy, 408
Rotational hysteresis, 394, 415
Rotation magnetization, 125f, 136f, 147f, 246, 275f
  incoherent, 415
RRu_2 (R: rare earth), 455
Russell-Saunders coupling, 56, 56f
Rutile-type compound, 445

Samarium, 447
Satellite reflections, 452
Saturation magnetization, 11, 15, 245
  approach to, 246, *274,* 278f
  of metals, 19t, 451, 494t
  of ferrite, 100t, 498
  of ferroxplana, 332t
  of powder, 498t
Saturation magnetostriction, 161, 163
Saturation moment, *63*
  of ferrites, 91f, 93f, 100t
  of ilmenites, 101f
  of garnets, 105t
  of rare earth metals, 448t
Scattering vector, 461
Scratch technique, 223f, 224, 224f
Sendust, 18f, *499*
Sendust powder, 498t
Sheared magnetization curve, 22
Shearing correction, 23f, 248
Short range order-Coarse slip type (S. C. type), *383,* 384, 384t, 386
Si-Fe alloys, 117, 120f, 125f, 132f, 145f, 174f, 202f, 221f, 223f, 231f, 295, 324f, 424f, *492,* 493f, 494t

Silmanal, 502t, 504
Single crystal films, 408, 409
Single domain particles, 281, 291, 292f
Single domain structure, 118f, *240,* 411, 414, 504, 505
critical radius, 240, 415
Singlet, 154, 155
Skin depth, 322, 343
Slater-Pauling curve, 72, 73f, 76, 496, 497
Slip bands, 380f
Slip deformation, 381f, 385f, 386f
Slip density, *378,* 384, 385
Snoek's model of magnetic aftereffect, 307f
Soft magnetic materials, 18, 494f
Solenoid, *3,* 32
Spatial quantization, 53, 53f, 58, 63, 69
Specific heat, 71, 71f
Spectroscopic notation, 59
Spectroscopic splitting factor, 52
Spinel, 89, 89f, 156f
Spin, 40
motion of, 335f, 338f
Spin angular momentum of nucleus, *see* Nuclear spin
Spin clusters, 70
Spin dynamics, 333
Spin magnetic moment, 57
Spin maps, 471, 471f
Spin momentum quantum number, 53, 449f
Spin-orbit interaction, 57, 158
Spin-orbit parameter, 57, 155t, 156
Spin state, 447f
Spin wave, 70, 343, 345, 472
Spin wave resonance, 344f, 410
Spiral structure, *see* Helical spin configuration
Spontaneous magnetization, 9, 103, 397, 411
field dependence, 279
temperature dependence, 69f, 70f, 98f, 104f, 107f, 413f
$Sr^{2+}Co^{4+}O_3$, 102
$Sr^{2+}Mn^{4+}O_3$, 102
Stainless steel, 394
13 stainless steel, 437
18-8 stainless steel, 437
State density, 75f
Strain gauge technique, 161, 173
Strain tensor, 168
Stress-strain curve, 435f
Striations, 222
Sucksmith ring balance, 29, 30f
Superexchange interaction, *81,* 90
Superinvar, 496
Superlattice, 361, 362f, 381f
Superlattice magnets, 504
Supermalloy, 18f, 19t, 47t, 49f, 51t, 65, 238, 301t, 494t
Superparamagnetism, *314,* 411, 412f, 413, 444
Superstructure, *see* Superlattice
Surface anisotropy, 343, *372*
Surface energy of the wall, *see* Domain wall
Surface tension of the wall, 288
Switching of magnetization, 411f
Switching time, 336, 337f, 508
Switchover effect, 388

TbAs, 456
TbN, 455
TbP, 456
TbSb, 456
Technical magnetization, 123, *245*
of thin film, 410
Tensor, 509
Terbium, 450, 453, 482t
Tetrahedral sites, *89,* 154f, 155t
Thermal fluctuation aftereffect, *see* Magnetic aftereffect
Thermal neutrons, 463
Thermoperm, 494t, 496
Thin film memory, 338, *508,* 509f
Thin films, 338, 338f, 343, *397,* 508, 509f
domain structure, 397, 398f
nickel, 397
Permalloy, 405, 407f, 408
spontaneous magnetization, 397
Thin wires, 338, 397, 411, 415
Thulium, 453
$Ti_xFe^{2+}_yFe^{3+}_zO_4$, 144t, 176t
TiS, 106t
TiSb, 106t

TiSe, 106t
TiTe, 106t
$T^{3/2}$ law, 71, 450
Torque magnetometer, 130f, 131
Torque reversal, 134
Total angular momentum quantum number, 53
Total quantum number or principal quantum number, 53, 54
Total susceptibility, 16
Transition element, 55, 55t, 155t
Tree patterns, 231, 231f, 232f
Triangular spin arrangement, 440
Triplet, 154
Twistor, 509

UMn_2, 394
Uniaxial anisotropy, 128, 260, 261t
Unidirectional anisotropy, *391*
Uniform mode, *338,* 342

V alloys, 76
VCl_2, 86t
VCl_3, 86t
Vector model, *52*
Velinvar, 437
Vertical magnetic pendulum, 30, 30f
Verwey order, 387, 388, 389, 389f, 390f, 469
Vibralloy, 437
Vibrating-specimen magnetometer, 31, 31f
Vicalloy, 502t, 503
VO, 106t
V_2O_3, 86t
V_2O_4, 86t
Volume magnetostriction, *176,* 181f
VS, 106t
VSb, 106t
VSe, 106t
VTe, 106t

Walker mode, 342, 343f
Wasp-waisted hysteresis loop, *see* Hysteresis
Wavy-type domain, 211, 212f, 399, 400f
Weber, 3
Weiss theory, 64
Whiskers, 203, 203f, 415, 416f
Wiedemann effect, 433
W steel, 18f, 502t, 503

YIG, *see* Ferrimagnetic garnets
Ytterbium, 446, 449

Zeeman effect, 40
Zigzag angle, 202, 202f, 213
Zigzag wall, 201, 201f, 202f
$Zn_xFe_yO_4$ or Zn ferrite, 91f, 92, 93f, 143t, 440, 470
  mixed, 93f, 440, 444